SOLAR ENERGY
Fundamentals, Design, Modelling and Applications

SOLAR ENERGY
Fundamentals, Design,
Modelling and Applications

G.N. Tiwari

Alpha Science International Ltd.
Pangbourne England

G.N. Tiwari
Centre for Energy Studies
Indian Institute of Technology, Delhi
New Delhi - 110 016

Reprint - 2004

Alpha Science International Ltd.
P.O. Box 4067, Pangbourne RG8 8UT, UK

ISBN 1-84265-106-4

Printed in India.

My Respected Teacher and Guru Ji
Padamashree Mahendra Singh Sodha
On his seventieth Birthday (February 8, 2002)

Preface

Sun is a source of one of renewable energy, known as solar energy. Solar energy is a basic need of living plants and human being on the earth. It is intermittent in nature, eco-friendly and nonpolluting energy. It is freely available throughout the world, particularly in India. Solar energy can be used for direct conversion into electricity (by photovolatic conversion) and into thermal energy. Further, thermal energy conversion can be classified into three category namely (i) low temperature range ($< 10\,^\circ$C) (ii) medium temperature range (10–$150\,^\circ$C) and (iii) high temperature range ($> 150\,^\circ$C).

The objective of this book is to provide a platform to disseminate the knowledge regarding fundamentals of solar energy and design of solar appliances namely,

- fundamental of solar energy and design of solar appliances
- basic heat transfer
- thermal modelling of solar appliances and its applications and
- basic economics of solar systems etc.

to undergraduate and postgraduate students, learners, professional, practioners and designers. To understand the above objectives about three hundred thirty figures, one hundred eighty solved examples and fifty-five tables have been provided. At the end of each chapter, problems/exercises (two hundred twenty six) have also been given along with hint to solve them.

I have drawn the materials for inclusion in the book with references at appropriate places needs to be mentioned. These includes: Solar Engineering of Thermal Processes by Duffie and Backman; Solar Thermal Engineering by Peter J. Lunde; Design and Installation of Solar Heating and Hot Water Systems by J. Richard Williams; Heat Transfer by Donald R. Pitts and Leighton E. Sissom; Jelen's Cost and Optimisation Engineering by Kennrth K. Humphreys; Heat Transfer by J.P. Holman; Solar Passive Building by M.S. Sodha, N.K. Bansal, A. Kumar, P.K. Bansal and Solar Distillation by M.A.S. Malik, G.N. Tiwari, A. Kumar and M.S. Sodha etc.

Solar Thermal Engineering Systems by G.N. Tiwari and Sangeeta Suneja provides the foundation to write this book with the majority of topics.

The present book has been written in fourteen chapters to study a basic knowledge of sun's structure and its radiation and applications of solar energy. Chapter I describes the basic correlation between sun's radiation, atmosphere and the earth angles, hourly variation of solar radiation, daily and monthly average radiation, sunshine hours etc. have also been highlighted in this chapter. Chapter II describes the basic knowledge of various solar devices (Chapters III–XIII). Chapters III and IV discussed in detail about design and performance of flat plate collector (FPC) and evacuated tabular collector respectively. Testing of FPC and basic heat transfer have also been dealt with examples. A steady state thermal model of solar water heating system, heat exchanger and its application for space heating has been discussed in Chapter V. Solar air heater and crop drying system have been covered in Chapter VI and VII respectively. The solar radiation received from the sun can also be concentrated on receiving surface with help of concentrators for high temperature ($\phi > 150\,°C$) applications (Chapter VIII). Purification of brackish/saline water by solar distillation has been discussed in Chapter IX. This also includes other design of solar still and parametric studies. Various space heating/cooling concepts and their steady state analysis have been described in Chapter X. Other application of solar energy (convective and non-convective water heater, solar cooker, solar refrigeration, swimming pool heating, biogas thermal heating and solar fraction etc.) and thermal energy storage have been dealt in Chapter XI and XII respectively. Chapter XIII has been used to discuss about the basic of photovoltaic system with examples. Last chapter deals with economics of solar thermal systems with and without depreciation, effect of tax has also been considered.

S.I. units have been used throughout the book. Appendix has been given in the end of book. Appendix includes conversion units, climatic data's, physical and chemical properties of metal and non-metal and steam tables. Appendix has been frequently used in the text particularly in solved examples.

This book has been aimed to provide a great insight in the subject particularly to the learning students/professionals doing self-study. In spite of my best efforts, some errors might have been crept in the text. I welcome the suggestions and comments, if any from all readers for further improvement of the book in the next edition.

I am highly obliged by Prof. P.K. Kaw, Director, IPR, Ahmedabad who released the book on the February 8, 2002 on the occasion of seventieth birthday of Professor M.S. Sodha (FNA).

I feel immense pleasure to express my heart-felt gratitude to Prof. R.S. Sirohi, Director, IIT Delhi for his constant and consistent encouragement and utmost co-operation at every stage which culminated in successful completion of this book.

I am highly obliged by Prof. S.K. Dube, Dean, Industrial Research and Development and Prof. S.C. Mullick, Head, Centre for Energy Studies, and Prof. P.C. Sinha, Coordinator CEP of IIT Delhi for constant encouragement provided during preparation of the book. I am also obliged by Prof. K.L. Chopra, Ex-Director, IIT Kharagpur, Prof. Chandrasekhar, IDDC, Prof. V.K. Srivastava, Chemical Engineering, Prof. S.C. Kaushik and Prof. A. Chandra, CES, IIT Delhi for their moral support.

I convey my sincere thanks to Prof. S.K. Joshi, Former DG, CSIR, Prof. B.N. Gupta, Director, NSIT, New Delhi, Dr. N.S.L. Srivastava ADG (Engg.), ICAR New Delhi and Dr. D.N. Singh, Executive Director, SCL, Mohali, Chandigarh for rending full help and support, in making this endeavor successful.

I acknowledge with thanks the financial support by the curriculum Development cell, IIT Delhi for preparation of the book.

The patience of my spouse, Mrs. Kamalawati Tiwari in bearing with this lengthy project in good humor is greatly appreciated. My special thanks go out to my children, Arvind, Ghanshyam, and Gopika for willingly giving up their valuable time due to them, used in the preparation of this book.

I acknowledge for moral support and encouragement extended by Sh. Prem Shanker Mishra, Mrs. Ram Kumari Mishra, Sh. V.V. Pandey, Mrs. Shakuntala Pandey, Sh. Shivji Tiwari, Mrs. Umrawati Tiwari, Sh. Nagesh Kumar Tiwari and Mrs. Manju Tiwari during the course.

I am thankful to Dr. J.C. Joshi, Dr. Sanjeev Kumar, Mr. S.K. Dubey and Mr. Dilip Jain for preparation of some of the figures on computer.

Thanks are also due to Sh. Lakhmi Chand and Sh. P.N. Prasad for their help during preparation of excellent figures for the book.

Full credit is due to the publishers for producing a nice print of the book. Special thanks are due to Mr. M.S. Sejwal, General Manager, Narosa Publishing House, New Delhi for editing of manuscript.

My list of acknowledgements would be incomplete without mention of my UG/PG students of non-conventional energy sources, solar energy utilization and economics and planning courses running at CES, IIT Delhi who are my constant critics and are partly responsible for the preparation of the various chapters in the book.

Last but not least, I express my deep gratitude to my respected parents Smt. Bhagirathi Tiwari and Late Sh. Pt. Bashisht Tiwari (21-03-2001) for their blessing which helped me to reach my target.

G.N. Tiwari

Contents

CHAPTER 1

Introduction

1.1 THE SUN

The sun, which is the largest member of the solar system with other members revolving around it, is a sphere of intensely hot gaseous matter with a diameter of 1.39×10^9 m and, on an average, at a distance of 1.5×10^{11} m from the earth. As observed from the earth, the sun rotates on its axis about once every four weeks, though, it does not rotate as a solid body. The equator takes about 27 days and the polar regions take about 30 days for each rotation.

With an effective black body temperature T_s of 5777 K, the sun is, effectively, a continuous fusion reactor. Several fusion reactions have been suggested to supply the energy radiated by the sun; the most important being in which hydrogen (i.e. four protons) combines to form helium (i.e. helium nucleus); the mass of the helium nucleus is less than that of four protons, mass having been lost in the reaction and converted to energy. The reaction is as follows:

$$4(_1H^1) \longrightarrow {}_2He^4 + 26.7\,\text{MeV} \tag{1.1}$$

This energy is produced in the interior of the solar sphere, at temperatures of many millions of degrees. The produced energy must be transferred out to the surface and then be radiated into space ($E = \varepsilon\sigma T_s^4$; ε and σ are respectively the emissivity of surface and Stefan–Boltzmann constant).

It is estimated that 90 percent of the sun's energy is generated in the region 0 to 0.23R (R being the radius of the sun); the average density ρ and temperature T in this region are 10^5 kg/m^3 and about 8–40×10^6 K respectively. At a distance of about 0.7R from the center, the temperature drops to about 1.3×10^5 K and the density to 70 kg/m^3. Hence for $r > 0.7R$ convection begins to be important and the region $0.7R < r < R$ is known as the convective zone. The outer layer of this zone is called the photosphere. The edge of the photosphere is sharply defined, even though it is of low density. Above the photosphere is a layer of cooler gases several hundred kilometers deep called the reversing layer. Outside that, is a layer referred to as the chromosphere, with a depth of about 10,000 km. This is a gaseous layer with temperatures somewhat higher than that of the photosphere and with lower density. Still further out is the corona, of very low density and of very high temperature (10^6 K). A schematic diagram of the structure of the sun is shown in Figure 1.1.

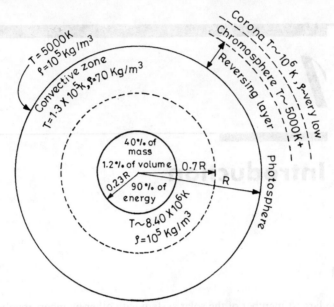

Figure 1.1 The structure of the sun.

1.2 THE EARTH

The earth came into existence some 4.6×10^9 years ago. The earth's inner core is a solid mass made of iron and nickle. Its outer core is made up of these two metals but in a melted state. Covering the outer core is the earth's mantle made up of solid rock. The outermost crust that covers the mantle is made up of rock too. The oldest rocks of sedimentary origin appear to date from about 3.7×10^9 years ago covering maximum area with liquid water. The fossil (remains of blue-green algae and bacteria) have been found in rocks/water at least 3×10^9 years ago. The existence of blue-green algae marks the beginnings of photosynthesis. As a result of photosynthesis, the level of O_2 and O_3 in the atmosphere is increased which blocks UV solar radiation coming from the sun. This phenomenon of blocking of UV radiation happened about 420 million years ago and this allowed the plants to grow on earth.

The earth is almost round in shape having a diameter of about 13000 kilometers. It revolves around the sun once in about a year. Nearly 70 percent of the earth is covered by the water and remaining 30 percent is land. Half of the earth is lit by sunlight at a time. It reflects one third of the sunlight that falls on it. This is known as earth's albedo. The earth is spinning about its axis constantly. Its axis is inclined at an angle of 23.5°. As a result, the lengths of days and nights keep changing. The diameter of earth is 12.75×10^6 m.

1.3 SOLAR SPECTRUM

The simplified picture of the sun, its physical structure, temperature and density gradients indicate that the sun, in fact, does not function as a blackbody radiator at a fixed temperature. Rather, the emitted solar radiation is the composite result of the several layers that emit and absorb radiation of various wavelengths. The photosphere is the source of most solar radiation and is essentially opaque, as the gases, of which it is composed, are strongly ionized and able to absorb and emit a continuous spectrum of radiation. In addition to the total energy in the solar spectrum (i.e., the solar constant) it is useful to know the spectral distribution of the extraterrestrial radiation, that is, the radiation that would be received in the absence of the atmosphere.

As shown in Figure 1.2, the maximum spectral intensity occurs at about 0.48 μm wavelength (λ) in the green portion of the visible spectrum. About 6.4 percent of the total energy is contained in ultraviolet region (λ < 0.38 μm); another 48 percent is contained in the visible region (0.38 μm < λ < 0.78 μm) and the remaining 45.6 percent is contained in the infrared region (λ > 0.78 μm).

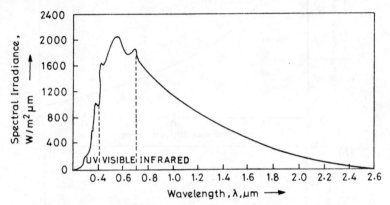

Figure 1.2 Spectral solar irradiance (*Thekaekara, 1977*).

The solar irradiance from the black body, in the present case either sun or earth, as a function of wavelength (μm) can be governed by Planck's law of radiation given by

$$E_{\lambda b} = \frac{C_1}{\lambda^5 [\exp\{C_2/(\lambda T)\} - 1]} \tag{1.2}$$

where $E_{\lambda b}$ represents the energy emmited per unit area per unit time per unit wavelength (μm) interval at a given wavelength, $C_1 = 3.742 \times 10^8$ W · μm^4/m^2(3.7405 × 10^{-16} m^{2}W) and $C_2 = 14387.9$ μm K (0.0143879 mK) (For details see Section 2.7.3).

The variation of $E_{\lambda b}$ with wavelength in μm is shown in Figure 1.3.

It is clear from Figures 1.3(*a*) and (*b*) that the wavelength of solar radiation emitted from the sun at about 6000 K and from earth at 288 K (15 °C) lies in the range of short wavelength and long wavelength range respectively as reported earlier. The comparison of these radiation from the sun and earth is shown in Figure 1.3(*c*).

The total emitted radiation from zero to any wavelength λ from the sun can be obtained from Equation 1.2 as

$$E_{0-\lambda,b} = \int_0^\lambda E_{\lambda b} d\lambda \tag{1.3}$$

If above equation is divided by σT^4 ($\sigma = 5.67 \times 10^{-8}$ W/m^{2}k^4) then integral can be made to be only a function of λT as follows:

$$f_{0-\lambda T} = \frac{E_{0-\lambda T}}{\sigma T^4} = \int_0^{\lambda T} \frac{C_1 d(\lambda T)}{\sigma (\lambda T)^5 [\exp(C_2/\lambda T) - 1]} \tag{1.4}$$

The value of $f_{0-\lambda T}$ for different λT, μmK, have been given in the Table 1.1.

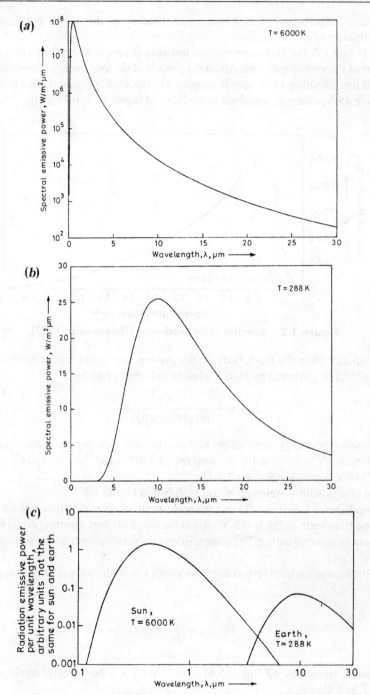

Figure 1.3 Effect of temperature of blackbody on emissive power
(a) $T = 6000\,K$, (b) $T = 288\,K$ and (c) comparison.

EXAMPLE 1.1

Calculate the total energy contained in ultra violet region $(0 < \lambda < 0.40\,\mu m)$, in the visible region $(0.40\,\mu m < \lambda < 0.70\,\mu m)$, in the infrared region $(0.70\,\mu m < \lambda < \infty)$ (Fig. 1.2) by using either Equation 1.4 or Table 1.1.

TABLE 1.1 Value of $f_{0-\lambda T}$ for different λT (μmK) for even increment of λT

λT, μmK	$f_{0-\lambda T}$	λT, μmK	$f_{0-\lambda T}$	λT, μmK	$f_{0-\lambda T}$
1000	0.0003	4500	0.5643	8000	0.8562
1100	0.0009	4600	0.5793	8100	0.8601
1200	0.0021	4700	0.5937	8200	0.8639
1300	0.0043	4800	0.6075	8300	0.8676
1400	0.0077	4900	0.6209	8400	0.8711
1500	0.0128	5000	0.6337	8500	0.8745
1600	0.0197	5100	0.6461	8600	0.8778
1700	0.0285	5200	0.6579	8700	0.8810
1800	0.0393	5300	0.6693	8800	0.8841
1900	0.0521	5400	0.6803	8900	0.8871
2000	0.0667	5500	0.6909	9000	0.8899
2100	0.0830	5600	0.7010	9100	0.8927
2200	0.1009	5700	0.7107	9200	0.8954
2300	0.1200	5800	0.7201	9300	0.8980
2400	0.1402	5900	0.7291	9400	0.9005
2500	0.1613	6000	0.7378	9500	0.9030
2600	0.1831	6100	0.7461	9600	0.9054
2700	0.2053	6200	0.7451	9700	0.9076
2800	0.2279	6300	0.7618	9800	0.9099
2900	0.2506	6400	0.7692	9900	0.9120
3000	0.2730	6500	0.7763	10000	0.9141
3100	0.2958	6600	0.7831	11000	0.9318
3200	0.3181	6700	0.7897	12000	0.9450
3300	0.3401	6800	0.7961	13000	0.9550
3400	0.3617	6900	0.8022	14000	0.9628
3500	0.3829	7000	0.8080	15000	0.9689
3600	0.4036	7100	0.8137	16000	0.9737
3700	0.4238	7200	0.8191	17000	0.9776
3800	0.4434	7300	0.8244	18000	0.9807
3900	0.4624	7400	0.8295	19000	0.9833
4000	0.4829	7500	0.8343	20000	0.9855
4100	0.4987	7600	0.8390	30000	0.9952
4200	0.5160	7700	0.8436	40000	0.9978
4300	0.5327	7800	0.8479	50000	0.9988
4400	0.5488	7900	0.8521	$\propto$	1.0000

Duffie and Beckman, 1991.

Solution

By using the value of Table 1.1,

i. The value of λT at 5777 K for ultra-violet region $= 0.4 \times 5777 = 2310.8\ \mu mK$
   The energy contained in fraction between 0–2310.8 μm K $= 0.1222\ (12.22\%)$
   The energy contained in W/m^2 between 0–2310.8 μm K $= 0.1222 \times 1367 = 167$

ii. The value of λT at 0.70 μm for visible region $= 0.7 \times 5777 = 4043.9\ \mu m$ K
   The energy contained in fraction between $0 - 4043.9\ \mu m$ K $= 0.4898\ (48.98\%)$
   The energy contained in W/m^2 between $0 - 0.4898\ \mu m$ K $= 0.4898 \times 1367 = 669.6$
   The energy contained in W/m^2 between $2310.8 - 4043.9\ \mu m$ K $= 669.6 - 167 = 502.6 (36.76\%)$

iii. The remaining 51.02 percent (697.4 W/m^2) is contained in the infrared region.
   It is clear from above calculation that there is significant change in energy contained by changing the range of ultra-violet, visible and infrared regions (Fig. 1.2).

According to Wein's displacement law, the wavelength corresponding to the maximum of solar irradiance from the sun can be obtained from

$$\lambda_{max} \cdot T = 2897.6\ \mu m\ K \tag{1.5}$$

EXAMPLE 1.2

Obtain Wein's displacement law by using Planck's law of radiation.

Solution

Differentiate Equation (1.2) with respect to λ and equate it to zero to get λ_{max}. After differentiation of Equation (1.2), one has,

$$\frac{\partial E_{\lambda b}}{\partial \lambda} = \frac{\partial}{\partial \lambda} \left[C_1 \lambda^{-5} (e^{C_2/\lambda T} - 1)^{-1} \right] = 0$$

$$= C_1 \left[(-5\lambda^{-6})(e^{C_2/\lambda T} - 1)^{-1} + \lambda^{-5}(-1)(e^{C_2/\lambda T} - 1)^{-2} e^{C_2/\lambda T} \left(-\frac{C_2}{\lambda^2 T} \right) \right] = 0$$

$$= \frac{C_1 \left[-5\lambda^{-6}(e^{C_2/\lambda T} - 1) + \lambda^5 (e^{C_2/\lambda T})(C_2/\lambda^2 T) \right]}{(e^{C_2/\lambda T} - 1)^2} = 0$$

or, $\quad -5(1 - e^{-C_2/\lambda T}) + \dfrac{C_2}{\lambda T} = 0$

Since $\quad \dfrac{C_2}{\lambda T} \gg 1,\ $ then $e^{-C_2/\lambda T} \to 0.$ Now

$$\frac{C_2}{\lambda_{\max}.T} = 5 \quad \text{or} \quad \lambda_{\max}.T = \frac{C_2}{5} = \frac{14387.9}{5} = 2877.58 \ \mu\text{m K}$$

This value is very close to the value given by Equation (1.5).

EXAMPLE 1.3

Calculate the maximum monochromatic emissive power at 288 K.

Solution

From the above example $\lambda_{\max} = 2897.6/288 = 10.06\ \mu\text{m}$
By using Equation (1.2), the maximum monochromatic power will be

$$E_{\lambda b} = \frac{3.742 \times 10^8}{(10.06)^5[\exp(14387.9/2897.6) - 1]} = 25.53 \ \text{W/m}^2 \cdot \mu\text{m}$$

The value of $E_{\lambda b}$ is the same as reported in Figure 1.3b at $\lambda_{\max} = 10.06\ \mu\text{m}$.

1.4 SOLAR RADIATION

The orientation of the earth's orbit around the sun is such that the sun–earth distance varies only by 1.7 percent and since the solar radiation outside the earth's atmosphere is nearly of fixed intensities, the radiant energy flux received per second by a surface of unit area held normal to the direction of sun's rays at the mean earth–sun distance, outside the atmosphere, is practically constant throughout the year. This is termed as the solar constant I_{sc} and its value is now adopted to be 1367 W/m^2. However, this extraterrestrial radiation suffers variation due to the fact that the earth revolves around the sun not in a circular orbit but follows an elliptic path, with sun at one of the foci. The intensity of extraterrestrial radiation I_{ext} measured on a plane normal to the radiation on the nth day of the year is given in terms of solar constant (I_{sc}) as follows (Duffie and Beckman, 1991):

$$I_{ext} = I_{sc}[1.0 + 0.033 \ \cos(360n/365)] \tag{1.6a}$$

The variation of extraterrestrial radiation with nth day of the year is shown in Figure 1.4.
For December 21, 1995 $n = 355$, $I_{ext} = 1411$ W/m^2 (Equation (1.6a))
For June 22, 1996 (Leap Year), $n = 174$, $I_{ext} = 1322$ W/m^2 (Equation (1.6a))

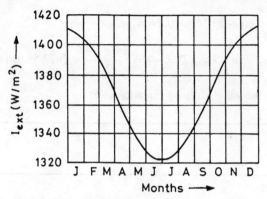

Figure 1.4 Variation of extraterrestrial solar radiation with time of the year.

EXAMPLE 1.4

Determine the temperature of the sun in the month of June for the following data: $I_{ext} = 1320\,\text{W/m}^2$, sun's diameter $(2R_s) = 1.39 \times 10^9$ m.
Mean sun–earth's distance $(L_{se}) = 1.5 \times 10^{11}$ m and $\sigma = 5.67 \times 10^{-8}$ W/m^{2}K^4.

Solution

Amount of solar radiation emitted by the surface of the sun $= \sigma T_s^4 (4\pi R_s^2)$
The total emitted radiation from the sun will be received by a sphere having a radius equal to mean sun–earth distance $= I_{ext} \times (4\pi L_{se}^2)$. Hence,

$$\sigma T_s^4 (4\pi R_s^2) = I_{ext} \times (4\pi L_{se}^2)$$

or,
$$T_s = \left[\frac{I_{ext}(4\pi L_{se}^2)}{\sigma(4\pi R_s^2)} \right]^{\frac{1}{4}} = 5738.5\,\text{K}.$$

1.4.1 Terrestrial and Extraterrestrial Regions

Solar radiations while passing through the earth's atmosphere are subjected to the mechanisms of atmospheric absorption and scattering. A fraction of the radiation reaching the earth's surface is reflected back into the atmosphere and is subjected to these atmospheric phenomenon again; the remainder is absorbed by the earth's surface. Figure 1.5(a) shows the position of terrestrial and extraterrestrial regions. The atmospheric absorption is due to ozone (O_3), oxygen (O_2), nitrogen (N_2), carbon dioxide (CO_2), carbon monoxide (CO) and water vapor (H_2O) and the scattering is due to air molecules, dust and water droplets. The x-rays and extreme ultra-violet radiations of the sun are absorbed highly in the ionosphere by nitrogen, oxygen and other atmospheric gases; ozone and water vapors largely absorb ultraviolet

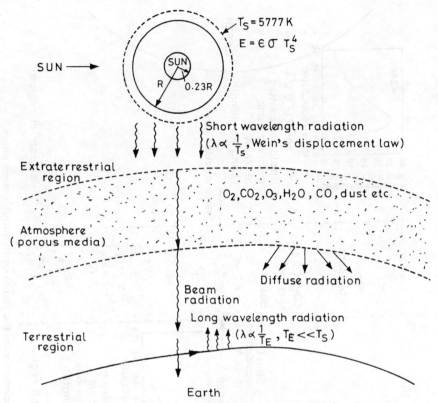

Figure 1.5(a) Position of terrestrial and extraterrestrial region.

($\lambda < 0.40 \, \mu m$) and infrared radiations ($\lambda > 2.3 \, \mu m$) respectively. There is almost complete absorption of short wave radiations ($\lambda < 0.29 \, \mu m$) in the atmosphere. Hence, the energy in wavelength radiation below 0.29 μm and above 2.3 μm, of the spectra of the solar radiation, incident on the earth's surface is negligible. Scattering by air molecules, water vapors and dust particles results in the attenuation of radiation. The range of wavelength radiation emitted from the sun, attenuation of its amplitude during propagation from the sun to atmosphere and further attenuation of radiation in the atmosphere has been shown in Figure 1.5(b) which also shows the long wavelength radiation emitted from the earth. The atmospheric attenuation is characterized by the term called *air mass*. It is defined as the ratio of the optical thickness of the atmosphere through which beam radiation passes to the optical thickness if the sun were at zenith. Large value of air mass indicates that solar radiation travel greater distance in atmosphere. Hence is prone to attenuation. An expression for air mass (referring to Figure 1.6(a)) is given by

$$\text{Air mass} = \frac{AB}{AC} = \sec \Psi \qquad (1.6b)$$

At noon, $\Psi = 0$.

The variation of air mass with time of the day for the latitude of Delhi for different number of days of the year is shown in Figure 1.6(b). It is clear from the figure that the sunshine hour is shorter and air mass is higher for the month of December 21, in comparison with other days as expected.

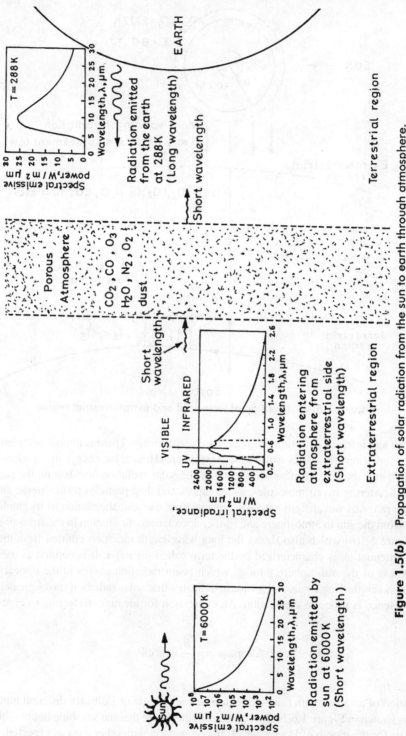

Figure 1.5(b) Propagation of solar radiation from the sun to earth through atmosphere.

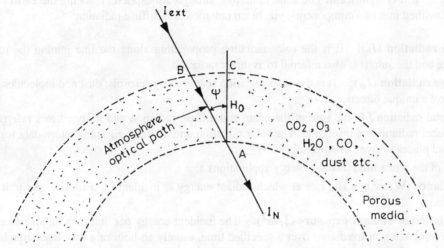

Figure 1.6(a)   Direction of sun's ray with respect to atmosphere.

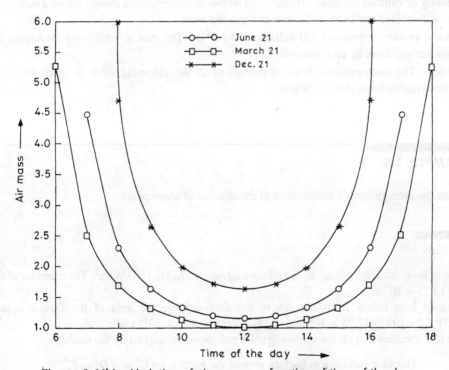

Figure 1.6(b)   Variation of air mass as a function of time of the day.

Thus, from the view of terrestrial applications of solar energy, only radiation of wavelength between 0.29 and 2.3 μm is significant. The solar radiation, through atmosphere, reaching the earth's surface can be classified into two components, viz. beam radiation and diffuse radiation.

Beam radiation (I_b):   It is the solar radiation propagating along the line joining the receiving surface and the sun. It is also referred to as direct radiation.

Diffuse radiation (I_d):   It is the solar radiation scattered by aerosols, dust and molecules. It does not have a unique direction.

The total radiation I is the sum of the beam and diffuse radiation and is sometimes referred to as the global radiation or insolation. The solar radiation and ambient air temperature data for certain selected places is given in Appendix II.

Some of the terms used in solar energy applications are:

Irradiance (W/m^2):   The rate at which radiant energy is incident on a surface, per unit area of surface.

Irradiation or radiant exposure (J/m^2):   The incident energy per unit area on a surface, found by the integration of irradiance over a specified time, usually an hour or a day. Insolation is a term applied specifically to solar energy irradiation. The symbol H is used for insolation for a day and I for insolation for an hour. Both H and I can represent beam, diffuse or total and can be on surfaces of any orientation.

Radiosity or radiant exitance (W/m^2):   The rate at which radiant energy leaves a surface per unit area, by combined emission, reflection and transmission.

Emissive power or radiant self-exitance (W/m^2):   The rate at which radiant energy leaves a surface per unit area, by emission only.

Albedo: The earth reflects about 30 percent of all the incoming solar radiation back to extra-terrestrial region through atmosphere.

EXAMPLE 1.5

Calculate the average earth's temperature in the absence of atmosphere.

Solution

The rate of solar radiation per m^2 in extraterrestrial region (I_{sc}) is 1367 W/m^2. The diameter of the earth (D_e) is 12.75×10^6 m.

The total heat inflow from the sun to the earth = Projected area of the earth × Solar flux = $\pi(D_e/2)^2 I_{sc} = (\pi/4)(12.75 \times 10^6 \text{ m})^2 \cdot 1367 \text{ W/m}^2 = 1.75 \times 10^{14} \text{ kW}$.

Here it is assumed that all the incoming solar radiation is absorbed by the earth.

$$\text{Outward radiation} = \text{Surface area of the earth} \times \varepsilon\sigma T_e^4 = \pi D_e^2 \varepsilon\sigma T_e^4$$

$$= \pi(12.75 \times 10^6 \text{ m})^2 \times 5.672 \times 10^{-11}(\text{kW/m}^2\text{K}^4) \cdot T_e^4$$

If the radiation due to nuclear decay and tidal friction with the moon on the earth, which is about 0.1

percent of the solar energy inflow, is neglected then

$$\text{All the incoming solar radiation} = \text{Outward radiation}$$

or, $\quad \pi (12.75 \times 10^6 \, \text{m})^2 \times 5.672 \times 10^{-11} (\text{kW/m}^2 \text{K}^4) \cdot (T_e^4) = 1.75 \times 10^{14} \, \text{kW}$

or, $\quad\quad\quad\quad\quad\quad\quad\quad\quad T_e = 278.7 \, \text{K} = 5.7 \, ^\circ\text{C},$

where $\varepsilon = 1$.

If the total heat flow in from the sun to the earth is multiplied by 0.7 due to the earth's albedo then $T_e = 255 \, \text{K} = -18 \, ^\circ\text{C}$, a frozen world. This indicates that existing atmosphere between the sun and the earth blocks maximum outgoing radiation so that the average earth's temperature has been raised to about 15 $^\circ$C.

EXAMPLE 1.6

What is the fraction of the outgoing radiation from the earth which is blocked by the atmosphere?

Solution

Assume that 30 percent of incoming radiation from the sun is reflected back to extraterrestrial region through atmosphere. This means that 70 percent of incoming radiation is absorbed by the earth. If we assume average earth's surface about 15 $^\circ$C $= 288 \, \text{K}$, then

$$\text{Fraction emitted} = \frac{0.7 \, (\text{total solar radiation inflow from the sun})}{\pi D^2 \sigma T^4}$$

$$= \frac{0.7 \times (1.75 \times 10^{14})}{3.14 \times (12.75 \times 10^6)^2 \times 5.672 \times 10^{-11} \times (288)^4}$$

$$= 0.606$$

The atmospheric outward transmission of radiant energy $= 0.606/0.7 = 0.86$.
Eighty six percent of the inward transmission of solar energy is blocked by atmosphere.

1.4.2 Solar Time

Time based on the apparent angular motion of the sun across the sky, with solar noon denoting the time the sun crosses the meridian of the observer. The difference in minutes between solar time and standard time is

$$\text{Solar time} - \text{Standard time} = 4(L_{\text{st}} - L_{\text{loc}}) + E \tag{1.7}$$

where L_{st} is the standard meridian for the local time zone (for India the value is 81°54'), L_{loc} is the longitude of the location in question (in degrees west) (Table 1.2) and E the equation of time (in minutes) (Table 1.3) is given as

$$E = 229.2 \, (0.000075 + 0.001868 \cos B - 0.032077 \sin B - 0.014615 \cos 2B - 0.04089 \sin 2B) \tag{1.8}$$

where $B = (n - 1)360/365$ and $n = $ day of the year.

TABLE 1.2 Latitude, longitude and elevation for different places in India

Place	Latitude (Φ)	Longitude (L_{loc})	Elevation (E_0)
Bangalore	12°58′ N	77°35′ E	921 m above msl
Bombay	18°54′ N	72°49′ E	11 m above msl
Jodhpur	26°18′ N	73°01′ E	224 m above msl
Mt. Abu	24°36′ N	72°43′ E	1195 m above msl
New Delhi	28°35′ N	77°12′ E	216 m above msl
Simla	31°06′ N	77°10′ E	2202 m above msl
Srinagar	34°05′ N	74°50′ E	1586 m above msl
Calcutta	22°32′ N	88°20′ E	6 m above msl
Madras	13°00′ N	80°11′ E	16 m above msl

The equation of time (minutes : seconds) for typical days for different months for Delhi (longitude 77°12′ E) is given in Table 1.3.

TABLE 1.3 The sun's equation of time (E) (minutes : second)

Month	1	8	15	22
January	−(3 : 16)	−(6 : 26)	−(9 : 12)	−(11 : 27)
February	−(13 : 34)	−(14 : 14)	−(14 : 15)	−(13 : 41)
March	−(12 : 36)	−(11 : 04)	−(9 : 14)	−(7 : 12)
April	−(4 : 11)	−(2 : 07)	−(0 : 15)	(1 : 19)
May	2 : 50	3 : 31	3 : 44	3 : 30
June	2 : 25	1 : 15	−(0 : 09)	−(1 : 40)
July	−(3 : 33)	−(4 : 48)	−(5 : 45)	−(6 : 19)
August	−(6 : 17)	−(5 : 40)	−(4 : 35)	−(3 : 04)
September	−(0 : 15)	2 : 03	4 : 29	6 : 58
October	10 : 02	12 : 11	13 : 59	15 : 20
November	16 : 20	16 : 16	15 : 29	14 : 02
December	11 : 14	8 : 26	5 : 13	1 : 47

EXAMPLE 1.7

Determine the solar time (ST) corresponding to 12.00 noon Indian Standard Time (IST) (Longitude 81°54′ E) on May 8, 1995 for Delhi.

Solution

Equation of time for May 8 is 3 min. 31 sec. (from Table 1.3). The longitude correction, in this case, would be negative as Delhi is west of standard meridian. Using Equation (1.7), we have

$$ST = 12\,hr\ 0\,min\ 0\,sec + 3\,min\ 31\,sec - 4\,min\ (81°54′ - 77°12′)$$

$$= 11\,hr\ 45\,min\ 23\,sec$$

Here, $1° = 60′$ and 4 min. (Appendix I)

1.5 INSTRUMENTS

Instruments for measuring solar radiation are basically of two types. The accepted terms for these are:

 i. *Pyrheliometer*:   An instrument using a collimated detector for measuring solar radiation from the sun and from a small portion of the sky around the sun (i.e. beam radiation) at normal incidence.

 ii. *Pyranometer*:   An instrument for measuring total hemispherical solar (beam + diffuse) radiation, usually on a horizontal surface. If shaded from the beam radiation by a shade ring or disc, a pyranometer measures diffuse radiation.

In addition, the terms solarimeter (which is same as pyranometer) and actinometer (which usually refers to a pyrheliometric instrument) can be used.

1.5.1 Pyrheliometer

In this instrument, two identical blackened manganin strips are arranged so that either one can be exposed to radiation at the base of collimating tubes by moving a reversible shutter. Each strip can be electrically heated and each is fitted with a thermocouple as shown in Figure 1.7.

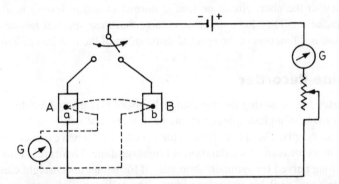

Figure 1.7   Circuit diagram for the thermoelectric type pyrheliometer.

With one strip shaded and one strip exposed to radiation, a current is passed through the shaded strip to heat it to the same temperature as the exposed strip. When there is no difference in temperature, the electrical energy to the shaded strip must equal the solar radiation absorbed by the exposed strip.

Solar radiation is then determined by equating the electrical energy to the product of incident solar radiation, strip area and absorptance. Then the position of the shutter is reversed, interchanging the electrical and radiation heating, and the second value is determined. Alternating the shade and the functions of the two strips compensates for minor differences in the strips, such as edge effects and lack of uniformity of electrical heating.

It may be noted that the solar intensity thus measured are the radiation from the sun and from a portion of the sky around the sun. Since the detectors do not distinguish between forward-scattered radiation, which comes from the circumsolar sky, and the beam radiation, the instruments are, in effect, defining beam radiation. Though it is difficult to generalize, but it appears that thin clouds or haze can affect the angular distribution of radiation within the field of view of standard pyrheliometers.

1.5.2 Pyranometer

The pyranometer is used for the measurement of global solar radiation received from the entire hemisphere on a horizontal surface. The principle of working is the same as pyrheliometer except for the fact that sensitive surface is exposed to total (beam + diffuse + reflected from earth and surroundings) radiations. The sensitive surface consists of a circular, blackened (hot-junction) multijunctions thermopile whose cold junctions are electrically insulated from basement. The temperature difference between hot and cold junctions is a function of the radiation falling on the surface. The sensitive surface is covered by two concentric hemispherical glass domes to shield it from wind and rain. This also reduces the convection currents.

Pyranometer, when provided with an occulting disc, measures the diffuse radiation. This disc or band, blocks the beam radiation from the surface. It may be noted that the pyranometers are calibrated so as to measure the solar radiation on horizontal surface. Therefore when tilted, the change in free convection regime within the glass dome may introduce an error in measurement.

A pyranometer (or pyrheliometer) produces voltage, as a function of the incident solar radiation, from the thermopile detectors. A potentiometer is required to detect and record this output. Radiation data usually must be integrated over some period of time, such as an hour or a day. Integration can be done by means of planimetry or electronic integrator. Pyranometers have also been based on photovoltaic (solar cell) detectors. Silicon cells are the most common for solar energy measurement, although cadmium sulfide and selenium cells have been used. Silicon solar cells have the property that their light current (approximately equal to the short-circuit current at normal radiation levels) is a linear function of the incident solar radiation. They have the disadvantage that their spectral response is not linear, so instrument calibration is a function of the spectral distribution of the incident radiation.

1.5.3 Sunshine Recorder

This instrument is used for measuring the duration, in hours, of bright sunshine during the course of the day. It essentially consists of a glass sphere mounted in a section of spherical brass bowl with grooves for holding the recorder cards. The sphere burns a trace on the card when exposed to the sun, the length of the trace being a direct measure of the duration of bright sunshine. There are sets of grooves for taking three sets of cards, long curved for summer, short curved for winter and straight cards at equinoxes.

Solar radiation data are available in various forms. The following information about the radiation data is useful in its application:

1. whether they are instantaneous measurements or values integrated over some period of time.
2. the time or time period of the measurements.
3. whether the measurements are of beam, diffuse or total radiation, and the instruments used.
4. the receiving surface orientation (usually horizontal, sometimes inclined at a fixed slope, or normal to the beam radiation).
5. if averaged, the period over which they were averaged (e.g. monthly average of daily radiation).

1.6 SUN–EARTH ANGLES

The energy flux of beam radiation on a surface with arbitrary orientation can be obtained from the knowledge of flux either on a surface perpendicular to the sun rays or on a horizontal surface.

If θ be the angle of incidence of a beam of flux I, incident on a plane surface then the flux incident on the plane surface is $I \cos \theta$. The angle θ can, in general, be evaluated from a general equation with the knowledge of angles shown in Figures 1.8(a), (b) and (c) as follows:

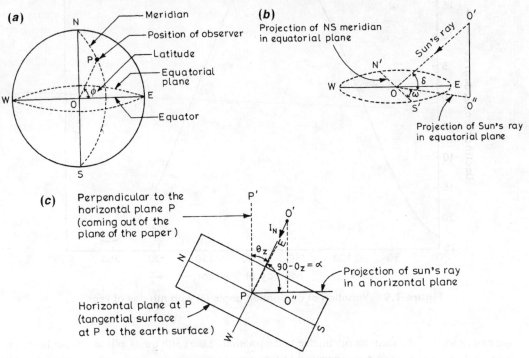

Figure 1.8(a), (b) and (c) View of different angles.

Latitude (ϕ): The latitude of a location is the angle made by the radial line, joining the given location to the center of the earth, with it's projection on the equatorial plane (Fig. 1.8(a)). The latitude is positive for the northern hemisphere and negative for southern hemisphere. The latitude for some places in India is given in Table 1.2.

Declination (δ): Declination may be defined as the angle between the line joining the centers of the sun and the earth and it's projection on the equatorial plane (Fig. 1.8(b)). Declination is due to the rotation of the earth about an axis which makes an angle of $66\frac{1}{2}°$ with the plane of it's rotation around the sun. The declination varies from a maximum value of 23.45° on June 21 to a minimum value of $-23.45°$ on December, 21. It may be calculated by the following relation (Cooper, 1969),

$$\delta = 23.45 \sin \left[\frac{360}{365} (284 + n) \right] \tag{1.9}$$

The variation of declination angle with nth day of year has been depicted in Figure 1.9.

Hour angle (ω): It is the angle through which the earth must be rotated to bring the meridian of the plane directly under the sun (Fig. 1.8(b)). In other words, it is the angular displacement of the

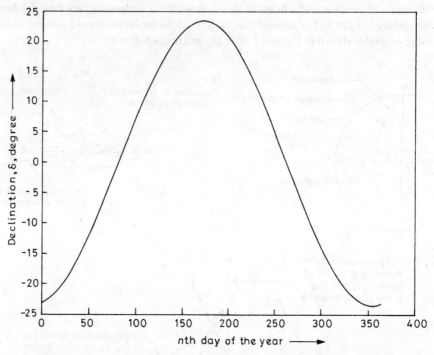

Figure 1.9 Variation of declination angle (δ) with nth day of year.

sun east or west of the local meridian, due to the rotation of the earth on its axis at 15° per hour. The hour angle is zero at solar noon, negative in the morning and positive in the afternoon (Table 1.4) for northern hemisphere (India) and vice-versa for southern hemisphere (Australia). Expression for hour angle is given by

$$\omega = (ST - 12)\,15° \qquad (1.10)$$

where ST is local solar time.

TABLE 1.4 The value of hour angle with time of the day (for northern hemisphere)

Time of the day (hours)	6	7	8	9	10	11	12
Hour angle (degree)	−90	−75	−60	−45	−30	−15	0
Time of the day (hours)	12	13	14	15	16	17	18
Hour angle (degree)	0	+15	+30	+45	+60	+75	+90

EXAMPLE 1.8(a)

Calculate the declination on 23 September 1995.

Solution

For 23 September 1995, $n = 266$. From Equation (1.9), we have

$$\delta = 23.45 \sin \left(\frac{360}{365} \right) (284 + 266) = -1.01°$$

Zenith (θ_z): It is the angle between sun's ray and perpendicular line to the horizontal plane (Fig. 1.8(c)).

Altitude (α): It is defined as the angle between sun rays and a horizontal plane (Fig. 1.8(c)). Also, $\alpha = 90 - \theta_z$.

Solar altitude angle (α_s): It is the angle between sun ray and its projection in a horizontal plane (Fig. 1.10) i.e. $\alpha = \alpha_s$.

Slope (β): It is the angle between the plane surface, under consideration, and the horizontal. It is taken to be positive for surface sloping towards south and negative for surfaces sloping towards north (Fig. 1.10).

Surface azimuth angle (γ): It is the angle in the horizontal plane, between the line due south and the projection of the normal to the surface(inclined plane) on the horizontal plane (Fig. 1.10). By

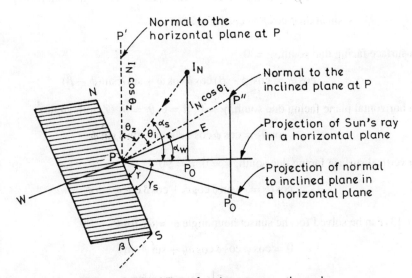

Figure 1.10 View of various sun-earth angles.

convention, the angle will be taken negative, if the projection is east of south and positive if west of south for Northern hemisphere (India) and vice-versa for southern hemisphere (Australia). The value of γ for some orientations are given in Table 1.5.

TABLE 1.5 Surface azimuth angle (γ) for various orientations in the Northern hemisphere

Surface orientation	γ
Sloped towards South	0°
Sloped towards North	−180°
Sloped towards East	−90°
Sloped towards West	+90°
Sloped towards South-East	−45°
Sloped towards South-West	+45°

Solar azimuth angle (γ_s):   It is the angle in a horizontal plane, between the line due south and the projection of beam radiation on the horizontal plane (Fig. 1.10). By convention, the angle is taken to be positive and negative, respectively, if the projection is east of south and west of south for Northern hemisphere (India) and vice-versa for Southern hemisphere (for details see Duffie and Beckman, 1991).

Solar wall azimuth angle (α_w):   It is the angle between normal to the inclined plane and projection of sun's ray in a horizontal plane (Fig. 1.10).

Angle of incidence (θ_i):   It is the angle between beam radiation on a surface and the normal to that surface (Fig. 1.10).

In general, the angle of incidence can be expressed as,

$$\cos\theta_i = (\cos\phi\cos\beta + \sin\phi\sin\beta\cos\gamma)\cos\delta\cos\omega + \cos\delta\sin\omega\sin\beta\sin\gamma$$
$$+ \sin\delta(\sin\phi\cos\beta - \cos\phi\sin\beta\cos\gamma) \tag{1.11}$$

i. For a surface facing due south, $\gamma = 0$

$$\cos\theta_i = \cos(\phi - \beta)\cos\delta\cos\omega + \sin\delta\sin(\phi - \beta) \tag{1.12}$$

ii. For a horizontal plane facing due south, $\gamma = 0$, $\beta = 0$, $\theta_i = \theta_z$ (Zenith angle)

$$\cos\theta_z = \cos\phi\cos\delta\cos\omega + \sin\delta\sin\phi \tag{1.13}$$

iii. For a vertical surface facing due south, $\gamma = 0$, $\beta = 90°$

$$\cos\theta_i = -\sin\delta\cos\phi + \cos\delta\cos\omega\sin\phi \tag{1.14}$$

Equation (1.13) can be solved for the sunset hour angle $\omega = \omega_s$ for $\theta_z = 90°$

$$0 = \cos\phi\cos\delta\cos\omega_s + \sin\phi\sin\delta$$

or,

$$\omega_s = \cos^{-1}(-\tan\phi\tan\delta)$$

Total angle between sunrise and sunset is given by

$$2\omega_s = 2\cos^{-1}(-\tan\phi\tan\delta) \tag{1.15}$$

Since $15° = 1$ hour, the number of daylight (sunshine) hours (N) is given by

$$N = \frac{2}{15}\cos^{-1}(-\tan\phi\tan\delta) \tag{1.16}$$

The variation of N with latitude (ϕ) for different n (nth day of the year) is given in Figure 1.11. It is clear from the figure that the total sunshine hour is 12 hours for March and September, 21 and it is independent of latitude. However, it is significantly varying with latitude for other days. The maximum total sunshine hour is for June 21 and minimum for December 21 at latitude more than 60°.

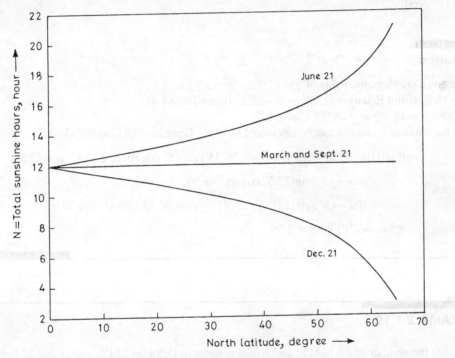

Figure 1.11 Variation of N with latitudes for different nth day of the year.

EXAMPLE 1.8(b)

Derive an expression for air mass in terms of zenith angle (θ_z) and altitude angle.

Solution

In equation (1.6b), substitute $\psi = \theta_z$ (both are same) hence

$$\text{air mass} = \sec\theta_z = \frac{1}{\cos\theta_z} \quad (\text{Equation}(1.13))$$

since $\alpha = 90 - \theta_z$, hence

$$\text{air mass} = \operatorname{cosec} \alpha.$$

EXAMPLE 1.9

Calculate the angle of incidence of beam radiation on a surface located at New Delhi at 1:30 (Solar time) on 16 February 1995, if the surface is tilted 45° from the horizontal and pointed 30° west of south $(1° = 60')$.

Solution

In the given problem, the value of n is 47.
$\delta = -13.0°$ (from Equation (1.9)); $\omega = +22.5°$ (from Table 1.4).
$\gamma = 30°$; $\beta = 45°$; $\phi = +28°35'$ (Table 1.2)
Now, the angle of incidence can be calculated by using Equation (1.11) as follows:

$$\cos \theta_i = (\cos(28°35')\cos 45° + \sin(28°35')\sin 45° \cos 30)\cos(-13°)\cos(22.5°)$$

$$+ \cos(-13°)\sin(22.5°)\sin 45° \sin 30°$$

$$+ \sin(-13°)(\sin(28°35')\cos 45° - \cos(28°35')\sin 45° \cos 30°)$$

$$\theta_i = \cos^{-1}(0.999) = 2.56°.$$

EXAMPLE 1.10

Calculate the number of daylight hours (sunshine hour) in Delhi on 22 December and 22 June 1995.

Solution

Here $\phi = 28°35'$ (Table 1.2). For 22 December 1995, $n = 356$ and $\delta = -23.44°$
From Equation (1.16), we have

$$N = \frac{2}{15} \cos^{-1}[-\tan(-23.44°)\tan(28°35')]$$

$$N = \frac{2}{15} \cos^{-1}[-(-0.434)(0.545)]$$

$$= (2/15)\cos^{-1}[0.237] = 10.18 \text{ hours}$$

Similarly, for 22 June 1995, $n = 173$; $\delta = 23.45°$ (Equation (1.9))
From Equation (1.16), we have

$$N = \frac{2}{15} \cos^{-1}(-\tan 23.45° \tan 28°35') = 13.82 \text{ hours}$$

EXAMPLE 1.11

Calculate zenith angle of the sun at New Delhi at 2.30PM on 20 February 1995.

Solution

We have, $n = 51$; $\phi = 28°35'$ (Table 1.2)
$\delta = -11.58°$ (Equation (1.9)); $\omega = 37.5°$ (Table 1.4)
From Equation (1.13), we have,

$$\cos \theta_z = \cos(28°35') \cos(-11.58°) \cos(37.5°) + \sin(-11.58) \sin(28°35')$$

$$= 0.587$$

$$\theta_z = \cos^{-1}(0.587) = 54.03°$$

1.7 AVAILABILITY OF SOLAR RADIATION ON AN INCLINED SURFACE

The total solar radiation incident on a surface consists of: (i) beam solar radiation, (ii) diffuse solar radiation and (iii) solar radiation reflected from the ground and the surroundings.

Normally the beam radiation (I_b) and diffuse radiation (I_d) on a horizontal surface are recorded. In the case of non availability of datas for beam and diffuse radiation, the following expression for beam and diffuse radiation on the horizontal surface can be used:

$$I_b = I_N \cos \theta_z \qquad (1.17a)$$

and

$$I_d = (1/3)[I_{\text{ext}} - I_N] \cos \theta_z \qquad (1.17b)$$

where an expression for I_N is given problem 1.6.

After knowing beam and diffuse radiation on horizontal surface, Liu and Jordan (1962) have given a formula to evaluate total radiation on a surface of arbitrary orientation.

$$I_T = I_b R_b + I_d R_d + \rho R_r (I_b + I_d) \qquad (1.18)$$

where R_b, R_d and R_r are known as conversion factors for beam, diffuse and reflected components respectively and ρ is the reflection coefficient of the ground ($= 0.2$ and 0.6 for ordinary and snow covered ground respectively). The expressions for these are as follows:

i. **R_b:** It is defined as the ratio of flux of beam radiation incident on an inclined surface to that on a horizontal surface.

The flux of beam radiation incident on a horizontal surface (I_b) is given by

$$I_b = I_N \cos\theta_z$$

and that on an inclined surface (I'_b) is

$$I'_b = I_N \cos\theta_i$$

where θ_z and θ_i are the angles of incidence on the horizontal and inclined surfaces, respectively, (Fig. 1.10) and I_N is the intensity of beam radiation. Now, R_b, for beam radiation can be obtained as

$$R_b = \frac{I'_b}{I_b} = \frac{\cos\theta_i}{\cos\theta_z} \tag{1.18a}$$

Depending on the orientation of inclined surface, the expression for $\cos\theta_i$ and $\cos\theta_z$ can be written from Equations (1.11) and (1.13) respectively.

The variation of R_b with nth days of the year for different latitude, inclination and hour angles is shown in Figure 1.12. Figure 1.12(a) shows that conversion factor for radiation has been significant effect at higher latitude and low value of n. Also R_b depends significantly at higher inclination of the plane (β) and becomes less significant at lower value of β (Fig. 1.12(b)). The maximum value of R_b is observed at higher β at low value of n as per expectation. Similar effect has been observed in Figure 1.12(c) for hour angle, however R_b tends to minimum value at higher n irrespective of any value of ω which is accordance with motion of sun at an early morning or at late evening.

ii. **R_d:** It is the ratio of the flux of diffuse radiation falling on the tilted surface to that on the horizontal surface.

This conversion factor depends on the distribution of diffuse radiation over the sky and on the portion of sky seen by the surface. But a satisfactory method of estimating the distribution of diffuse radiation over the sky is yet to be found. It is, however, widely accepted that sky is an isotropic source of diffuse radiation. If $(1 + \cos\beta)/2$ be the radiation shape factor for a tilted surface with respect to sky, then

$$R_d = \frac{1 + \cos\beta}{2} \tag{1.18b}$$

iii. **R_r:** The reflected component comes mainly from the ground and other surrounding objects. If the considered reflected radiation is diffuse and isotropic, then the situation is opposite to that in the above case.

$$R_r = \left(\frac{1 - \cos\beta}{2}\right) \tag{1.18c}$$

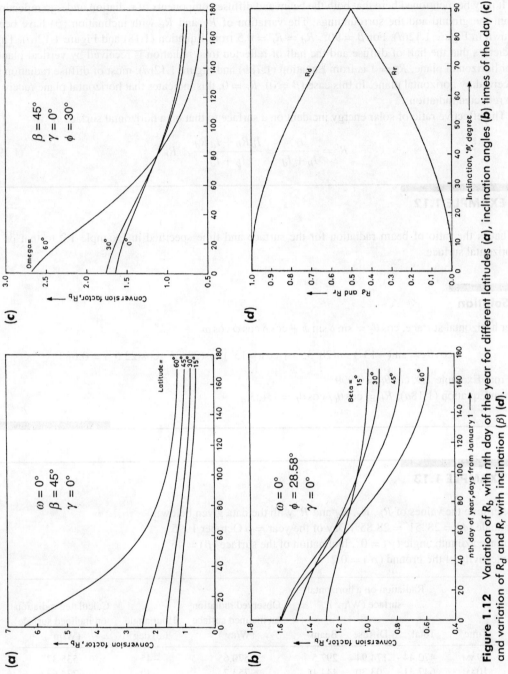

Figure 1.12 Variation of R_b with nth day of the year for different latitudes (**a**), inclination angles (**b**) times of the day (**c**) and variation of R_d and R_r with inclination (β) (**d**).

It may be mentioned here that both the beam and diffuse components of radiation undergo reflection from the ground and the surroundings. The variation of R_d and R_r with inclination (β) have been shown in Figure 1.12(d). For $\beta = 90°$, $R_d = R_r = 0.5$ from Equation (1.18) and Figure 1.12(d). This indicates that the half of diffuse and the half of reflected total radiation is received by vertical plane. For horizontal plane, $R_d = 1$ is from Equation (1.18b) and Figure 1.12(d), most of diffuse radiation is received by horizontal plane. In this case ($\beta = 0$), $R_r = 0$, this indicates that horizontal plane receives no reflected radiation.

The effective ratio of solar energy incident on a surface to that on a horizontal surface R' is

$$R' = \frac{I_t}{I_b + I_d} = \frac{I_b R_b + I_d R_d}{I_b + I_d} + R_r \tag{1.19}$$

EXAMPLE 1.12

What is the ratio of beam radiation for the surface and time specified in Example 1.9 to that on a horizontal surface.

Solution

For horizontal surface, $\cos\theta_z = \sin\delta\sin\phi + \cos\delta\cos\phi\cos\omega$

$$\cos\theta_z = \sin(-13°)\sin(28°35') + \cos(-13°)\cos(28°35')\cos(22.5°) = 0.683$$

From Example 1.9, $\cos\theta_i = 0.999$
From Equation (1.18a); $R_b = \cos\theta_i / \cos\theta_z = 1.463$

EXAMPLE 1.13

Determine the values of R_b, R_r, R_d and R' with the data given below:
Latitude (ϕ) = $28°51' = 28.85°$, Day of the year = 6 October 1995
Surface azimuth angle (γ) = $0°$, Inclination of the surface (β) = $45°$
Reflectivity of the ground (ρ') = 0.2

Time	Radiation on a horizontal surface (W/m²)			Observed radiation on inclined surface (W/m²)	Hour angle (ω) deg	Calculated radiation on inclined surface (W/m²)
	Total	Diffuse	Beam			
9AM	472.44	174.94	297.5	570.65	−45	535.37
10AM	647.41	203.30	444.11	753.7	−30	734.63
11AM	752.40	222.22	530.18	839.83	−15	851.607
12 noon	769.7	231.0	538.70	832.9	0	867.93
1PM	752.40	236.4	516.00	872.9	15	846.60

Solution

Here, $n = 279$, $\delta = -6.18°$ (from Equation (1.9))

Calculation for R_b:

   From Equation (1.11), $\cos\theta_i = 0.985$

   From Equation (1.13), $\cos\theta_z = 0.819$

   From Equation (1.18a), $R_b = \cos\theta_i / \cos\theta_z = 1.203$

   Calculation for R_d and R_r:

   From Equations (1.18b) and (1.18c),

$$R_d = \frac{1 + \cos 45°}{2} = 0.8536 \quad \text{and} \quad R_r = \rho\left(\frac{1 - \cos 45°}{2}\right) = 0.029$$

Hence, using the data of the above table, for 12 noon, the total solar radiation on an inclined surface can be obtained from Equation (1.18) and is given by

$$I_T = I_b \cdot R_b + I_d \cdot R_d + R_r(I_b + I_d) = 648.19 + 197.18 + 22.56 = 867.93 \text{ W/m}^2$$

The above calculated value (867.93) is in close agreement with the observed value (832.9) given in the table.

Calculation for R':

From Equation (1.19) $$R' = \frac{648.19 + 197.18}{769.7} + 0.029 = 1.127$$

The calculations done above can be repeated for other hours given in the table.

1.7.1 Monthly Average Daily Radiation on Sloped Surfaces

In the earlier section, we have discussed the calculation of total hourly radiation on sloped surfaces from measurements on a horizontal surface. For use in solar process design procedures we also require the monthly average daily radiation, H_T, on the tilted surface. It may be noted that as per our convention I denotes the hourly radiation, H denotes the daily radiation and $\bar{H}$ denotes the monthly average of the daily radiation. The procedure for calculating H_T is parallel to that for I_T, that is, by summing the contributions of the beam radiation ($H_b = H - H_d$), the components of the diffuse radiation (H_d), and the radiation reflected from the ground ($\rho' H$). If the diffuse and ground-reflected radiation are each assumed to be isotropic, then the monthly mean daily solar radiation on an unshaded tilted surface can be expressed as (Klein, 1977):

$$\bar{H}_T = (\bar{H} - \bar{H}_d)\bar{R}_b + \bar{H}_d\left(\frac{1 + \cos\beta}{2}\right) + \bar{H}\rho'\left(\frac{1 - \cos\beta}{2}\right) \tag{1.20a}$$

$$\bar{R} = \frac{\bar{H}_T}{\bar{H}} = \left(1 - \frac{\bar{H}_d}{\bar{H}}\right)\bar{R}_b + \frac{\bar{H}_d}{\bar{H}}\left(\frac{1 - \cos\beta}{2}\right) + \rho'\left(\frac{1 - \cos\beta}{2}\right) \tag{1.20b}$$

where $\bar{R}_b$ is the conversion factor for average daily beam radiation.

1.8 SOLAR RADIATION ON HORIZONTAL SURFACE

At any point of time, the solar radiation outside the atmosphere (I_0) incident on a horizontal plane is

$$I_0 = I_{sc} \left[1.0 + 0.033 \cos \left(\frac{360n}{365} \right) \right] \cos \theta_z \qquad (1.21a)$$

where I_{sc} is the solar constant and n is the day of the year. The recommended average days for each month and values of n by months are given in Table 1.6.

TABLE 1.6 Recommended Average Days for Months and Values of n by Months

Month	n for ith day of month	For the average day of the month		
		Date	Day of year n	Declination δ
January	i	17	17	−20.9
February	$31 + i$	16	47	−13.0
March	$59 + i$	16	75	−2.4
April	$90 + i$	15	105	9.4
May	$120 + i$	15	135	18.8
June	$151 + i$	11	162	23.1
July	$181 + i$	17	198	21.2
August	$212 + i$	16	228	13.5
September	$243 + i$	15	258	2.2
October	$273 + i$	15	288	−9.6
November	$304 + i$	14	318	−18.9
December	$334 + i$	10	344	−23.0

From Duffie and Beckman, 1991.

Substitution of the expression for $\cos \theta_z$, from Equation (1.13), in (1.21(a)) gives,

$$I_0 = I_{sc} \left[1.0 + 0.33 \cos \left(\frac{360n}{365} \right) \right] (\cos \phi \cos \delta \cos \omega + \sin \delta \sin \phi) \qquad (1.21b)$$

It is often required to have the integrated daily extraterrestrial radiation on a horizontal surface, H_0, for the calculations of daily solar radiation. This can be obtained by integrating Equation (1.21b) over the period from sunrise to sunset.

If I_{sc} is in Watts/m^2 and H_0 in Joules/m^2 then;

$$H_0 = \frac{24 \times 3600}{\pi} I_{sc} \left[1 + 0.033 \cos \frac{360n}{365} \right]$$

$$\times \left(\cos \phi \cos \delta \sin \omega_s + \left(\frac{2\pi \omega_s}{360} \right) \sin \phi \sin \delta \right) \qquad (1.22)$$

where ω_s is the sunset hour angle, in degrees.

$\bar{H}_0$, the monthly average daily extraterrestrial radiation can be obtained from the above expression by using day numbers from Table 1.6. The extraterrestrial radiation on a horizontal surface for an hour period can be calculated by integrating Equation (1.21b) for a period defined by hour angles ω_1 and ω_2 which define an hour (where ω_2 is larger):

$$I_0 = \frac{12 \times 3600}{\pi} I_{sc} \left[1 + 0.033 \cos\left(\frac{360n}{365} \right) \right]$$

$$\times \left[\cos\phi \cos\delta(\sin\omega_2 - \sin\omega_1) + \frac{2\pi(\omega_2 - \omega_1)}{360} \sin\phi \sin\delta \right] \qquad (1.22a)$$

EXAMPLE 1.14

Calculate the day's solar radiation on a horizontal surface (H_0), in the absence of the atmosphere at latitude 30°N on 31 May 1995.

Solution

We have, $n = 151$ and $\delta = 21.90°$ (from Equation (1.9)), $\phi = 30°$
From Equation (1.15), $\omega_s = \cos^{-1}(-\tan\phi\tan\delta) = 103.41°$
Now, the daily solar radiation on a horizontal surface can be obtained from Equation (1.22) and is given by

$$H_0 = \frac{24 \times 3600 \times 1367}{\pi} \left[1 + 0.033 \cos\left(\frac{360 \times 151}{365} \right) \right]$$

$$\times \left[\cos 30° \cos 21.90° \sin 103.41° + \frac{\pi \times 103.41}{180} \sin 30° \sin 21.90° \right]$$

or,
$$H_0 = 40.87 \, \text{MJ/m}^2$$

EXAMPLE 1.15

What is the solar radiation on the surface given in Example 1.14 between the hours of 10 and 11?

Solution

Given: $\omega_1 = -30°$ and $\omega_2 = -15°$ (Table 1.4)

First Method
From Equation (1.22a), we have

$$I_0 = (12 \times 3600/\pi) \times 1367(1 + 0.033 \cos(360 \times 151/365))[\cos(30°)\cos(21.90°)$$

$$\times(\sin(-15°) - \sin(-30°)) + \{\pi(-15 - (-30))/180\} \sin 30° \sin 21.90°]$$

$$= 4.43 \, \text{MJ/m}^2$$

Second Method

The hourly extraterrestrial radiation can also be approximated by using the relation (Equation (1.21a))

$$I_0 = 3600 I_{sc}\{1 + 0.033 \cos(360n/365)\}[\cos\phi\cos\delta\cos\omega + \sin\phi\sin\delta]$$

where ω is evaluated at the midpoint of the hour i.e. $\omega = (\omega_1 + \omega_2)/2 = (-30 + (-15))/2 = -22.5°$.
Now, $I_0 = 3600 \times 1367 \times (0.972)(0.743 + 0.187) = 4.44\,\text{MJ/m}^2$

Differences between the hourly radiation calculated by the above two methods will be slightly larger at times near sunrise and sunset.

1.8.1 Estimation of Average Solar Radiation

The regression equation relating monthly average daily radiation (H) to clear day radiation at the location in question, and average fraction of possible sunshine hours is given by:

$$\frac{\bar{H}}{\bar{H}_c} = a' + b'\frac{\bar{n}}{\bar{N}} \tag{1.23}$$

where

$\bar{H}$ = the monthly average daily radiation on a horizontal surface,
$\bar{H}_c$ = the monthly average clear sky radiation for the location and month in question,
a', b' = empirical constants,
$\bar{n}$ = the monthly average daily hours of bright sunshine, and
$\bar{N}$ = monthly average of the maximum possible daily hours of bright sunshine
     (i.e. the day length of the average day of the month).

$\bar{N}$ can be calculated from Equation (1.16) by using the value of δ from Table 1.6, for a given month.

A basic difficulty with the above equation lies in the ambiguity of the terms $\frac{\bar{n}}{N}$ and $\bar{H}_c$. The former is an instrumental problem (records from sunshine recorders can be interpreted in different ways), whereas the latter arises from the uncertainty in the definition of a clear sky.

The equation is modified to use extraterrestrial radiation on a horizontal surface, rather than on clear sky radiation (Page, 1964)

$$\frac{\bar{H}}{\bar{H}_0} = a + b\frac{\bar{n}}{\bar{N}} \tag{1.24}$$

where $\bar{H}_0$ is the monthly average of the radiation outside of the atmosphere for the same location and a and b are modified constants. Their values are determined by regression. The values of a and b obtained for New Delhi (India) are 0.341 and 0.446 respectively.

1.8.2 Distribution of Clear and Cloudy Days and Hours

$\bar{K}_T$, the monthly average clearness index, is the ratio of monthly average radiation on a horizontal surface to the monthly average daily extraterrestrial radiation.

$$\bar{K}_T = \frac{\bar{H}}{\bar{H}_0} \tag{1.25a}$$

K_T, a daily clearness index, is the ratio of a particular day's radiation to the extraterrestrial radiation for that day.

$$K_T = \frac{H}{H_0} \qquad (1.25b)$$

k_T the hourly clearness index, can be defined as

$$k_T = \frac{I}{I_0} \qquad (1.25c)$$

The data, $\bar{H}$, H and I are from measurements of total solar radiation on a horizontal surface. $\bar{H}_0$, H_0. and I_0 can be calculated by the methods given in Section 1.8. After knowing the value of clearness index, the diffuse component of the radiation can be obtained by the known empirical relations, which are being discussed in the following sections.

1.8.3 Estimation of Beam and Diffuse Components of Hourly Radiation

The split of total solar radiation on a horizontal surface into its beam and diffuse components is important as methods for calculating total radiation on surfaces of other orientation from data on a horizontal surface require separate treatments of beam and diffuse radiation.

I_d/I, the fraction of the hourly radiation on a horizontal plane which is diffuse, can be correlated to the hourly clearness index, k_T, the ratio of the total radiation to the extraterrestrial radiation for the hour.

The equations for the correlation (Orgill and Hollands, 1977 correlation) are

$$I_d/I = \begin{cases} 1.0 - 0.249 k_T & \text{for} & k_T < 0.35 \\ 1.557 - 1.84 k_T & \text{for} & 0.35 < k_T < 0.75 \\ 0.177 & \text{for} & k_T > 0.75 \end{cases} \qquad (1.26)$$

1.8.4 Estimation of Beam and Diffuse Components of Daily Radiation

Studies of available daily radiation indicate that the average fraction which is diffuse, H_d/H, is a function of K_T. The equations for the correlation (Collares-Pereira and Rabl, 1979 correlation) are:

$$\frac{H_d}{H} = \begin{cases} 0.99 & \text{for} & K_T \leq 0.17 \\ \left.\begin{array}{l} 1.188 - 2.272 K_T + 9.473 K_T^2 \\ -21.865 K_T^3 + 14648 K_T^4 \end{array}\right\} & \text{for} & 0.17 < K_T < 0.75 \\ -0.54 K_T + 0.632 & \text{for} & 0.75 < K_T < 0.80 \\ 0.2 & \text{for} & K_T \geq 0.80 \end{cases} \qquad (1.27)$$

1.8.5 Estimation of the Monthly Average of Daily Total Radiation on a Horizontal Surface

For many applications, one requires the knowledge of the monthly average of daily solar radiation available on a horizontal surface. The correlation for estimating the monthly average daily total radiation on a horizontal surface can be given as,

$$\frac{\bar{H}}{\bar{H}_0} = a + b\left(\frac{\bar{n}}{\bar{N}}\right) \tag{1.28}$$

where

$\bar{H}$ = the monthly average daily total radiation on a horizontal surface.
$\bar{H}_0$ = the monthly average of daily extraterrestrial radiation on a horizontal surface.
$\bar{n}$ = monthly average daily hours of bright sunshine obtained from actual records.
$\bar{N}$ = monthly average of the maximum possible daily hours of bright sunshine
    (i.e. the day length of the average day of the month).

a and b are constants.

The regression parameters a and b can be given as

$$a = -0.309 + 0.539 \cos\phi - 0.0693 E_0 + 0.290(\bar{n}/\bar{N}) \tag{1.29a}$$

and

$$b = 1.527 - 1.027 \cos\phi + 0.0926 E_0 - 0.359(\bar{n}/\bar{N}) \tag{1.29b}$$

where ϕ is the latitude and E_0 is the elevation of the location above sea level in kilometers (Table 1.2) and $\left(\frac{\bar{n}}{N}\right)$ is the percent possible sunshine.

These coefficients can be used in Equation (1.28) to estimate the global radiation on a horizontal surface with an accuracy of about 10 percent

The regression coefficients can be computed from latitude and sunshine hours

$$a = -0.110 + 0.235 \cos\phi + 0.323(\bar{n}/\bar{N}) \tag{1.30a}$$

$$b = 1.449 - 0.553 \cos\phi - 0.694(\bar{n}/\bar{N}) \tag{1.30b}$$

It may be noted that the correlation are valid for the following ranges:

$0.2 \leq (\bar{n}/\bar{N}) \leq 0.6$ for monsoon period
$0.4 \leq (\bar{n}/\bar{N}) \leq 0.9$ for pre-monsoon and post-monsoon period

1.8.6 Estimation of the Monthly Average of Daily Diffuse Radiation on a Horizontal Surface

Investigations relating to estimation of monthly average of daily diffuse radiation on a horizontal surface have been less in number than that of global radiation. Mainly two types of correlations have been developed. One type refers to the correlation between $\bar{H}_d/\bar{H}$, the ratio of monthly average of daily diffuse radiation to the monthly average of daily global radiation and K_T, the ratio of monthly average of daily global radiation to the monthly average of daily extraterrestrial radiation.

The other type refers to correlation between $\bar{H}_d/\bar{H}$ and the number of bright sunshine hours. On the basis of the measured data for Madras, New Delhi and Pune the following correlation has been proposed

(Gopinathan, 1988):

$$\frac{\bar{H}_d}{\bar{H}} = 1.194 - 0.838\bar{K}_T - 0.0446\frac{\bar{n}}{\bar{N}} \tag{1.31a}$$

The prediction based on the above equation has been compared with that obtained using two other linear correlations, one based on K_T and the other based on $\bar{n}/\bar{N}$:

$$\frac{\bar{H}_d}{\bar{H}} = 1.403 - 1.672\bar{K}_T \tag{1.31b}$$

$$\frac{\bar{H}_d}{\bar{H}} = 0.931 - 0.814(\bar{n}/\bar{N}) \tag{1.31c}$$

The agreement between the experimental and estimated data is shown to be better when Equation (1.31a) is used. It would be possible to calculate the monthly mean daily diffuse radiation with an accuracy of about 10 percent and it would be desirable to include both K_T and $\left(\frac{\bar{n}}{N}\right)$ in the correlation.

1.8.7 Estimation of Hourly Radiation from Daily Data

Statistical studies of the time distribution of total radiation on horizontal surfaces through the day, using monthly average data from a number of places, have led to generalized charts of r_t, the ratio of hourly to daily total radiation, as a function of day length and the hour in question. The hours are designated by the time for the midpoint of the hour, and days are assumed to be symmetrical about solar noon. Day length can be calculated from Equation (1.16). Thus from a knowledge of day length (a function of latitude φ and declination δ) and daily total radiation, the hourly total radiation for symmetrical days can be estimated by the equation (Collares-Pereira and Rabl, 1979):

$$r_t = \frac{I(t)}{\Sigma I(t)} = \frac{I(t)}{H_0} \tag{1.32}$$

or

$$I(t) = r_t H_0 \, \text{J/m}^2/\text{hr} = (r_t H_0)/3600 \, \text{W/m}^2$$

where

$$r_t = \frac{\pi}{24}(a + b\cos\omega)\frac{\cos\omega - \cos\omega_s}{\sin\omega_s - (2\pi\omega_s/360)\cos\omega_s} \tag{1.32a}$$

The coefficients 'a' and 'b' are given by

$$a = 0.409 + 0.5016\sin(\omega_s - 60) \tag{1.32b}$$

$$b = 0.6609 - 0.4767\sin(\omega_s - 60) \tag{1.32c}$$

where, ω is the hour angle in degrees for the time in question and ω_s is the sunset hour angle.

EXAMPLE 1.16

Calculate the monthly average clearness index, for 16 March 1995 at a surface placed at latitude 30 °C. The monthly average daily terrestrial radiation on a horizontal surface is 28.1 MJ/m².

Solution

Given, for 16 March 1995 $n = 75$ (Table 1.6); $\delta = -2.4°$ (from Equation (1.9))
$\omega_s = \cos^{-1}(-\tan 30° \tan(-2.4°)) = 88.61°$ (from Equation (1.15))
By using the above given data, the monthly average daily extraterrestrial radiation is given by

$$\bar{H}_0 = \{24 \times 3600 \times 1367/\pi\}(1.01)(0.832) = 31.57 \, \text{MJ/m}^2 \, (\text{from Equation (1.22)})$$

Now, the monthly average clearness index is given by

$$\bar{K}_T = \frac{\bar{H}}{\bar{H}_0} = \frac{28.1}{31.57} = 0.890$$

EXAMPLE 1.17

Estimate the fraction of the average May, 1995 radiation on a horizontal surface in Madras, that is diffuse and beam. The May average radiation, H for Madras is $22.69 \, \text{MJ/m}^2$.

Solution

Here, $\delta = 18.86°$ (from Equation (1.9) and Table 1.6), $\phi = 13°$
Now, $\omega_s = \cos^{-1}(-\tan 13° \tan 18.86°) = 94.5°$ (from Equation (1.15))
The monthly average daily extraterrestrial radiation (from Equation (1.22)) is given by

$$\bar{H}_0 = \frac{24 \times 3600 \times 1367}{\pi} \left[1 + 0.033 \cos \left(\frac{360 \times 135}{365} \right) \right]$$

$$\times \left[\cos 13° \cos 18.9° \sin 94.5° + \frac{\pi \times 94.5}{180} \sin 13° \sin 18.9° \right]$$

$$= 38.18 \, \text{MJ/m}^2$$

The monthly average clearness index is given by

$$\bar{K}_T = \frac{\bar{H}}{\bar{H}_0} = \frac{22.69}{38.18} = 0.594$$

Using the above value of K_T in Equation (1.31b), we get the diffuse component of monthly average daily radiation as

$$\frac{\bar{H}_d}{\bar{H}} = 1.403 - 1.672 \bar{K}_T = 0.410$$

$$\bar{H}_d = 0.410 \times 22.69 \, \text{MJ/m}^2 = 9.30 \, \text{MJ/m}^2$$

Further, the beam component of monthly average daily radiation can be calculated as,

$$\bar{H}_b = \bar{H} - \bar{H}_d = 22.69 - 9.30 = 13.39 \, \text{MJ/m}^2$$

EXAMPLE 1.18

Using the isotropic diffuse model, estimate the beam, diffuse and general reflected components of solar radiation and the total radiation on a surface sloped 45° toward the south at a latitude of 40°N for the hour 2 to 3 PM on 20 February 1995. Here, $I = 1.04 \, MJ/m^2$ and $\rho_g = 0.60$.

Solution

Given $n = 51$; $\delta = -11.6°$ (from Equation (1.9)); $\phi = 40°N$

$\omega_2 = 45$ and $\omega_1 = 30$ (from Table 1.4)

The hourly extraterrestrial radiation is given by,

$$I_0 = \frac{12 \times 3600 \times 1367}{\pi} \left[1 + 0.033 \cos \frac{360 \times 51}{365} \right]$$

$$\times \left(\cos 40° \cos(-11.6°)(\sin 45° - \sin 15°) + \frac{\pi \times (45-30)}{180} \sin 40° \sin(-11.6°) \right)$$

$$I_0 = (12 \times 3600 \times 1367/\pi)(1.02)(0.3026) = 5.807 \, MJ/m^2$$

The hourly clearness index is given by,

For the above value of $k_T = \frac{I}{I_0} = \frac{1.04}{5.807} = 0.18$, the diffuse component can be evaluated from Equation (1.26) and is given by

$$\frac{I_d}{I} = 1.0 - 0.249 \times 0.18 = 0.955$$

For a given value of $I (= 1.04 \, MJ/m^2)$

$$I_d = 0.955 \times 1.04 = 0.9936 \, MJ/m^2$$

$$I_b = 0.045 \times 1.04 = 1.04 - 0.9936 = 0.0468 \, MJ/m^2, \text{ here } 0.0468 = 1.04 - 0.9936$$

The conversion factor for beam radiation at midpoint of the hour ($\omega = 37.5°$) is given by,

$$R_b = \frac{\cos(40-45)\cos(-11.6°)\cos 37.5° + \sin(-11.6°)\sin(40-45)}{\cos 40° \cos(-11.6°)\cos 37.5° + \sin(-11.6°)\sin 40°} = 1.697 \approx 1.70$$

The total radiation is given as

$$I_T = 0.0468 \times 1.70 + 0.9936 \times \frac{(1 + \cos 45)}{2} + 1.04 \times 0.60 \frac{(1 - \cos 45)}{2}$$

$$I_T = 0.0795 + 0.8481 + 0.091 = 1.0186 \, MJ/m^2$$

The beam contribution is $0.447 \, MJ/m^2$; the diffuse component is $0.663 \, MJ/m^2$ and the ground reflected component is $0.091 \, MJ/m^2$.

EXAMPLE 1.19

Estimate the global radiation on a horizontal surface at New Delhi in May, 1995 if the monthly average daily hours of bright sunshine observed (n) is 12.1 hours.

Solution

Given $\delta = 18.86°$ (from Equation (1.9) and Table 1.6), $\phi = 28°35' = 28.58°$ and
$\omega_s = \cos^{-1}(-\tan 28.58° \tan 18.86°) = 100.74°$ (from Equation (1.15)).
The average daily extraterrestrial radiation is given by

$$\bar{H}_0 = \frac{24 \times 3600 \times 1367}{\pi}\left(1 + 0.033\cos\frac{360 \times 135}{365}\right)[\cos(28.58)\cos(18.9)\sin(100.74)$$

$$+(\pi \times 100.74/180)\sin(28.58)\sin(18.9)] = 39.98\,\text{MJ/m}^2$$

The number of daylight hours (from Equation (1.16) and Table 1.6) is given by

$$N = (2/15)\cos^{-1}(-\tan 28.58° \tan 18.16°) = 13.43$$

The values of a and b (from Equations (1.30a) and (1.30b)) can be given as

$$a = -0.110 + 0.235 \times \cos(28.58) + 0.323 \times (12.1/13.43) = 0.387$$

$$b = 1.449 - 0.553\cos(28.58) - 0.694(12.1/13.43) = 0.338$$

The above values of a and b can be used in Equation (1.28) to get the monthly average daily solar radiation incident on the horizontal surface

$$\frac{\bar{H}}{\bar{H}_0} = a + b\left(\frac{\bar{n}}{\bar{N}}\right) = 0.387 + 0.338\left(\frac{12.1}{13.43}\right) = 0.692$$

or,
$$\bar{H} = 27.66\,\text{MJ/m}^2$$

EXAMPLE 1.20

Calculate the ratio of hourly to daily total radiation r_t on a sshorizontal surface in Delhi in the month of May, 1995 at 2PM. Also calculate the hourly radiation.

Solution

Given $\phi = 28°35' = 28.56°$ (Table 1.2); $\delta = 18.86°$ (from Equation (1.9) and Table 1.6)
$\omega_s = 100.74°$ (from Example 1.19) and $\omega = 30°$ (from Table 1.4)
From Equations (1.32b) and (1.32c), the values of a and b can be evaluated as

$a = 0.409 + 0.5016 \sin(100.74 - 60) = 0.736$
$b = 0.6609 - 0.4767 \sin(100.74 - 60) = 0.3498$
Knowing the values of a and b we get from Equation (1.32a)

$$r_t = \frac{\pi}{24}(0.736 + 0.350 \cos 30°)\left[\frac{(\cos 30 - \cos 100.74)}{\sin(100.74) - \left(\frac{2\pi \times 100.74}{360}\right)\cos 100.74°}\right] = 0.109$$

Further, using Equation (1.32) and for a given value of $H_0 = 39.98 \, \text{MJ/m}^2$ (from Example 1.19), the total radiation in one hour (1.30–2.30) is given as

$$I(t) = r_t \times \bar{H}_0 = 0.109 \times 39.98 \, \text{MJ/m}^2 = 4.36 \, \text{MJ/m}^2$$

PROBLEMS

1.1 Determine the temperature of the sun (T_S), in the month of January, for the following given data:

Solar constant $= 1367 \, \text{W/m}^2$

Sun diameter $(2R_S) = 1.39 \times 10^9 \, \text{m}$

Sun–Earth distance $(L_{se}) = 1.5 \times 10^{11} \, \text{m}$.

Hint $I_{ext} = \sigma T_S^4 (4\pi R_S^2)/(4\pi L_{se}^2)$

σ = Stefan-Boltzmann constant $= 5.67 \times 10^{-8} \, \text{W/m}^2\text{K}^4$.

1.2 Calculate the declination angle (δ) for March 31 in a leap year.

Hint Use Equation (1.9).

1.3 Calculate the hour angle (ω) at 2.30 PM.

Hint Use $\omega = 15 \, (ST - 12 \, \text{hours})$, ST is in hours.

1.4 Find out the daily variation of the extraterrestrial solar intensity (I_{ext}) and the declination angle (δ) for the month of June, 1996.

Hint Use Equations (1.6a) and (1.9), respectively.

1.5 Calculate air mass for sun's position at noon.

Hint Air mass $= m/H = \sec\psi$ (see figure given below).

At noon; $\psi = 0$. Here $\psi = \theta_z$ (Equation (1.13))

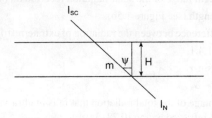

Also $m = H \sec\theta_z$ or $H = m \cos\theta_z$.

1.6 Calculate the hourly direct radiation on horizontal surface (I_b) and on inclined surface with an inclination of 45° on 15 January 1995 for city condition in terrestrial region.

Hint $I_N = I_{ext} \exp[-T_R/(0.9 + 9.4 \sin \alpha)]$, $\alpha = 90 - \theta_z$

For horizontal surface $I_b = I_N \cos \theta_z$; for inclined surface $I'_b = I_N \cos \theta_i$

The turbidity factor (T_R) for different months is as follows:

Month → Region↓	1	2	3	4	5	6	7	8	9	10	11	12
Mountain	1.8	1.9	2.1	2.2	2.4	2.7	2.7	2.7	2.5	2.1	1.9	1.8
Flat land	2.2	2.2	2.5	2.9	3.2	3.4	3.5	3.3	2.9	2.6	2.3	2.2
City	3.1	3.2	3.5	3.9	4.1	4.2	4.3	4.2	3.9	3.6	3.3	3.1

For cloudy condition $T_R = 10.0$.

1.7 Derive an expression for the number of daylight (sunshine) hours, N.

Hint Use $\theta_z = 90°$ (sunset/sunrise) in Equation (1.13)

with $\omega = \omega_s$ and $15° = 1$ hour, $N = 2\omega_s$.

1.8 Derive an expression for the extraterrestrial radiation on a horizontal surface (H_0).

Hint

Use,
$$H_0 = \frac{12}{\pi} \times 3600 \int_{-\omega_s}^{\omega_s} I_{ext} \cos \theta_z \, d\omega.$$

1.9 Calculate the number of daylight hours at Delhi on December 21 and June 21 in a leap year.

Hint Use Equation (1.16).

1.10 Calculate the diffuse radiation (I_d) on a horizontal surface for Problem 1.6.

Hint Use, $I_d = (1/3)(I_{sc} - I_N) \sin \alpha$.

1.11 What is the conversion factor for beam and diffuse radiation for Delhi at noon, for a 45° inclined surface facing east–south on 16 February.

Hint Use Equation (1.18a) and (1.18b).

1.12 Derive an expression for extraterrestrial radiation on a horizontal surface for an hour period.

Hint Integrate between ω_1 and ω_2 of Problem 1.8.

1.13 Derive an expression for extraterrestrial radiation on a south facing ($\gamma = 0$) tilted surface (H_{OT}).

Hint

Use,
$$H_{OT} = \frac{12}{\pi} \times 3600 \int_{-\omega_s}^{\omega_s} I_{ext} \cos \theta \, d\omega.$$

1.14 What is the wavelength range of the solar radiation received on the earth.

Hint Short wavelength (see Figure 1.5b).

1.15 What is the basic difference between the radiation of extraterrestrial region and that of the terrestrial region.

Hint See Section 1.4.1.

1.16 Prove that $\cos \theta_z = \sin \alpha$.

Hint Use $\theta_z = 90 - \alpha$.

1.17 Calculate the percentage of the total radiation that falls in ultra-violet region (0.2–0.38 μm), visible region (0.38–0.78 μm) and infra-red region (0.78–3 μm).

Hint Use Example 1.1.

1.18 Calculate the solar altitude angle (α) for Problem 1.2 at 1 AM.

 Hint See Problem 1.16.

1.19 Find out the hourly variation of direct and diffuse radiation on a horizontal surface, for 11 March 1996 for the following locations:

 (a) Madras, (b) Bangalore, (c) Leh, and (d) Simla.

 Hint Use expressions given in Problem 1.6 and 1.10 respectively.

1.20 Calculate $\bar{N}$ for each month for Delhi location.

 Hint Use Equation (1.16) and Table 1.6 for δ and n.

1.21 Calculate the average earth's temperature without atmosphere and assuming 40 percent reflection losses from earth's surface (neglect heat capacity of the earth).

 Hint $0.7\pi(D_e/2)^2 I_{sc} = 4\pi(D_e)^2 \varepsilon\sigma(T_e)^4$.

1.22 Calculate an average earth's temperature without atmosphere and by considering heat capacity of the earth.

 Hint $\pi(D_e/2)^2 I_{sc} = (4/3)\pi(D_e/2)^3 \rho_e C_e \nabla T$.

1.23 Plot the variation of monochromatic emissive power with λ in μm for different temperatures of blackbody (6000 K, 3000 K, 1000 K and 288 K) and verify the Wien's displacement law.

 Hint Compute $E_{b\lambda}$ for different λ (Equation (1.2)) and the plot the variation and find λ_{max} for each temperature and verify the Wien's displacement law.

1.24 Calculate the beam and diffuse radiation on horizontal surface for 21 December, 21 March and 21 June 2000, for latitude of Delhi.

 Hint See Problem 1.6.

1.25 Plot the conversion factor for beam radiation (R_b) with latitude for different days from 1 January i.e. $n = 0$, 40, 80, 120 and 160 for a given hour angle (ω).

 Hint Use either Figure 1.12(a) or Equation (1.18a).

1.26 Plot the conversion factor for beam radiation (R_b) with hour angle (ω) for different days from 1 January i.e. $n = 0$, 40, 80, 120 and 160 for a given latitude ($\phi = 30°$).

 Hint See Problem 1.26.

1.27 Calculate the hourly direct radiation on horizontal surface (I_b) and on inclined surface ($\beta = 45°$) on 15 January 1995 for mountain and flat land conditions in terrestrial region. Compare the results with Problem 1.6.

 Hint See Problem 1.6.

1.28 Calculate diffuse radiation on a horizontal ($\beta = 0$) and inclined surfaces ($\beta = 45°$) for the Problems 1.6 and 1.27 for mountain, flat land and city conditions and compare the results.

 Hint See Problem 1.10.

1.29 Compute the Problems 1.6 and 1.27 for summer condition (15 June 1995) and discuss the results. Also answer the following questions:

 i. Which surface gives maximum beam radiation in winter (15 January 1995) and summer (15 June 1995) conditions respectively.

 ii. Which surface gives minimum beam radiation in both winter and summer conditions.

1.30 Repeat Problem 1.29 for diffuse radiation.

 Hint Formula for diffuse radiation on inclined surface is

$$I_{di} = I_{dh}\left[\frac{I_b' \cos\theta_i}{I_{sc}\sin\alpha} + \left(1 - \frac{I_b'}{I_{sc}}\right)\cos^2\frac{\beta}{2}\right]$$

and other terms are defined in Problem 1.6.

1.31 Calculate and plot the variation of air mass with hour angle for different nth day of the year for a given latitude.

    Hint See Problem 1.5 and consider $\psi = \theta_z$ (Equation (1.13)) and n corresponding 1 January and 30 June 1995.

1.32 Repeat Problem 1.31 and plot the variation of air mass with hour angle for different latitude for a given n.

    Hint See Problem 1.31.

1.33 Plot the variation of sunshine hour (N) for different latitude (Φ).

    Hint Use Equation (1.16).

1.34 Plot the variation of R_b with nth day of year from January 1 for a given $\bar{\Phi}$, β and ω.

    Hint See Figure 1.12.

2

Heat Transfer: Concepts and Definitions

2.1 INTRODUCTION

The transfer of heat energy occurs as a result of a driving force called temperature difference. Heat transfer is of great importance in modern technology and therefore the understanding of its basic principles and practical applications is of vital importance. In this chapter, some of the elementary fundamentals of heat transfer have been reviewed. Heat is transferred by conduction, convection or thermal radiation. These modes of heat transfer differ profoundly in nature and are governed by different laws.

Heat transfer by conduction takes place between bodies or particles of bodies that are in direct contact and at different temperatures. Heat conduction is a molecular process, that is, heat is transferred from molecule to molecule and there occurs negligible movement of particles of the body. Heat conduction may be through solids, liquids and gases.

The second mode of heat transfer occurs only in fluids, i.e. gases and liquids, and is called convection. This mode of heat transfer takes place when the entire mass of a non-uniformly heated liquid or gas is displaced and mixed. Heat transfer by convection takes place at a greater rate when the velocity of motion of the liquid or gas is higher, as in this case a greater number of fluid particles are displaced per unit time. In liquids and gases heat transfer by convection is always accompanied by conduction, since in the course of convection, particles at a different temperature are always in direct contact.

The combined mode of heat transfer by convection and conduction is called convective heat transfer. Convection can be either free (or natural) or forced. If the working medium is put into motion artificially (by means of a fan, compressor etc.) then such a convection is called as forced. But if the working medium begins to move due to the difference between the densities of individual parts of the fluid upon heating, then the mode of heat transfer is referred to as free, or natural convection.

The third mode of heat transfer is known as thermal radiation. This process of heat transfer, between two bodies separated by a medium which transmits radiation fully or partly, takes place in three stages: conversion of a fraction of internal energy of one of the bodies into the energy of electromagnetic waves, propagation of the electromagnetic waves in space, and absorption of radiant energy by the other body.

2.2 CONDUCTION

The phenomenon of heat conduction is a process of propagation of energy between the particles of a body which are in direct contact and have different temperatures.

2.2.1 Temperature Field

For propagation of heat in any body or space to take place, there must be a difference between the temperatures at different points of the body. This must also be so for heat transfer by conduction when the temperature gradient at various points of the body is not zero. Analytical investigation of heat conduction amounts to the study of space time variations of temperature T, namely, the determination of the temperature field, which is expressed as

$$T = f(x, y, z, t) \tag{2.1}$$

Equation (2.1) describes the temperature field in which temperature varies in time and space, a characteristic of the transient condition which is referred to as a transient temperature field. If the temperature of a body is a function of the spatial coordinates only and does not vary with time, the temperature field is referred to as a steady-state one, i.e.

$$T = f(x, y, z) \quad \text{or} \quad \frac{\partial T}{\partial t} = 0 \tag{2.2}$$

2.2.2 Fourier's Law

The basic equation for steady state heat conduction is known as Fourier's equation. According to this, the quantity of heat (dQ), passing through an isothermal surface (dA), per time interval (dt) is proportional to the temperature gradient $(\partial T/\partial n)$ and mathematically can be expressed as

$$dQ = -K \frac{\partial T}{\partial n} dA dt \tag{2.3}$$

The proportionality factor K in Equation (2.3) is the physical property of the substance, which defines the ability of the substance to conduct heat and is called the thermal conductivity. The heat flux q, defined as the rate of heat flow per unit area of the isothermal surface, is given by

$$\dot{q} = -K \frac{\partial T}{\partial n} \tag{2.4}$$

The direction of heat flux $\dot{q}$ is normal to the surface and is positive in the direction of decreasing temperature, which explains the negative sign on the right hand side of Equation (2.4).

2.2.3 Thermal Conductivity

As already stated, thermal conductivity is a physical property of a substance. The values of thermal conductivity of a few commonly used materials are given in Appendix III.

Thermal conductivity of gases, liquids and solids depends on temperature. Experimental studies have shown that for many materials the dependence of thermal conductivity on temperature can be assumed to be linear.

$$K = K_0[1 + \beta(T - T_0)] \tag{2.5}$$

where K_0 is the thermal conductivity at temperature T_0 and β is a constant for the material.

In general, an increase in temperature causes the conductivity of a gas to increase (positive β) and conductivity of a solid or liquid to decrease (negative β). However, there are some exceptions to this generalization.

Thermal conductivity is greatly influenced by moisture content of the substance. Experimental studies have shown that the thermal conductivity increases substantially with an increase in moisture content.

2.2.4 Differential Equation of Conduction

The rate of heat flow in x, y and z directions can be computed if the temperature gradient in these directions is known. The temperature gradient in any direction can be determined if the temperature distribution in the medium is known. The latter may be determined from the solution of the differential equation of heat conduction subject to appropriate boundary conditions.

In deriving the differential equation of conduction the following assumptions are made:

i. solid is homogenous and isotropic
ii. its physical parameters are constant
iii. deformation of the volume, caused by the temperature variations, is very small in comparison to the volume itself, and
iv. generally the inner heat sources are uniformly distributed in the solid.

Figure 2.1 shows a volume element with sides dx, dy, dz. Let the quantities of heat imparted to the sides of the elementary volume in time dt and in the directions OX, OY, OZ be denoted by dQ_x, dQ_y and dQ_z respectively. The quantity of heat removed through the opposite sides in the same directions be denoted respectively by dQ_{x+dx}, dQ_{y+dy}, dQ_{z+dz}.

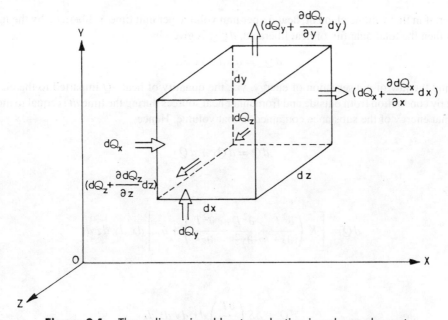

Figure 2.1 Three dimensional heat conduction in volume element.

Thus, the amount of heat imparted to the elementary volume in the OX direction is,

$$dQ_x = -K \, dy \, dz \frac{\partial T}{\partial x} dt \qquad (2.6)$$

The quantity of heat removed through the opposite side in the OX direction is,

$$dQ_{x+dx} = dQ_x + \left(\frac{\partial \, dQ_x}{\partial x} \right) dx \qquad (2.7)$$

Now,

$$dQ_{x1} = dQ_x - dQ_{x+dx} = -\left(\frac{\partial \, dQ_x}{\partial x} \right) dx = +K \, dx \, dy \, dz \, dt \left(\frac{\partial^2 T}{\partial x^2} \right) \qquad (2.8)$$

Similarly,

$$dQ_{y1} = K \left(\frac{\partial^2 T}{\partial y^2} \right) dx \, dy \, dz \, dt \qquad (2.9)$$

$$dQ_{z1} = K \left(\frac{\partial^2 T}{\partial Z^2} \right) dx \, dy \, dz \, dt \qquad (2.10)$$

Therefore, the quantity of heat, dQ_1, imparted to the considered volume by conduction is given by the expression:

$$dQ_1 = K \left(\frac{\partial^2 T}{\partial x^2} + \frac{\partial^2 T}{\partial y^2} + \frac{\partial^2 T}{\partial z^2} \right) dx \, dy \, dz \, dt \qquad (2.11)$$

However, if in this volume element, heat $\dot{q}_v$ per unit volume per unit time, is liberated by the inner heat sources then the total quantity of heat liberated, dQ_2, is given by

$$dQ_2 = \dot{q}_v \, dx \, dy \, dz \, dt \qquad (2.12)$$

Following the law of conservation of energy, viz. the quantity of heat dQ imparted to the elementary volume by conduction from outside and from inner heat sources during the time dt is equal to the change in internal energy of the substance contained in that volume. Hence,

$$dQ = dQ_1 + dQ_2$$

or,

$$dQ = \left[K \left(\frac{\partial^2 T}{\partial x^2} + \frac{\partial^2 T}{\partial y^2} + \frac{\partial^2 T}{\partial z^2} \right) + \dot{q}_v \right] dx \, dy \, dz \, dt \qquad (2.13)$$

where,

$$dQ = \rho S \left(\frac{\partial T}{\partial t} \right) dx \, dy \, dz \, dt \qquad (2.14)$$

gives the heat stored in the volume element in time dt. Here, ρ and S represent the density and the specific heat of element respectively. Substituting the value of dQ from Equation (2.14) in (2.13), we have,

$$\rho S \frac{\partial T}{\partial t} = K \left(\frac{\partial^2 T}{\partial x^2} + \frac{\partial^2 T}{\partial y^2} + \frac{\partial^2 T}{\partial z^2} \right) + \dot{q}_v$$

or,

$$\rho S \frac{\partial T}{\partial t} = K \Delta^2 T + \dot{q}_v \tag{2.15}$$

Equation (2.15) is the general differential equation of heat conduction. It reduces to Fourier's equation if the system contains no inner sources.

$$\dot{q}_v = 0$$

and,

$$\frac{\partial T}{\partial t} = \frac{K}{\rho S} \Delta^2 T = \alpha \Delta^2 T \tag{2.16}$$

where $\alpha = K/\rho S$ is called thermal diffusivity and is a physical property of a substance. It determines the rate at which a non-uniform temperature distribution approaches the equilibrium condition.

However, if the inner sources are present and the temperature field corresponds to a steady state condition, i.e. $T = f(x, y, z)$, then Equation (2.15) reduces to Poisson's equation as

$$\Delta^2 T = -\frac{\dot{q}_v}{K} \tag{2.17}$$

Finally, for steady state condition in the absence of inner heat sources, Equation (2.15) takes the form of Laplace's equation

$$\Delta^2 T = 0 \tag{2.18}$$

2.2.5 Solution of Heat Conduction in a Medium

As discussed earlier, heat conduction equation governs the temperature at any point in the given region and the explicit determination of temperature requires a set of initial and boundary conditions. However, to know it's qualitative behavior one needs to solve this heat conduction equation.

We can solve this equation analytically, if the boundaries of the region are regular, i.e., when the boundary surfaces coincide with the coordinate surface of an orthogonal coordinate system. Various mathematical techniques, for example, separation of variables, Laplace transform, integral transform, complex variables, are available for analytical solution. However, in the case of solids having complicated geometries and boundary conditions, numerical methods based on the finite difference techniques are used.

In this section, we will outline the method of separation of variables which is the most straightforward approach to solve the equation. However, in most of the solar energy applications one dimensional heat conduction equation is sufficient to govern the temperature distribution. From Equation (2.15), the equation governing the temperature distribution $T(x, t)$ is

$$\frac{K}{\rho S} \frac{\partial^2 T}{\partial x^2} + \frac{\dot{q}_v}{\rho S} = \frac{\partial T}{\partial t} \tag{2.19}$$

If the boundary conditions, meteorological parameters etc. are assumed to be periodic, then the temperature distribution will also be periodic and may be written as,

$$T(x, t) = T_0(x) + Re \sum_{n=1}^{\infty} T_n(x) \exp(in\,\omega t) \tag{2.20}$$

where Re denotes the real part. Substituting $T(x, t)$ from Equation (2.20) in (2.19), the expressions for T_0 and T_n can be obtained so that solution for temperature distribution is known.

The expression for general solution of temperature distribution, can be determined in two cases:

(i) Without a heat source in the medium:

$$T(x, t) = A \cdot x + B + Re \sum_{n=1}^{\infty} \{C_n \exp(-\beta_n x) + D_n \exp(\beta_n x)\} \exp(in\,\omega t) \tag{2.21}$$

(ii) With a heat source in the medium:

The temperature distribution cannot be known unless explicit expression for $\dot{q}_v$ is used. In solar energy systems, $\dot{q}_v$ can be given as,

$$\dot{q}_v = -\frac{\partial I}{\partial x} \tag{2.22}$$

where I is the solar intensity available at a distance x away from the surface and is given by

$$I(x, t) = I_0(1 - r) \sum_{j=1}^{m} [\exp(-\mu_j x)] \left(\frac{\Delta E}{E_b} \right) \tag{2.23}$$

where μ_j is the extinction coefficient and $(\Delta E/E_b)$ is the emissive power. The variation of emissive power with extinction coefficient is given in Appendix IV.

Solar intensity I being periodic in nature can be expressed as a Fourier series in time

$$I_0 = I_0' + Re \sum_{n=1}^{\infty} I_n' \exp(in\,\omega t) \tag{2.24}$$

The solution for the temperature distribution is, therefore, given as

$$T(x, t) = Ax + B - \frac{I_0'}{K}(1 - r) \sum_{j=1}^{m} \frac{\exp(-\mu_j x)}{\mu_j} \left(\frac{\Delta E}{E_b} \right) + Re \sum_{n=1}^{\infty} \{C_n \exp(-\beta_n x) + D_n \exp(\beta_n x)$$

$$- \frac{1-r}{K} \sum_{j=1}^{m} I_n' \frac{\exp(-\mu_j x)}{(\mu_j^2 - \beta_n^2)} \mu_j \frac{\Delta E}{E_b} \} \exp(in\,\omega t) \tag{2.25}$$

where

$$\beta = \sqrt{\frac{n\omega\rho S}{2K}}(1 + i)$$

and the constants appearing in the solution can be determined by the boundary conditions.

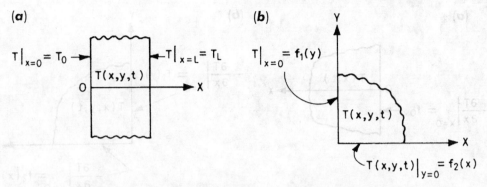

Figure 2.2 Illustration of boundary conditions of the first kind (**a**) one dimensional (**b**) two dimensional temperature distribution.

2.3 BOUNDARY CONDITIONS

In order to solve the differential equation of heat conduction for the purpose of determining the temperature distribution in a medium, a set of boundary conditions and an initial condition is required.

The initial condition specifies the temperature distribution in the medium at the origin of the time coordinate ($t = 0$) and is required only for time dependent problems. The boundary conditions specify the temperature or the heat flow situation at the boundaries of the region. For example, at a given boundary surface, the temperature distribution, or the heat flux distribution, and the heat exchange by convection etc., with a medium at a given temperature may be described. These may be classified as boundary conditions of the first kind, the second kind and the third kind, discussed as follows:

I. When the distribution of temperature is prescribed at a boundary surface, the boundary condition is said to be of the first kind. For example, for the slab geometry of Figure 2.2 the boundary condition

$$T(x, t)|_{x=0} = T_0 \quad \text{and}, \quad T(x, t)|_{x=L} = T_L \tag{2.26a}$$

denotes the temperatures of boundary surfaces $x = 0$ and $x = L$, respectively.

In the case of two dimensional geometry (Fig. 2.2(b)) the boundary condition

$$T(x, y, t)|_{x=0} = f_1(y) \tag{2.26b}$$

denotes the temperature of boundary surface at $x = 0$, which is a function of y.

II. When the flux at a boundary is given, it is called boundary condition of the 'second kind'. In case of one dimensional region, Figure 2.3(a), the heat flux entering the medium through the boundary surface at $x = 0$ is given by

$$\dot{q}_0 = -K \frac{\partial T(x, t)}{\partial x}\bigg|_{x=0} \tag{2.27a}$$

$$-\frac{\partial T}{\partial x}\bigg|_{x=0} = \frac{\dot{q}_0}{K} = f_0 \tag{2.27b}$$

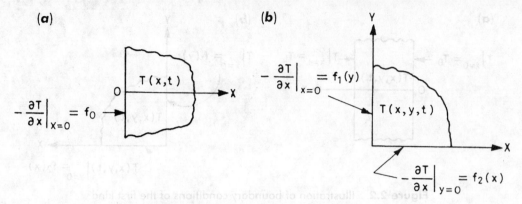

Figure 2.3 Illustration of boundary conditions of the second kind
(a) one dimensional **(b)** two dimensional temperature distribution.

In the case of two dimensional region (Fig. 2.3(b)) the heat flux entering the region through the boundary surface at $x = 0$ may be written as

$$-\frac{\partial T}{\partial x}\Big|_{x=0} = \frac{q_1(y)}{K} = f_1(y) \tag{2.27c}$$

III. When the boundary surface is subjected to a convective heat transfer into the medium at a particular temperature, the boundary condition is said to be of the 'third kind'.

In order to demonstrate the physical significance of this boundary condition, let us consider a slab geometry (Fig. 2.4) in which convection takes place from both the boundaries into the environment at a temperature T_A. If the heat transfer coefficients at the boundary surfaces $x = 0$ and $x = L$ are h_1 and h_2 respectively, then the energy balance at these surfaces may be written as:

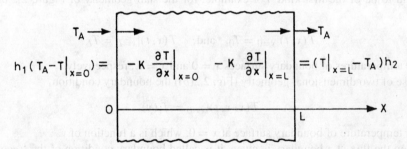

Figure 2.4 Illustration of the boundary condition of the third kind.

The boundary condition at $x = 0$ is

Heat entering by convection = Heat leaving by conduction

or

$$h_1(T_A - T|_{x=0}) = -K\frac{\partial T}{\partial x}\Big|_{x=0} \tag{2.28a}$$

for a unit area on the boundary surface $x = 0$.

Similarly, for the boundary surface at $x = L$,

Heat entering by conduction = Heat leaving by convection

or
$$-K \frac{\partial T}{\partial x}\bigg|_{x=L} = h_2 (T|_{x=L} - T_A) \qquad (2.28b)$$

2.4 OVERALL HEAT TRANSFER

Most practical problems in heat transfer involve a medium composed of several different parallel layers each having different thermal conductivity or involve two or more of the heat transfer modes, viz. conduction, convection and radiation. In such cases, the concept of overall heat transfer coefficient (U) is applied, to predict the one dimensional steady state heat transfer rate.

Figures 2.5–2.7 show typical configurations of such composite layers for slabs, cylinders and spheres and the value of U in each case is determined as follows:

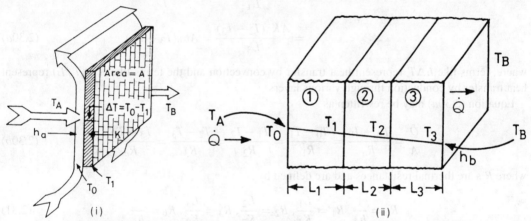

Figure 2.5(a) One dimensional heat flow through parallel slabs in perfect (i) single and (ii) triple thermal contact.

2.4.1 Single and Parallel Slabs

Let us consider a single slab as shown in Figure 2.5(a i). One side of the slab is maintained at temperature T_0 by flowing hot air over it and other side is maintained at temperature T_1 due to conduction of heat through slab. T_A and T_B are temperatures of air on both side of the slab. The rate of heat transfer in a steadily state condition can be written as

$$\dot{Q} = Ah_a(T_A - T_0) = A\frac{K}{L}(T_0 - T_1) = Ah_b(T_1 - T_B) \qquad (2.29a)$$

The first and last terms are heat transfer rate due to convection and the middle is due to conduction. The above equation can be written as

$$\frac{\dot{Q}}{A} = \frac{T_A - T_0}{R_A} = \frac{T_0 - T_1}{R} = \frac{T_1 - T_B}{R_B} \qquad (2.29b)$$

After simplification,

$$\dot{Q} = U A (T_A - T_B) \tag{2.29c}$$

where

$$U = \left[\frac{1}{h_a} + \frac{L}{K} + \frac{1}{h_b} \right]^{-1} = [R_A + R + R_B]^{-1}$$

with

$$R_A = \frac{1}{h_a}, R = \frac{L}{K} \quad \text{and} \quad R_b = \frac{1}{h_b}.$$

Consider a composite wall, Figure 2.5(a ii), through which heat is transferred from the hot fluid at temperature T_A to the cold fluid at temperature T_B. Assuming steady state, i.e. the heat transfer rate, $\dot{Q}$, per unit area of the structure is the same through each layer, we may write,

$$\dot{Q} = A h_a (T_A - T_0) = \frac{A K_1 (T_0 - T_1)}{L_1} = \frac{A K_2 (T_1 - T_2)}{L_2}$$

$$= \frac{A K_3 (T_2 - T_3)}{L_3} = A h_b (T_3 - T_B) \tag{2.30a}$$

where, terms like $h \Delta T$ represent heat transfer by convection and the terms like $K(\Delta T / L)$ represent heat transfer by conduction through various layers.

Equation (2.30a) may be rewritten as

$$\frac{\dot{Q}}{A} = \frac{T_A - T_0}{R_a} = \frac{T_0 - T_1}{R_1} = \frac{T_1 - T_2}{R_2} = \frac{T_2 - T_3}{R_3} = \frac{T_3 - T_B}{R_b} \tag{2.30b}$$

where R's are thermal resistances and are defined by

$$R_a = \frac{1}{h_a}, R_1 = \frac{L_1}{K_1}, R_2 = \frac{L_2}{K_2}, R_3 = \frac{L_3}{K_3}, R_b = \frac{1}{h_b} \tag{2.31}$$

Eliminating interface temperatures T_0, T_1, T_2 and T_3, we obtain the total heat transfer rate through the composite layer given by

$$\frac{\dot{Q}}{A} = \frac{T_A - T_B}{R}$$

where

$$R = R_a + R_1 + R_2 + R_3 + R_b \tag{2.32}$$

Hence, the total thermal resistance R in the path of heat flow from temperature T_A to temperature T_B consists of the sum of several different thermal resistances (Equation (2.32)).

The overall heat transfer coefficient, U, which if characterized as the unit conductance of a composite layer, is related to the total thermal resistance R of the composite wall by,

$$\frac{1}{U} = \frac{1}{h_a} + \frac{L_1}{K_1} + \frac{L_2}{K_2} + \frac{L_3}{K_3} + \frac{1}{h_b} \tag{2.33a}$$

or

$$U = 1/R$$

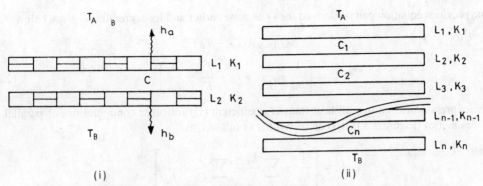

Figure 2.5(b) Configuration of parallel slabs with air cavities
(i) single cavity and (ii) multiple cavity.

2.4.2(a) Parallel Slabs with Air Cavity (Sodha et al. 1979b)

Consider a concrete wall/roof with air cavity of air-conductance 'C' as shown in Figure 2.5(b). The variation of thermal air conductance with thickness of air gap has been given in Figure 2.5(c).

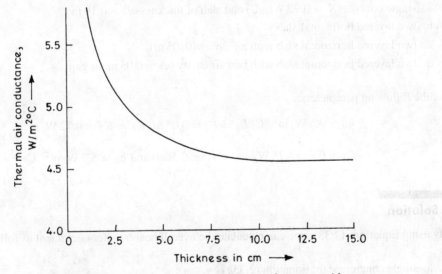

Figure 2.5(c) Variation of thermal air conductance with air gap thickness.

The heat is transfer from the hot surface at temperature T_A to the cold surface temperature T_B. Assuming steady state condition, i.e. the rate of heat tansfer per unit area of structure will be same at each layer. Then we can write (Figure 2.5 (b i))

$$\dot{Q} = Ah_a(T_A - T_0) = \frac{AK_1(T_0 - T_1)}{L_1} = AC(T_1 - T_2)$$

$$= \frac{AK_2(T_2 - T_3)}{L_2} = Ah_b(T_3 - T_B)$$

The previous equation can be derived for $\dot{Q}$ as done earlier and its expression is given below:

$$\dot{Q} = UA(T_A - T_B)$$

where

$$U = \left[\frac{1}{h_a} + \frac{L_1}{K_1} + \frac{1}{C} + \frac{L_2}{K_2} + \frac{1}{h_b} \right]^{-1} \qquad (2.33b)$$

The expression for an overall heat transfer coefficient (U) for other configurations of parallel slab with air cavities (Figure 2.5 (bii)) can be written as follows:

$$U = \left[\frac{1}{h_a} + \sum_i \frac{L_i}{K_i} + \sum_i \frac{1}{C_i} + \frac{1}{h_b} \right]^{-1} \qquad (2.33c)$$

EXAMPLE 2.1(a)

Calculate the overall heat transfer coefficient U for

a. single concret ($K = 0.72$ W/mC) and slab of thickness $L = 0.10$ m
b. two layered horizontal slab
c. two layered horizontal slab with air cavity (0. 05 m)
d. two layered horizontal slab with two air cavity (each 0.05 m air gap)

for the following parameters:

$$h_a = 9.5 \text{ W/m}^2\,{}^\circ\text{C}, L_1 = L_2 = 0.05 \text{ m}, K_1 = K_2 = 0.72 \text{ W/m}^\circ\text{C}$$

$$C_1 = C_2 = 4.75 \text{ W/m}^2\,{}^\circ\text{C} \text{ (Fig. 2.5}(c)\text{) and } h_b = 5.7 \text{ W/m}^2\,{}^\circ\text{C}$$

Solution

By using Equations (2.33), one can calculate the overall heat transfer coefficient as follows:

a. single concrete slab, Equation (2.33a)

$$U = [(1/9.5) + (0.10/0.72) + (1/5.7)]^{-1} = 2.38 \text{ W/m}^2\,{}^\circ\text{C}$$

b. two layered horizontal slab, Equation (2.33a)

$$U = [(1/9.5) + (0.05/0.72) + (0.05/0.72) + (1/5.7)]^{-1} = 2.38 \text{W/m}^2\,{}^\circ\text{C}$$

c. two layered horizontal slab with air cavity, Equation (2.33 b)

$$U = [(1/9.5) + (0.05/0.72) + (1/4.75) + (0.05/0.72) + (1/5.7)]^{-1}$$
$$= 1.59 \text{W/m}^{2\circ}\text{C}$$

d. two layered horizontal slab with two air cavity gap, Equation (2.33c)

$$U = [(1/9.5) + (0.05/0.72) + (1/4.75) + (1/4.75) + (0.05/0.72) + (1/5.7)]^{-1}$$
$$= 1.19\,\text{W/m}^2\,°\text{C}$$

From above calculation, it is clear that introduction of air cavity reduces the overall heat transfer coefficient from 2.38 W/m^2 °C to 1.59 W/m^2 °C. Further, an increase of the number of air cavity from one to two reduces U from 1.59 W/m^2 °C to 1.19 W/m^2 °C. It is also important to note that there is no change in U from one layer slab to two layer slab for same thickness of wall and of the same material. However, there will be a change in numerical value if material in the two slabs are different.

2.4.2(b)   Heat Transfer in Parallel Paths

Consider Figure 2.5 (d) for the case of heat transfer in parallel path unlike previous cases (Figs. 2.5(a)–2.5(c)). In this case, the heat transfer will be added. Therefore,

$$\dot{Q} = \dot{Q}_0 + \dot{Q}_1$$

or
$$U\,\Delta T = h_0 \Delta T + \frac{K}{L}\Delta T$$

or
$$U = h_0 + \frac{K}{L} \qquad (2.33d)$$

or
$$\frac{1}{R} = \frac{1}{R_0} + \frac{1}{R_1}$$

which is the equation of total thermal resistance in the case of parallel path unlike the case of series path derived earlier. This type of situation arose when more then one mode of heat transfer occurs from the same surface for example internal heat loss in case of solar distillation.

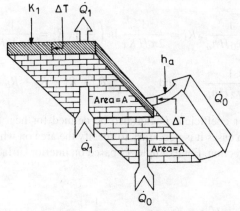

Figure 2.5(d)   One dimensional heat flow through parallel slabs in perfect thermal contact.

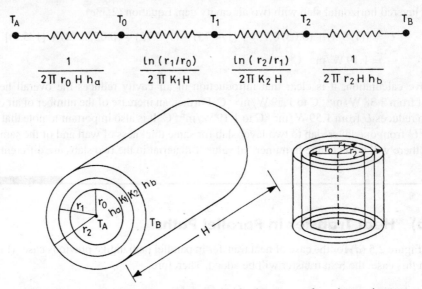

Figure 2.6 Heat flow through coaxial cylinders in perfect thermal contact (H is the height of cylinder).

2.4.3 Coaxial Cylinders

Consider a composite coaxial cylinder as illustrated in Figure 2.6. A hot fluid at temperature T_A flows inside the tube along height 'H' and the exterior surface of the tube is in contact with a cold fluid at temperature T_B. Then, at steady state,

$$\dot{Q} = \frac{T_A - T_0}{R_a} = \frac{T_0 - T_1}{R_1} = \frac{T_1 - T_2}{R_2} = \frac{T_2 - T_B}{R_b} \tag{2.34a}$$

where, various thermal resistances are defined as

$$R_a = \frac{1}{2\pi r_0 H h_a}, \; R_1 = \frac{1}{2\pi H K_1} \ln\left(\frac{r_1}{r_0}\right), \; R_2 = \frac{1}{2\pi H K_2} \ln\left(\frac{r_2}{r_1}\right),$$

$$R_b = \frac{1}{2\pi r_2 H h_b} \quad \text{and} \quad R = R_a + R_1 + R_2 + R_b$$

An overall heat transfer coefficient U can also be defined for heat transfer through a composite cylinder. However, in such cases it is necessary to specify the area on which U is based as the area of a cylinder varies in axial direction. For example, U based on interior surface A of the cylinder is defined as

$$U = \frac{1}{\frac{1}{h_a} + \frac{r_0}{K_1}\ln\left(\frac{r_1}{r_0}\right) + \frac{r_0}{K_2}\ln\left(\frac{r_2}{r_1}\right) + \frac{r_0}{r_2}\frac{1}{h_b}} \tag{2.34b}$$

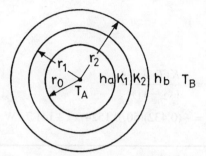

Figure 2.7 Heat flow through concentric spheres in perfect thermal contact.

2.4.4 Concentric Spheres

Figure 2.7 shows a composite sphere consisting of two concentric layers. The interior and exterior surfaces are subjected to heat exchange by convection with fluids at constant temperatures T_A and T_B, with heat transfer coefficients h_a and h_b, respectively. The total radial heat flow through the sphere is given by

$$\dot{Q} = \frac{T_A - T_B}{R}$$

where
$$R = R_a + R_1 + R_2 + R_b, \qquad R_a = \frac{1}{4\pi r_0^2 h_a}, \qquad R_b = \frac{1}{4\pi r_2^2 h_b}$$

$$R_1 = \frac{1}{4\pi K_1} \frac{r_1 - r_0}{r_1 r_0}, \qquad \text{and} \qquad R_2 = \frac{1}{4\pi K_2} \frac{r_2 - r_1}{r_2 r_1}$$

The overall heat transfer coefficient based on the exterior surface area of the sphere is defined as,

$$U = \frac{1}{\frac{r_2^2}{r_0^2}\left(\frac{1}{h_a}\right) + \frac{r_2^2}{K_1}\left(\frac{r_1 - r_0}{r_1 r_0}\right) + \frac{r_2^2}{K_2}\left(\frac{r_2 - r_1}{r_2 r_1}\right) + \frac{1}{h_b}} \tag{2.35}$$

EXAMPLE 2.1(b)

A plane wall 15.24 cm thick, is of a homogenous material and has a steady and uniform surface temperatures $T_1 = 21.1°C$ and $T_2 = 71.1°C$. Determine the heat transfer rate in the positive x direction. Given $K = 0.432\,\text{W/mK}$.

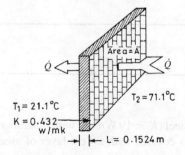

$T_1 = 21.1°C$

$K = 0.432 \text{ w/mk}$

Area = A

$T_2 = 71.1°C$

$L = 0.1524\,m$

Solution

Using Equation 2.29a, we have

$$\frac{\dot{Q}}{A} = \frac{+K(T_2 - T_1)}{(x_2 - x_1)}$$

$$= +(0.432)(50/0.1524) = +141.73 \,\text{W/m}^2$$

EXAMPLE 2.2

A hollow cylinder having inner and outer radii r_1 and r_2, respectively, is subjected to a steady heat transfer resulting in constant surface temperatures T_1 and T_2. If the thermal conductivity can be expressed as $K = K_0(1 + b\theta)$ where $\theta = T - T_{\text{ref}}$, obtain an expression for the heat transfer rate per unit length of the cylinder.

Solution

In terms of θ, Fourier's law is
$$\dot{Q} = -K A_r \frac{d\theta}{dr}$$

where A_r is the area normal to r.

Substituting the value of K and A_r in the above equation we get,

$$\dot{Q} = -K_0(1 + b\theta)2\pi r L \frac{d\theta}{dr}$$

or
$$\frac{\dot{Q}}{L}\frac{dr}{r} = -2\pi K_0(1 + b\theta)d\theta$$

where L is the length of the cylinder.

In the steady state $\frac{\dot{Q}}{L}$ is constant. Hence, integrating the above equation, we get

$$\frac{\dot{Q}}{L} = -2\pi K_0 \left[1 + \frac{b}{2}(\theta_2 + \theta_1)\right] \frac{\theta_2 - \theta_1}{\ln\left(\frac{r_2}{r_1}\right)}$$

EXAMPLE 2.3

A thick walled tube of stainless steel ($K = 19$ W/mK) with inner diameter (ID) and outer diameter (OD) as 3 cm and 5 cm respectively, is covered with a 4 cm layer of asbestos insulation ($K = 0.2$ W/mK).

If the inside wall temperature is maintained at 600°C, calculate the heat loss per meter of length if the outer temperature is 100°C.

The figure given below explains the structure. The equivalent thermal circuit is shown below:

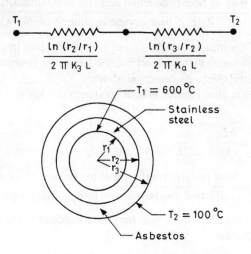

The rate of heat loss per meter of length is given by

$$\frac{\dot{Q}}{L} = \frac{2\pi(T_1 - T_2)}{\ln(r_2/r_1)/K_3 + \ln(r_3/r_2)/K_a}$$

After substituting the given parameters, we have

$$\frac{\dot{Q}}{L} = \frac{2\pi(500)}{\ln(0.05/0.03)/19 + \ln(0.09/0.05)/0.2} = 1058.7\,\text{W/m}$$

2.5 DIMENSIONLESS HEAT-CONDUCTION PARAMETERS

The number of variables in a heat conduction problem may be reduced by introducing dimensionless variables. The heat conduction equation, Equation (2.15) and various initial and boundary conditions involve the following parameters:

$$Bi = \frac{hL}{K} = \text{Biot number} \tag{2.36a}$$

$$Fo = \frac{\alpha t}{L^2} = \text{Fourier number} \tag{2.36b}$$

These are important dimensionless parameters and are frequently used in heat conduction problems. The physical significance of the Fourier number can be seen when it is arranged in the following form:

$$Fo = \frac{\alpha t}{L^2} = \frac{K(1/L)L^2}{\rho S L^3/t} \tag{2.36c}$$

or

$$Fo = \frac{\text{Rate of heat conduction across } L \text{ in volume } L^3, \text{ W/}^\circ\text{C}}{\text{Rate of heat storage in volume } L^3, \text{ W/}^\circ\text{C}}$$

Thus, the Fourier number is a measure of the rate of heat conduction in comparison with the rate of heat storage in a given volume element. Therefore, the larger the Fourier number, the deeper is the penetration of heat into a solid over a given period of time.

The physical significance of Biot number is better understood when it is rearranged in the following form:

$$Bi = \frac{hL}{K} = \frac{h}{K/L} \tag{2.36d}$$

or

$$Bi = \frac{\text{heat transfer coefficient at the surface of the solid}}{\text{internal conductance of solid across length } L}$$

That is, the Biot number is the ratio of the heat transfer coefficient to the unit conductance of a solid over the characteristic dimension.

2.6 CONVECTION

Thermal convection is the transfer of heat from one part of a fluid to another part, at a lower temperature, by mixing of fluid particles. Heat is transported simultaneously during the process by convection and by conduction. In most of the solar energy applications, convection is accompanied by conduction.

The temperature difference between the fluid and the contact surface causes the density variation in the fluid resulting in buoyancy. The fluid motion thus produced is 'free convection'. However, if the motion of the fluid is caused by forces arising from externally imposed pressure difference and is independent of the temperature difference in the fluid, it is referred to as 'forced convection'. These two processes may sometimes occur simultaneously. The rate of heat transfer by convection between the fluid and the boundary surface may be evaluated by the formula,

$$\dot{Q} = hA\Delta T \tag{2.37}$$

where h is the local heat transfer coefficient and the rate of heat flow, $\dot{Q}$ at the fluid and body interface is related to the temperature difference between the surface of the body concerned and it's surroundings.

The heat transfer coefficient, h, is a complicated function of the fluid flow, thermophysical properties of the fluid and geometric arrangements of the system. However, the appropriate mathematical form of h in a limited domain of applicability is evaluated from empirical equations obtained by correlating experimental results with the method of dimensional analysis.

The unit of h can be expressed either in W/m^{2}K or in W/m^2 $^\circ$C. The value of h in both units will be same.

2.6.1 Dimensionless Heat-Convection Parameters

The convection equations contain following dimensionless terms with dissimilar physical parameters:

$$Nu = \frac{hX}{K} \quad \text{Nusselt number} \tag{2.38a}$$

$$Re = \frac{\rho v_0 X}{\mu} = \frac{v_0 X}{v} \quad \text{Reynold number} \tag{2.38b}$$

$$Pr = \frac{\mu C_p}{K} \quad \text{Prandtl number} \tag{2.38c}$$

$$Gr = \frac{g\beta' \rho^2 X^3 \Delta T}{\mu^2} = \frac{g\beta' X^3 \Delta T}{v^2} \quad \text{Grashof number} \tag{2.38d}$$

$$Ra = Gr\,Pr = \frac{g\beta' \rho^2 X^3 C_p \Delta T}{\mu K} \quad \text{Rayleigh number} \tag{2.38e}$$

where $\qquad\qquad v = \text{Kinematic Viscosity} = \dfrac{\mu}{\rho}$

and X is characteristic dimension of the system and is given in Table 2.1 for most of the cases except rectangular shape. Reynold (Re), Prandtl (Pr) and Grashof (Gr) numbers should be calculated by using fluid physical properties at the average temperatures (T_f) of the hot surface (T_1) and surrounding air (T_2), i.e.

$$T_f = (T_1 + T_2)/2$$

The thermal expansion coefficient β should be calculated at:
  (a) temperature of surrounding air T_2 for exposed surface i.e.

$$\beta' = 1/(T_2 + 273) \tag{2.38f}$$

  (b) average temperature T_f for parallel plate i.e.

$$\beta' = 1/(T_f + 273) \tag{2.38g}$$

The characteristic dimension X of other shape is given by

$$X = A/P \tag{2.38h}$$

where A and P are the area and perimeter of the surface.
  Some times, for a rectangular horizontal surface($L_0 \times B_0$), the characteristic dimension can also be calculated by

$$X = (L_0 + B_0)/2 \tag{2.38i}$$

These terms, referred to as criteria of similarity, are called after the scientists who made great contributions to the development of hydro dynamics and heat transfer.
  The **Nusselt number**, Nu, characterizes the process of heat transfer at the wall-fluid boundary. It is the ratio of convective heat transfer to heat transfer by conduction in the fluid and is usually unknown in problems of convection, since it involves the heat transfer coefficient, h, which is an unknown parameter. X is a characteristic dimension of the system and K, the thermal conductivity. Although the Nusselt

number resembles the Biot number, the two are essentially different. The Biot number includes the thermal conductivity of a solid, the Nusselt number that of a fluid.

The **Reynold's number**, Re, characterizes the relation between the forces of inertia and viscosity. It is the ratio of the fluid dynamic force (ρv_0^2) to the viscous drag force ($\mu\, v\,_0/X$) where, ρ is the density and μ the dynamic viscosity. It indicates the flow behavior in forced convection and serves as a criterion for the stability of laminar flow.

The third dimensionless number, the **Prandtl number**, Pr, is the ratio of momentum diffusivity (μ/ρ) to the thermal diffusivity ($K/\rho C_p$); C_p is the specific heat at constant pressure. It represents the relation of heat transfer to fluid motion.

The **Grashof number**, Gr, is the ratio of the buoyancy force to the viscous force; β' is the coefficient of volumetric thermal expansion, g the gravitational acceleration and T the temperature. It represents the effects of hydrostatic lift force and viscous force of the fluid in free convection.

The **Rayleigh number**, Ra, is the ratio of the thermal buoyance to viscous inertia.

With the help of these dimensionless numbers, one can describe a large number of convection processes. Furthermore, similar processes would be described by a single formula as discussed below.

The various dimensionless numbers given by Equation (2.38) can be obtained by using the physical properties of dry and moist air and water given in Appendix III.

2.6.2 Bulk Temperature

In tube flow, the convection heat-transfer coefficient is usually defined as,

$$\text{Local heat flux} = \dot{q} = h(T_w - T_b) \tag{2.39a}$$

where T_w is the wall temperature and T_b is the so-called bulk temperature, or energy-average fluid temperature T across the tube, which may be calculated from the expression,

$$T_b = \bar{T} = \frac{\int_0^{r_0} \rho 2\pi\, r\, dr\, u\, C_p T}{\int_0^{r_0} \rho 2\pi\, r\, dr\, u\, C_p} \tag{2.39b}$$

where u is the instantaneous velocity, r the radius of the circular section considered and r_0 the radius of the tube.

The reason for using the bulk temperature in the definition of heat-transfer coefficients for tube flow may be explained as follows: in a tube flow there is no easily discernable free-stream condition as is present in the flow over a flat plate. Even the centerline temperature T_c is not easily expressed in terms of the inlet flow variables and the heat transfer. For most tube or channel flow heat transfer problems the topic of central interest is the total energy transferred to the fluid in either an elemental length of the tube or over the entire length of the channel. At any position x, the temperature that is indicative of the total energy of the flow is an integrated mass-energy average temperature over the entire flow area. The numerator of Equation (2.39b) represents the total energy flow through the tube, and the denominator represents the product of mass flow and specific heat integrated over the flow area. The bulk temperature is thus representative of the total energy of the flow at the particular location. For this reason, the bulk temperature is sometimes referred to as the 'mixing cup' temperature, since it is the temperature the fluid would assume if placed in a mixing chamber and allowed to come to equilibrium.

2.6.3 Free Convection

In the event of free or natural flow, the origin of motion in the considered volume of fluid is due to the heterogeneity of the mass forces acting upon the volume. If this natural flow is not confined within a space by solid boundaries, it is referred to as free convection. However, in general, no distinction is made between them. For free convection, therefore, the terrestrial gravitational field, acting on the fluid with a non-uniform density distribution owing to the temperature difference between the fluid and the contacting surface, causes the fluid motion. Other types of body forces, such as centrifugal forces and coriolis forces also have influence on free convective heat transfer, particularly, in rotating systems.

The coefficient of heat transfer, h, usually incorporated with Nusselt number, depends on whether the flow is laminar or turbulent, free or forced.

For free convection,

$$Nu = C'(Gr\ Pr)^n K' \tag{2.40}$$

where the relationship is obtained by the method of dimensional analysis. The constants C' and n are determined by the correlation of experimental data of geometrically similar bodies. The correlation factor, K', is introduced to represent the entire physical behavior of the problem and it also increases the range of the parameters. An extensive summary of the equations for convective heat transfer, for different geometries is given in ASHRAE Handbook of fundamentals (1967). The free convective heat transfer, for some geometrical systems is given in the Table 2.1.

In addition to Table 2.1 some other empirical relations for free convection can be used.

2.6.4 Empirical Relations

It has been found that average free-convection heat transfer coefficients can be represented in the following functional form for a variety of configurations:

$$\bar{Nu}_f = C(Gr_f Pr_f)^m \tag{2.41}$$

where the subscript f indicates that the properties in the dimensionless groups are evaluated at the film temperature $T_f = (T_\infty + T_w)/2$; T_w is the wall temperature and T_∞ is the free-stream temperature.

The characteristic dimension to be used in the Nusselt and Grashof numbers depends on the geometry of the problem. For a vertical plate it is the height of the plate L; for a horizontal cylinder it is the diameter, and so forth.

Table 2.2 gives the values of the constants C and m to be used, with Equation (2.41) for different conditions.

2.6.5 Free Convection from Inclined Surfaces

Extensive experiments have been conducted by Fuji and Imura (1972) for heated plates in water at various angles of inclination. The angle which the plate makes with the vertical is designated θ, with positive angles indicating that the heated surface faces downward. For the inclined plate facing downward with approximately constant heat flux, the correlation obtained for the average Nusselt number was (Fig. 2.8)

$$\bar{Nu}_e = 0.56(Gr_e Pr_e \cos\theta)^{1/4} \quad \text{for} \quad \theta < 88° \quad \text{and} \quad 10^5 < Gr_e Pr_e \cos\theta < 10^{11} \tag{2.42}$$

TABLE 2.1 Free Convective heat transfer of various systems (after Wong, 1977)

System	Schematic	C'	n	K'	Operating Conditions
Horizontal cylinder		0.47	1/4	1	Laminar flow
		0.1	1/3	1	Turbulent flow
Vertical plate and vertical cylinder with large diameter		0.8	1/4	$\left[1+\left(1+\frac{1}{\sqrt{Pr}}\right)^2\right]^{-1/4}$	Laminar flow to obtain local Nu, use $C'=0.6$, $X=x$, formula applicable to vertical cylinder when $D/L \gg 38(Gr)^{-1/4}$
		0.0246	2/5	$[Pr^{1/6}/(1+0.496\,Pr^{2/3})]^{2/3}$	Turbulent flow; to obtain local Nu use $C'=0.0296$ and $X=x$
Vertical cylinder with small diameter		0.686	1/4	$[Pr/(1+1.05Pr)]^{1/4}$	Laminar flow $\bar{N}u_{local} = \bar{N}u + 0.52(L/D)$
Heated horizontal plate facing upward		0.54	1/4	1	Laminar flow $(10^5 < Gr\,Pr < 2\times10^7)$, $X=(L_0+B_0)/2$
		0.14	1/3	1	Laminar flow $(10^7 < Gr\,Pr < 10^{11})$, $X=A/P$ for circular disc of diameter, D, use $X=0.9D$
					Turbulent flow $(2\times10^7 < Gr\,Pr < 3\times10^{10})$, $X=(L_0+B_0)/2$
		0.15	1/3	1	Turbulent flow $(10^7 < Gr\,Pr < 10^{11})$, $X=A/P$
Heated horizontal plate facing downward		0.27	1/4	1	Laminar flow only

TABLE 2.1 (Continued)

System	Schematic	C'	n	K'	Operating Conditions
Sphere		0.49	1/4	1	Laminar flow (air)
Moderately inclined plane		0.8	1/4	$\left[\dfrac{\cos\theta}{1+\left(1+\frac{1}{\sqrt{Pr}}\right)^2}\right]^{1/4}$	Laminar flow (multiply Gr by $\cos\theta$ in the formula for vertical plate)
Two vertical parallel plates at the same temperature		0.04	1	$(d/L)^3$	Air layer
Hollow vertical cylinder with open ends		0.01	1	$(d/L)^3$	Air column
Two horizontal parallel plates hot plate uppermost		0.27	1/4	1	Pure conduction $\dot{q} = K(T_h - T_c)/d$ Laminar (air) $3 \times 10^5 < Gr.Pr < 3 \times 10^{10}$
Two concentric cylinders		0.317	1/4	$\left[X^3\left(\dfrac{1}{d_i^{3/5}} + \dfrac{1}{d_o^{3/5}}\right)^5\right]^{-1/4}$	Laminar flow

(Contd.)

TABLE 2.1 (Continued)

System	Schematic	C'	n	K'	Operating Conditions
Two vertical parallel plates of different temperatures (h for both surfaces)		0.18	1/4	$(L/d)^{-1/9}(Pr)^{-1/4}$	Laminar (air) $2 \times 10^4 < Gr < 2 \times 10^5$
		0.065	—	$(L/d)^{-1/9}(Pr)^{-1/3}$	Turbulent (air) $2 \times 10^5 < Gr < 10^7$
Two inclined parallel plates				$\overline{Nu} = \tfrac{1}{2}\left[\overline{Nu}_{\text{vert}}\cos\theta + \overline{Nu}_{\text{horz}}\sin\theta\right]$	
Two horizontal parallel plates cold plate uppermost		0.195	1/4	$Pr^{-1/4}$	Laminar (air) $10^4 < Gr < 4 \times 10^5$
		0.068	—	$Pr^{-1/3}$	Turbulent (air) $Gr > 4 \times 10^5$
Two concentric spheres		0.61	1/4	$\left[\dfrac{1}{2(d_o+d_i)}\left[X^3\left(\dfrac{1}{d_i^{7/5}} + \dfrac{1}{d_o^{7/5}}\right)^5\right]\right]^{-1/4}$	Laminar flow

TABLE 2.2 The value of C and m for different geometries

Geometry	$Gr_f Pr_f$	C	m	Ref.
Vertical planes and cylinders	$10^4 - 10^9$	0.59	1/4	McAdams (1954)
	$10^9 - 10^{13}$	0.10	1/3	Bayley (1955)
Horizontal cylinder	$10^4 - 10^9$	0.53	1/4	McAdams (1954)
	$10^9 - 10^{12}$	0.13	1/3	McAdams (1954)
	$10^{-10} - 10^{-2}$	0.675	0.058	Morgan (1975)
	$10^{-2} - 10^2$	1.02	0.148	Morgan (1975)
	$10^2 - 10^4$	0.850	0.188	Morgan (1975)
Upper surface of heated plates or lower surface of cooled plates	$2 \times 10^4 - 8 \times 10^6$	0.54	1/4	Fuji & Imura (1972)
	$8 \times 10^6 - 10^{11}$	0.15	1/3	Fuji & Imura (1972)
Lower surface of heated plates or upper surface of cooled plates	$10^5 - 10^{11}$	0.27	1/4	Fuji & Imura (1972) Clifton & Chapman (1969)
Vertical cylinder height = diameter	$10^4 - 10^6$	0.775	0.21	Sparrow & Ansari (1983)
For constant heat flux condition:				
Heated surface facing upward	$< 2 \times 10^8$	0.13	1/3	Fuji & Imura (1972)
	$2 \times 10^8 - 10^{11}$	0.16	1/3	
Heated surface facing downward	$10^6 - 10^{11}$	0.58	1/5	Fuji & Imura (1972)

In the above equation, all properties except β are evaluated at a reference temperature T_e defined by

$$T_e = T_w - 0.25(T_w - T_\infty) \tag{2.43}$$

where T_w is the mean wall temperature and T_∞ the free-stream temperature. β is evaluated at a temperature of $T_\infty + 0.5(T_w - T_\infty)$.

For almost horizontal plates facing downwards, i.e. $88° < \theta < 90°$ an additional relation was obtained as

$$\bar{Nu}_e = 0.58(Gr_e Pr_e)^{1/5}, \qquad 10^6 < Gr_e Pr_e < 10^{11} \tag{2.44}$$

For an inclined plate with heated surface facing upward the empirical relations become complicated for angles between $-15°$ and $-75°$ a suitable correlation is

$$\bar{Nu}_e = 0.14[(Gr_e Pr_e)^{1/3} - (Gr_c Pr_c)^{1/3}] + 0.56(Gr_e Pr_e \cos\theta)^{1/4};$$

$$10^5 < Gr_e Pr_e \cos\theta)10^{11} \tag{2.45}$$

The quantity Gr_c is a critical Grashof relation indicating when the Nusselt number starts to separate

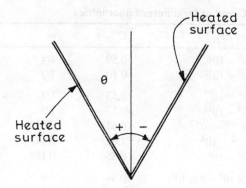

Figure 2.8 Inclined plates.

from the laminar relation, Equation (2.42) and is given as

θ	Gr_c
-15	5×10^9
-30	2×10^9
-60	10^8
-75	10^6

For inclined cylinder, the laminar heat transfer, under constant heat flux condition, may be calculated by the relation

$$\bar{Nu}_L = \left[0.60 - 0.488(\sin\theta)^{1.03}\right]\left[(Gr_L Pr)^{1/4+1/12(\sin\theta)^{1.75}}\right]$$

for
$$Gr_L Pr < 2 \times 10^8 \tag{2.46}$$

TABLE 2.3   Simplified equations for free convection from various surfaces to air at atmospheric pressure (Heat transfer, J.P. Holman 1992)

Surface	Laminar $10^4 < Gr_f Pr_f < 10^9$	Turbulent $Gr_f Pr_f > 10^9$
Vertical plane or cylinder	$h = 1.42(\Delta T/L)^{1/4}$	$h = 1.31(\Delta T)^{1/3}$
Horizontal cylinder	$h = 1.32(\Delta T/d)^{1/4}$	$h = 1.24(\Delta T)^{1/3}$
Horizontal plate: Heated plate facing upward or cooled plate facing downward	$h = 1.32(\Delta T/L)^{1/4}$	$h = 1.52(\Delta T)^{1/3}$
Heated plate facing downward or cooled plate facing upward	$h = 0.59(\Delta T/L)^{1/4}$	

where θ is the angle the cylinder makes with the vertical. Properties are evaluated at the film temperature, except β, which is evaluated at ambient conditions. Further, the simplified equations for free convection from various surfaces to air at atmospheric pressure are given in Table 2.3.

EXAMPLE 2.4(a)

Estimate the rate of heat loss from a horizontal rectangular surface ($1.0\,\mathrm{m} \times 0.8\,\mathrm{m}$) which is maintained at $134°C$ and it is exposed toward the upward direction to an environment at $20°C$ as shown in the following figure:

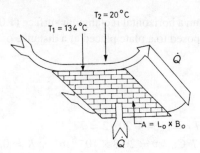

Solution

The average film temperature, $T_f = (134 + 20)/2 = 77°C$

From Appendix IIId, at $T_f = 77°C$; $\nu = 20.8 \times 10^{-6}\,\mathrm{m^2/s}$; $K = 0.030$ W/m.K
$Pr = 0.697$ and $\beta' = 1/(20 + 273)$ due to the exposure of surroundings.
For the characteristic dimension of
(i) $X = A/P = 0.8/3.6 = 0.222$ m

$$Gr_L\, Pr = \frac{(9.8)(134 - 20)(0.222)^3(0.697)}{(293)(2.08 \times 10^{-5})^2} = 6.72 \times 10^7$$

Using Table 2.1 for hot surface facing upward and turbulent flow condition, the heat transfer coefficient can be calculated as

$$h = (K/L)0.15(Gr_L\, Pr)^{0.333}$$
$$= (0.03/0.222)(0.15)(6.72 \times 10^7)^{0.333} = 8.23\,\mathrm{W/m^2°C}$$

Hence, the rate of heat loss from the hot plate to the surroundings is

$$\dot{Q} = hA(T_1 - T_2) = 8.23(1.0 \times 0.8)(134 - 20) = 751\,\mathrm{W}$$

(ii) $X = (L_0 + B_0)/2 = (1.0 + 0.8)/2 = 0.9$

$Gr_L\, Pr = 4.47 \times 10^9$

and $\qquad h = (0.03/0.9)(0.14)(4.47 \times 10^9)^{1/3} = 7.74\,\mathrm{W/m^2°C}$

Hence, the rate of heat loss from the hot plate to the surroundings is

$$\dot{Q} = hA(T_1 - T_2) = 7.742(1.0 \times 0.8)(134 - 20) = 706\,\text{W}$$

From above calculation, it is seen that there is a little variation with rate of heat transfer by considering different characteristic length in cases (a) and (b) respectively.

EXAMPLE 2.4(b)

Estimate the rate of heat loss from a horizontal rectangular surface ($1.0\,\text{m} \times 0.8\,\text{m}$) which is maintained at $134°\text{C}$. The hot surface is exposed to a plate placed at a distance of $0.10\,\text{m}$ above it and temperature of plate is maintained at $20°\text{C}$.

Solution

The average film temperature, $T_f = (134 + 20)/2 = 77°$ C
From Appendix IIId, at $T_f = 77°\text{C}$; $\quad \nu = 20.8 \times 10^{-6}\,\text{m}^2/\text{s}$; $K = 0.030\,\text{W/m.K}$
$Pr = 0.697$ and $\beta' = 1/(77 + 273) = 2.857 \times 10^{-3}\,\text{K}^{-1}$ due to exposure to another plate and the characteristic dimension ($X = d$) $= 0.10\,\text{m}$

$$Gr_L = \frac{(9.8)(2.857 \times 10^{-3})(134 - 20)(0.1)^3}{(20.8 \times 10^{-6})^2} = 7.377 \times 10^6$$

Using Table 2.1 for turbulent flow condition for two horizontal parallel plates (cold plate uppermost), the heat transfer coefficient can be calculated as

$$h = \frac{K}{L}(0.068)(Gr_L\,Pr)^{1/3}\,Pr^{-1/3}$$

$$= \frac{0.030}{0.10}(0.068)(7.377 \times 10^6)^{1/3} = 3.97\,\text{W/m}^2\text{K}$$

Hence, the rate of heat loss from a horizontal surface is

$$\dot{Q} = hA(T_s - T_\infty) = (3.97)(1.0 \times 0.8)(134 - 20) = 362.14\,\text{W}$$

From this example one can observe that there is a change in numerical value of convective heat transfer coefficient from $7.47\,\text{W/m}^2\text{K}$ to $3.97\,\text{W/m}^2$ K due to change in condition from open to closed.

EXAMPLE 2.4(c)

Estimate the rate of heat transfer from a horizontal rectangular surface ($1.0\,\text{m} \times 0.8\,\text{m}$) which is maintained at $134°\text{C}$. The hot surface is exposed to water at $20°\text{C}$.

Solution

The average film temperature, $T_f = (134 + 20)/2 = 77°C$
From Appendix IIIe, water thermal properties at $T_f = 77°C$

$\mu = 3.72 \times 10^{-4}$ kg/ms; $K = 0.668$ W/m.K, $\rho = 973.7$ kg/m³, $Pr = 2.33$ and $\beta' = 1/(77 + 273) = 2.857 \times 10^{-3}$K^{-1}. Consider the characteristic dimension $(X = d) = (1.0 + 0.8)/2 = 0.90$ m.
The Grashof number is

$$Gr_L = \frac{g\beta'\rho^2(\Delta T)X^3}{\mu^2}$$

$$= \frac{(9.8)(2.857 \times 10^{-3})(973.7)^2(114)(0.9)^3}{(3.72 \times 10^{-4})^2} = 1.594 \times 10^{13}$$

This is a turbulent flow, hence using Table 2.1 for heated plate facing upward for $X = L = (L_0 + B_0)/2$, consider $C = 0.14$ and $n = 1/3$. Now, a convective heat transfer coefficient can be calculated as

$$h = \frac{K}{L}(0.14)(Gr_L\,Pr)^{1/3} = \frac{0.668}{0.9}(0.14)(1.594 \times 10^{13} \times 2.33)^{1/3}$$

$$= 3467 \text{ W/m}^2\text{K}$$

Hence, the rate of heat loss from a vertical wall is

$$\dot{Q} = hA(T_s - T_\infty) = (3467)(1.0 \times 0.8)(134 - 20) = 316190.4 \text{ W} = 316.19 \text{ kW}.$$

If this result has to be compared with the result of Example 2.4(a), one can see that the change in convective heat transfer coefficient from 7.74 W/m²°C to 3467 W/m²°C is due to change of fluid from air to water.

EXAMPLE 2.5(a)

Estimate the rate of heat loss from a vertical wall exposed to air at 1 atm and at 15°C. The wall is 1.80 m high and 2.40 m wide, and is maintained at 49°C.

Solution

The average film temperature, $T_f = (15 + 49)/2 = 32°C$
At $T_f = 32°C$; $\nu = 16.19 \times 10^{-6}$ m²/s; $K = 0.0264$ W/m.K, $Pr = 0.707$
(see Appendix IIId) giving a Grashof number (from Equation (2.38d)), of

$$Gr_L = \frac{(9.8)(3.472 \times 10^{-3})(34)(1.8)^3}{(16.19 \times 10^{-6})^2} = 2.573 \times 10^{10}$$

In this case, $L = X = 1.8$ m and $\beta' = 1/(Ta + 273) = 1/(15 + 273) = 3.472 \times 10^{-3}$ K^{-1}

Using Table 2.2 (vertical planes), the heat transfer coefficient can be calculated as follows:

$$h = \frac{K}{L}(0.1)(Gr_L Pr)^{1/3} = \frac{0.0264}{1.80}(0.1)(1.819 \times 10^{10})^{1/3} = 3.857 \, \text{W/m}^2\text{K}$$

Hence, the rate of heat loss from a vertical wall is

$$\dot{Q} = hA(T_s - T_\infty) = (3.857)(1.8 \times 2.40)(34) = 566.52 \, \text{W}$$

EXAMPLE 2.5(b)

Estimate the rate of heat loss from a vertical wall exposed to a plate placed at a distance of $d = 0.10$ m from vertical wall which is maintained at 15°C. The wall is 1.80 m high and 2.40 m wide, and is maintained at 49°C.

Solution

The average film temperature, $T_f = (15 + 49)/2 = 32°C$
At $T_f = 32°C$; $v = 16.19 \times 10^{-6} \, \text{m}^2/\text{s}$; $K = 0.0264 \, \text{W/m.K}$; $Pr = 0.707$.
Since, $L/d = 1.8/0.10 = 18 > 3$, hence characteristic dimension $(X = L) = d = 0.10$ m and $\beta' = 1/(T_f + 273) = 1/305 = 3.278 \times 10^{-3} \text{K}^{-1}$ (see Appendix IIId) giving a Grashof number (from Equation (2.38d)), of

$$Gr_L = \frac{(9.8)(3.278 \times 10^{-3})(34)(0.1)^3}{(16.19 \times 10^{-6})^2} = 4.1667 \times 10^6$$

Now, it is turbulent flow condition hence using Table 2.1 for two vertical parallel plates of different temperature, the heat transfer coefficient can be calculated as

$$h = \frac{K}{L}(0.065)(Gr_L Pr)^{1/3} \left(\frac{L}{d}\right)^{-1/9} Pr^{-1/3}$$

$$= \frac{0.0264}{0.10}(0.065)(4.1667 \times 10^6)^{1/3}(18)^{-0.1111} = 2.00 \, \text{W/m}^2\text{K}$$

Hence, the rate of heat loss from a vertical wall is

$$\dot{Q} = hA(T_s - T_\infty) = (2.00)(1.8 \times 2.40)(34) = 294.22 \, \text{W}.$$

By comparing the results of Examples 2.5(a) and 2.5(b), one can observe that the rate of heat loss is significantly reduced due to change in condition from open to closed.

EXAMPLE 2.5(c)

Estimate the rate of heat loss from a horizontal plate exposed to a plate at 15°C and placed at distance of 0.10 m. The plate is 1.80 m long and 2.40 m wide, and is maintained at 49°C.

Solution

Here, the characteristic dimension $(X = L) = 0.1$ m and $\beta' = 1/(T_f + 273) = 1/305 = 3.278 \times 10^{-3} \text{K}^{-1}$ and other physical properties will remain the same as considered in the above example.

$$Gr_L = \frac{(9.8)(3.278 \times 10^{-3})(34)(0.1)^3}{(16.19 \times 10^{-6})^2} = 4.1667 \times 10^6$$

Using Table 2.1 for turbulent flow condition for two horizontal parallel plates (cold plate upermost), the heat transfer coefficient can be calculated as

$$h = \frac{K}{L}(0.068)(Gr_L Pr)^{1/3} Pr^{-1/3}$$

$$= \frac{0.0264}{0.10}(0.068)(4.1667 \times 10^6)^{1/3} = 2.9 \text{ W/m}^2\text{K}$$

Hence, the rate of heat loss from a horizontal plate is

$$\dot{Q} = hA(T_s - T_\infty) = (2.9)(1.8 \times 2.40)(34) = 425.95 \text{ W}$$

EXAMPLE 2.5(d)

Estimate the rate of heat loss from a horizontal plate exposed to ambient air at 15°C. The plate is 1.80 m long and 2.40 m wide, and is maintained at 49°C.

Solution

In this, characteristic dimension $(X = L) = (1.8 + 2.4)/2 = 2.1$ m and $\beta' = 1/(T_a + 273) = 1/288 = 3.472 \times 10^{-3}\text{K}^{-1}$ and other physical properties will remain the same as considered in the above example.

The Grashof number is

$$Gr_L = \frac{(9.8)(3.472 \times 10^{-3})(34)(2.1)^3}{(16.19 \times 10^{-6})^2} = 4.087 \times 10^{10}$$

Using Table 2.1 for turbulent flow condition for heated horizontal plates facing upward , the heat transfer coefficient can be calculated as

$$h = \frac{K}{L}(0.14)(Gr_L \, Pr)^{1/3} = \frac{0.0264}{2.1}(0.14)(4.087 \times 10^{10} \times 0.707)^{1/3} = 5.4 \text{ W/m}^2\text{K}$$

Hence, the rate of heat loss from a horizontal plate is

$$\dot{Q} = hA(T_s - T_\infty) = (5.4)(1.8 \times 2.40)(34) = 793.15 \text{ W}$$

Here h is the heat transfer coefficient in W/m^{2}K; $\Delta T = T_w - T_\infty$°C; L is the vertical or horizontal dimension, m; d is the diameter, m.

From Examples 2.5(a–d) it is evident that there is a significant change in the values of heat transfer coefficients and corresponding the rate of heat transfer with change in position of plates from vertical to horizontal. Hence, in order to calculate accurate heat transfer coefficient it is necessary to choose proper characteristic dimension and use proper expression for thermal expansion coefficient β'.

2.6.6 Forced Convection

In the forced convection, the fluid motion is artificially induced, say with a pump or a fan that forces the fluid flow over the surface. Also considered as forced convection is the flow of the investigated volume of fluid under the action of an internal homogeneous field of mass forces. The external energy is supplied to maintain the process in which two types of forces are in operation: the fluid pressure related to flow velocity $(1/2)\rho v^2$ and the frictional force produced by viscosity $(\mu.dv/dy)$. Their relative importance in heat transfer is signified by the non-dimensional Reynold number. It also controls the flow, laminar or turbulent, in the boundary layer with which the rate of heat transfer is closely connected. The heat transfer by forced convection is represented by the following Nusselt equation

$$Nu = C(Re \, Pr)^n K \tag{2.47a}$$

where C and n are the constants for a given type of flow and geometry and K the correction factor (shape factor) added to obtain greater accuracy.

The empirical relation for forced convective heat transfer through cylindrical tubes may be represented as

$$\bar{Nu} = \frac{hD}{K_{th}} = C Re^m \, Pr^n K \tag{2.47b}$$

where $D = 4A/P$ is the hydraulic diameter (m); P the perimeter of the section (m) and K_{th} the thermal conductivity (W/m K).

The values of C, m, n and K for various conditions are given in Table 2.4 (after Wong, 1977).

In addition to the above table, some other empirical relation may also be used.

For turbulent flow in smooth tubes the following relation is recommended by Dithus and Boelter (1930)

$$Nu = 0.023 Re^{0.8} \, Pr^n \tag{2.47c}$$

where $n = 0.4$ for heating and $n = 0.3$ for cooling the fluid.

TABLE 2.4 The value of constants for forced convection

Cross-section	D	C	m	n	K	Operating conditions
	d	1.86	1/3	1/3	$(d/l)^{1/3}(\mu/\mu_w)^{0.14}$	Laminar flow short tube $Re < 2000$, $Gz > 10$
	d	3.66	0	0	1	Laminar flow long tube $Re < 2000\ Gz < 10$
	d	0.023	0.8	0.4	1	Turbulent flow of gases $Re > 2000$
	d	0.027	0.8	0.33	$(\mu/\mu_w)^{0.14}$	Turbulent flow of highly viscous liquids $0.6 < Pr < 100$

In the entrance region of the tube where the flow is not developed, Nusselt (1931) recommended the relation

$$Nu = 0.036\, Re^{0.8}\, Pr^{1/3} \left(\frac{d}{L}\right)^{0.055} \quad \text{for} \quad 10 < \frac{L}{d} < 400 \tag{2.47d}$$

where L is the length and d the diameter of the tube.

For fully developed laminar flow in tubes at constant wall temperature, Hausen (1943) gave the relation

$$Nu_d = 3.66 + \frac{0.0668(d/L)Re\,Pr}{1 + 0.04[(d/L)Re\,Pr]^{2/3}} \tag{2.47e}$$

The heat transfer coefficient calculated from this relation is the average value over the entire length of the tube. When the tube is sufficiently long the Nusselt number approaches a constant value of 3.66. For the plate heated over it's entire length, Nusselt number can be obtained by integrating the equation given below over the length of the plate,

$$Nu_x = 0.332(Pr)^{1/3} Re_x^{1/2} \tag{2.47f}$$

Now,

$$\bar{h} = \frac{1}{L}\int_0^L \frac{K}{x}(0.332)(Pr^{1/3})(Re_x^{1/2})dx = \frac{K}{L}(0.332)(Pr^{1/3})\int_0^L \frac{1}{x}\left(\frac{v_0 x}{\nu}\right)^{1/2} dx$$

$$= \frac{K}{L}(0.332)(Pr^{1/3})(2)\left(\frac{v_0 L}{\nu}\right)^{1/2} = \frac{K}{L}(0.664)(Pr^{1/3})(Re_L^{1/2})$$

Thus,

$$\bar{Nu}_L = 0.664\, Re_L^{1/2}\, Pr^{1/3} \tag{2.47g}$$

For laminar flow on an isothermal flat plate, Churchill and Ozoe (1973) gave the following relation

$$Nu_x = \frac{0.3387\, Re_x^{1/2}\, Pr^{1/3}}{\left[1 + \left(\frac{0.0468}{Pr}\right)^{2/3}\right]^{1/4}} \quad \text{for } Re\,Pr > 100 \tag{2.47h}$$

For constant heat flux, the relation is

$$Nu_x = \frac{0.4637 Re^{1/2} Pr^{1/3}}{\left[1 + \left(\frac{0.0207}{Pr}\right)^{2/3}\right]^{1/4}} \tag{2.47i}$$

For solar air heaters, the forced convective heat transfer coefficient between two plates can be determined (Kays and Grawford, 1980), from

$$Nu = 0.0158 \, Re^{0.8} \text{ for } Re > 2100 \text{ and } \frac{L}{Dn} \text{ is large} \tag{2.47j}$$

In this case, one side is heated and the other is insulated. The above relation is valid for fully developed turbulent flow.

The characteristic length (Dn) is the hydraulic diameter (twice two plate spacing).

2.6.7 Combined Free and Forced Convection

A number of practical situations occur where the convective heat transfer is neither 'forced' nor 'free' in nature. The circumstances arise when a fluid is forced over a heated surface at a rather low velocity. Coupled with the forced-flow velocity is a convective velocity which is generated by the buoyant forces resulting from a reduction in fluid density near the heated surface.

A large value of Reynold number implies a large forced-flow velocity and hence less influence of free-convection currents. The larger the value of the Grashof-Prandtl number product, the more one would expect free convection effects to prevail.

Brown and Gauvin (1965) developed a correlation for the mixed convection, laminar flow region of flow through horizontal tubes as

$$Nu = 1.75 \left(\frac{\mu_b}{\mu_w}\right)^{0.14} [Gz + 0.012(Gz \, Gr^{1/3})^{4/3}]^{1/3} \tag{2.48}$$

where $Gz = Re \, Pr \, d/L$ is the Graetz number, μ_b the fluid viscosity at bulk temperature and μ_w the fluid viscosity at wall temperature.

The general notion which is applied in combined-convection analysis is that the predominance of a heat-transfer mode is governed by the fluid velocity associated with that mode. A forced-convection situation involving a fluid velocity of 30 m/s, for example, would be expected to overshadow most free convection effects encountered in ordinary gravitational fields because the velocities of the free-convection currents are small in comparison with 30 m/s. On the other hand, a forced-flow situation at very low velocities (~ 0.3 m/s) might be influenced appreciably by free-convection currents. A general criterion is that when $Gr/Re^2 > 10$, free convection is of primary importance.

2.6.8 Convective Heat Transfer Due to Wind

The heat transfer from a flat plate, exposed to outside winds, has been analyzed by several workers. McAdams (1954) reports the following equation for convection coefficient

$$h_c = 5.7 + 3.8 \, V \qquad \text{for} \qquad 0 \leq V \leq 5 \, \text{ms}^{-1} \tag{2.49a}$$

where V is the wind speed, m/s. The above equation for zero wind speed gives heat loss by natural convection. It may be mentioned here that the process taking place is not as simple as it appears, as the wind may not always be blowing parallel to the surface.

It is probable (Duffie and Beckman,1991) that in this equation the effect of free convection and radiation are included. For this reason Watmuff *et al.* (1977) reported that this equation should be

$$h_c = 2.8 + 3.0\,V \quad \text{for} \quad 0 \le V \le 7\,\text{ms}^{-1} \tag{2.49b}$$

This can be justified from Example 2.5 (*c* and *d*), where h_c are coming close to the above value for $V = 0.0$.

EXAMPLE 2.6(a)

Air at 27°C approaches a 0.91 m long and 0.61 m wide flat plate with an approach velocity $v_\infty = 4.57$ m/s. Determine the local heat transfer coefficient at a distance 0.457 m from the leading edge, for a plate surface temperature of 127°C. Also calculate the total rate of heat transfer from plate to the air.

Solution

The average film temperature, $T_f = (27 + 127)/2 = 77°C = 350$ K. At this value of T_f, the values of other constants are as follows:
$v = 2.079 \times 10^{-5}\text{m}^2/\text{s}, Pr = 0.697, K = 3.001 \times 10^{-2}$ W/mK (see Appendix III). The value of Reynold number is given by

$$Re_x = \frac{v_\infty x}{v} = \frac{(4.57)(0.457)}{2.079 \times 10^{-5}} = 1 \times 10^5 < 5 \times 10^5$$

Thus, the flow is laminar at this x-value, and

$$h_x = (0.332)\left(\frac{K}{x}\right) Re_x^{1/2} Pr^{1/3} \quad \text{(from Equation(2.47f))}$$

$$= (0.332)\left(\frac{3.001 \times 10^{-2}}{0.458}\right)(1 \times 10^5)^{1/2}(0.697)^{1/3} = 6.11\,\text{W/m}^2\text{K}$$

The Reynold number for overall heat transfer calculation is

$$Re_L = \frac{v_\infty L}{v} = \frac{(4.57)(0.91)}{2.079 \times 10^{-5}} = 200000$$

The flow is laminar, therefore

$$\bar{h} = \frac{K}{L}\bar{N}u = \frac{K}{L}(0.664)Re_L^{1/2} Pr^{1/3} \quad \text{(from Equation(2.47g))}$$

$$= \frac{3.001 \times 10^{-2}}{0.91}(0.664)(2 \times 10^5)^{1/2}(0.697)^{1/3} = 8.68\,\text{W/m}^2\text{K}$$

Then, from Equation (2.37), $\dot{Q} = \bar{h}A\Delta T = 8.68(0.91)(0.61)(100) = 482$ W.

EXAMPLE 2.6(b)

Water flows over a smooth flat plate with a free stream velocity of 0.20 m/s. The plate is 1.0 m long and the water properties at the fluid film temperature are: $v = 7.66 \times 10^{-7} \text{m}^2/\text{s}$, $K = 0.621 \text{W/mK}$, $Pr = 5.13$.

Determine the average rate of heat transfer to the water per square meter of plate surface if the plate temperature is 27.8° C above the free stream fluid temperature.

Solution

The maximum Reynold number (from Equation (2.38b)) is

$$Re_L = \frac{(0.20)(1.0)}{7.66 \times 10^{-7}} = 261096$$

The flow is laminar. Therefore, from Equation (2.47 g), we have

$$\bar{h} = \frac{K}{L}(0.664)Re_L^{1/2} \, Pr^{1/3}$$

$$= (0.621/1.0)(0.664)(261096)^{1/2}(5.13)^{1/3} = 363.45 \text{ W/m}^2\text{K}$$

The average heat flux to the water (using Equation (2.37)) is

$$\frac{\dot{Q}}{A} = \bar{h}\Delta T = (363.45)(27.8) = 10.104 \text{ k W/m}^2.$$

EXAMPLE 2.7

Saturated liquid water at bulk temperature, $T_b = 20°C$ flows at 0.61 m/s in a 25.4 mm ID tube; the tube wall is heated by solar irradiation to $T_s = 26°C$, constant. Find the average heat transfer coefficient in the fully developed velocity and temperature profile region assuming that $L/D > 60$.

Solution

At $T_b = 20°C$, $v_b = 1.006 \times 10^{-6} \text{m}^2/\text{s}$, $K_b = 0.597$ W/mK, $Pr_b = 7.02$ (see Appendix III).
The Reynold number, (from Equation (2.38b), for the present case $x = D$), is

$$Re = \frac{Dv_b}{v_b} = \frac{(0.0254)(0.61)}{1.006 \times 10^{-6}} = 1.54 \times 10^4, \text{ turbulent}$$

In this case, the Nusselt number (from Equation (2.47c)) is given by

$$Nu_D = (0.023)(1.54 \times 10^4)^{0.8} \times (7.02)^{0.4} = 112 \text{ (from Table 2.4)}$$

The heat transfer coefficient, from Equation (2.37), can be given as

$$h = (0.597/0.0254)(112) = 2.632\,\text{kW/m}^2\text{K}$$

EXAMPLE 2.8

Water at average bulk temperature 80°C flows inside a 0.020 m ID circular tube with average bulk velocity 2.0 m/s. The pipe wall temperature is at each point 40°C below the local value of T_b (= 80°C), and the cooling takes place at constant heat flux. Estimate the rate of energy loss from a 4 m length of tubing.

Solution

The fluid properties at $T_b = 80°C$ and $T_s = 40°C$ are
$\nu_b = 3.64 \times 10^{-7}\text{m}^2/\text{s}$, $Pr_b = 2.22$, $K_b = 0.6676$ W/mK
$\nu_s = 6.58 \times 10^{-7}\text{m}^2/\text{s}$, $\rho_b = 974$ kg/m^2, $\rho_s = 995$ kg/m^3 (see appendix III)
The Reynold number (from Equation (2.38b) and using $x = D$) is

$$Re = \frac{D\nu_b}{\nu_b} = \frac{(0.020)(2.0)}{3.64 \times 10^{-7}} = 1.10 \times 10^5, \text{ highly turbulent}$$

Since $T_s - T_b = 40°C$, and noting that, for $\frac{L}{D} = \frac{4}{0.02} = 200 > 60$, we have

$$\bar{h} = \frac{K_b}{D}(0.027)(Re)^{0.8}(Pr)^{1/3}\left(\frac{\mu_b}{\mu_s}\right)^{0.14} \quad \text{(from Table 2.4)}$$

$$= (0.6676/0.02)(0.027)(1.10 \times 10^5)^{0.8}(2.22)^{1/3}[(3.64/6.58)(974/995)]^{0.14}$$

$$= 11.65\,\text{kW/m}^2\text{K}$$

Now, $$\dot{Q}_{\text{conv}} = \bar{h}\pi DL(T_s - T_b) \quad \text{(from Equation(2.37))}$$

or, $$\dot{Q}_{\text{conv}} = (11.65)\pi(0.020)(4)(-40) = -117.06\,\text{kW}$$

2.7 RADIATION

Thermal radiation is the transfer of heat from a body at higher temperature to another at lower temperature by electromagnetic waves (0.1 to 100 μm). Temperature is transmitted in the space in the form of electromagnetic waves. Thermal radiation is in the infrared range and obeys all the rules as that of light, viz. travels in straight line through a homogenous medium, is converted into heat when it strikes any body which can absorb it and is reflected and refracted according to the same rule as that of light.

2.7.1 Radiation Involving Real Surfaces

When radiant energy falls on a body, a part of it is reflected, another is absorbed and the rest is transmitted through it. The conservation of energy states that the total sum must be equal to the incident radiation, thus,

$$I_T = I_r + I_a + I_t$$

$$1 = \rho' + \alpha' + \tau' \tag{2.50}$$

where, ρ', α' and τ' are reflectivity, absorptivity and transmissivity of the intercepting body, respectively. The ratio of the energy reflected to that which is incident is called *reflectivity*. The ratios of the energy absorbed and the energy transmitted to that which is incident are the absorptivity and transmissivity respectively.

For an opaque surface, $\tau' = 0$, therefore $\rho' + \alpha' = 1$. However, when $\rho' = \tau' = 0; \alpha = 1$, that is, the substance absorbs the whole of the energy incident on it. Such a substance is called the *black body*. Similarly, for a white body which reflects the whole of the radiation falling on it, $\alpha' = \tau' = 0$, $\rho' = 1$.

The energy which is absorbed is converted into heat and this heated body, by virtue of it's temperature, emits radiation. The radiant energy emitted per unit area of a surface in unit time is referred to as the *emissive power* (E_λ). However, if defined as the amount of energy emitted per second per unit area perpendicular to the radiating surface in a cone formed by a unit solid angle between the wavelengths lying in the range $d\lambda$, it is called *spectral emissive power* (e_λ). Further, emissivity, defined as the ratio of the emissive power of a surface to the emissive power of a black body of the same temperature, is the fundamental property of a surface.

2.7.2 Kirchoff's Law

It states that for a body in thermal equilibrium, the ratio of it's emissive power to that of a black body at the same temperature is equal to it's absorptivity, i.e.

$$\frac{e}{e_b} = \alpha' \quad \text{or,} \quad \varepsilon = \alpha' \tag{2.51}$$

Thus, at a given temperature, a body can absorb as much incident radiation as it can emit. However, it may not be valid if the incident radiation comes from a source at different temperature. Further, it applies to surfaces bearing the grey surface characteristics, viz. radiation intensity is taken to be a constant proportional to that of a black body. The radiative properties α_λ, ε_λ and ρ_λ are assumed to be uniform over the entire wavelength spectrum.

2.7.3 Laws of Thermal Radiation

Laws of thermal radiation have been obtained for black bodies and conditions of thermodynamic equilibrium.

i. Planck's law

The emission of energy with respect to wavelength is not uniform and depends on temperature. Planck's law establishes the relation of the spectral emissive power, wavelength and temperature as follows:

$$E_{b\lambda} = \frac{C_1}{(\lambda)^5} \frac{1}{\exp[C_2/(\lambda T)] - 1} \tag{2.52}$$

where $$C_1 = 3.742 \times 10^8 \, \text{W}\mu\text{m}^4/\text{m}^2 (= 3.7405 \times 10^{-16} \, \text{W}\text{m}^2)$$

and $$C_2 = 1.4387 \times 10^4 \, \mu\text{mK} \, (= 0.01439 \, \text{mK})$$

are called Planck's first and second radiation constants respectively. Planck's law has two limiting cases depending on the relative value of C_2 and λT:

(a) when $\lambda T \gg C_2$

$$E_{b\lambda} = \frac{C_1}{(\lambda)^5} \frac{\lambda T}{C_2} \text{ Rayleigh-Jean's law}$$

(b) when $\lambda T \ll C_2$

$$E_{b\lambda} = \frac{C_1}{(\lambda)^5} \exp\left(-\frac{C_2}{\lambda T}\right)$$

ii. Wein's displacement law
The wavelength corresponding to the maximum intensity of black body radiation for a given temperature T is given by this law:

$$\lambda_{\max} T = C_3 \tag{2.53}$$

where $C_3 = 2897.6 \, \mu\text{m K}$. Hence, an increase in temperature shifts the maximum black body radiation intensity towards the shorter wavelength.

iii. Stefan-Boltzmann law
This law relates the hemispherical total emissive power, viz. total energy and temperature. By integrating Planck's law over all wavelengths, the total energy emitted by a black body is found to be,

$$E_b = \int_0^\infty E_{b\lambda} d\lambda = \sigma T^4 \tag{2.54}$$

where $\sigma = 5.6697 \times 10^{-8} \text{W}/\text{m}^2\text{K}^4$ is the Stefan-Boltzmann constant.

iv. Sky radiation
In order to evaluate radiation exchange between a body and the sky, certain equivalent black-body sky temperature is defined. This accounts for the fact that the atmosphere is not at a uniform temperature and that it radiates only in certain wavelength regions. Thus, the net radiation to a surface with emittance ε and temperature T is

$$\dot{Q} = A\varepsilon\sigma(T_{\text{sky}}^4 - T^4) \tag{2.55}$$

In order to express the equivalent sky temperature T_{sky}, in terms of ambient air temperature, various expressions have been given by different people. These relations, although simple to use, are only approximations. Swinbank (1963) relates sky temperature to the local air temperature by the relation,

$$T_{\text{sky}} = 0.0552 \, T_a^{1.5} \tag{2.56}$$

where T_{sky} and T_a are both in Kelvin.

Another commonly used relation is that given by Whillier (1967):

$$T_{\text{sky}} = T_a - 6 \tag{2.57a}$$

or $$T_{\text{sky}} = T_a - 12 \tag{2.57b}$$

2.7.4 Radiative Heat Transfer Coefficient

As seen in Equation (2.54), the radiant heat exchange between two infinite parallel surfaces at temperatures T_1 and T_2 may be given as

$$\dot{q}_r = \varepsilon\sigma(T_1^4 - T_2^4) \tag{2.58a}$$

Equation (2.58a) may be rewritten as

$$\dot{q}_r = h_r(T_1 - T_2) \tag{2.58b}$$

where

$$h_r = \varepsilon\sigma(T_1^2 + T_2^2)(T_1 + T_2) = \varepsilon(4\sigma\bar{T})^3 \quad \text{for} \quad \bar{T}_1 \cong \bar{T}_2$$

and

$$\varepsilon = \frac{1}{\varepsilon_1} + \frac{1}{\varepsilon_2} - 1, \text{ for two parallel surfaces}$$

$$= \varepsilon, \text{ for surface exposed to atmosphere}$$

ε_1 and ε_2 are the emissivities of the two surfaces. By writing radiant heat exchange as Equation (2.58b), one retains the simplicity of a linear equation. However, when one of the surfaces is sky, Equation (2.58a) becomes

$$\dot{q}_r = \varepsilon\sigma(T_1^4 - T_{sky}^4) \tag{2.58c}$$

The equation may be modified as

$$\dot{q}_r = \varepsilon\sigma(T_1^4 - T_a^4) + \varepsilon\sigma(T_a^4 - T_{sky}^4)$$

or

$$\dot{q}_r = h_r(T_1 - T_a) + \varepsilon\Delta R \tag{2.58d}$$

where $\Delta R = \sigma[(T_a + 273)^4 - (T_{sky} + 273)^4]$ is the rate of long wavelength radiation exchange between ambient air and sky. The reduction of $\dot{q}_r$ in the form of Equation (2.58d) will enable one to find exact closed form solution for T_1. It may be noted here that this solution is based on the assumption that T_a and T_{sky} are constant.

Also, the radiative heat transfer coefficient is given by

$$h_r = \frac{\varepsilon\sigma[T_1^4 - T_a^4]}{(T_1 - T_a)} \tag{2.58e}$$

EXAMPLE 2.9(a)

Calculate the rate of long wavelength radiation exchange ΔR between the ambient air and sky temperature for $T_a = 15°C$.

Solution

Using

$$\Delta R = \sigma[(T_a + 273)^4 - (T_{sky} + 273)^4]$$

from Equations (2.56) and (2.57), we get

$$T_{sky} = 0.0552(15)^{1.5} = 3.2°C$$

$$= 15 - 6 = 9°C$$

$$= 15 - 12 = 3°C$$

Now,
$$\Delta R = 5.67 \times 10^{-8}[(15+273)^4 - (3.2+273)^4] = 60.11 \, \text{W/m}^2$$
$$= 5.67 \times 10^{-8}[(15+273)^4 - (9.0+273)^4] = 31.50 \, \text{W/m}^2$$
$$= 5.67 \times 10^{-8}[(15+273)^4 - (3.0+273)^4] = 61.06 \, \text{W/m}^2$$

Here, it is important to mention that the value of T_{sky} is nearly same for two cases hence the numerical value of ΔR should be considered as 60 W/m^2.

EXAMPLE 2.9(b)

Calculate the radiative heat transfer coefficient between the surface of a wall at 25°C and room air temperature at $T_a = 24$°C.

Solution

Since the temperatures are same, we have
$$h_r = 4\varepsilon\sigma T^3 = 4 \times 5.67 \times 10^{-8}(25+273)^3 = 6 \, \text{W/m}^2\text{°C}$$

2.7.5 Radiation Shape Factor (Pitts and Sissom, 1991)

Let us consider two black surfaces A_1 and A_2 (Fig. 2.9). We want to obtain a general expression for the energy exchange between these surfaces when they are maintained at different temperatures. The problem boils down to the determination of the amount of energy which leaves one surface and reaches the other. To solve this problem, a quantity known as *radiation shape factor* is defined. Other names for the radiation shape factor are view factor, angle factor and configuration factor. If, F_{12} is the fraction of energy leaving surface 1 which reaches surface 2 and F_{21}, the fraction of energy leaving surface 2 which reaches surface 1, then the energy leaving surface 1 and arriving at surface 2 is $E_{b1}A_1F_{12}$ and the energy leaving surface 2 and arriving at surface 1 is $E_{b2}A_2F_{21}$. Since the surfaces are black, all the incident radiation will be absorbed, and the net energy exchange is

$$E_{b1}A_1F_{12} - E_{b2}A_2F_{21} = \dot{Q}_{12}$$

If both surfaces are at the same temperature, there can be no heat exchange, that is $\dot{Q}_{12} = 0$, or

$$E_{b1} = E_{b2} \quad \text{and} \quad A_1F_{12} = A_{21}F_{21} \tag{2.59}$$

The net exchange is

$$\dot{Q}_{12} = A_1F_{12}(E_{b1} - E_{b2}) = A_2F_{21}(E_{b1} - E_{b2}) \tag{2.60a}$$

In general, for any two surfaces m and n,

$$A_mF_{mn} = A_nF_{nm} \tag{2.60b}$$

Equation (2.59) is known as a *reciprocity relation*; and though derived for black surfaces, holds for other surfaces also as long as diffuse radiation is involved. We now wish to determine a general relation for F_{12}. For this, let us consider the elements of area dA_1 and dA_2 in Figure (2.9). The angles ϕ_1 and ϕ_2 are measured between a normal to the surface and the line drawn between the area elements r. The projection of dA_1 on the line between centres is $dA_1 \cos\phi_1$. We assume that the surfaces are diffuse, i.e. the intensity of the radiation is the same in all directions. The intensity is the radiation emitted per unit area and per unit of solid angle in a certain specified direction. In order to obtain the energy emitted by the element of area dA_1 in a certain direction, we must multiply the intensity by the projection of dA_1 in the specified direction. Thus the energy leaving dA_1 in the direction given by the angle ϕ_1 is

$$I_b \, dA_1 \cos \phi_1$$

where I_b is the black body intensity.

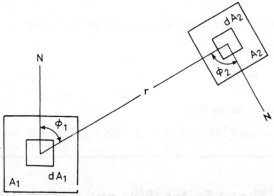

Figure 2.9 Sketch showing area elements used in the derivation of shape factor.

The radiation arriving at some area element dA_n at a distance r from A_1 would be

$$d\dot{Q}_{12} = I_b \, dA_1 \cos\phi_1 \frac{dA_n}{r^2} \qquad (2.60c)$$

where dA_n is constructed normal to the radius vector. The quantity dA_n/r^2 represents the solid angle subtended by the area dA_n. From Figure 2.10,

$$dA_n = r^2 \sin \phi \, d\psi \, d\phi$$

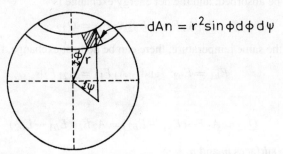

Figure 2.10 Spherical coordinate system used in derivation of radiation shape factor.

The intensity may be obtained in terms of the emissive power by integrating expression over a hemisphere enclosing the element of area dA_1,

$$E_b \, dA_1 = I_b \, dA_1 \int_0^{2\pi} \int_0^{\pi/2} \sin\phi \, \cos\phi \, d\phi \, d\psi = \pi I_b \, dA_1$$

or
$$I_b = E_b / \pi \tag{2.60d}$$

The area element dA_n is given by,

$$dA_n = \cos\phi_2 \, dA_2 \tag{2.60e}$$

(substitute Equations (2.60d) and (2.60e) in Equation (2.60c))
The energy leaving dA_1 which arrives at dA_2 is,

$$d\dot{Q}_{12} = E_{b1} \cos\phi_1 \cos\phi_2 \frac{dA_1 dA_2}{\pi r^2}$$

Similarly, the energy leaving dA_2 and arriving at dA_1 is,

$$d\dot{Q}_{21} = E_{b2} \cos\phi_2 \cos\phi_1 \frac{dA_2 dA_1}{\pi r^2}$$

Now, the net energy exchange is,

$$\dot{Q}_{net1-2} = (E_{b1} - E_{b2}) \int_{A2} \int_{A1} \cos\phi_1 \cos\phi_2 \frac{dA_1 dA_2}{\pi r^2} \tag{2.61}$$

The integral is either $A_1 F_{12}$ or $A_2 F_{21}$, according to Equation (2.59). To evaluate the integral, the specific geometry of the surfaces A_1 and A_2 must be known.

Radiant energy exchange between surfaces involves not only the emission, absorption and reflection characteristics of the surfaces but also their geometrical arrangement. The energy emitted by a body propagates in space with various intensities. The law giving the dependence of radiation intensity on direction is known as the *Lambert's cosine law*. According to this law, the amount of energy radiated from an elementary area dF_1 in the direction of area dF_2 is proportional to the product of the energy emitted along the normal $d\dot{Q}_n$, the solid angle $d\omega$, and $\cos\phi$, where ϕ is the angle enclosed between the direction of radiation and the normal.

$$d\dot{Q}_\phi = d\dot{Q}_n d\omega \cos\phi = E_n \, dF_1 d\omega \cos\phi \tag{2.62}$$

where E_n is the radiant energy emitted in the direction of the normal.

To determine the quantity E_n, Equation (2.62) must be integrated over the surface of the hemisphere lying on top of elementary area dF_1 (Fig. 2.11). The solid angle $d\omega$ is that at which the elementary area dF_2 on the surface of a sphere with radius r is seen from any point of elementary area dF_1.

$$dw = \frac{dF_2}{r} \quad \text{or} \quad dw = \sin\phi \, d\psi d\phi$$

where ψ is the angle of latitude and ϕ the angle complimentary to the angle of latitude.

$$d\dot{Q}_\phi = E_n dE_1 \sin\phi \, d\psi d\phi \cos\phi$$

Integrating the above expression over the surface of a hemisphere, i.e. between the limits of variation of angle ϕ from 0 to $\pi/2$ and of angle ψ from 0 to 2π, the energy emitted by the area dF_1, EdF_1, is

$$E \, dF_1 = E_n dF_1 \int_0^{2\pi} d\psi \int_0^{\pi/2} \sin\phi \cos\phi \, d\phi = \pi E_n dF_1 \tag{2.63}$$

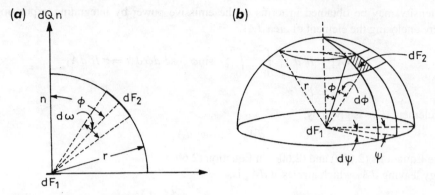

(a) dQ n

(b)

Figure 2.11 Radiant heat exchange between two small black surface elements.

The Lambert cosine law is true for black or grey bodies and for bodies where there is diffused radiation in the range $\phi = 0 - 60°$ only. The Lambert cosine law is not applicable to polished surfaces.

i. Radiation exchange between surfaces separated by a nonabsorbing medium

Suppose, we want to determine the radiant heat exchange between the horizontal and vertical surfaces as shown in Figure 2.12. We consider the radiant exchange between any small element of area ΔA_1 in surface A_1 with any similar size element ΔA_2 in surface A_2. We assume that both surfaces are black.

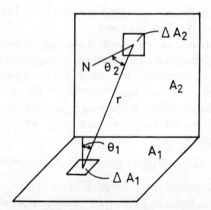

Figure 2.12 Different geometries for calculation of shape factor.

The rate at which radiant energy leaves ΔA_1 in the direction of ΔA_2 would be

$$\Delta \dot{Q}_{\theta 1} = \frac{\dot{Q}}{\Delta A_1 d\omega \cos \theta_1} \Delta A_1 \cos \theta_1$$

where θ_1 is the angle between the normal to ΔA_1 and the line r connecting it with ΔA_2. The amount of this energy which is intercepted by ΔA_2 depends on the solid angle which ΔA_2 subtends with center at ΔA_1

$$\Delta \omega_2 = \frac{\Delta A_2 \cos \theta_2}{r^2}.$$

Thus the amount of energy $\Delta \dot{Q}_{12}$ which leaves ΔA_1 and impinges on ΔA_2 is given as:

$$\Delta \dot{Q}_{12} = \frac{\dot{Q} \cos \theta_1 \cos \theta_2 \Delta A_1 \Delta A_2}{\Delta A_1 d\omega_1 \cos \theta_1 r^2}$$

Since A_2 is black, all of this energy is absorbed. In turn, however, ΔA_2 is radiating towards ΔA_1. The amount intercepted by ΔA_1 is given as

$$\Delta \dot{Q}_{21} = \frac{\dot{Q} \cos \theta_1 \cos \theta_2 \Delta A_1 \Delta A_2}{\Delta A_2 d\omega_2 \cos \theta_2 r^2}$$

This will be absorbed by ΔA_1, so that the net exchange $\Delta \dot{Q}$ equals:

$$\Delta \dot{Q} = \left(\frac{\dot{Q}}{\Delta A_1 d\omega_1 \cos \theta_1} - \frac{\dot{Q}}{\Delta A_2 d\omega_2 \cos \theta_2} \right) \frac{\cos \theta_1 \cos \theta_2 \Delta A_1 \Delta A_2}{r^2}$$

or,

$$\frac{\dot{Q}}{\Delta A_1 d\omega_1 \cos \theta_1} = \frac{E_1}{\pi} = \sigma T_1^4$$

Thus,

$$\Delta \dot{Q} = \frac{\sigma}{\pi} (T_1^4 - T_2^4) \frac{\cos \theta_1 \cos \theta_2 \Delta A_1 \Delta A_2}{r^2}$$

In order to determine the total energy radiated from surface 1 to 2, it is necessary to sum the radiation from each ΔA_1 to all of A_2. This is accomplished by integrating over A_1 and A_2. The net exchange between surfaces 1 and 2 is given as

$$\dot{Q} = \sigma (T_1^4 - T_2^4) \frac{1}{\pi} \int_{A_1} \int_{A_2} \frac{\cos \theta_1 \cos \theta_2}{r^2} dA_1 dA_2 \qquad (2.64a)$$

$$\dot{Q} = \sigma (T_1^4 - T_2^4) F_{12} \qquad (2.64b)$$

ii. Geometrical factor

The total energy emitted from surface 1 is $\sigma T_1^4 A_1$. Since only part of this will be intercepted by surface 2, it is obvious that

$$\frac{1}{\pi} \int_{A_1} \int_{A_2} \frac{\cos \theta_1 \cos \theta_2}{r^2} dA_1 dA_2$$

is less than A_1. Further, we note that this quantity involves only the geometry of the system; consequently the fraction of the total energy emitted by A_1 intercepted by A_2 is the same for all temperatures. This suggests that the radiation from 1 to 2 can be conveniently calculated from an expression of the type

$$\dot{Q}_{12} = \sigma T_1^4 A_1 F_{12}$$

where

$$F_{12} = \frac{1}{A_1} \frac{1}{\pi} \int_{A_1} \int_{A_2} \frac{\cos \theta_1 \cos \theta_2}{r^2} dA_1 dA_2 \qquad (2.65)$$

F_{12} is defined as the *geometrical factor* of surface 1 with respect to surface 2, and represents that fraction of the total energy emitted per unit area of A_1 which is intercepted by A_2 and F_{12} is also known as the *view factor* or *shape factor*.

$$F_{12} + F_{13} + F_{14} + \dots = 1$$

where $F_{13}, F_{14} \dots$ are the geometrical factors for other surfaces which are seen by surface 1.

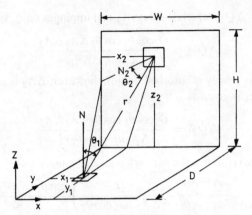

Figure 2.13 Coordinates for evaluating the geometrical factor for adjacent rectangles in perpendicular planes.

iii. Determination of geometrical factors

The actual evaluation of a geometrical factor may be simple or difficult depending on the individual circumstances. Let us consider two mutually perpendicular surfaces (Fig. 2.13).

$$F_{12} = \frac{1}{WD\pi} \left[\frac{1}{\pi}(D^2 - W^2 + H^2)\ln(D^2 + H^2 + W^2) - (D^2 + H^2)\ln(D^2 + H^2) \right.$$

$$-(D^2 - W^2)\ln(D^2 + W^2) - (H^2 - W^2)\ln(H^2 + W^2) + D^2\ln\,D^2 + H^2\ln\,H^2 - \frac{W^2}{n}$$

$$\left. -HW\,\tan^{-1}\frac{W}{H} - DW\,\tan^{-1}\frac{W}{D} - W\sqrt{D^2 + H^2}\,\tan^{-1}\frac{W}{\sqrt{(D^2 + H^2)}} \right]$$

Letting $y = D/W$, $z = H/W$ and dividing and multiplying by W^2

$$F_{12} = \frac{1}{\pi} \left[\frac{1}{4}\ln\left(\frac{(1+y^2+z^2)^{y-(1/y)+(z^2/y)}(y^2)^y(z^2)^{z^2/y}}{(1+y^2)^{y-(1/y)}(1+z^2)^{(z^2/y)-(1/y)}(y^2+z^2)^{y+(z^2/y)}}\right) \right.$$

$$\left. + \tan^{-1}\frac{1}{y} + \frac{z}{y}\tan^{-1}\frac{1}{z} - \sqrt{1+\frac{z^2}{y^2}}\,\tan^{-1}\frac{1}{\sqrt{y^2+z^2}} \right] \qquad (2.66)$$

EXAMPLE 2.10

Calculate the radiative heat transfer for parameter given in the following configuration:

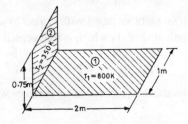

Solution

By the reciprocity theorem

$$\dot{Q}_{12} = F_{12}A_1\sigma T_1^4 - F_{21}A_2\sigma T_2^4$$
$$= F_{12}A_1\sigma(T_1^4 - T_2^4) \quad \text{since } F_{12}A_1 \simeq F_{21}A_2$$

From Figure of Problem 2.7

$$\beta = \frac{2}{1} = 2.0 \quad \text{and} \quad \gamma = \frac{a}{c} = \frac{0.75}{1} = 0.75$$

For $\beta = 2.0$ and $\gamma = 0.75$, $F_{12} \cong 0.1$. Thus,

$$\dot{Q}_{12} = 0.1(2 \times 1)(5.67 \times 10^{-8})(800^4 - 350^4)$$
$$= 4474.69\,\text{W}$$

EXAMPLE 2.11

Determine the radiative heat transfer from surface 2 to surface 1 for parameters given in the following configuration:

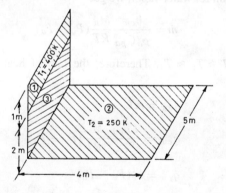

Solution

Here $F_{21} = F_{2\to(1,3)} - F_{23}$

From Figure of Problem 2.7, (i) for $F_{2\to(1,3)}$, $\beta = \frac{4}{5} = 0.8$, $\gamma = \frac{3}{5} = 0.6 \Rightarrow F_{2-(1,3)} = 0.20$ and (ii) for F_{2-3}, $\beta = \frac{4}{5} = 0.8$, $\gamma = \frac{2}{5} = 0.4 \Rightarrow F_{2-3} = 0.16$. Thus, $F_{21} = 0.21 - 0.16 = 0.04$.

Then by using reciprocity theorem

$$\dot{Q}_{21} = 0.04(4 \times 5)(5.67 \times 10^{-8})(250^4 - 400^4) = -984.03\,\text{W}.$$

2.8 HEAT AND MASS TRANSFER

In a moving single component medium, heat is transferred by conduction and convection; the process is known as *convective heat transfer*. By analogy, the process of molecular and molar transport of matter; in a moving heterogenous medium, is called *convective mass transfer*. The surface of the liquid phase plays a part similar to that of a solid wall in heat transfer process without accompanying diffusion. The process of heat and mass transfer are of practical interest in evaporation, condensation etc. The heat transfer is based on the Newton's law, viz.

$$\dot{q} = h(T_w - T_a) \tag{2.67}$$

where T_w and T_a are the fluid (water) and the surrounding air temperatures respectively. Mass transfer rate is based on a similar equation:

$$\dot{m} = h_D(\rho_w^0 - \rho_a^0) \tag{2.68}$$

where $\dot{m}$ is the rate of mass flow per unit area, (kg/m^{2}s), h_D the mass transfer coefficient [(kg/s) (sqm/kg/m^3)], ρ^0 the partial mass density of water vapor, kg/m^3. The subscripts w and a indicate the mass density directly on the phases interface and at a distance from it.

According to Lewis relation for air and water vapor mixture

$$\frac{h}{h_D \rho^0 C_p} = 1 \tag{2.69}$$

That is, the ratio of the heat transfer coefficient to mass transfer coefficient is equal to the specific heat per unit volume at constant pressure of the mixture. Substituting for h_D from Equation (2.69) into (2.68) and using perfect gas equation for water vapor, we get

$$\dot{m} = \frac{h_{cw}}{\rho_a^0 C_{pa}} \frac{M_w}{RT}(P_w - P_a) \tag{2.70}$$

where we have assumed $T \approx T_w \approx T_a$. Therefore, the rate of heat transfer on account of mass transfer is

$$\dot{q}_{ew} = \dot{m}L$$

$$\dot{q}_{ew} = \frac{Lh_{cw}}{\rho_a^0 C_{pa}} \frac{M_w}{RT}(P_w - P_a)$$

$$\dot{q}_{ew} = h_e(P_w - P_a) \tag{2.71}$$

where

$$\frac{h_e}{h_{cw}} = \frac{L}{\rho_a^0} C_{pa} \frac{M_w}{RT} \tag{2.72}$$

Using the perfect gas equation for air, i.e. $P_a \frac{M_a}{\rho_a^0} = RT$, Equation (2.72) can be written as

$$\frac{h_e}{h_{cw}} = \frac{L}{C_{pa}} \frac{M_w}{M_a} \frac{1}{P_T} \tag{2.73}$$

as, for small P_w, $P_T \approx P_a$. Substitution of the appropriate values for the parameters yields,

$$h_e = 0.013 \, h_{cw} \tag{2.74}$$

The value of h_e/h_{cw} given in the above equation is smaller than that obtained by Bowen (1926) and Dunkle (1961). It is seen that the best representation of the mass-heat transfer phenomenon is obtained if the value of the h_e/h_{ew} is taken to be 16.273×10^{-3}. Hence,

$$\dot{q}_{ew} = 16.273 \times 10^{-3} h_{cw}(P_w - P_a) \tag{2.75}$$

The expressions for variation of the saturated vapor pressure with temperature have been given in Appendix Va and Appendix Vb gives the steam table for saturated vapor pressure. Equation (2.75) can be applied for a horizontally wetted surface exposed to atmosphere to give the expression,

$$\dot{q}_{ew} = 16.273 \times 10^{-3} h_{cw}(P_w - \gamma P_a) \tag{2.76}$$

where γ is the relative humidity.

After linearisation , the above equation can be written in terms of evaporative heat transfer coefficient as follows:

$$\dot{q}_{ew} = h_{ew}(T_w - T_a),$$

where
$$h_{ew} = \frac{16.276 \times 10^{-3} h_{cw}|\bar{P}_w - \gamma \bar{P}_a|}{(\bar{T}_w - \bar{T}_a)} \tag{2.77}$$

In the case of small value of $(\bar{T}_w - \bar{T}_a)$, h_{ew} becomes very large. In this situation the partial vapor should be expressed as

$$P(T) = R_1 T + R_2$$

where R_1 and R_2 can be obtained from linear regression analysis by using stream table data of appendix (Vb).

EXAMPLE 2.12

Calculate the rate of evaporation for exposed wetted surface having a temperature of 35°C and a relative humidity of 50 percent. Ambient air temperature is 15°C. Also calculate evaporative heat transfer coefficient.

Solution

The pressures may be calculated as

$$P(T_w) = \exp\left(25.317 - \frac{5144}{273 + 35}\right) = 5517.6 \text{ N/m}^2$$

and
$$P(T_a) = \exp\left(25.317 - \frac{5144}{273 + 15}\right) = 1730 \text{ N/m}^2$$

Using $h_c = 2.7$ and substituting the values in Equation (2.76) we have the rate of evaporation as:

$$\dot{q}_{ew} = 16.273 \times 10^{-3} \times 2.7(5517.6 - 0.5 \times 1730) = 204.42 \text{ W/m}^2$$

The evaporative heat transfer coefficient can be calculated as

$$h_{ew} = \frac{\dot{q}_{ew}}{(T_w - T_a)} = \frac{204.42}{(35 - 15)} = 10.22 \, W/m^2 \, {}^\circ C$$

PROBLEMS

2.1 Derive an expression for interface temperature of two material composite wall in a steady state condition.

Hint

$$\dot{Q} = -K_a A \frac{T_2 - T_1}{L_a} = -K_b A \frac{T_3 - T_2}{L_b}.$$

2.2 Obtain an analytical expression for temperature distribution $T(x)$ in the plane wall having uniform temperature T_1 and T_2 at x_1 and x_2 respectively. The thermal conductivity of the plane wall varies linearly with temperature, $K = K_0(1 + bT)$.

Hint Use $\dot{Q} = -KA(dT/dx)$

$$\frac{\dot{Q}}{A} \int_{x_1}^{x_2} dx = -K_0 \int_{T_1}^{T_2} (1 + bT) dT.$$

After algebraic manipulation divide the equation by $T_m = (T_1 + T_2)/2$.

2.3 Derive an expression for the rate of heat transfer in a spherical shell of radii r_1 and r_2 if the surfaces are held at temperatures T_1 and T_2 respectively. The thermal conductivity of shell material varies as $K = K_0 T^2$.

Hint Integrate equation $\dot{Q} = -(K_0 T^2)(4\pi r^2)(dT/dr)$ between r_1 and r_2 and T_1 and T_2.

2.4 A 0.30 m brick wall with thermal conductivity of 0.69 W/m°C used in an office building in Leh has the following measured parameters: inside air temperature, 20°C, outside air temperature, -10°C, inside surface temperature (T_1), 16°C and outside surface temperature $(T_2) = -5$°C. Estimate the average value of the inner and outer heat transfer coefficient h_i and h_o alongwith the U-value. Also calculate the rate of heat transfer using U-value.

Hint Calculate $(\dot{Q}/A) = -K(\Delta T/\Delta x)$ and then $\dot{Q}/A = h_i(20 - T_i)$ also, $\dot{Q}/A = h_0(T_2 - (-10))$ and use Equation (2.33a).

2.5 A mild steel pipe with inside and outside radius of 5 cm and 5.5 cm respectively is covered with 10 cm thick glass wool. Steam at 100 °C flows through it. The inside and outside heat transfer coefficient are 30 W/m² °C and 5.0 W/m² °C respectively. Determine the overall heat transfer coefficient (U) and the rate of heat transfer per unit length, if the surrounding air temperature is 20 °C. The length of the pipe is 50 cm. The thermal conductivity of the pipe and the insulation are 36 W/m °C and 0.04 W/m °C respectively.

Hint Use $\dot{Q}/L = U_0(A_0/L)(\Delta T)_{overall}$ and Equation (2.34b).

2.6 A black rectangular box of 10 m × 8 m × 10 m has been kept 560 K at bottom and 450 K at top and 480 K at other surfaces. Find out the net heat transfer rate.

Hint Use figure given below and Equation (2.64b)

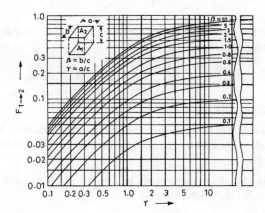

(From Schaum's Solved Problems Series, Heat Transfer by Donald R. Pitts and Leighton E. Sissom).

2.7 Assuming two black surfaces 1 m × 0.75 m and 2 m × 1 m are at 460 K and 820 K respectively. The two surfaces being at right angles to each other, determine the radiative heat transfer from one surface to the other surface.

Hint Use Figure given below and Equation (2.64b)

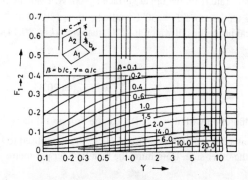

(From Schaum's Solved Problems Series, Heat Transfer by Donald R. Pitts and Leighton E. Sissom).

2.8 Calculate the shape factor for problem 2.7 by using Equation (2.66) and compare the results with the figure given in problem 2.7.

2.9 Calculate the convective heat transfer coefficient for an insulated hot plate at 60°C facing upward with water column at 40°C. The dimension of hot plate are 1 m × 1 m.

Hint Use Equation (2.40) and Table 2.1.

2.10 Calculate the evaporative heat transfer coefficient for wetted surface at 60°C and the surrounding temperature 35°C and relative humidity of about 60 percent.

Hint From Equation (2.76) $\dot{q}_{ew} = 0.016h_c[P(T_w) - \gamma P(T_a)]$.

After linearisation $\dot{q}_{ew} = h_{ew}(T_w - T_a)$

Here $h_{ew} = 0.016h_c[P(T_w) - \gamma P(T_a)]/(T_w - T_a)$ with $h_c = 2.7 + 3.8V$, take $V = 0.0$

and
$$P(T) = \exp\left(25.317 - \frac{5144.0}{T + 273.15}\right).$$

2.11 Calculate the radiative heat transfer coefficient for problem 2.10 and compare it with the evaporative heat transfer coefficient with justification.

$$h_{rw} = \frac{\varepsilon\sigma\left[(T_w + 273)^4 - (T_{sky} + 273)^4\right]}{T_w - T_a}.$$

Hint After linearisation of Equation (2.58c).

2.12 Compare the radiative, convective and evaporative heat transfer coefficients for wetted surface (Problems 2.10 and 2.11) and find out the total heat transfer coefficient for the same surface.

Hint See problems 2.10 and 2.11.

2.13 Estimate the evaporative heat transfer coefficient for different relative humidity ($\gamma = 20\%, 40\%, 60\%$, and 80%), in problem 2.10 and explain the variation, if any.

Hint See problem 2.10.

2.14 A horizontal plate, 1 m × 1 m, with it's upper surface heated to 50°C is exposed to an ambient air at 30°C at atmospheric pressure. Estimate the heat transfer rate.

Hint $(hL/k) = 0.14(Gr.Pr)$, evaluate Gr and Pr for thermal properties of air at 30°C.
$L = (L_1 + L_2)/2$ (Table 2.1 and Equation (2.40)).

2.15 A horizontal plate, with an effective area of 0.3 m × 0.3 m, with it's upper surface heated, is exposed to water at 20°C. The heat transfer rate from the upper surface is 900 W. Evaluate the upper surface temperature.

Hint Assume surface temperature $T_s = 30°C$ and take $T_f = (30 + 20)/2 = 25°C$. Calculate Gr and Pr by using physical properties of water at 25°C, then $h = (k/L)(0.14)(Gr.Pr)^{1/3}$
Then,

$$T_s^1 - T_\infty = \frac{\dot{Q}}{hA} \text{ with } T_\infty = 20°C$$

Repeat the computation with new T_s^1. Repeat the process till the value of T_s becomes constant.

2.16 A 3 m vertical plate is at a constant temperature of 60°C and is exposed to still air at 10°C on one side and still water at 20°C on the other side. Determine the heat transfer per metre width.

Hint Use,

$$Nu = \frac{hL}{k} = 0.68 + \frac{(0.67)(Gr.Pr)^{1/3}}{\left[1 + (0.492/Pr)^{0.5625}\right]^{0.44}} \text{ for } 10^4 < Ra < 10^9$$

and calculate Gr and Pr separately for both the sides for evaluating convective heat transfer coefficient.

2.17 Air, at 30°C, approaches a 1 m long and 0.5 m wide flat plate with a velocity of 5 m/s. Determine the rate of heat transfer from the plate at 120°C to the air.

Hint Use Equation (2.47f) as $(hL/k) = 0.664(Re)^{1/2}(Pr)^{1/3}$.

2.18 Water, at 30°C, is flowing at 2 m/min over a flat plate which is maintained at 50°C. What is the heat transfer coefficient for 2 m length.

Hint From Equation (2.47g) $Nu = 0.664(Re)^{1/2}(Pr)^{1/3}$.

2.19 Air, at 20°C and atmospheric pressure, is passing through a 2 m long cylinder with a diameter of 0.5 m, at a speed of 3 m/s. The cylinder is maintained at a temperature of 70°C. Calculate the heat transfer coefficient.

Hint If the maximum boundary layer thickness ($\delta_{max} = 5.0L/(Re)^{1/2}$) is small compared to the diameter (D) of the cylinder then assumption for flat plate will be valid. First check δ_{max}/D should be small, use $Nu = 0.664(Re)^{1/2}(Pr)^{1/3}$.

2.20 Prove that $h_r = 4\varepsilon\sigma T^3$.

Hint Use $T_1 = T_2$ in Equation 2.58b.

2.21 Estimate long wavelength radiation exchange between ambient air temperature, and sky temperature for $Ta = 5°C, 10°C, 20°C$ and $25°C$. Find out the variation.

Hint See Example 2.9 (a).

2.22 Calculate radiative heat transfer coefficient from the surface having $30°C$ surface temperature and exposed to various ambient air temperature given in problem 2.21.

Hint See Example 2.9(b).

2.23 Determine the rate of heat transfer per square metre between the two surfaces separated by 0.125 m of air. The upper and lower surfaces are maintained at the temperature of 300 K and 350 K respectively.

Hint From Table 2.1,

$$\overline{Nu} = \frac{\overline{h}L}{K} = \begin{cases} 0.195Gr^{1/4} & 10^4 < Gr < 4 \times 10^5 \\ 0.068Gr^{1/3} & 4 \times 10^5 < Gr \end{cases}$$

$$L = 0.125 \text{ m}, \ \beta' = \frac{1}{T} \quad \text{and appendix III } d.$$

2.24 (a) Determine the rate of heat transfer from an inclined square plate, having an inclination of $20°$ from the vertical, and temperature of $60°C$, to an environment air at $16°C$. Area of the square plate is 0.30 cm × 30 cm. (b) Repeat part (a) with water instead of air, $\beta = 0.153 \times 10^{-3} \text{K}^{-1}$ at $16°C$.

Hint (i) For inclined plate multiply Grashof number by $\cos\theta$ and use vertical plate constant $c = 0.59$ and $n = \frac{1}{4}$ for air and $c = 0.13$ and $n = \frac{1}{3}$ for water.

2.25 Determine configuration factor for the following figure:

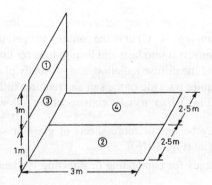

Hint $F_{(1,3)\to(2,4)} = \frac{1}{2}\left[F_{1\to(2,4)} + F_{3\to(2,4)}\right] = \frac{1}{2}\left[2F_{1-2} + F_{3\to(2,4)}\right]$ and $F_{1-2} = F_{(1,3)\to(2,4)} - \frac{1}{2}F_{3\to(2,4)}$

2.26 Prove the Stefan-Boltzman law by using Plank's law of radiation.

Hint Integrate Equation (2.54) after substituting expression for $E_{b\lambda}$ from Equation (1.2).

CHAPTER 3

Flat-Plate Collectors

3.1 INTRODUCTION

Solar energy can be used by three technological processes namely (i) heliochemical, (ii) helioelectrical and (iii) heliothermal. Heliochemical process, through photosynthesis, maintains life on earth by producing food and converting CO_2 to O_2. Helioelectrical process, using photovoltaic converters, provides power for spacecraft and is used in many terrestrial applications. Heliothermal process can be used to provide much of the thermal energy required for solar water heating and building heating. The most important part of the system, the collector, will be discussed in this chapter.

3.2 FLAT-PLATE COLLECTOR

The flat-plate collector is the heart of any solar energy collection system designed for operation in the low temperature range (ambient-$60°C$) or in the medium temperature range (ambient-$100°C$). It is used to absorb solar energy, convert it into heat and then to transfer that heat to a stream of liquid or gas. It absorbs both the beam and the diffuse radiation, and is usually planted on the top of a building or other structures. It does not require tracking of the sun and requires little maintenance.

A flat-plate collector (Fig. 3.1(*a*)) usually consists of the following components:

i. glazing, which may be one or more sheets of glass or some other diathermanous (radiation transmitting) material (Fig. 3.1(*b*))

ii. tubes, fins or passages for conducting or directing the heat transfer fluid from the inlet to the outlet.

iii. absorber plate which may be flat, corrugated or grooved with tubes, fins or passages attached to it.

iv. header or manifolds, to admit and discharge the fluid.

v. insulation, which minimizes heat loss from the back and sides of the collector.

vi. container or casing, which surrounds the various components and protects them from dust, moisture etc.

The complete view of the collector is shown in Figure 3.1(*c*).

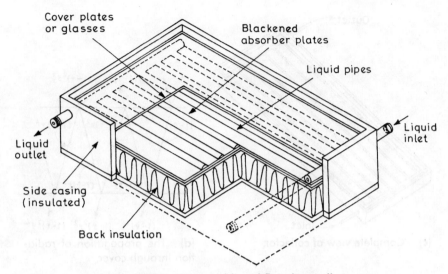

Figure 3.1(a) A typical liquid flat plate collector.

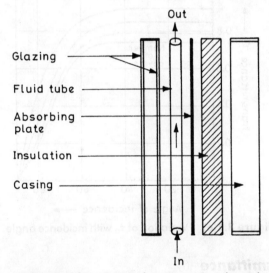

Figure 3.1(b) Exposed cross-section through double-glazed flat-plate collector.

3.2.1 Glazing Materials

The role of glazing is to admit the maximum possible radiation and to minimize the upward loss of heat. The most commonly used glazing material is glass as it can transmit upto 90 percent of the incident short wave radiations while its transmittance to the long wave heat radiation (5.0 to 50 μm), emitted by the absorber plate, is negligible.

Plastic films and sheets may also be used for the purpose of glazing as they possess high transmittance to short wave solar radiation, but transmission bands in the middle of thermal radiation spectrum and dimensional changes in this temperature range restricts their use as a good glazing surface.

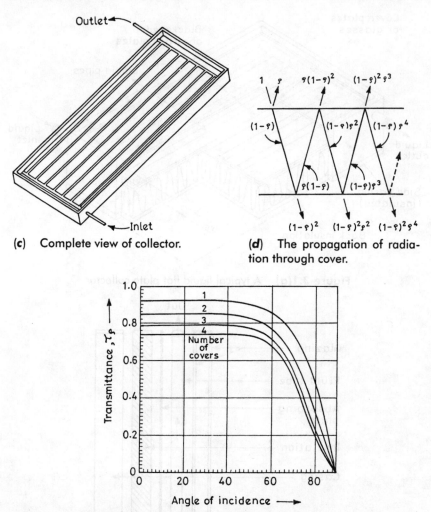

(c) Complete view of collector.

(d) The propagation of radiation through cover.

Figure 3.1(e) Variation of τ_ρ with incidence angle.

3.2.1.1 The transmittance

The propagation of radiation through one non-absorbing cover is shown in Figure 3.1(d). Considering only the perpendicular component of polarization of incoming radiation, $(1 - \rho_\perp)$ of incident radiation reaches the second interface. The $\rho_\perp$ part of $(1 - \rho_\perp) = \rho_\perp(1 - \rho_\perp)$ is reflected back to first interface from the second interface and remaining $= (1 - \rho_\perp) - \rho_\perp(1 - \rho_\perp) = (1 - \rho_\perp)^2$ passes through the interface and so on. Summing the transmitted terms, we get

$$\tau_\perp = (1 - \rho_\perp)^2 \sum_{n=0}^{n=\infty} \rho_\perp^{2n} = \frac{(1 - \rho_\perp)^2}{(1 - \rho_\perp^2)} = \frac{1 - \rho_\perp}{1 + \rho_\perp}. \tag{3.1a}$$

The same expression can be obtained for the parallel component of polarization of incoming radiation. The $\rho_\perp$ and $\rho_\parallel$ are not equal except at normal incidence and the transmittance of initially unpolarized radiation is the average of the two components and it is given by

$$\tau_\rho = \frac{1}{2}\left[\frac{1-\rho_\|}{1+\rho_\|} + \frac{1-\rho_\perp}{1+\rho_\perp}\right]$$

where the subscript ρ indicates that only reflection losses has been considered for transmission of radiation.

Similarly, an expression for $\tau_{\rho N}$, for a system of N covers, can be derived and can be written as

$$\tau_{\rho N} = \frac{1}{2}\left[\frac{1-r_\|}{1+(2N-1)r_\|} + \frac{1-r_\perp}{1+(2N-1)r_\perp}\right] \qquad (3.1b)$$

where,

$$r = \frac{I_\rho}{I_i} = \frac{1}{2}(\rho_\perp + \rho_\|) = \left(\frac{n-1}{n+1}\right)^2$$

Here one medium is air, i.e. a refractive index of nearly unity.

The variation of τ_ρ with angle of incidence for different glass cover has been shown in Figure 3.1(e). It is clear that transmittance decreases as number of glass cover increases as per expectation.

3.2.1.2 Absorption by glazing

According to Bouguer's law, the absorbed radiation is proportional to the intensity and the distance traveled (x) in the medium and can be expressed as

$$dI = -IK\,dx$$

where K is proportionality constant (the extinction coefficient which is assumed to be constant in the solar spectrum). The value of K varies from $4\,m^{-1}$ for "water white" glass to $32\,m^{-1}$ for poor glass (which appears greenish when viewed on the edge).

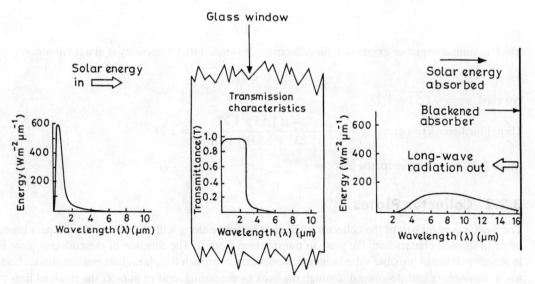

Figure 3.1(f) Propagation of solar radiation through glass window.

Integrating from zero to actual path length $(L/\cos\theta_2)$ in the medium, we get

$$\tau_a = \frac{I_{\text{transmitted}}}{I_{\text{incidence}}} = \exp\left(-\frac{KL}{\cos\theta_2}\right) \tag{3.1c}$$

The subscript 'a' indicates that the transmission is due to absorption only. The absorptance of a solar collector cover can be written as

$$\alpha = 1 - \tau_a$$

The reflectance of a single cover can also be written as

$$\rho = 1 - \alpha - \tau = \tau_a - \tau$$

Now, transmittance of single becomes

$$\tau = \tau_a \tau_r$$

The propagation of short wavelength solar radiation through a glass window has been shown in Figure 3.1(f). The transmission characteristic of a window glass cover has also been shown in the same figure. After transmission it is absorbed by the blackened surface. The blackened surface emits the long wavelength radiation. Due to its properties, the window glass cover behaves as a opaque for long wave length radiation and hence it blocked the emitted long wavelength radiation from the blackened surface.

3.2.1.3 *Effect of dielectric*

If a clear of a second dielectric (light-transmitting) material is coated on the glass cover in a thickness of several microns, the reflectivity of the glass cover is reduced. It is given by

$$r = 1 - \frac{4n_1 n_2}{(n_2^2 + n_1)(n_1 + 1)} \tag{3.1d}$$

If $n_1 = n_2$, the above equation reduces to

$$r = \left(\frac{n-1}{n+1}\right)^2$$

which is similar to earlier expression for reflectivity. It is noted that reflectivity is at a minimum for

$$n_2 = \sqrt{n_1}$$

For glass, $n_2 = \sqrt{1.53} = 1.23$.

Then, Equation (3.1d) gives $\quad f = 1 - \dfrac{4 \times 1.53 \times 1.23}{(1.23^2 + 1.53)(1.53 + 1)} = 2.21$

which is 50 percent of uncoated glass.

3.2.2 Collector Plates

The most important part of the collector is the absorber plate along with the pipe or duct to pass liquid or air in thermal contact with the plate to transfer heat from it. The function of the collector plate is to absorb maximum possible solar radiation incident on it through the glazing, to emit minimum heat, to the atmosphere and downward, through the back of the casing, and to transfer the retained heat to the fluid. Materials generally used for collector plates, in decreasing order of cost and conductance, are copper, aluminum and steel. The surface coating of the plate should be such that it has high absorptivity and poor emissivity for the required temperature range. Selective surfaces are particularly important when the collector surface temperature is much higher than the ambient air temperature. For domestic water heating system the plate is normally painted black. The energy absorbed by the plate is extracted by circulating a fluid, through a network of tubes in good thermal contact with the plate or directly in contact with the plate. The bottom and sides of the collector are covered with insulation to reduce the conductive heat loss. The collector is placed inclined at suitable angle to receive the maximum solar radiation.

3.3 CLASSIFICATION

In steady-state condition, the rate of energy absorbed by plate per unit area should be equal to the sum of the rate of the useful energy ($\dot{q}_u$) transferred to the fluid and the rate of energy lost ($\dot{q}_L$) per unit area by the plate to the surrounding. Thus

$$\dot{q}_{ab} = \dot{q}_L + \dot{q}_u \qquad (3.2a)$$

where
$$\dot{q}_{ab} = (\tau_0\alpha_0)I(t) \text{ and,}$$

$$\dot{q}_L = U_L(T_p - T_a)$$

Equation (3.2a) can be rewritten as

$$\dot{q}_u = (\tau_0\alpha_0)I(t) - U_L(T_p - T_a) \qquad (3.2b)$$

Further,
$$\eta_i = \frac{\dot{q}_u}{I(t)} = \tau_0\alpha_0 - U_L\frac{T_p - T_a}{I(t)} \qquad (3.2c)$$

where τ_0 and α_0 is the transmissivity and absorptivity, of the glazing surface, respectively. $I(t)$ is the incident solar radiation in the plane of absorber (W/m^2), T_p and T_a are the plate and ambient temperature (°C) respectively and U_L is the heat loss coefficient for collector (W/m^2 °C).

As shown in Figure 3.2, the useful energy decreases with increasing temperature difference. The thermal loss to the surroundings is an important factor in the determination of performance of a collector, the higher the losses the lower is the useful energy output and the heat loss depends upon ($T_p - T_a$). Hence, for different ranges of temperature difference, different types of collectors have been designed to minimize $\dot{q}_L$ and optimize $\dot{q}_u$, as attempts to decrease $\dot{q}_L$ also decreases $\dot{q}_{ab}$.

The collectors, according to their shape, are divided into two categories, namely, evacuated tubular collectors and flat-plate collectors.

3.3.1 Evacuated Tubular Collectors (Charters and Window, 1978)

In order to increase the useful energy one can use the selective surface, which reduces the radiative heat loss, but a further reduction in the convective loss is required in order to fully utilize the potential of such surfaces. This can be achieved by using more glazing covers but at the cost of some incident radiation, due to an increase in reflective losses (Figure 3.2).

This can also be achieved by removing the air over the absorbing surface, i.e. by evacuation. The collector, based on this concept, developed at the University of Sydney, is shown in Figure 3.3.

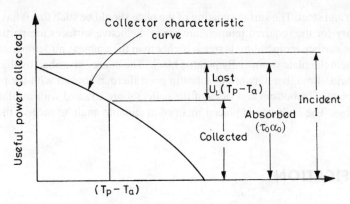

Figure 3.2 Collector characteristic curve.

The simplest modules are made of two connected tubes, assembled as shown in Figure 3.3. The selective surface on the outer side of the inner tube is vacuum-encased (less than 10^{-5} of atmospheric pressure), and the fluid to extract heat must be passed in and out at the same end of the inner tube. One major disadvantage of such collectors is the necessity of using cylindrical shape. It is difficult to arrange the position so that the entire surface always receives the full solar flux. In commercial unit, the tubes are spaced apart, with some type of reflector behind them, to reflect the radiation passing between them on to the back of the tubes. As a result, these collectors are not as efficient as flat-plate collectors at low extraction temperatures and high solar flux, but perform much better, because of low losses, at temperatures above $100°C$ and in low solar flux. The details of evaluated-tube collector are discussed in Chapter 4.

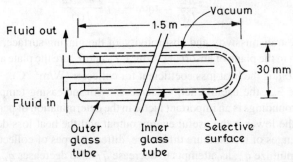

Figure 3.3 An all-glass evacuated-tube collector module (*Charters and Window, 1978*).

The group in School of Physics, University of Sydney, developed a sputtered iron/iron carbide on copper, a selective surface which can be deposited directly onto glass, and made collectors with diffuse reflector, which gave temperatures upto $320°C$ in full solar flux. Extraction efficiencies more than 50% at $100°C$ can be obtained with improved reflectors.

3.3.2 Type of Flat-plate Collectors

In flat-plate collectors, the heat loss by convection is more important in the determination of their performance. The convective heat loss may be decreased by using double glazing, but the radiation

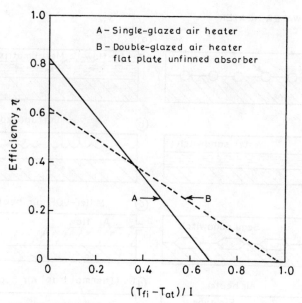

Figure 3.4 Efficiency vs. $(T_{fi} - T_{at})/I$ (ASHRAE handbook, 1967), characteristic curve.

reaching the absorber is reduced due to double reflection. Hence, at low temperatures where this loss is small, use of single glazing gives a better efficiency than the double one while at higher temperature difference the use of double glazing is advisable for better performance. Figure 3.4 compares the efficiencies of single and double glazing flat plate collectors in different temperature ranges.

Flat plate collectors are basically divided into two categories according to their use, (i) water or liquid heaters and (ii) air heaters. The collectors meant for these uses are sub-divided as follows:

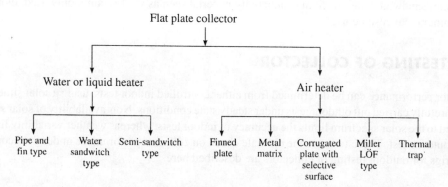

The schematic diagram of all these collectors, with single glazing are shown in Figure 3.5. The conditions, in which the various types of flat-plate collectors are used for water-heating purpose are:

i. full pipe and fin, with comparatively low wetted area and water capacity, should be used with fin of highly conducting material, i.e. copper or aluminum. It is used for high temperature domestic water heating,

WATER HEATER

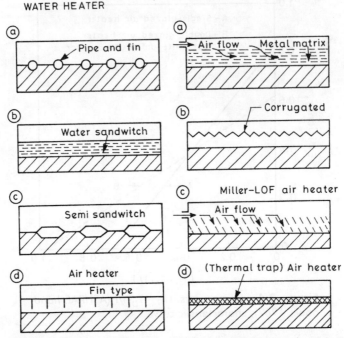

Figure 3.5 Various types of flat-plate collectors/air heaters.

ii. full water sandwich type, where both the wetted area and the water capacity are high. As the thermal conduction is only across the skin thickness (short distance), low conductivity materials may be employed. Both plastic and steel have been used. It is commonly used for heating swimming pools with plastic panel, and

iii. semi-sandwich type, medium conductivity material such as steel is commonly used, though aluminum may also be used.

3.4 TESTING OF COLLECTOR

Collector performance can be determined from either controlled indoor tests, using a solar simulator, or from carefully carried out outdoor tests under steady state conditions. Non-availability of solar simulator matched to the solar spectrum limits the accuracy of indoor tests, whereas weather variability limits that of the outdoor test. These tests are the ultimate check on collector performance under field conditions. Three rigs for outdoor testing of collectors are described here.

3.4.1 Orientable Test Rig (Charters and Window, 1978)

The standard procedure for collector testing is to operate the collector on the test stand, under conditions in which operation is nearly steady, i.e. the radiation and other conditions are essentially constant for a time long enough for the outlet temperature and useful gain to become steady.

In order to achieve the above conditions, water at constant temperature is supplied to the collector at constant flow rate from a fixed-head tank; as the collector is rotated to face the sun to reduce the angle of incidence and capacitance effects as shown in Figure 3.6. This adjustment feature extends the duration of the day available for constant steady input of radiation to the system. The efficiency is then calculated from the carefully measured incident radiation, inlet and outlet temperatures of the fluid, the ambient temperature and the flow rate of the fluid through the collector.

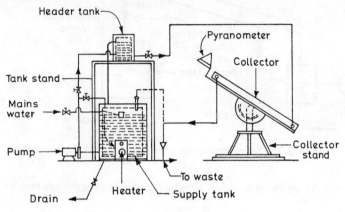

Figure 3.6 Orientable test rig.

The useful energy of the collector is given by

$$\dot{Q}_u = \dot{m}\, C_f (T_{fo} - T_{fi}) \tag{3.3}$$

where $\dot{m}$ is the flow rate of fluid through the collector, C_f is the specific heat of the fluid, T_{fo}, T_{fi} are the outlet and inlet fluid temperatures respectively.

The instantaneous efficiency of the collector is given by

$$\eta_i = \frac{\dot{Q}_u}{A_c I(t)} \tag{3.4}$$

where A_c is the collector area and $I(t)$ is the incident radiation on the collector (W/m^2).

3.4.2 Series-connected Test Rig

Using a series-connected bank of collectors exposed to the same atmospheric conditions it is possible to carry out simultaneous testing of several collectors under different operating conditions. In this process, inlet water temperature is maintained with a controlled flow rate. The inlet and outlet temperatures are continuously monitored for each of the collector separately. Although all collectors are exposed to the same environmental conditions, each collector is at a successively higher operational temperature level. It is thus possible to generate a characteristic curve of collector performance for the mean radiant load over the test period.

3.4.3 Testing of Solar Collector with Intermittent Output

This test rig was developed by Huang and Lu (1982) to overcome the limitations of previous methods, in which steady state condition, i.e. constant solar radiation and ambient temperature, is required for a long time. In this model the output water from the collector is not continuous but intermittent, controlled by a thermal switch which operates only when the outflow temperature reaches a pre-set value. The

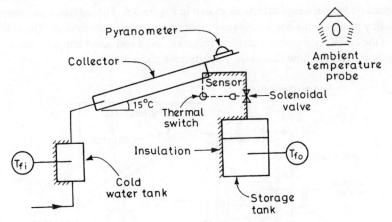

Figure 3.7 Schematic diagram of experimental apparatus.

outflow of hot water is driven by inlet cold water which comes from a storage tank on the roof or higher level.

The complete test model, shown in Figure 3.7, consists of a cold water tank from which inlet water at temperature T_{fi} is fed to the collector, a collector to be tested at some known inclination depending upon the latitude of the testing place, a pyranometer to measure the incident radiation on the collector, thermal switch and storage tank. The thermal switch controls the solenoidal valve with a sensor immersed inside the outflow header tube of the collector. When the water inside the collector attains a temperature equal to a pre-set value, the solenoidal valve opens and hot water is forced out. As the sensor is cooled down below the pre-set value, by the inflow of cold water, the valve is shut down and the solar water heating process is repeated again. The temperatures are measured with copper constantan thermocouple. The thermal sensor used by Huang and Lu was of mercury type immersed inside the upper header tube. As the temperature distribution inside the absorber during the heating period is not uniform due to the natural convection inside the tilted collector, the outflow temperature may not equal the pre-set temperature. Therefore, in practical applications it is more convenient to measure the average temperature, T_{fo}, of the total discharged and well-mixed fluid collected in an insulated storage tank during the discharging period.

Huang and Lu (1982) measured the quantity of water discharged to the storage tank by graduated cylinder in each cycle and found it to be different from the total fluid in the absorber. They correlated this measured quantity of discharged hot water in each cycle with a parameter Y as

$$M_{fo} = \begin{matrix} 3.74Y^{-0.507} \pm 0.6 \, \text{kg} \, (0 < \theta < 45°) \\ 4.37Y^{-0.33} \pm 0.3 \, \text{kg} \, (45° < \theta < 60°) \end{matrix} \qquad (3.5)$$

where $Y = \ln X$ and

$$X = \frac{E - (\bar{T}_{f0} - T_a)/I(t)}{E - (\bar{T}_f - T_a)/I(t)} \qquad (3.6)$$

where E is the equilibrium constant and it's values have been determined for different values of inclination and has been found to be approximately independent of solar incident angle. The values of E for different ranges of incident solar angle is given in Table 3.1.

TABLE 3.1 Determination of constant E

	0–10°	10°–20°	20°–30°	30°–40°	40°–50°	50°–60°
E (°C-m^2/kW)	87.8 ± 2.5	82.8 ± 2.7	79.3 ± 2.5	82.6 ± 3.0	78.1 ± 2.3	82.7 ± 1.7

The average efficiency of energy collection in each cycle is given by

$$\eta_{av.} = \frac{M_{f0} C_f (\bar{T}_{f0} - T_{fi})}{A_c \tau I(t)} \tag{3.7}$$

where τ is the time period of discharge of fluid M_{fo}.

The measured efficiency has been correlated with X by Huang and Lu as

$$\eta = \frac{0.762 X^{0.514} \pm 0.03 \ (0° < \theta < 45°)}{0.614 X^{0.356} \pm 0.03 \ (45° < \theta < 60°)} \tag{3.8}$$

Using Equations (3.7) and (3.8) with Table 3.1, we can test the performance of a collector.

3.4.4 The ASHRAE Method

ASHRAE 93-77 gives three standard procedures for liquid heaters and one for air heater. The essential features of all the procedures can be summarized as follows:

i. arrangement is provided to feed the collector with fluid at a controlled inlet temperature and tests are made over a range of inlet temperatures.

ii. solar radiation is measured by a pyranometer on the plane of the collector.

iii. means for measuring flow rate, inlet and outlet fluid temperatures, and ambient conditions are provided.

iv. means are provided for measurements of pressure and pressure drop across the collector.

The ASHRAE method for air collectors includes the essential features of those for liquid heaters, with the addition of detailed specifications of conditions relating to air flow, air mixing, air temperature measurements and pressure drop measurements.

The general test method is to operate the collector in the test facility under nearly steady conditions (Figure 3.8(a)), measure the data to determine $\dot{Q}_u$ from Equation (3.3), and measure $I(t)$, T_{fi} and T_a, which are required in the analysis based on Equation (3.4). This indicates that outdoor tests are done in the mid-day hours on clear days when beam radiation is high and nearly normal to the collector. Thus, the transmittance-absorptance product for these test conditions is approximately the normal incidence value and is written as $(\tau\alpha)_n$.

Tests are done for various values of inlet temperature. In order to minimize the effects of heat capacity of collectors, tests are usually done in nearly symmetrical pairs, one before and one after solar noon, with results of the pairs averaged. Instantaneous efficiencies are determined from the expression given in Equation (3.4) for the averaged pairs, and are plotted as a function of $(T_{fi} - T_a)/I(t)$. A sample plot of data taken at five test sites under conditions meeting ASHRAE 93-77 specifications is shown in Figure 3.8(b).

For constant U_L, F_R and $(\tau\alpha)_n$, the plots of η_i versus $(T_{fi} - T_a)/I(t)$ would be straight lines with intercept $F_R(\tau\alpha)_n$ and slope $-F_R U_L \cdot U_L$ is a function of temperatures and wind speed, the dependence

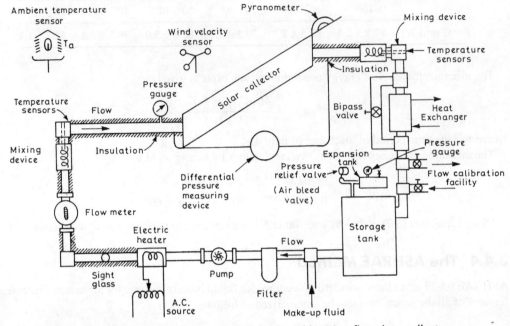

Figure 3.8(a) Closed-loop test setup for liquid heating flat-plate collectors (*Duffie and Beckman, 1991*).

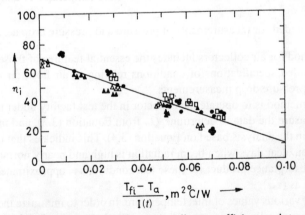

Figure 3.8(b) Experimental collector efficiency data measured for a liquid heating flat-plate collector with one cover and a selective absorber. Sixteen points are shown for each of five test sites. Curve represents the theoretical characteristics derived from points calculated for the test conditions (*Duffie and Beckman, 1991*).

decreasing, as the number of cover increase. F_R is a weak function of temperature (Section 3.8.5). Some changes in the relative proportions of beam, diffuse and ground-reflected components of solar radiation also occur. Thus the scatter from straight line data is to be expected due to solar radiation variation, wind effects, and angle of incidence variations. In spite of these deviations, long time performance estimates of many solar heating systems, collectors can be characterized by the intercept $F_R(\tau\alpha)_n$ and slope $(F_R U_L)$.

3.5 HEAT TRANSFER COEFFICIENTS

The thermal loss to the surroundings is an important factor in the study of the performance of a solar flat-plate collector. Heat is lost to the surroundings, from (a) the plate through the glass cover (referred as top loss) and (b) the plate through the insulation (referred as bottom loss). These losses took place by conduction, convection and radiation. These losses have also been shown in Figure 3.9(a). The equivalent losses in terms of thermal resistance circuit have been shown in Figure 3.9(b). Here, it is important to mention that the thermal resistance (r) is inversely proportional to heat transfer coefficient (h). The knowledge of heat transfer coefficient corresponding to the phenomenon is essential for the analysis of the performance of the collector. A discussion of these heat transfer coefficients is given in the following sections.

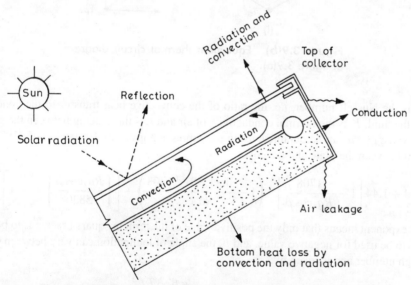

Figure 3.9(a)   Various heat losses from absorber to ambient.

3.5.1 Top Loss Coefficients

3.5.1.1 Convective heat transfer coefficient

i. From plate to cover ($h_{1c} = 1/r_3$)
The convective heat transfer coefficient between plate and cover, parallel to each other and inclined at an angle β to the horizontal can be expressed as

$$h_{1c} = Nu\,K/L \tag{3.9}$$

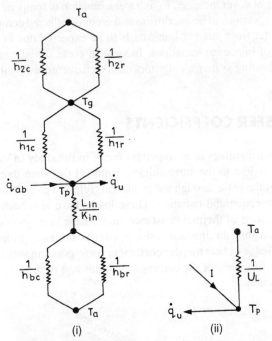

Figure 3.9(b) Equivalent thermal circuit diagram of figure 3.9(a).

where Nu is the Nusselt number, i.e. the ratio of the convective heat transfer to the conductive heat transfer in the fluid, K is the thermal conductivity of air and L is the spacing between the plate and the cover. The value of Nu can be obtained, using the expression given by Holland et al. (1976) for air as the medium between the plate and the cover:

$$Nu = 1 + 1.44 \left[1 - \frac{1708}{Ra \cos \beta} \right]^{+} \left(1 - \frac{\sin(1.8\beta)^{1.6} 1708}{Ra \cos \beta} \right) + \left[\left\{ \frac{Ra \cos \beta}{5830} \right\}^{1/3} - 1 \right]^{+} \quad (3.10)$$

The '+' exponent means that only the positive value of the term in square bracket is to be considered and zero is to be used for negative value, and β, the angle of inclination can vary between 0–75°, Ra is the Rayleigh number and is given by:

$$Ra = Gr\, Pr = \frac{g\, \beta'\, \Delta T L^3}{\nu \alpha} \quad (3.10a)$$

If, $75° \leq \beta \leq 90°$, then

$$Nu = \left[1, 0.288 \left(\frac{\sin \beta\, Ra}{A} \right)^{1/4}, 0.039(\sin \beta\, Ra)^{1/3} \right]_{\max} \quad (3.11a)$$

where the subscript 'max' indicates that for a given value of Rayleigh number, the maximum of the three quantities separated by commas should be taken. A is the ratio of length of collector plate inclined to spacing between cover and absorbing plate.

Buchberg *et al.* (1976) have also recommended the following three region correlation for a convective heat transfer losses for an inclined flat plate collector:

$$Nu = 1 + 1.446 \left[1 - \frac{1708}{Ra \cos \beta} \right] \quad \text{for } 1708 < Ra \cos \beta < 5900 \tag{3.11b}$$

$$Nu = 0.229(Ra \cos \beta)^{0.252} \quad \text{for } 5900 < Ra \cos \beta < 9.23 \times 10^4 \tag{3.11c}$$

$$Nu = 0.157(Ra \cos \beta)^{0.285} \quad \text{for } 9.23 \times 10^4 < Ra \cos \beta < 10^6 \tag{3.11d}$$

EXAMPLE 3.1(a)

Find the convective heat transfer coefficient between two parallel plates separated by 20 mm with a 40° tilt, the lower plate is at 50°C and the upper plate is at 30°C.

Solution

At the mean air temperature of 40°C, air properties are $K = 0.0272$ W/mK; $T = 313$ K so $\beta' = 1/T = 1/313$; $\nu = 1.70 \times 10^{-5}$ m^2/s; $\alpha = 2.40 \times 10^{-5}$ m^2/s (see Appendix III).
Substituting the values of various parameters in Equation (3.10a), we get the value of Rayleigh number as,

$$Ra = \frac{9.81 \times 20 \times (0.020)^3}{313 \times 1.70 \times 10^{-5} \times 2.40 \times 10^{-5}} = 1.23 \times 10^4$$

Substituting this value of Ra in Equation (3.10), we get the value of Nusselt number as 2.481. The heat transfer coefficient is found from the relation,

$$h = Nu \, K/L = (2.481 \times 0.0272)/0.020 = 3.37 \, \text{W/m}^2\text{K}$$

ii. From cover to ambient ($h_{2c} = 1/r_1$)

The convective heat loss coefficient from the top glazing to the ambient is given by (Watmuff *et al.* 1977)

$$h_{2c} = 2.8 + 3.0 \, V \tag{3.12}$$

where V is the wind speed in m/s, over the collector.

3.5.1.2 *Radiative heat transfer coefficient*

i. From plate to cover ($h_{1r} = 1/r_4$)

The amount of heat transferred from plate to cover, through radiation, can be given by the expression:

$$\dot{q}_{rad} = h_{1r}(T_p - T_g) \tag{3.13}$$

where T_p and T_g are the plate and glazing temperatures in °C, respectively, and h_{1r} is the coefficient of radiative heat loss from collector plate to the cover and is expressed as

$$h_{1r} = \varepsilon_{\text{eff}}\sigma\frac{[(T_p + 273)^4 - (T_g + 273)^4]}{T_p - T_g} \tag{3.14}$$

where $\sigma(= 5.67 \times 10^{-8}\,\text{W/m}^2\text{K}^4)$ is the Stefan's constant and ε_{eff} is the effective emissivity of plate-glazing system, given by

$$\varepsilon_{\text{eff}} = \left[\frac{1}{\varepsilon_p} + \frac{1}{\varepsilon_g} - 1\right]^{-1} \tag{3.15}$$

ii. From glazing cover to ambient ($h_{2r} = 1/r_2$)

The coefficient of radiative heat loss from cover to the ambient depends upon the radiation exchange with sky at sky temperature T_{sky} given by

$$T_{\text{sky}} = T_a - 6 \tag{3.16}$$

The radiative heat transfer coefficient, h_{2r}, can thus be expressed as

$$h_{2r} = \varepsilon_g\sigma\frac{[(T_g + 273)^4 - (T_s + 273)^4]}{T_g - T_a} \tag{3.17}$$

3.5.1.3 Top loss coefficient

The total heat transfer coefficient from collector plate to cover can be expressed as the sum of h_{1c} and h_{1r}, i.e.

$$h_1 = h_{1c} + h_{1r} = \frac{1}{r_3} + \frac{1}{r_4} \tag{3.17a}$$

and that from the cover to ambient as

$$h_2 = h_{2c} + h_{2r} = \frac{1}{r_1} + \frac{1}{r_2} \tag{3.17b}$$

The effective top heat transfer coefficient from plate to ambient is given by

$$U_t = [(1/h_1) + (1/h_2)]^{-1} = \left[\left(\frac{1}{r_3} + \frac{1}{r_4}\right)^{-1} + \left(\frac{1}{r_1} + \frac{1}{r_2}\right)^{-1}\right]^{-1} \tag{3.18}$$

and the rate of heat loss from the top per unit area can be given as

$$\dot{q}_{\text{loss,top}} = U_t(T_p - T_a) \tag{3.19}$$

The value of U_t is calculated by the method of iteration because in general the cover temperature, T_g is not known. An arbitrary value of T_g is assumed and then at the mean value of temperature, h_{1c}, h_{1r}, h_{2r} are calculated, and then using these values, the top loss coefficient is calculated. Substituting this value of U_t in the energy balance condition,

$$h_1(T_p - T_g) = U_t(T_p - T_a) \tag{3.20a}$$

A new value of T_g is obtained from above equation as

$$T_g = T_p - \frac{U_t}{h_1}(T_p - T_a) \tag{3.20b}$$

With this value of T_g the above calculations are repeated till the two consecutive values of T_g obtained are nearly the same. This is referred as iterative process which is as follows:

First step: Chose a initial appropriate value of T_g

Second step: Calculate the value of U_t for known value of T_p, T_g and T_a as in Example 3.1(b)

Third step: After calculating the value of U_t, calculate new value of T_g from the above equation

Fourth step: Repeat the same calculations with T_p, T_a and new value of T_g till the two consecutive values of T_g obtained are nearly the same.

EXAMPLE 3.1(b)

Calculate the overall top loss coefficient from the absorber to ambient through the glass cover (U_t) for the following collector specifications:

Absorber-to-cover distance (L)	0.025 m
Absorber plate emittance (ε_p)	0.95
Glass cover emittance (ε_g)	0.88
Inclination of collector (β)	45°
Wind heat transfer coefficient (h_{2c})	10 W/m²°C
Ambient air temperature (T_a)	16°C
Mean absorber plate temperature (T_p)	100°C

Solution

Here, $\varepsilon_{\text{eff}} = \left[\frac{1}{0.95} + \frac{1}{0.88} - 1\right]^{-1} = 0.84$ (Equation (3.15))

Let us assume, $T_g = 35°C$ and $T_{\text{sky}} = 16 - 6 = 10°C$ (Equation (3.16))

From Equations (3.14) and (3.17), we get

$$h_{1r} = 0.84 \times 5.6 \times 10^{-8} \frac{[(100 + 273)^4 - (35 + 273)^4]}{100 - 35} = 7.496 \text{ W/m}^2 \, °C$$

and $\quad h_{2r} = 0.88 \times 5.6 \times 10^{-8} \frac{[(35 + 273)^4 - (10 + 273)^4]}{35 - 16} = 6.705 \text{ W/m}^2 \, °C$

In order to calculate the convective heat transfer coefficient between the absorber and the cover, the air properties will be considered at mean temperature between the absorber and the glass cover, i.e. 67.5°C. Air properties at 67.5°C are $\nu = 1.96 \times 10^{-5} \text{ m}^2/\text{s}$, $K = 0.0293$ W/m°C, $T = 340.5 \, K$ and $Pr = 0.7$. The Rayleigh number is

$$Ra = Gr \cdot Pr = \frac{9.81(100 - 35)(0.025)^3 \times (0.7)}{340.5(1.96 \times 10^{-5})^2} = 5.33 \times 10^4$$

From Equation (3.10), the Nusselt number is 3.19. The convective heat transfer coefficient is

$$h_{1c} = Nu \cdot \frac{K}{L} = 3.19 \times \frac{0.0293}{0.025} = 3.73 \, \text{W/m}^2 \, ^\circ \text{C}$$

The first estimate of U_t can be obtained from Equation (3.18) as

$$U_t = \left(\frac{1}{7.496 + 3.73} + \frac{1}{6.705 + 10} \right)^{-1} = 6.714 \, \text{W/m}^2 \, ^\circ \text{C}$$

With $U_t = 6.714 \, \text{W/m}^2 \, ^\circ \text{C}$, a new value of T_g is obtained from Equation (3.20b) as

$$T_g = 100 - \frac{6.714}{(7.496 + 3.73)} (100 - 16) = 49.76 ^\circ \text{C}$$

With this value of T_g, the new values of h_{1r} and h_{2r} will be

$$h_{1r} = 0.84 \times 5.6 \times 10^{-8} \frac{[(100 + 273)^4 - (49.76 + 273)^4]}{100 - 49.76} = 7.96 \, \text{W/m}^2 \, ^\circ \text{C}$$

and

$$h_{2r} = 0.88 \times 5.6 \times 10^{-8} \frac{[(49.76 + 273)^4 - (10 + 273)^4]}{49.76 - 16} = 6.48 \, \text{W/m}^2 \, ^\circ \text{C}$$

Air properties at mean temperature of T_p and new T_g, i.e. $74.88 \cong 75 ^\circ \text{C}$ are $\nu = 2.076 \times 10^{-5} \, \text{m}^2/\text{s}$, $K = 0.03 \, \text{W/m}^\circ \text{C}$, $T = 348 \, K$ and $Pr = 0.697$ (Appendix Table IIId). The Rayleigh number is

$$Ra = Gr \cdot Pr = \frac{9.81(100 - 49.76)(0.025)^3 \times (0.697)}{348 \times (2.076 \times 10^{-5})^2} = 3.58 \times 10^4$$

In this case, the Nusselt number is 2.9. Hence

$$h_{1c} = Nu \cdot \frac{K}{L} = 2.9 \times \frac{0.03}{0.025} = 3.48 \, \text{W/m}^2 \, ^\circ \text{C}$$

and second estimate of U_t is

$$U_t = \left[\frac{1}{3.48 + 7.96} + \frac{1}{6.48 + 10} \right]^{-1} = 6.75 \, \text{W/m}^2 \, ^\circ \text{C}$$

With above value of U_t, further new value of T_g will be

$$T_g = 100 - \frac{6.75}{11.44} (100 - 16) = 50.4 ^\circ \text{C}$$

Now $T_g = 46.6 ^\circ \text{C}$ can be used to get h_{1r} and h_{2r} for third estimate of U_t as

$$h_{1r} = 7.98 \, \text{W/m}^2 \, ^\circ \text{C}$$
$$h_{2r} = 6.48 \, \text{W/m}^2 \, ^\circ \text{C}$$

At mean temperature $= 75.2 ^\circ \text{C}$, approximately same as considered earlier at $75 ^\circ \text{C}$. Now, Rayleigh number is 3.533×10^4. The Nusselt number is given by

$$Nu = 2.894$$

and the convective heat transfer is given by

$$h_{1c} = 2.894 \times \frac{0.03}{0.025} = 3.47 \, \text{W/m}^2 \, ^\circ\text{C}$$

Third estimate of U_t is

$$U_t = \left[\frac{1}{7.98 + 3.47} + \frac{1}{6.47 + 10} \right]^{-1} = 6.76 \, \text{W/m}^2 \, ^\circ\text{C}$$

The new estimate of T_g is

$$T_g = 100 - \frac{6.76}{11.45}(100 - 16) = 50.4$$

which is similar to the value calculated earlier. Hence, the same value of U_t will be obtained if calculations are repeated further. This mease the exact value of U_t will be 6.76 W/m^2 °C.

3.5.2 Back Loss Coefficient

Heat is lost from the plate to the ambient by conduction through the insulation and then subsequently by convection and radiation from the bottom surface casing. The bottom loss coefficient can be given by,

$$U_b = \left[\frac{L_{in}}{K_{in}} + \frac{1}{h_b} \right]^{-1} \tag{3.21}$$

where h_b is the heat loss coefficient from the bottom, and is the sum of convective heat loss coefficient h_{bc} and radiative heat loss coefficient h_{br}. The value of h_{br}, h_{bc} can be calculated as in the case of top cover. The magnitude of K_{in} and h_b are such that the second term in the Equation (3.21) is negligible as compared to the first one, thus,

$$U_b \simeq \frac{K_{in}}{L_{in}}$$

The suffix 'in' indicates insulation.

3.5.3 Edge Loss Coefficient

The energy lost from the side of the collector casing is exactly the same as that from the back if the thickness of the edge insulation is the same as that of the back insulation (Tabor, 1958). The edge loss, generally expressed in terms of the area of collector and the back loss coefficient, is given as

$$U_e = U_b(A_e/A_c) \tag{3.22}$$

where A_e is the edge area.

3.5.4 Overall Heat Loss Coefficient

The overall heat loss coefficient U_L is the sum of the top, bottom and edge loss coefficient, i.e.,

$$U_L = U_t + U_b + U_e \tag{3.23}$$

and the overall heat lost by the absorber to the ambient per unit area per unit time can be expressed as,

$$\dot{q}_L = U_L(T_p - T_a) \tag{3.24}$$

where q_L is the same as used in Equation 3.2a.

EXAMPLE 3.2

Determine the overall heat transfer coefficient (U_L) for a flat-plate collector (FPC) system inclined at 45° to the horizontal and facing south. The average ambient air temperature for the day is 20°C and the observed glass and absorber plate temperatures are 45°C and 69°C respectively. The system is provided with 6 cm thick insulation (glass wool) at the bottom. The thermal conductivity of the insulation being 0.04 W/m°C. The air space between the absorber plate and glass cover has optimum thickness of 7.5 cm and the emissivity of the glass and plate is 0.88 and 0.95 respectively.

Solution

The average air temperature in the space between glass and plate is 57°C. The physical properties of air at this temperature are:

$$\nu = 1.88 \times 10^{-5}\,\text{m}^2/\text{s}; \alpha = 2.69 \times 10^{-5}\,\text{m}^2/\text{s}; K = 0.028\,\text{W/m°C}$$

$$\sigma = 5.6697 \times 10^{-8}\,\text{W/m}^2\,\text{K}^4; g = 9.81\,\text{m/s}^2(\text{see Appendix III})$$

Now, the Prandtl number (from Equation (2.38c)) is

$$Pr = \frac{\nu}{\alpha} = \frac{1.88 \times 10^{-5}}{2.69 \times 10^{-5}} = 0.7$$

Here, $\Delta T = 24°C$; $L = 0.075$ m; $Pr = 0.7$; $\beta' = 1/330\,\text{K}^{-1}$.

Then, the Rayleigh number (from Equation (3.10a)) is

$$Ra = \frac{9.81 \times 24(0.075)^3 \times 0.7}{330 \times (1.88 \times 10^{-5})^2} = 5.961 \times 10^5$$

Substituting the values in Equation (3.10), we have

$$Nu = 1 + 1.44\left[1 - \frac{1708}{5.961 \times 10^5 \cos 45}\right]^+ \left(1 - \frac{\sin(1.8 \times 45)^{1.6}}{17085.961 \times 10^5 \cos 45}\right)$$

$$+ \left[\left(\frac{5.961 \times 10^5 \cos 45}{5830}\right)^{1/3} - 1\right]^+ = 5.595$$

From Equation 3.9, we get

$$h_{1c} = \frac{5.595 \times 0.028}{0.075} = 2.089\,\text{W/m}^2\,°\text{C}$$

From Equation (3.12), we get $h_{2c} = 17.8$ W/m^2 °C (for $V = 5$ m/s). In this problem,

$$\varepsilon_{\text{eff}} = \left[\frac{1}{0.88} + \frac{1}{0.95} - 1\right]^{-1} = 0.841$$

Now, the radiative heat transfer coefficient from plate to cover, is given by

$$h_{1r} = \frac{0.841 \times 5.6697 \times 10^{-8}[(69 + 273)^4 - (45 + 273)^4]}{69 - 45} = 6.85 \text{ W/m}^2 \text{ °C}$$

The radiative heat transfer coefficient from glass cover to sky is

$$h_{2r} = 0.88 \times 5.6697 \times 10^{-8}\left[\frac{(45 + 273)^4 - (14 + 273)^4}{45 - 20}\right] = 6.868 \text{ W/m}^2 \text{ °C}$$

Now, $\qquad\qquad h_1 = h_{1c} + h_{1r} = 2.089 + 6.85 = 8.939$ W/m^2 °C

and $h_2 = h_{2c} + h_{2r} = 17.8 + 6.868 = 24.668$ W/m^2 °C; by using Equation (3.17a and b).
The overall heat transfer coefficient can be obtained by using Equation (3.18)

$$U_t = \left[\frac{1}{8.939} + \frac{1}{24.668}\right]^{-1} = 6.57 \text{ W/m}^2 \text{ °C}$$

Bottom loss coefficient (from Equation (3.21)) $U_b = 0.04/0.06 = 0.666$ W/m^2 °C. From Equation (3.22), the overall heat loss coefficient is

$$U_L = U_t + U_b = 6.567 + 0.666 = 7.23 \text{ W/m}^2 \text{ °C}$$

EXAMPLE 3.3

For a collector with a top loss coefficient of 6.6 W/m^2 °C, calculate the overall loss coefficient using the following data:
Back insulation thickness = 45 mm, Insulation conductivity = 0.04 W/m °C
Collector bank length = 8 m, Collector bank width = 2.5 m
Collector thickness = 80 mm, Edge insulation thickness = 20 mm

Solution

The bottom loss coefficient (from Equation (3.21)) is given by

$$U_b = \frac{K_{in}}{L_{in}} = \frac{0.04}{0.045} = 0.889 \text{ W/m}^2 \text{ °C}$$

The edge loss coefficient (from Equation (3.22)) (for 21 m perimeter) is

$$U_e = \frac{(0.04/0.020)(21)(0.080)}{8 \times 2.5} = 0.17 \text{ W/m}^2 \text{ °C}$$

The collector overall loss coefficient is

$$U_L = 6.6 + 0.889 + 0.17 \approx 7.659 \approx 7.7 \, \text{W/m}^2 \, ^\circ\text{C}$$

The edge loss for this $20 \, \text{m}^2$ ($8 \, \text{m} \times 2.5 \, \text{m}$) collector array is about two percent of the total losses. However, if this collector were $1 \times 2 \, \text{m}$, the edge losses would increase to over six percent. Thus edge losses for well-constructed large collector arrays are usually negligible, but for small arrays or individual modules the edge losses may be significant.

3.5.5 Film Heat Transfer Coefficient (Ong, 1974)

The Nusselt number, for the film heat transfer coefficient between the fluid in the collector tube and the tube wall, is given by (Baker, 1967)

$$Nu = 1.75 \left(\frac{\mu_w}{\mu'_w} \right)^{0.14} \left[Gz + 0.0083(Ra)^{0.75} \right]^{1/3} \tag{3.25}$$

where μ_w is the viscosity of fluid at the mean system temperature, μ'_w the viscosity of fluid at the temperature of the inner wall of the tube and Gz the Greatz number given by

$$Gz = \frac{Re_D \, Pr}{x/D} \tag{3.25a}$$

where Re_D is the value of Reynold number at the diameter D and x is the characteristic dimension. The temperature of water and the inner wall of the tube are, in general, time dependent but their average values may be used for the calculation of heat transfer coefficients.

The thermal conductance from the inner wall of heat exchanger to the cold water, U_c, and thermal conductance from hot water to the outer wall of the heat exchanger, U_h, can be calculated from the formula (Wong, 1977):

$$Nu = \frac{U_h \, D_2}{K_w} = C \, Re^m \, Pr^n \, K_R \tag{3.26}$$

where Re, Pr are the Reynold and the Prandtl numbers, respectively and D_2 the outer diameter of the tube. For laminar flow and long tube, $m = n = 0$, $K_R = 1$ and $C = 3.66$ (from Table 2.4). Hence

$$U_h = 3.66 \, K_w / D_2 \tag{3.27}$$

Similarly,

$$U_c = 3.66 \, K_w / D_1 \tag{3.28}$$

where D_1 is the inner diameter of the tube (Figure 3.10).

The overall heat conductance, U_k, can be calculated from the formula (Equation (2.34b), $D = 2r$)

$$\frac{1}{U_k} = \frac{1}{\frac{D_1}{D_2}(U_h)} + \frac{1}{U_c} + \frac{1}{\frac{2K_s}{D_1 \ln(D_2/D_1)}} \tag{3.29}$$

Assuming $D_2 \approx D_1$, the above equation becomes

$$\frac{1}{U_k} = \frac{1}{U_h} + \frac{1}{U_c} \tag{3.30}$$

If h is the heat transfer coefficient per unit length of tube, of length L_0, then h can be given as

$$h = U_k / L_0 \tag{3.31}$$

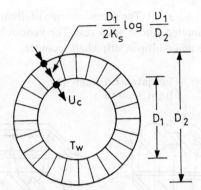

Figure 3.10 Cross-section of the tube.

3.6 OPTIMIZATION OF HEAT LOSSES

There is about 33–50 percent heat loss in most commercial flat-plate collectors, the break-up of which can be given as 22–30 percent convective and 5–7 percent radiative loss from the back surface. In order to improve the collector efficiency the heat losses should be minimized. A method suggested by Madhusudan *et al.* (1981) to minimize the heat loss is being discussed in brief.

To minimize the heat loss from the back surface, a highly reflective coating is used on the back surface of the collector and another highly reflective surface supported by glass wool insulation placed below it as shown in Figure 3.11. In this arrangement, the radiation loss from the back surface of the absorber plate is minimized in first step by reflective coating then in the second step it is minimized due to reflecting surface. Mainly convective and radiative losses take place from the front space of the absorbing plate. These two losses are minimized by altering the spacing between cover and plate, and using absorbing surface of different emissivities. The optimum gap width is different for different absorbing surface. The convective loss is further reduced to almost zero by illuminating the absorbing plate from below in the reverse flat-plate configuration. In this configuration the hot air sticks to the plate because of a favorable density gradient, hence, the convection is avoided.

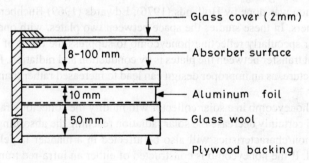

Figure 3.11 Cross-section of collector assembly.

A comparative study of the losses from normal and reverse flat-plate has been made by Madhusudan *et al.* (1981). The source of radiation used in this experiment, were two 1000 watt cylindrical tungsten halogen lamps mounted in Phillips NV-51 (2 × 15) reflectors, which uniformly illuminates the absorber plate kept in a leak proof cell (Figure 3.12). The measurements of illumination intensity was done from the short circuit current of a calibrated silicon solar cell. The various temperatures were measured by copper-constantan thermocouple and multipen strip chart recorder.

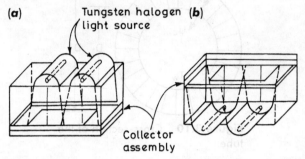

Figure 3.12 Experimental setup for flat plate collector in (*a*) normal configuration and (*b*) reverse configuration (*Madhusudan et al., 1981*).

The theoretical and experimental values have been compared for black paint and cobalt oxide ($\alpha_s = 0.88$, $\varepsilon(100°C) = 0.15$) coated absorbing surfaces in normal and reverse configurations. The optimum gap width for black painted surface is ~2.5 cm whereas for selective surface, in normal and reverse configuration it's value is ~5 cm and no effect of intensity of incident radiation is observed on this optimum gap width. Although the higher gap widths show equivalent or better performance in the case of selective surfaces, smaller gap is preferred to avoid shading of absorber surface by the collector walls.

From this study it is evident that a highly reflective surface, supported on 3 mm plyboard, spaced behind the collector plate is equivalent to the glass wool insulation of about 5 cm thickness and upto 90°C rise in temperature. At higher temperature the performance of glass wool is better. The two employed together are observed to be superior at all plate temperatures.

3.6.1 Transparent Insulating Material (Honeycomb)

A primary objective in designing solar collectors is to reduce the heat loss through the covers. Studies of convection suppression was done by Hollands (1976), Edwards (1969) Buchberg *et al.* (1976), Meyer *et al.* (1978) and others. In these studies the space between two plates, with one plate heated, is filled with a transparent or specularly reflecting honeycomb to suppress the onset of fluid motion. Without fluid motion, the heat transfer between the plates is by conduction and radiation. Full care must be taken in designing the structure, as an improper design can lead to increased rather than decreased convection losses.

The addition of a honeycomb in a solar collector will modify the collector's radiative characteristics. The honeycomb will certainly decrease the solar radiation reaching the absorbing plate of the collector. The infra-red radiation characteristics will also be affected in a manner largely dependent upon the honeycomb material. If the honeycomb is constructed of either an infra-red transparent material or an infra-red specularly reflective material, then the infra-red radiative characteristics of the collector will not be significantly changed. If the honeycomb material is constructed of a material that is opaque in the infra-red, then the radiative characteristics of the collector will approach that of a blackbody. This, of course, is undesirable.

3.6.2 Selective Surface

Solar collectors must have a high absorptance for radiation in the solar energy spectrum. These collectors lose energy by a combination of mechanisms (convection, radiation etc.) including thermal radiation from the absorbing surface. Thus it is required to have long-wave emittance of the surface as low as possible to reduce the losses. It is possible to devise surfaces having high solar absorptance and low long-wave emittance, that is, selective surfaces.

The concept of a selective surface is shown in Figure 3.13. This ideal surface is called a semigray surface, since it can be considered gray in the solar spectrum (i.e. at $\lambda < 3.0\ \mu m$) and also gray, but with different properties in the infrared spectrum ($\lambda > 3.0\ \mu m$). For this idealized surface, the reflectance below the cutoff wavelength is very low. At wavelengths greater than λ_c the reflectance is nearly unity and since $\varepsilon_\lambda = \alpha_\lambda = 1 - \rho_\lambda$, the emittance in this range is low. For normal operation of flat-plate solar collectors, the temperature is low enough so that practically all energy will be emitted at wavelengths greater than 3 μm.

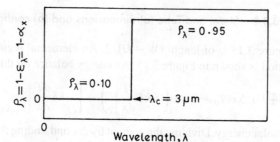

Figure 3.13 A hypothetical selective surface with the cut-off wavelength at 3 μm.

3.7 DETERMINATION OF FIN EFFICIENCY

Heat transfer takes place from a rod or fin of uniform cross-sectional area protruding from a flat wall. In practical application, fins may have varying cross-sectional areas and may be attached to circular surfaces.

In order to indicate the effectiveness of a fin in transferring a given quantity of heat, a new parameter, called *fin efficiency*, is defined as

$$\text{Fin efficiency} = \frac{\text{Actual heat transferred}}{\text{Heat which would be transferred if entire fin area were at base temperature}}$$

$$= F \tag{3.32}$$

The temperature distribution between two tubes can be derived if we temporarily assume the temperature gradient in the flow direction as negligible. Let us consider the sheet and tube dimensions shown in Figure 3.14. The distance between the tubes is W, the tube diameter is D, and the sheet has a small thickness δ. The sheet material being a good conductor, the temperature gradient through the sheet is negligible. The sheet above the bond is assumed to be at some local base temperature T_b. The region between the centerline separating the tubes and the tube base can be considered as a fin problem.

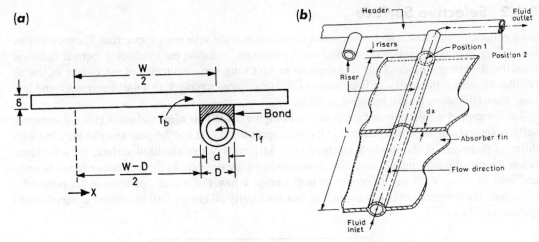

Figure 3.14 Sheet and Tube (*a*) dimensions and (*b*) configuration.

The fin, shown in Figure 3.14 is of length $(W - D)/2$. An elemental region of width Δx and unit length in the flow direction is shown in Figure 3.15. An energy balance on this element gives

$$I\Delta x + U_L \Delta x (T_a - T) + \left(-k\delta \frac{dT}{dx}\bigg|_x \right) - \left\{ -k\delta \frac{dT}{dx}\bigg|_{x+\Delta x} \right\} = 0 \tag{3.33}$$

where, I is the absorbed solar energy. Dividing throughout by Δx and finding the limit as Δx approaches zero yields

$$\frac{d^2 T}{dx^2} = \frac{U_L}{K\delta} \left(T - T_a - \frac{I}{U_L} \right) \tag{3.34}$$

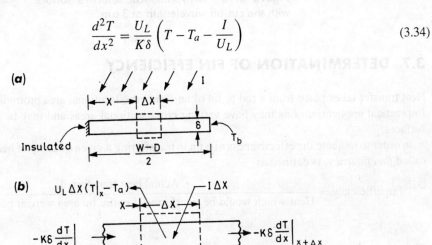

Figure 3.15 Energy balance on fin element.

The two boundary conditions required to solve this second-order differential equation are, symmetry at the centerline and known root temperature:

$$\frac{dT}{dx}\bigg|_{x=0} = 0 \quad \text{and} \quad T\,|_{x=(W-D)/2} = T_b \tag{3.35}$$

If we define

$$m^2 = U_L/K\delta \quad \text{and,} \qquad \psi = T - T_a - I/U_L \tag{3.35a}$$

Equation (3.34) becomes

$$\frac{d^2\psi}{dx^2} - m^2\psi = 0 \tag{3.36}$$

The general solution of the above equation can be given as

$$\psi = C_1 \sinh mx + C_2 \cosh mx \tag{3.37}$$

Constants C_1 and C_2 can be determined with the boundary conditions

$$\frac{d\psi}{dx}\bigg|_{x=0} = 0 \quad \text{and} \quad \psi|_{x=(W-D)/2} = T_b - T_a - \frac{I}{U_L} \tag{3.38}$$

Thus,

$$\frac{T - T_a - I/U_L}{T_b - T_a - I/U_L} = \frac{\cosh mx}{\cosh m(W-D)/2} \tag{3.39}$$

The energy conducted to the region of the tube per unit of length in the flow direction can be found by evaluating Fourier's law at the fin base.

$$\dot{q}_{fin} = -K\delta \frac{dT}{dx}\bigg|_{x=(W-D)/2}$$

or

$$\dot{q}_{fin} = \frac{K\delta}{U_L}[I - U_L(T_b - T_a)]\tanh m(W-D)/2 \tag{3.40}$$

The above equation accounts for the energy collected on only one side of a tube; for both sides, the energy collection is

$$\dot{q}_{fin} = (W - D)[I - U_L(T_b - T_a)]\frac{\tanh m(W-D)/2}{m(W-D)/2} \tag{3.41}$$

where

$$F = \frac{\tanh m(W-D)/2}{m(W-D)/2} \tag{3.42}$$

is the standard fin efficiency for straight fins with rectangular profile.

3.8 THERMAL ANALYSIS OF FLAT-PLATE COLLECTORS

3.8.1 Basic Energy Balance Equation

The useful energy output per unit time of a collector of area A_c is the difference between the absorbed solar radiation, $\dot{q}_{ab}$, and the thermal loss and is given by,

$$\dot{Q}_u = A_c\dot{q}_u = A_c[\dot{q}_{ab} - U_L(T_p - T_a)] \tag{3.43a}$$

where

$$\dot{q}_{ab} = (\tau_0\dot{\alpha}_0)I(t)$$

The instantaneous efficiency, η_i, is given as

$$\eta_i = \frac{\dot{Q}_u}{A_cI(t)} = \frac{\dot{q}_{ab}}{I(t)} - \frac{U_L(T_p - T_a)}{I(t)} = \tau_0\alpha_0 - U_L\frac{T_p - T_a}{I(t)} \tag{3.43b}$$

where $I(t)$ is the intensity of incident radiation, τ_0 and α_0 are the transmissivity of the glass cover, and absorptivity of an absorber respectively. The expression for U_L, the overall heat loss coefficient is given in section (3.5).

The collection efficiency, defined as the ratio of the useful gain to the incident solar energy over the same period of time, is given by

$$\eta_c = \frac{\int \dot{Q}_u \, dt}{A_c \int I(t) dt} \tag{3.44}$$

3.8.2 Effective Transmittance-Absorptance Product $(\tau_0 \alpha_0)_e$

The value of $(\tau_0 \alpha_0)_e$ is usually 1 to 2 percent greater than $(\tau_0 \alpha_0)$. Since surface radiation properties are seldom known to be within 1 percent, the equivalent transmittance-absorptance product can be approximated by (Duffie and Beckman, 1991),

$$(\tau_0 \alpha_0)_e = 1.02(\tau_0 \alpha_0) \tag{3.45a}$$

For a collector with low iron (water white) (no absorption) glass, $(\tau_0 \alpha_0)_e$ and $(\tau_0 \alpha_0)$ are nearly identical, i.e.

$$(\tau_0 \alpha_0)_e = 1.01(\tau_0 \alpha_0) \tag{3.45b}$$

In order to determine $(\tau \alpha)$ as a function of angle of incidence, Klein (1977) developed a relationship between $(\tau \alpha)/(\tau \alpha)_n$ and θ, where, $(\tau \alpha)_n$ is the value of $(\tau \alpha)$ at normal incidence. The variation of $(\tau \alpha)/(\tau \alpha)_n$ with angle of incidence is shown in Figure 3.16.

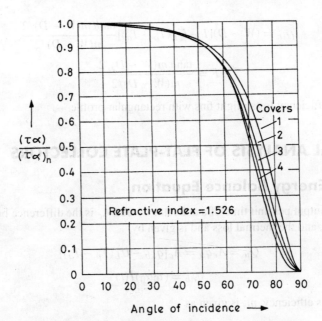

Figure 3.16 Typical $(\tau \alpha)/(\tau \alpha)_n$ curves for 1 to 4 covers (*Duffie and Beckman, 1991*).

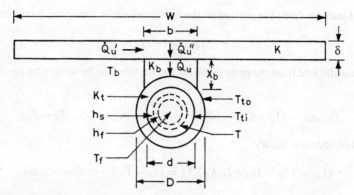

Figure 3.17 Schematic representation of a section through a typical finned-tube (*Ong, 1974*).

3.8.3 Collector Efficiency Factor F'

The collector efficiency factor, F', is defined as the ratio of actual useful heat collection rate to the useful heat collection rate when the collector absorbing plate (T_p) is at the local fluid temperature (T_f), i.e.

$$F' = \frac{\dot{Q}_{\text{useful}}}{\dot{Q}_u|_{T_p=T_f}} = \frac{\dot{Q}_{\text{useful}}}{A_c[\dot{q}_{ab} - U_L(T_f - T_a)]} \tag{3.46}$$

or

$$\dot{Q}_{\text{useful}} = F' A_c[\dot{q}_{ab} - U_L(T_f - T_a)] \tag{3.46a}$$

A general expression for F' can be obtained with the following simplifying assumptions (Figure 3.17) (Ong, 1974):

  i. the incident solar radiation is absorbed only by the plate.

  ii. the bottom surface of the plate and tubes are perfectly insulated and the heat loss occurs only from the top surface of the plate.

  iii. the tubes are brazed onto the plate as shown in Figure 3.17 and the bond has a thermal conductivity K_b which also accounts for the effects of a poor joint.

  iv. thermal inertia effects can be neglected, i.e. a steady state analysis is valid.

The rate of heat transfer from the base of the fin (T_b) to the tube is given by,

$$\dot{Q}'_u = (W - b)L_r F[\dot{q}_{ab} - U_L(T_b - T_a)] \tag{3.47}$$

where, F, the fin efficiency factor (Equation (3.42)) is given as,

$$F = \frac{\tanh m[(W - D)/2]}{[m(W - D)/2]} \quad \text{and} \quad m^2 = \frac{U_L}{K\delta} \tag{3.48}$$

where W is the tube pitch (m), b the bond breadth (m) and L_r the length of riser (m).

The rate of heat transfer from the portion of the fin immediately above the tube bond is,

$$\dot{Q}_u'' = b\, L_r[\dot{q}_{ab} - U_L(T_b - T_a)] \qquad (3.49)$$

The total heat transfer rate from the plate to the fluid in the tube can be written by combining Equations (3.47) and (3.49)

$$\dot{Q}_{useful} = \dot{Q}_u' + \dot{Q}_u'' = L_r[(W-b)F + b][\dot{q}_{ab} - U_L(T_b - T_a)] \qquad (3.50)$$

From simple heat conduction theory

$$\dot{Q}_{ub} = (T_b - T_{t0})b\, L_r K_b/X_b = (T_b - T_{t0})/r_b; \text{ across bond} \qquad (3.51)$$

$$\dot{Q}_{ut} = (T_{t0} - T_{ti})\pi\, A_{mw} L_r K_t/X_t = (T_{t0} - T_{ti})/r_t; \text{ across tube wall} \qquad (3.52)$$

$$\dot{Q}_{us} = (T_{ti} - T_s)\pi\, d_i L_r h_s = (T_{ti} - T_s)/r_s; \text{ across scale} \qquad (3.53)$$

$$\dot{Q}_{uf} = (T_s - T_f)\pi\, d_i L_r h_f = (T_s - T_f)/r_f; \text{ across film} \qquad (3.54)$$

The mean wall area A_{mw} is given by,

$$A_{mw} = \frac{D-d}{\ln(D/d)}$$

where T, K, X, D, d, r and h are temperature, thermal conductivity, thickness, outer and inner diameter of riser, thermal resistance and heat transfer coefficient respectively. From Equations (3.51) to (3.54) one obtains,

$$\dot{Q}_{useful}(r_b + r_f + r_s + r_f) = (T_b - T_f) \qquad (3.55)$$

where $\dot{Q}_{ub} = \dot{Q}_{ut} = \dot{Q}_{us} = \dot{Q}_{uf} = \dot{Q}_{useful}$ in a steady state condition

Defining the fin resistance, r_f' as

$$r_f' = \{L_r[(W-b)F + b]U_L\}^{-1} \qquad (3.56)$$

Equation (3.55) may be expressed as

$$\dot{Q}_{useful} = \frac{1}{U_L r_f'}[\dot{q}_{ab} - U_L(T_b - T_a)] \qquad (3.57)$$

Eliminating T_b from Equations (3.50) and (3.55), we get

$$\dot{Q}_{useful} = W\, L_r F'[\dot{q}_{ab} - U_L(T_f - T_a)] \qquad (3.58)$$

The expression for η_i can be given as

$$\eta_i = \frac{\dot{Q}_{useful}}{A\, I(t)} = F'\left[\frac{\dot{q}_{ab}}{I(t)} - \frac{U_L(T_f - T_a)}{I(t)}\right]$$

or
$$\eta_i = F'\left[\tau_0\alpha_0 - U_L\frac{(T_f - T_a)}{I(t)}\right] \qquad (3.58a)$$

Equation (3.58) is the same as Equation (3.46a) and F' is called the collector efficiency factor, given by

$$F' = \{U_L \, W \, L_r (r_b + r_t + r_s + r_f + r_f')\}^{-1} \tag{3.59}$$

or

$$F' = \frac{1/U_L}{W \left[\frac{1}{U_L[(W-b)F+b]} + \frac{1}{(bK_b/X_b)} + \frac{1}{(\pi \, A_{mw} K_t/X_t)} + \frac{1}{\pi \, dh_s} + \frac{1}{\pi \, dh_f} \right]} \tag{3.60}$$

From the above equation, it is evident that the denominator is the heat transfer resistance from the fluid to the ambient air and is given as $1/U_0$. The numerator is the heat transfer resistance from the absorber plate to the ambient air.

$$F' = \frac{1/U_L}{1/U_0} = U_0/U_L \tag{3.61}$$

The collector efficiency factor, F', is essentially a constant for any collector design and fluid flow rate. F' decreases with an increase in centre to centre distance of the tube and increases with increase in material thickness and the thermal conductivity. Increasing the overall loss coefficient decreases F', while an increase in the fluid to tube heat transfer coefficient increases it. F' and F for some other configurations of collector have been shown in Figure 3.18.

(a)

$$U_L = U_t + U_b$$

$$F' = \frac{1}{\frac{WU_L}{\pi \, Dh} + \frac{WU_L}{C_{\text{bond}}} + \frac{W}{D+(W-D)F}}$$

$$F = \frac{\tanh[m(W-D)]/2}{[m(W-D)]/2}$$

$$m^2 = U_L/k\delta$$

(b)

$$U_L = U_t + U_b$$

$$F' = \frac{1}{\frac{WU_L}{\pi \, Dh} + \left[\frac{1}{\left[\frac{D}{W} + \frac{1}{(WU_L/C_{\text{bond}})} + \frac{W}{(W-D)F} \right]^{-1}} \right]}$$

$$F = \frac{\tanh[m(W-D)]/2}{[m(W-D)]/2}$$

(c)

$$U_L = U_t + U_b$$

$$F' = \frac{1}{\frac{WU_L}{\pi \, Dh} + \frac{W}{D+(W-D)F}}$$

$$F = \frac{\tanh[m(W-D)]/2}{[m(W-D)]/2}$$

(d)

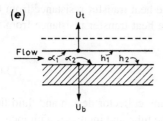

$$U_L = \frac{(U_b + U_t)(h_1 h_2 + h_1 h_r + h_2 h_r) + U_b U_t (h_1 + h_2)}{h_1 h_r + h_2 U_t + h_2 h_r + h_1 h_2} = U_t + U_b$$

$$F' = \frac{h_r h_1 + h_2 U_t + h_2 h_r + h_1 h_2}{(U_t + h_r + h_1)(U_b + h_2 + h_r) - h_r^2}$$

$$h_r = \frac{\sigma \left(T_1^2 + T_2^2\right)(T_1 + T_2)}{(1/\varepsilon_1) + (1/\varepsilon_2) - 1}$$

(e)

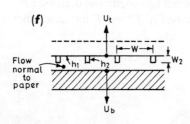

$$U_L = U_t + U_b$$

$$F' = \frac{1}{1 + \dfrac{U_L}{h_1 + (1/h_2 + 1/h_r)^{-1}}}$$

$$h_r = \frac{\sigma \left(T_1^2 + T_2^2\right)(T_1 + T_2)}{(1/\varepsilon_1 + 1/\varepsilon_2 - 1)^{-1}}$$

(f)

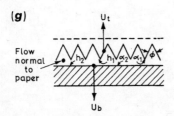

$$F_0' = \frac{1}{1 + \dfrac{U_L}{h_1 + (1/h_2 + 1/h_r)^{-1}}}$$

$$F' = F_0' \left[1 + \frac{(1 - F_0')}{(F_0'/F_p) + (W h_1 / 2 W_2 h_2 F_F)} \right]$$

F_p = fin efficiency of plate

F_f = fin efficiency of fin

(g)

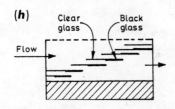

$$U_L = U_t + U_b$$

U_t is based on projected area

$$F' = \frac{1}{1 + \dfrac{U_L}{\dfrac{h_1}{\sin \phi / 2} + \dfrac{1}{1/h_2 + 1/h_r}}}$$

(h)

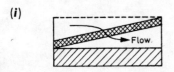

See Selcuk (1971)

(i)

See Hamid and Backman (1971)

(j)

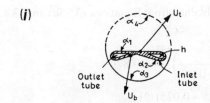

$U_L = U_t + U_b$

$F' = 1/(1 + U_L/h)$

Valid only with negligible heat transfer between inlet and outlet fluid tubes.

(k)

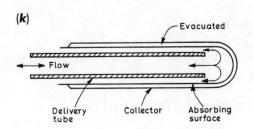

See Eberlain (1976)

(l)

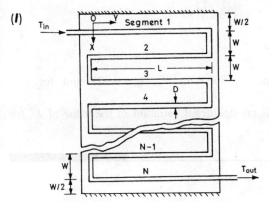

See Abdel–Khalik (1976)

Figure 3.18 An expression of F' and F for various configurations of collector.

EXAMPLE 3.4

A flat-plate collector system has aluminum absorber plate ($K_p = 211$ W/m °C) of thickness 0.35 mm and area 1.5 m² and it has ten riser tubes of diameter 0.025 m each. The length of the tubes being l m, find out the collector efficiency factor F' for this collector, if the convective heat transfer coefficient from inner tube surface to water is 50,100 and 500 W/m² °C. The overall loss coefficient is 7.2 W/m² °C.

Solution

The width of the spacing between the two riser tubes is

$$W = (1.5 - 0.025 \times 10)/10 = 0.125 \text{ m}$$

The value of m and fin efficiency factor (F) can be obtained from Equations (3.35a) and (3.42) respectively as,

$$m = \left(\frac{7.2}{211 \times 0.35 \times 10^{-3}}\right)^{1/2} = 9.87$$

$$F = \frac{\tanh[9.87(0.125 - 0.025)/2]}{9.87 \times (0.125 - 0.025)/2}$$

$$= \frac{\tanh 0.4935}{0.4935} = 0.926$$

The collector efficiency factor (F') (Equation (3.60)), for $h = 50 \text{ W/m}^2 \,^\circ\text{C}$ and $b = D = 0.025 \text{ m}$ is

$$F' = \frac{1/7.2}{0.125\left[\frac{1}{7.2[(0.125-0.025)\times0.926+0.025]} + \frac{1}{3.14\times0.025\times50}\right]}$$

$$= \frac{1}{\frac{0.125}{(0.125-0.025)\times0.926+0.025} + \frac{0.125\times7.2}{3.14\times0.025\times50}}$$

$$= \frac{1}{1.0629 + 0.2293} = 0.774$$

Similarly, for $h = 100 \text{ W/m}^2 \,^\circ\text{C}$, $F' = 0.849$; for $h = 500 \text{ W/m}^2 \,^\circ\text{C}$, $h = 0.921$ and for $h = 1000 \text{ W/m}^2 \,^\circ\text{C}$, $F' = 0.931$

It can be seen from above calculation that there is no significant variation in the value of F' for $h \geq 500 \text{ W/m}^2 \,^\circ\text{C}$.

EXAMPLE 3.5

Calculate the fin efficiency factor and the collector efficiency factor for the following data.

Overall loss coefficient $= 6 \text{ W/m}^2 \,^\circ\text{C}$, Tube Spacing $= 100 \text{ mm}$, Tube diameter $= 8 \text{ mm}$, Plate thickness $= 0.45 \text{ mm}$, Thermal conductivity $= 385 \text{ W/m} \,^\circ\text{C}$, Heat transfer coefficient inside tubes $= 100 \text{ W/m}^2 \,^\circ\text{C}$, Bond resistance $= 0$.

Solution

From Equations (3.35a) and (3.42), we have

$$m = \left(\frac{6}{385 \times 4.5 \times 10^{-4}}\right)^{1/2} = 5.88$$

and,

$$F = \frac{\tanh[5.88(0.10 - 0.008)/2]}{5.88 \times (0.10 - 0.008)/2} = 0.976$$

The collector efficiency factor F' (from Equation (3.60)) is

$$F' = \frac{1/6}{0.10 \left[\frac{1}{6[(0.10-0.008)0.976+0.008]} + \frac{1}{\pi(0.008)\times 100} \right]} = 0.800$$

Also, calculate the collector efficiency factor for the value of heat transfer coefficient inside the tubes as 300 W/m^2 °C and 1000 W/m^2 °C.

The collector efficiency for $h_{fi} = 300$ W/m^2 °C, is given as

$$F' = \frac{1/6}{0.10 \left[\frac{1}{6[(0.10-0.008)0.976+0.008]} + \frac{1}{\pi(0.008)\times 300} \right]} = 0.91$$

Similarly for $h_{fi} = 1000$ W/m^2 °C, we have $F' = 0.96$ and, for $h_{fi} = 2000$ W/m^2 °C, $F' = 0.97$. We see that as the heat transfer coefficient inside the tubes is increased the collector efficiency factor increases. However, not much increase in efficiency is observed when the value of h_{fi} is increased beyond 1000 W/m^2 °C.

EXAMPLE 3.6

Calculate the fin efficiency factor and the collector efficiency factor for the data given below:
Tube spacing = 100 mm, Tube diameter (inside) = 8 mm,
Plate thickness = 0.45 mm, Plate thermal conductivity = 385 W/m°C,
Heat transfer coefficient inside tubes = 300 W/m^2 °C.
$U = 2, 4$ and 8 W/m^2 °C.

Solution

For $U = 2$ W/m^2 °C
From Equation (3.35a), we have

$$m = \left(\frac{2}{385 \times 4.5 \times 10^{-4}} \right)^{1/2} = 3.40$$

From Equation (3.42) we have,

$$F = \frac{\tanh[3.40(0.10-0.008)/2]}{3.40 \times (0.10-0.008)/2} = 0.99$$

Further, F' is given by Equation (3.60) and its value will be

$$F' = \frac{1/2}{0.10 \left[\frac{1}{2[(0.10-0.008)0.99+0.008]} + \frac{1}{\pi(0.008)\times 300} \right]} = 0.96$$

For $U = 4$, m, F and F' are given by

$$m = \left(\frac{4}{385 \times 4.5 \times 10^{-4}} \right)^{1/2} = 4.81$$

$$F = \frac{\tanh[4.81(0.10 - 0.008)/2]}{4.81 \times (0.10 - 0.008)/2} = 0.98$$

and

$$F' = \frac{1/4}{0.10 \left[\frac{1}{4[0.099]} + \frac{1}{\pi(0.008) \times 300} \right]} = 0.94$$

For $U = 8$, m, F and F' are given by

$$m = \left(\frac{8}{385 \times 4.5 \times 10^{-4}} \right)^{1/2} = 6.795$$

$$F = \frac{\tanh[6.795(0.10 - 0.008)/2]}{4.81 \times (0.10 - 0.008)/2} = 0.968$$

and

$$F' = \frac{1/8}{0.10 \left[\frac{1}{8[0.097]} + \frac{1}{\pi(0.008) \times 300} \right]} = 0.879$$

Hence, we see that on increase of an overall loss coefficient decreases the collector efficiency factor F'.

3.8.4 Temperature Distribution in Flow Direction

The useful heat gain per unit flow length, transferred to the flowing fluid, is given by

$$\dot{q}_u = W F'[\dot{q}_{ab} - U_L(T_f - T_a)] \tag{3.62}$$

From Figure 3.19(a), the energy balance on the flowing fluid along x-direction through a single tube of length Δx can be written as

$$\left(\frac{\dot{m}}{n_0} \right) C_f T_f \big|_x - \left(\frac{\dot{m}}{n_0} \right) C_f T_f \big|_{x+\Delta x} + \dot{q}_u \Delta x = 0 \tag{3.63}$$

where $\dot{m}$ is the total collector flow rate, n_0 is the number of riser tubes and C_f is the specific heat of fluid. Substituting the value of $\dot{q}_u$ from Equation (3.62) in (3.63) we get

$$\dot{m} C_f \frac{dT_f}{dx} - n_0 W F'[\dot{q}_{ab} - U_L(T_f - T_a)] = 0 \tag{3.64}$$

Assuming F' and U_L to be constant, the solution of Equation (3.64) with the boundary condition $T_f = T_{fi}$ at $x = 0$ is

$$\frac{T_f - T_a - (\dot{q}_{ab}/U_L)}{T_{fi} - T_a - (\dot{q}_{ab}/U_L)} = \exp \left[\frac{-U_L n_0 W F' x}{\dot{m} C_f} \right] \tag{3.65}$$

The outlet fluid temperatures T_{fo} at $x = L_r$, can be obtained as

$$T_{fo} = T_f|_{x=L_r} = [(\dot{q}_{ab}/U_L) + T_a] + [T_{fi} - T_a - (\dot{q}_{ab}/U_L)]$$
$$\exp\left[-A_c U_L F'/(\dot{m} C_f)\right] \tag{3.66}$$

where $A_c(= n_0 W L_r)$ is the collector area and L_r the length of riser in the flow direction.

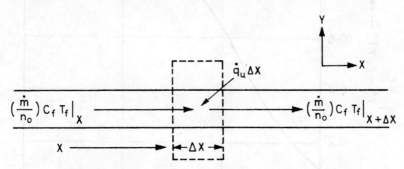

Figure 3.19(a) Energy balance on water element.

3.8.5 Collector Heat Removal Factor, F_R

The collector heat removal factor is defined as the ratio of the actual useful energy gain to the useful energy gain if the entire collector were at the fluid inlet temperature T_{fi} (forced circulation flow) and can be expressed as,

$$F_R = \frac{\dot{m} C_f (T_{fo} - T_{fi})}{A_c[\dot{q}_{ab} - U_L(T_{fi} - T_a)]} \tag{3.67}$$

The above equation can also be written as

$$F_R = [\dot{m} C_f/(A_c U_L)]\left[1 - \frac{T_{fo} - T_a - (\dot{q}_{ab}/U_L)}{T_{fi} - T_a - (\dot{q}_{ab}/U_L)}\right] \tag{3.68}$$

From Equations (3.65) and (3.68), F_R can be written as,

$$F_R = [\dot{m} C_f/(A_c U_L)][1 - \exp(-A_c U_L F'/(\dot{m} C_f))] \tag{3.69a}$$

In order to avoid heat removal factor (F_R) [Equation (3.69a)] graphically, it is convenient to define the collector flow factor F'' as

$$F'' = \frac{F_R}{F'} = \frac{\dot{m} C_p}{A_c U_L F'}\left[1 - \exp\left(-\frac{A_c U_L F'}{\dot{m} C_f}\right)\right] \tag{3.69b}$$

The variation of collector flow factor (F'') is shown in Figure 3.19(b). It is inferred that F'' becomes constant after $\dot{m} C_f/A_c U_L F'$, the dimensionless collector capacitance rate, becomes more than 10.

From Equation (3.67), the actual useful energy collected by fluid is given by

$$\dot{Q}_{useful} = A_c F_R[\dot{q}_{ab} - U_L(T_{fi} - T_a)] \tag{3.70}$$

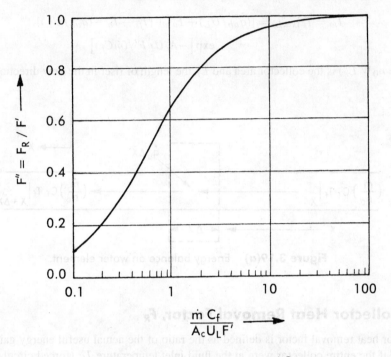

Figure 3.19(b) Variation F'' with $\dfrac{\dot{m}C_p}{A_c U_L F'}$.

EXAMPLE 3.7(a)

Calculate the net rate of useful energy per m^2 for the following parameters:

i. The overall heat loss coefficient $(U_L) = 6.0$ W/m^2 °C and $F' = 0.8$ (Example 3.5)
ii. $\dot{m} = 0.35$ kg/s and $C_f = 4190$ J/kg°C
iii. $I(t) = 500$ W/m^2 and $\alpha_0 \tau_0 = 0.8$
iv. $T_{fi} = 60$°C and $T_a = 40$°C

Solution

The heat removal factor is given by

$$F_R = [\dot{m}C_f/(A_c U_L)][1 - \exp(-A_c U_L F'/(\dot{m}C_f))]$$

$$= [0.35 \times 4190/(1 \times 6)][1 - \exp(-1 \times 6 \times 0.8/\{0.35 \times 4190\})] = 0.7986$$

The net rate of useful energy per m^2 can be calculated from Equation (3.70) as

$$\dot{q}_u = F_R[\alpha_0 \tau_0 I(t) - U_L(T_{fi} - Ta)]$$

$$= 0.7986[0.8 \times 500 - 6(60 - 40)] = 223.6 \, \text{W/m}^2$$

EXAMPLE 3.7(b)

Find out the rate of useful energy per m^2 for Example 3.7(a) with the mass flow rate of 0.035 kg/sec.

Solution

The flow rate factor can be evaluated as

$$F_R = \frac{(0.035 \times 4190)}{(1 \times 6)} \left[1 - \exp\left(-\frac{1 \times 6 \times 0.8}{0.035 \times 4190}\right)\right] = 0.787$$

The net rate of useful energy per m^2 will be

$$\dot{q}_u = 0.787[0.8 \times 500 - 6(60 - 40)]$$

$$= 220.36 \, \text{W/m}^2$$

It is clear that the change in flow rate have no effect on $\dot{q}_u$ for a given design and climatic parameters of a collector.

The factor F_R reduces the useful energy gain from the expected value, had the whole collector absorber plate been at fluid inlet temperature, to what it actually is. The temperature rise through the collector decreases as the mass flow rate through the collector increases. This results in lower losses and consequently the actual useful energy gain goes up. Heat removal factor F_R can never exceed the collector efficiency factor F'.

The models given by Hottel and Whiller (1958), Bliss (1959) and Whiller (1953) give the value of F_R as

$$F_R = F'[1 - \exp(-N_0)]/N_0 \tag{3.71}$$

where $F' = (U_0/U_L)$ is the efficiency factor of the collector and $N_0 = A_c U_0/(\dot{m} C_f)$ is a dimensionless design parameter.

The analysis leading to Equation (3.71) neglects axial conduction in both the fluid and the receiver. In the case of the fluid, this assumption seems reasonable as the fluids commonly used in collectors are of low conductivity. However, the collector plates are of good conducting material, as such, it does not seem appropriate to neglect axial conduction in the receiver under all conditions. Axial conduction in the collector plate causes a flow of heat in the direction opposite to the flow of fluid, thus reducing the fluid outlet temperature and the overall efficiency of the system. In some cases, the prediction made by Hottel, Whiller and Bliss model is as much as 30 percent high.

Taking the axial conduction of heat into consideration, Phillips (1979) derived analytically an expression for F_R as

$$F = F'[1 - \exp(-N_0)]/N_0 \text{ when } K = 0$$

and,

$$F_R = \frac{F'[1 - \exp[-N_0/(1 - F')]]}{N_0 + F'[1 - \exp[N_0/(1 - F')]]} \text{when } K \to \infty$$

For $N_0 = \infty$, $F_R = F'/N_0$ and for $N_0 = 0$, $F_R = F'$

Phillips also developed empirically an approximate expression for F_R over the full range of parameters N_0, F' and K as,

$$F_R = \frac{F'[1 - \exp[-N_o/(1 - F'K')]]}{N_0 + F'K'[1 - \exp[N_0/(1 - F'K')]]} \tag{3.72a}$$

where $K' = K/(K + 0.11)$.

The error due to neglecting axial conduction, in the Hottel, Whiller and Bliss models can be represented as, conduction error and its value is,

$$\text{Conduction error} = \frac{F''(N_0, 0) - F''(N_0, F', K')}{F''(N_0, F', K')} \tag{3.72b}$$

where $F''(N_0 F' K') = F_R/F$.

The practical upper limit on the error for collectors with $F'K' = 0.8$ which could result from neglecting axial conduction is about 22 percent. For collectors having $F'K'$ products in the range of 0.1–0.5, the error is not more than 12 percent, by neglecting axial conduction.

From Equation (3.70), an instantaneous thermal efficiency of a flat plate collector can be defined as follows:

$$\eta_i = \frac{\dot{Q}_{\text{useful}}}{A_c I(t)} = F_R \left[(\alpha\tau) - U_L \frac{T_{fi} - T_a}{I(t)} \right] \tag{3.73a}$$

The variation of η_i with $\frac{T_{fi} - T_a}{I}$ has been shown in Figure 3.8(b). Equation (3.73a) is known as the Hottel-Whiller-Bliss equation of flat plate collector. This is also known as characteristic equation of a flat plate collector.

3.8.6 Threshold Radiation Flux

The threshold radiation flux (I_{th}) is the lowest radiation needed to allow the collector to operate for thermal heating. This will be achieved by equating Equation (3.43a) is exactly zero

$$I_{\text{th}} = \frac{U_L(T_p - \bar{T}_a)}{\alpha_0 \tau_0} \tag{3.73b}$$

This gives the minimum level of solar flux to operate the collector for thermal heating. If solar radiation is lower than I_{th}, no thermal heating will take place.

The threshold solar flux (I_{th}) mainly depends on

   i. the value of U_L, it should be as low as possible

  ii. the value of product $\alpha_0 \tau_0$, it should be as high as possible.

EXAMPLE 3.8

Find out threshold radiation flux for Example 3.1(b) for $\alpha_0 \tau_0 = 0.80, 0.60, 0.40$ and 0.20.

Solution

From Example 3.1(b), $T_p = 100°C$, $T_a = 16°C$ and $U_L = 6 \text{ W/m}^2 °C$.
Therefore, the threshold radiation flux levels from Equation (3.73b) are

$$I_{th} = \frac{6(100 - 16)}{0.8} = 630 \text{ W/m}^2 \quad \text{for } \alpha_0 \tau_0 = 0.8$$

$$= 840 \text{ W/m}^2 \quad \text{for } \alpha_0 \tau_0 = 0.6$$

$$= 1260 \text{ W/m}^2 \quad \text{for } \alpha_0 \tau_0 = 0.4$$

$$= 2520 \text{ W/m}^2 \quad \text{for } \alpha_0 \tau_0 = 0.2$$

This indicates that solar radiation can't be used for thermal heading for $\alpha_0 \tau_0 = 0.2$ and 0.4 due to higher value of I_{th}.

3.9 CONFIGURATION OF FPC CONNECTION

A large bank of collectors can be formed by series, parallel, or mixed combination of collectors in one or multiple rows.

3.9.1 Collectors Connected in Series

It is convenient to define array characteristics for the series connected modules that are analogues to $F_R(\tau_0 \alpha_0)$ and $F_R U_L$ for single module.

If two sets of collector (each set having m collectors in parallel) 1 and 2, are connected in series, the useful heat output of the combination is,

$$\dot{Q}_{um1} + \dot{Q}_{um2} = A_{m1} F_{Rm1}[(\tau_0 \alpha_0)_1 I(t) - U_{Lm1}(T_{fi} - T_a)]$$

$$+ A_{m2} F_{Rm2}[(\tau_0 \alpha_0)_2 I(t) - U_{Lm2}(T_{fo1} - T_a)] \tag{3.74}$$

where
$$F_{Rm1} = \frac{\dot{m}C_f}{A_{cm1}U_{Lm1}}\left[1 - \exp\left(-\frac{A_{cm1}U_{Lm1}F'}{\dot{m}C_f}\right)\right]$$

and
$$F_{Rm2} = \frac{\dot{m}C_f}{A_{cm2}U_{Lm1}}\left[1 - \exp\left\{-\frac{A_{cm2}U_{Lm2}F'}{\dot{m}C_f}\right\}\right]$$

A_{m1}, A_{m2} are the area of each set of collectors and T_{fi} is the inlet fluid temperature for the first set of collectors, T_{fo1} is the outlet temperature of the first set of collectors and is also the inlet temperature to the second set of collectors.

$$T_{fo1} = T_{fi} + [\dot{Q}_{um1}/\dot{m}C_f] \tag{3.75}$$

By eliminating T_{fo1} from Equations (3.74) and (3.75), the useful heat output of the combination can be given as

$$\dot{Q}_{um,1+2} = [A_{m1}F_{Rm1}(\tau_0\alpha_0)_1(1 - K_K) + A_{m2}F_{Rm2}(\tau_0\alpha_0)_2]I(t)$$

$$-[A_{m1}F_{Rm1}U_{Lm1}(1 - K_K) + A_{m2}F_{Rm2}U_{Lm2}](T_{fi} - T_a) \tag{3.76}$$

where
$$K_K = (A_{m2}F_{Rm2}U_{Lm2})/(\dot{m}C_f) \tag{3.77}$$

The form of Equation (3.76) suggests that the combination of the two sets can be considered as a single collector with the characteristics

$$A = A_{m1} + A_{m2} \tag{3.78}$$

$$F_R(\tau_0\alpha_0) = \frac{A_{m1}F_{Rm1}(\tau_0\alpha_0)(1 - K_K) + A_{m2}F_{Rm2}(\tau_0\alpha_0)_2}{A} \tag{3.79}$$

and
$$F_RU_L = \frac{A_{m1}F_{Rm1}U_{Lm1} + A_{m2}F_{Rm2}U_{Lm2}}{A} \tag{3.80}$$

If three or more collector modules are placed in series, these equations can be used for the first two modules to define a new equivalent first collector. The equations are applied again with this equivalent first collector and the third module becoming the second module. The process can be repeated for as many modules as desired.

If the two sets of collectors are identical, the above equations become

$$F_R(\tau_0\alpha_0) = F_{Rm1}(\tau_0\alpha_0)_1[1 - K_K/2] \tag{3.81}$$

$$F_RU_L = F_{Rm1}U_{Lm1}[1 - K_K/2] \tag{3.82}$$

For N identical set of collectors in series, Oonk *et al.* (1979) have shown that repeated applications of Equations (3.79) and (3.80) give,

$$F_R(\tau_0\alpha_0) = F_{Rm1}(\tau_0\alpha_0)_1\left[\frac{1 - (1 - K_K)^N}{NK_K}\right] \tag{3.83}$$

and
$$F_RU_L = F_{Rm1}U_{Lm1}\left[\frac{1 - (1 - K_K)^N}{NK_K}\right] \tag{3.84}$$

EXAMPLE 3.9

Calculate flow rate factor for two identical collectors connected in series for same configuration of Example 3.7.

Solution

From Equation (3.77),

$$K_K = \frac{A_{m2} F_{Rm2} U_{Lm2}}{\dot{m} C_f} = \frac{1 \times 0.7986 \times 6}{0.35 \times 4190} = 0.00326$$

From Equation (3.82), one has

$$F_{R2} = F_{R1} \left[1 - \frac{K_{K'}}{2} \right] \text{ since } U_L = U_{L1} = U_{L2}$$

$$= 0.7986 \left[1 - \frac{0.00326}{2} \right] = 0.7973$$

There is not much change in flow rate factor for two identical collectors connected in series.

3.9.2 Outlet Temperature at Nth Collector

Equation (3.66) can be expressed as

$$T_{fo1} = [\dot{q}_{ab}/U_{L1} + T_a] - [\dot{q}_{ab}/U_{L1} + T_a] \exp[-A_{c1} U_{L1} F'/\dot{m} C_f]$$

$$+ T_{fi} \exp[-A_{c1} U_{L1} F'/\dot{m} C_f] \tag{3.85}$$

For a number of collectors connected in series, the inlet temperature of the second will be the outlet temperature of the first, the inlet temperature of the third will be the outlet temperature of the second and so on. Hence, for a system of N-collectors connected in series, the outlet fluid temperature from the Nth collector can be expressed in terms of the inlet temperature of the first collector as given below.

From Equation (3.85), the outlet fluid temperature of the second collector T_{fo2}, can be given as,

$$T_{fo2} = [\dot{q}_{ab}/U_{L2} + T_a] - [\dot{q}_{ab}/U_{L2} + T_a] \exp[-A_{c2} U_{L2} F_2'/(\dot{m} C_f)]$$

$$+ T_{fi2} \exp[-A_{c2} U_{L2} F_2'/(\dot{m} C_f)]$$

As $T_{fi2} = T_{fo1}$,

$$T_{fo2} = [q_{ab}/U_{L2} + T_a][1 - \exp\{-(A_{c2} U_{L2} F_2')/\dot{m} C_f\}] + T_{fi1} \exp\{-(A_{c1} U_{L1} F_1' + A_{c2} U_{L2} F_2')\}/\dot{m} C_f$$

$$+ [(\dot{q}_{ab}/U_{L1} + T_a)[1 - \exp\{-(A_{c1} U_{L1} F_1')/\dot{m} C_f\}]] \exp\{-(A_{c2} U_{L2} F_2')/\dot{m} C_f\} \tag{3.86}$$

Similarly, the outlet fluid temperature for Nth collector can be given as

$$T_{foN} = [(\dot{q}_{ab}/U_{Ln}) + T_a][1 - \exp\{-A_{CN}U_{LN}F'_N/\dot{m}C_f\}]$$

$$+[(\dot{q}_{ab}/U_{LN-1}) + T_a][1 - \exp\{-A_{CN-1}U_{LN-1}F'_{N-1}/\dot{m}C_f\}]\exp\{-A_{CN}U_{LN}F'_N/\dot{m}C_f\}$$

$$+[(\dot{q}_{ab}/U_{L1}) + T_a][1 - \exp\{-A_{C1}U_{L1}F'_l/\dot{m}C_f\}]$$

$$\times \exp\{-(A_{C2}U_{L2}F'_2 + \cdots + A_{CN}U_{LN}F'_N)/\dot{m}C_f\}$$

$$+T_{fi}\exp\{-(A_{c1}U_{L1}F'_1 + \cdots A_{CN}U_{LN}F'_N)/\dot{m}C_f\} \tag{3.87}$$

If all the collectors are identical, i.e.

$$U_{L1} = U_{L2} = \ldots\ldots\ldots = U_{LN} = U_L$$

$$A_{C1} = A_{C2} = \ldots\ldots\ldots = A_{CN} = A_C$$

and,

$$F'_1 = F'_2 = \ldots\ldots\ldots = F'_N = F'$$

Then Equation (3.87) reduces to

$$T_{foN} = \left(\frac{\dot{q}_{ab}}{U_L} + T_a\right)\left\{1 - \exp\left\{-\frac{NA_cU_LF'}{\dot{m}C_f}\right\}\right\} + T_{fi}\exp\left\{-\frac{NA_cU_LF'}{\dot{m}C_f}\right\} \tag{3.88}$$

EXAMPLE 3.10

Calculate the outlet fluid temperature at outlet of two and four collector connected in series for same configuration of Example 3.7 with the following climatic and design parameters

$$I(t) = 500 \text{ W/m}^2, T_a = 40°C \quad \text{and} \quad \alpha_0\tau_0 = 0.8$$

Solution

(a) For two identical collectors connected in series
From Equation (3.88),

$$T_{fo2} = \left[\frac{0.8 \times 500}{6.0} + 40\right]\left\{1 - \exp\left(-\frac{2 \times 1 \times 6 \times 0.8}{0.35 \times 4190}\right)\right\}$$

$$+60\exp\left\{-\frac{2 \times 1 \times 6 \times 0.8}{0.35 \times 4190}\right\} = 60.30°C$$

(b) For four identical collectors connected in series

$$T_{fo4} = 60.6°C$$

This indicates that the outlet temperature at end of 4th collector is higher than two collectors connected in series. However, the rise in temperature is insignificant due to large value of $\dot{m}$. In this case, the water is not getting sufficient time for thermal heating.

EXAMPLE 3.11

Determine the convective heat transfer coefficient for air flow in a channel 1 m wide and 2 m long. The channel thickness is 10 mm and the air flow rate is 0.025 kg/s. The average air temperature is 35°C. What is the heat transfer coefficient if the channel thickness is halved? What is the heat transfer coefficient if the flow rate is halved?

Solution

At 35°C, the air properties are: $v = 1.88 \times 10^{-5}$ m²/s and $K = 0.0268$ W/mK (see Appendix III). The hydraulic diameter D_h is twice the plate spacing t and the Reynold number can be expressed in terms of the flow rate per unit width. The Reynold number (from Equation (2.38b)) is

$$Re = \frac{\rho v D_h}{v} = \frac{2\rho v t W}{wv} = \frac{2\dot{m}}{wv} = \frac{2 \times 0.025}{1 \times 1.88 \times 10^{-5}} = 2660$$

The flow is turbulent and the value of Nusselt number for fully developed turbulent flow with one side heated and the other insulated is given by

$$Nu = 0.0158 \, Re^{0.8} = 0.0158(2660)^{0.8} = 8.68$$

(a) The heat transfer coefficient (from Equation (2.38a)) is $h = Nu \, K / D_h$ is

$$h = 8.68 \times 0.0268/2 \times 1.0 \times 10^{-2} = 11.63 \, \text{W/m}^2 \, °\text{C}.$$

(b) If the channel thickness is halved, the Reynold number remains the same but the heat transfer coefficient doubles.

(c) If the flow rate is halved, the Reynold number becomes 1330 (laminar flow). Then,

$$Nu = 4.9 + \frac{0.0606(Re \, Pr \, D_h/L)^{1.2}}{1 + 0.0909(Re \, Pr \, D_h/L)^{0.7} Pr^{0.17}}$$

Here, $Re \, Pr. \, D_h/L = 1330 \times 0.7 \times 0.02/2 = 9.31$.
From Equation (2.38a), the heat transfer coefficient $h = Nu \, K/D_h$. The heat transfer coefficient is 7.40 W/m² °C.

3.9.3 Collectors Connected in Parallel

An experiment was conducted at IIT Delhi during the month of November/December, 1981 on twelve collectors connected in parallel with each collector having an effective area of 1.5 m². The temperature were measured at six points in upper header as shown in Figure 3.20(a). The temperature at inlet was also measured for the comparison. The data recorded at 9 AM, 12 noon and 3 PM shown in Figure 3.20(a). It is clear from the data that there is not much variation in temperature at different points at 9 AM due to low value of solar intensity and foggy condition. This may linear due to laminar flow in riser as well as header. However at 12 noon and 3 PM, there is sudden rise in upper header temperature after two collector and then decreases upto 8th collector and then rises. It is mainly due to turbulent flow in upper header at peak hour.

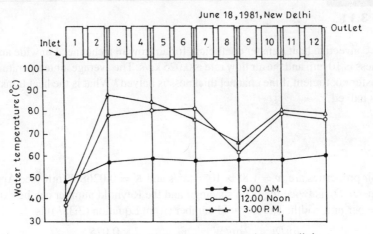

Figure 3.20(a) Collectors connected in parallel.

3.9.4 Collectors Connected in Mixed Mode

Table 3.2 gives the value of K_K, F_{Rm} and F_{Rmn} for collectors connected in mixed mode (Fig. 3.20(b)). From Table 3.2 we can infer that six panels (each panel consists of two collectors connected in parallel) connected in series gives $F_{Rm} = 0.8666$ and $F_{Rmn} = 0.8034$. This is in accordance with the present practices generally adopted for mixed mode.

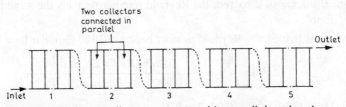

Figure 3.20(b) Collectors connected in parallel and series.

3.10 EFFECT OF HEAT CAPACITY IN FLAT-PLATE COLLECTOR

In order to see the effect of heat capacity of absorber plate and glass cover, the following assumptions have been made:

i. the temperatures of the absorber plate, the water in the tube, and the back insulation are the same($U_L \cong U_t$, back and edge losses are neglected)

ii. the glass cover is at uniform temperature

The energy balance of the absorber plate can be written (see Equation (3.43a)) as

$$(mC)_p \frac{dT_p}{dt} = A_c[\dot{q}_{ab} - h_1(T_p - T_g)] \tag{3.89}$$

where h_1 is the heat transfer coefficient from the plate to glass cover (see Example 3.2).

TABLE 3.2 Values of K_k, F_{Rm} and F_{Rmn} for collectors connected in mixed combination

Values of n'	Different combination of m and n	K_K	F_{Rm}	F_{Rmn}
	$m' = m = 12, n = 1$	–	0.8034	0.8034
	$m' = 12, m = 6, n = 2$	0.0884	0.8405	0.8033
$n' = 1$	$m' = 12, = m = 4, n = 3$	0.0598	0.8534	0.8034
	$m' = 12, m = 3, n = 4$	0.0452	0.8600	0.8034
	$m' = 12, m = 2, n = 6$	0.0304	0.8666	0.8034
	$m' = 12, m = 1, m = 12$	0.0152	0.8733	0.8034
	$m = 6, n = 1$	–	0.803	0.803
	$m = 3, n = 2$	0.0886	0.8404	0.8032
$n' = 2$	$m = 2, n = 3$	0.0599	0.8534	0.8037
	$m = 1, n = 6$	0.0927	0.8665	0.8032
	$m = 4, n = 1$	–	0.803	0.803
$n' = 3$	$m = 2, n = 2$	0.0885	0.8404	0.8033
	$m = 1, n = 4$	0.0435	0.8599	0.8031
$n' = 4$	$m = 3, n = 1$	–	0.803	0.803
	$m = 1, n = 3$	0.0599	0.8534	0.8036
$n' = 6$	$m = 2, n = 1$	–	0.803	0.803
	$m = 1, n = 2$	0.0885	0.8404	0.8036
$n' = 12$	$m = n = 1$	–	0.8032	0.8032

$m \to m$ collectors connected in parallel form one module
$n \to n$ module connected in series form one panel
$n' \to n'$ panel connected in parallel.

Similarly, the energy balance equation for cover can be written as

$$(mC)_c \frac{dT_c}{dt} = A_c[h_1(T_p - T_g) - h_2(T_g - T_a)] \tag{3.90}$$

where h_2 is convective and radiative heat transfer coefficients from glass cover (T_g) to ambient air (T_a). In a steady state condition, if we assume

$$U_L(T_p - T_a) = h_2(T_g - T_a) \tag{3.91}$$

Differentiating Equation (3.91) with respect to time, we have

$$\frac{dT_g}{dt} = \frac{U_L}{h_2} \frac{dT_p}{dt} \tag{3.92a}$$

After adding Equations (3.89) and (3.90), substitute Equation (3.92a), we have

$$\left[(mC)_p + \frac{U_L}{h_2}(mC)_c \right] \frac{dT_p}{dt} = A_c[\dot{q}_{ab} - U_L(T_p - T_a)] \tag{3.92b}$$

If the term in the square bracket is defined as an effective heat capacity, $(mC)_e$, of the collector, then above equation can be written as

$$\frac{dT_p}{[\dot{q} - U_L(T_p - T_a)]} = \frac{A_c}{(mC)_e} dt \tag{3.93}$$

or
$$\frac{\dot{q}_{ab} - U_L(T_p - T_a)}{\dot{q}_{ab} - U_L(T_{p0} - T_a)} = \exp\left(-\frac{A_c U_L}{(mC)_e}t\right) \tag{3.94a}$$

The solution of above equation is

$$T_p = \left(\frac{\dot{q}_{ab}}{U_L} + T_a\right) \times \left\{1 - \exp\left(-\frac{A_c U_L}{(mC)_e}\right)\right\} + T_{p0}\exp\left(-\frac{A_c U_L}{(mC)_e}t\right) \tag{3.94b}$$

where T_{p0} is the plate temperature at $t = 0$.

The expression for energy stored in time ($\Delta t = t - 0$) for the absorber plate, water and insulation can be written as

$$\dot{Q}_L = (mC)_e(T_p - T_{p0}) = (mC)_e\left\{\frac{\dot{q}_{ab}}{U_L} + T_a\right\}\left[1 - \exp\left(-\frac{A_c U_L}{(mC)_e}t\right)\right] \tag{3.95}$$

3.11 OPTIMUM INCLINATION OF FLAT-PLATE COLLECTOR

As we saw in Chapter 1 (Equation (1.18)) that there is a variation of solar radiation with inclination of the surface for a given latitude, orientation, the time of the day and day of the year. Hence, there should be an optimum inclination to receive maximum radiation. On the basis of literature survey, a thumb rule for Indian condition has been made. According to this rule, an optimum inclination of the surfaces receiving maximum radiation is given by

$$\beta_{optimum} = \varphi \pm 15° \tag{3.96}$$

where +ve sign refers for the winter condition and −ve refers for summer condition.

Since the design of collector is made from water heating point of view during the winter days hence the optimum inclination of collector for region around Delhi will be $28°35' + 15° = 43°35'$. For convenience, it is chosen as 45° from the horizontal facing south in the northern hemisphere. The surface should face the north direction in the southern hemisphere due to motion of sun.

For the year round performance, the optimum tilt is 0.9 times the latitude of the location.

3.12 EFFECT OF DUST IN FLAT-PLATE COLLECTOR

The deposition of the dust on the transparent cover reduces the transmittance of the cover. It is very difficult to generalize its effects due to its dependance on the type of the glazing, angle of incidence, wind speed, rainfall and place etc. The plastic cover attracts more dust due to its electrostatic nature than the window glass cover. It has been reported that the transmittance of glass cover is reduced by an average of 8–10 percent at an angle of 45° in the region of North India. The transmittance of the glass cover will be less affected for horizontal surface and clear sky condition of cold region (Srinagar, Shimla and Leh).

EXAMPLE 3.12

Calculate the hourly and daily useful gain and efficiency of 1 m^2 collector of the following design and climatical parameters:

(a) *Design parameters*: $\alpha_0\tau_0 = 0.8$, $\dot{m} = 0.035$ kg/s, $F' = 0.8$, $U_L = 6$ W/m^2 °C and $F_R = 0.787$ and $T_{fi} = 40$°C [Example 3.8(b)].

(b) *Climatic parameters*: The climatic data for January of Srinagar are given below:

Time	$I(t)$ (W/m^2)	$\dot{q}_{ab}$ (W/m^2)	T_a °C	$\dot{q}_L$ (W/m^2)	$\dot{q}_u$ (W/m^2)	q_u (MJ/m^2)	$\eta = \%$
7–8	270	216.0	−1.5	249	0	0	0
8–9	363	290.4	−0.8	244.8	35.88	0.12	0.09
9–10	486	388.8	0.1	239.4	117.57	0.42	0.242
10–11	566	452.8	1.2	232.8	173.14	0.62	0.306
11–12	593	474.4	2.4	225.6	195.80	0.72	0.330
12–13	566	452.8	3.2	220.8	182.23	0.651	0.322
13–14	486	388.8	4.0	216.0	135.99	0.49	0.280
14–15	363	290.4	4.3	214.2	59.97	0.215	0.165
15–16	207	216.0	4.3	214.2	1.42	0.005	0.007
16–17	100	80.0	3.9	216.6	0	0	
Total	4000				902	3.26	0.225

Solution

From Equation (3.43a), $\dot{q}_{ab}$ for 7–8 hrs is

$$\dot{q}_{ab} = 0.8 \times 270 = 216 \text{ W/m}^2$$

Similarly for other hours $\dot{q}_{ab}$ is calculated and it is given in the above table.

The hereby losses is based on an inlet temperature of 40°C and it can be calculated by using Equation (3.24) for 10–11 hrs as

$$\dot{q}_L = 6(40 - 1.2) = 232.8 \text{ W/m}^2$$

The useful energy gain ($\dot{q}_u$) in W/m^2 [Equations (3.2a) and (3.2b)] is

$$\dot{q}_u = 0.787[452.8 - 232.8] = 173.14 \text{ W/m}^2$$

and in MJ/m^2 is

$$q_u = 173.14 \times 3600 = 0.62 \times 10^6 \text{ J/m}^2 = 0.62 \text{ MJ/m}^2$$

The collector thermal efficiency for this hour (Equation (3.43b)) is

$$\eta = \frac{\dot{q}_u}{I(t)} = \frac{173.14}{566} = 0.306$$

The day-long overall collector thermal efficiency (Equation (3.44)) is

$$\eta_c = \frac{902 \times 3600}{4000 \times 3600} = 0.225$$

The daily useful gain $= 3.26$ MJ/m^2.

EXAMPLE 3.13

Calculate the hourly and daily useful gain for Example 3.12 by considering effect of dust and absorption of a single glass cover. The absorber plate has a flat black (non-selection) absorbing surface.

Solution

The glass absorbs about four percent of incident solar radiation and 27 percent of this is not lost. In other term, 27 percent of four percent means about one percent of incident solar radiation is retained by the glass cover. Thus

$$(\alpha_0 \tau_0)_e = 1.01(\alpha_0 \tau_0) \text{ (From Equation (3.45}b))$$

Due to dust, the absorbed solar radiation is reduced by eight percent at $45°$ inclination (Section 3.12). The net effect is to decrease $I(t)$ by nine percent. The table given below gives new value of $I(t)$. The hourly and daily useful gain is calculated as done in Example 3.11 with same value of $F_R = 0.787$ and $U_L = 6 \text{ W/m}^2$

Time	New $I(t) = 0.93$ of $I(t)$ of Example 3.11	T_a (°C)	$\dot{q}_{ab}$ (W/m²)	$\dot{q}_L$ (W/m²)	$\dot{q}_u$ (W/m²)	q_u (MJ/m²)
7–8	246.61	−1.5	197.29	249	0	
8–9	330.33	−0.8	264.26	244.8	15.32	0.07
9–10	442.26	0.1	353.81	239.4	90.04	0.41
10–11	515.06	1.2	412.05	232.8	141.07	0.65
11–12	539.63	2.4	431.71	225.6	162.21	0.74
12–13	515.06	3.2	412.05	220.8	150.51	0.69
13–14	442.26	4.0	353.81	216.0	108.46	0.496
14–15	330.33	4.3	264.26	214.2	39.39	0.18
15–16	188.37	4.3	150.69	214.2'	0	0
16–17	91.00	3.9	72.8	216.6	0	0
Total					707.00	3.24

In this case, the day-long overall collector thermal efficiency is

$$\eta_c = \frac{707.00 \times 3600}{4000 \times 3600} = 0.177(17.7\%)$$

The above calculation shows that the overall daily thermal efficiency of collector is reduced from 21.7 to 17.7 percent due to absorption of radiation by the glass cover and deposition of dust on the glass cover due to bad weather condition.

EXAMPLE 3.14

Estimate the reduction in useful energy gain due to heat capacity effect of the design and climatic parameters of collector of Example 3.12. The plate and tubes are copper. The collector has following specifications:

a. Plate thickness   0.0005 m (0.5 mm)
b. Inner diameter of the riser   0.01 m (10.0 mm)
c. Riser spacing   0.15 m (150.0 mm)
d. Thickness of the glasscover   0.035 m (3.5 mm)
e. Thickness of back insulation   0.05 m (50.0 mm)

The collector materials have following properties:

Materials	Specific heat (J/kg°C)	Density (kg/m³)
Copper	480	8800
Glass	800	2500
Insulation	800	50

Solution

The flat plate collector heat capacity includes glass plate, tube, water in tubes and insulation. These can be obtained for 1 m² collector area as follows:

i. The heat capacity of the glass $= 1 \times 0.0035 \times 2500 \times 800 = 7000 \, \text{J/m}^2 \, {}^\circ\text{C}$.

ii. The heat capacity of insulation $= 1 \times 0.05 \times 50 \times 800 = 2000 \, \text{J/m}^2 \, {}^\circ\text{C}$. The bottom of insulation is exposed to the ambient remains near ambient air temperature so that the effective insulation heat capacity is one-half of its actual value. Therefore, effective heat capacity of insulation $= 1000 \, \text{J/m}^2 \, {}^\circ\text{C}$.

iii. The heat capacity of absorber plate $= 1 \times 0.0005 \times 8800 \times 480 = 2112 \, \text{J/m}^2 \, {}^\circ\text{C}$.

iv. The heat capacity of one tube $= 2 \times 3.14 \times 0.005 \times 1 \times 0.0005 \times 8800 \times 480 = 66.3 \, \text{J/m}^2 \, {}^\circ\text{C}$. The heat capacity of 10 tubes $= 10 \times 66.3 = 663 \, \text{J/m}^2 \, {}^\circ\text{C}$

v. The heat capacity of the water in 10 tubes $= 10 \times 3.14 \times (0.005)^2 \times 1 \times 1000 \times 4190 = 3289 \, \text{J/m}^2 {}^\circ\text{C}$.

The effective collector heat capacity

$$(mc)_e = (mc)_p + \frac{U_L}{h_2}(mc)_c = [1000 + 2112 + 663 + 3289] + \frac{6}{10} \times 7000 = 7064 + 4200$$

$$= 11264 \, \text{J/m}^2 \, {}^\circ\text{C} \, (\text{assuming } U_L = U_t = 6 \text{ and } V = 1 \, \text{m/sec})$$

In the beginning at 8 AM, the collector temperature will be equal to ambient air temperature ($T_{po} = T_a$) and to calculate T_p at 9 AM, Equation (3.94a) reduces to

$$T_p(9\text{AM}) = T_a + \frac{\dot{q}_{ab}}{U_L}\left[1 - \exp\left(-\frac{A_c U_L t}{(mc)_e}\right)\right] = -0.8 + \frac{290.4}{6.0}\left[1 - \exp\left(-\frac{6 \times 3600}{11,264}\right)\right]$$

$$= 40.5°C$$

For second hour period, $T_{po} = 40.5°C$. The temperature at 10AM (Equation (3.94b)) becomes

$$T_p(10\text{AM}) = T_a + \frac{\dot{q}_{ab}}{U_L} - \left[\frac{\dot{q}_{ab}}{U_L} - (T_{po} - Ta)\right]\exp\left(-\frac{A_c U_L \cdot t}{(mc)_e}\right)$$

$$= 0.1 + \frac{388.8}{6} - \left[\frac{388.8}{6} - (40.5 - 0.1)\right]\exp\left(-\frac{6 \times 3600}{11264}\right)$$

$$= 64.9 - (24.4) \times 0.1469 = 61.3°C$$

By 9.00 AM, the collector has been heated to within 0.5°C of its operating temperature of 40°C. The reduction in useful gain is the energy required to heat the collector, i.e. 0.5°C or $0.5 \times 11264 = 5632\ \text{J/m}^2$.

Thus useful energy gain from 8.00 AM to 9.00 AM should be reduced from 0.12 MJ/m² to (120000 − 5632 = 114368 J/m²).

PROBLEMS

3.1 Derive an expression for an overall heat transfer coefficient (U_L) by using the thermal circuit analysis.

   Hint See Sections 2.4 and 3.5.

3.2 Draw the curve between fin efficiency factor (F) and thickness of an absorber (δ) for a given diameter (D) of pipe, and thermal conductivity (K) for a given collector specification.

   Hint See Section 3.7.

3.3 Calculate the collector efficiency factor (F') with ($x_b = 0$) and without ($x_b = 0$) bond conductance for a given collector.

   Hint See Section 3.8.3.

3.4 Calculate U-value for double glazed flat plate collector. The spacing between two glass cover is 5 cm.

$$\text{Use, } U = \left[\frac{1}{h_1} + \frac{1}{C} + \frac{1}{h_2}\right]^{-1}$$

   Hint air conductance for 5 cm thickness $C = 5\ \text{W/m °C}$ and see Example 2.1(a).

3.5 What is collector efficiency (η) and the rate of heat collection $\dot{q}_u$ for the following collector specifications and climatic conditions: $(\alpha_0\tau_0) = 0.8$, $U_L = 6\ \text{W/m}^2\ °C$, $I = 800\ \text{W/m}^2$, $T_p = 55°C$, $T_a = 20°C$, $W = 150\ \text{mm}$, $D_i = 7\ \text{mm}$, $D_o = 10\ \text{mm}$, $\delta = 0.5\ \text{mm}$, $x_b = 0$, $K = 385\ \text{W/m°C}$, $h_{fi} = 250\ \text{W/m}^2\ °C$, $A_c = 1\ \text{m}^2$.

Hint Use Equation (3.43a) and $\eta_i = \dot{Q}_u / I$.

3.6 Derive an expression for the threshold radiation level for a given operating temperature (T_o).

Hint Substitute $\eta = 0$ in Equation (3.43b) with $I(t) = I_{th}$.

3.7 How do you calculate the stagnation temperature of the collector.

Hint Obtain $f_c = (T_p - T_a)/I(t)$ at $\eta = 0$ in Figure 3.8(b) and evaluate $T_p(\text{stagnate}) = f_c I(t) + T_a$.

3.8 Derive an expression for total useful energy (Q_T) for a time interval of $0 - t_T$ time interval.

Hint

$$\text{Use } Q_T = \int_0^{t_T} \dot{Q}_u(t)dt \text{ with } \bar{I} = \frac{\int_0^{t_T} I(t)dt}{t_T} \text{ and } \bar{T}_a = \frac{\int_0^{t_T} T_a(t)dt}{t_T}$$

3.9 What is the lowest radiation level at which heat can be collected.

Hint See Problem 3.6.

3.10 Draw the curve between heat removal factor (F_R) and flow rate $\dot{m}$ for a given collector area (A_c) and collector efficiency factor.

Hint Use Equation (3.67).

3.11 Derive an expression for mean fluid temperature for a flat plate collector.

Hint

$$\text{Use, } T_{fm} = \frac{1}{L}\int_0^L T_f(x)dx$$

3.12 What are the basic properties of transparent materials (window glass cover).

Hint It transmits short wavelength and behaves as an opaque material for long wavelength (see figure given below).

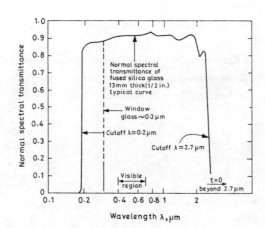

3.13 Draw the variation of transmittance of glass cover with angle of incidence of irradiance, for different number of glass plate.

Hint See Figure given below.

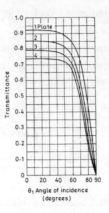

3.14 Find out an analytical expression for mean plate temperature.

Hint Equate Equations (3.43a) and (3.70) with $T_{pm} = T_p$.

3.15 Plot the curve between collector flow factor ($F'' = F_R/F'$) and $\dot{m}C_p/A_cU_LF$.

Hint Use Equation (3.69b).

3.16 Determine the mean fluid and the plate temperature for collector of Problem 3.5.

Hint See Problems 3.14 and 3.15.

3.17 Plot the variation between $F_R(\alpha\tau)$ and F_RU_L and the number of collectors connected in series.

Hint Use Equations (3.81) and (3.82).

3.18 Calculate the heat removal factor (F_{Rm}) for two collectors connected in parallel (one module) for the parameters of Example 3.7(a)

Hint See Example 3.7(a). Calculate F_R with $A_c = 2\,\text{m}^2$.

3.19 Calculate heat removal factor (F'') for Example 3.7(a).

Hint Use Equation (3.69b).

3.20 Plot the curve between heat removal factor (F_R) with collector area for a given flow rate ($\dot{m}$).

Hint See Problem 3.10.

3.21 Calculate the heat removal factor (F_{Rm}) for one panel of six module connected in series of Problem 3.18.

3.22 Discuss Equation (3.95) for two limiting cases of heat capacity, i.e. $(mC)_e \to 0$ and $(mC)_e \to \infty$.

Hint Substitute $(mC)_e \to 0$ and $(mC)_e \to \infty$ in Equation 3.95

CHAPTER 4

Evacuated Solar Collector

4.1 INTRODUCTION

As seen in Chapter 3, on flat plate collector that there is further scope of reducing convection heat losses from the absorber to glass cover. This can be achieved by completely removing the air between absorber and glass cover. The only heat loss mechanism remaining is radiation. The resulting stress on the cover plate due to outside air restricts the use of vacuum in flat plate collector. To avoid this the plate must be supported at frequent interval. Also, it is very difficult to maintain vacuum in flat plate collector and hence evacuated-tube collector was invented.

The classification of evacuated-tube collector has been given in Table 4.1.

The brief description and working principle of each has been discussed in the following sections.

TABLE 4.1   Classification of evacuated solar collector (*Williams, 1983*)

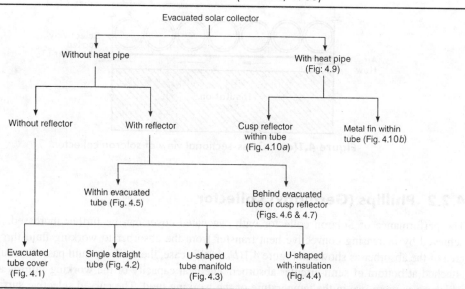

4.2 EVACUATED-TUBE COVER COLLECTOR

In evacuated-tube cover collector vacuum is created between the absorber and transparent glass cover as discussed below.

4.2.1 Solaron Collector

Figure 4.1(*a*) shows a coss-sectional view of Solaron collector with evacuated tube cover. In this case, evacuated tubes are arranged above the absorber so that there should not be any space left between consecutive tubes. The evacuated tubes provides a vacuum layer above the absorber. The vacuum layer suppressed the convection heat loss from the absorber to glass cover. Similar to the flat plate collector, incident solar radiation is absorbed by selectively coated absorber after transmission from the glass cover and transparent evacuated tubes. After absorption, most of the available thermal energy at the absorber will be first conducted and then convected to the working fluid below the absorber. Rest of absorbed thermal energy is lost to upper portion of evacuated tubes by radiative heat loss unlike conventional flat plate collector. Further, there will be a convective and radiative heat losses from the upper portion of evacuated tubes to glass cover. Since, the temperature of upper portion of evacuated tubes will be small, hence there will be a small heat losses. The working fluid may be either air or any liquid fluid (say water). The temperature of working fluid in this case will be more in comparison to the fluid temperature of conventional flat plate collector due to reduced upward heat loss.

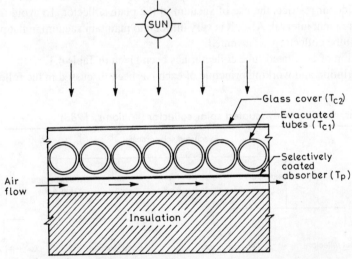

Figure 4.1(a)   Cross-sectional view of solaron collector.

4.2.2 Phillips (Germany) Collector

The performance of Solaron collector with evacuated cover can be further increased. This can be achieved by increasing convective heat transfer from the absorber to working fluid through surface area of the absorber as shown in Figure 4.1(*b*). In this case, the working fluid passes through the tubes attached at bottom of semi-circular absorber. The heat capacity of the working fluid is also reduced which gives more rise in the temperature of the working fluid. The curved selective surface absorber acts as a heat exchanger. The top surface of the evacuated tubes are directly exposed to solar radiation unlike to Solaron collector.

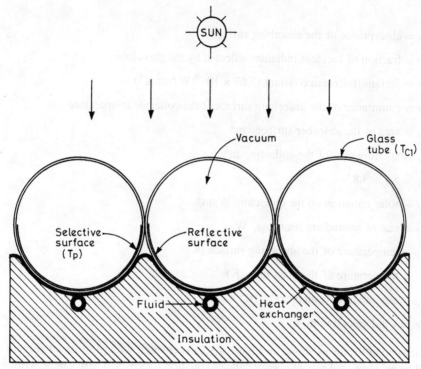

Figure 4.1(b)   Cross-sectional view of evacuated Phillips (Germany) collector.

Solar radiation is transmitted after reflection from the outer curved portion of the glass tube inside vacuum space. It is finally absorbed by curved selective surface after reflection. The reflected radiation is further transmitted to atmosphere through curved outer portion of the tube. Most of absorbed thermal energy is transferred to working fluid through conduction and convection. Rest is lost to atmosphere through radiation, conduction, convection, and radiation.

4.2.3 Thermal Efficiency

In order to determine the thermal efficiency of evacuated-tube cover collector, the question naturally arises as to what constitutes the aperture area of the collector. It is always easier and simple that solar collector efficiencies be determined by considering the outside area equal to the overall dimensions of collector.

The efficiency (instantaneous) of such evacuated cover collector (Fig. 4.1(b)) can be written as

$$\eta_i = \alpha(1 - \rho) - \frac{\varepsilon\sigma\left(T_p^4 - T_c^4\right)A_a}{I A_c} - \frac{\dot{Q}_f}{I A_c} \qquad (4.1)$$

and, the rate of energy balance for cover will be

$$\varepsilon\sigma \left(T_p^4 - T_c^4\right) A_a = h_c(T_c - T_a)A_a \tag{4.2}$$

where

α = absorptance of the absorbing surface

ρ = fraction of incident radiation reflected by the glass tube

σ = Stefan–Boltzman constant ($5.67 \times 10^{-8}\,\mathrm{W/m^2\,K^4}$)

ε = emmitance of the absorbing surface at the collector temperature

A_a = area of the absorber surface, $\mathrm{m^2}$

A_c = aperture area of the collector, $\mathrm{m^2}$

$h_c = 5.7 + 3.8\,V$

I = solar radiation on the collector, $\mathrm{W/m^2}$

$\dot{Q}_f$ = rate of nonradiant heat loss, W

T_p = temperature of the absorbing surface (K)

T_c = temperature of the glass tube (K)

T_a = temperature of ambient air(K)

V = wind velocity, m/s

If the evacuated collector as shown in Figure 4.1(b) is covered with glass cover (Figure 4.1(a)), then there will be an additional rate of energy balance equation for cover. These can be written as follows:

For cover I

$$\varepsilon\sigma \left(T_p^4 - T_{c1}^4\right) A_a = h_1(T_{c1} - T_{c2})A_c \tag{4.3}$$

For cover II

$$h_1(T_{c1} - T_{c2})Ac = h_c(\dot{T}_{c2} - T_a)A_c \tag{4.4}$$

where h_1 is the sum of convective and radiative heat transfer coefficients between cover I and cover II respectively.

In the remaining collector, each tube acts as a collector and then connected either in parallel or in series to make a module with effective area equal to a conventional flat plate collector for comparison.

EXAMPLE 4.1

Derive an expression for an instantaneous thermal efficiency (η_i) for an evacuated cover collector (Fig. 4.1(b)) in terms of absorber plate temperature T_p.

Solution

Equation (4.2) can be rewritten by linearising the left hand side as follows:

$$\varepsilon\sigma\left(T_p^4 - T_c^4\right)A_a = h_r(T_p - T_c)A_a = h_c(T_c - T_a)A_a$$

The above equation can be written in the form

$$h_r(T_p - T_c) = U_t(T_p - T_a)$$

where, $\qquad h_r = \varepsilon\sigma(T_p + T_c)\left(T_p^2 + T_c^2\right)$ and $\quad U_t = \dfrac{h_r h_c}{h_r + h_c}$

Substitute the value of $h_r(T_p - T_c)$ in Equation (4.1)

$$\eta_i = \alpha(1 - \rho) - \frac{U_t(T_p - T_a)A_a}{I A_c} - \frac{\dot{Q}_f}{I A_c}$$

If $\dot{Q}_f = U_b A_a(T_p - T_a)$, the overall bottom heat loss, then above equation reduces to

$$\eta_i = \alpha(1 - \rho) - \frac{U_L(T_p - T_a)A_a}{I A_c} = (\alpha\tau) - \left(U_L\frac{A_a}{A_c}\right)\left[\frac{T_p - T_a}{I}\right]$$

The above equation is similar to the characteristic equation of a flat-plate collector with intercept of $(\alpha\tau) = \alpha(1 - \rho)$ and the slope of $\left(-U_L\dfrac{A_a}{A_c}\right)$ as discussed in Sections 3.3 and 3.8 respectively. Here, $U_L = U_t + U_b$.

EXAMPLE 4.2

Derive an expression for η_i for Example 4.1 when it is covered with the glass cover (Figure 4.1(a)).

Solution

Eliminating T_{c2} from Equations (4.3) and (4.4), we get after algebraic manipulation

$$\varepsilon\sigma\left(T_p^4 - T_{c1}^4\right)A_a = h_r(T_p - T_{c1})A_a = U_{t1}(T_{c1} - T_a)A_c$$

where $\qquad\qquad\qquad U_{t1} = \dfrac{h_1 hc}{h_1 + hc}$

Follow similar procedure of Example 4.1 to obtain

$$\eta_i = (\alpha\tau) - \left(U_L\frac{A_a}{A_c}\right)\left[\frac{T_p - T_a}{I}\right]$$

where $U_L = U_t + U_b$ and $U_t = \dfrac{h_r U_{t1}}{h_r + U_{t1}}$.

EXAMPLE 4.3

Calculate the overall heat loss coefficient U_L for Figure 4.1 for the parameters: $\sigma = 5.6 \times 10^{-8} \, \text{W/m}^2\text{K}^4$, $T_a = 40\,°\text{C}$, $T_p = 160\,°\text{C}$, $T_{c1} = 140\,°\text{C}$, $T_{c2} = 80\,°\text{C}$, and $V = 1 \, \text{m/s}$ and $\varepsilon = 0.05$ (selective surface).

Solution

Here, the conductance of the glass material is neglected.
For Example 4.1 (Fig. 4.1(b)):

$$h_r = \varepsilon\sigma(T_p + T_c)\left(T_p^2 + T_c^2\right) = 0.05 \times 5.67 \times 10^{-8}(433 + 353)\,(18.74 \times 10^4 + 12.46 \times 10^4)$$

$$= 0.70 \, \text{W/m}^2\text{K}$$

and,

$$h_c = 5.7 + 3.8 = 9.5 \, \text{W/m}^2\text{K}$$

$$U_t = \frac{h_r h_c}{h_r + h_c} = \frac{0.70 \times 9.5}{0.7 + 9.5} = 0.65 \, \text{W/m}^2\text{K}$$

and,

$$U_b = \left[\frac{L_i}{K_i} + \frac{1}{h_i}\right]^{-1} = \left[\frac{0.10}{0.04} + \frac{1}{5.7}\right]^{-1} = 0.37 \, \text{W/m}^2\text{K}$$

So,

$$U_L = U_t + U_b = 0.65 + 0.37 = 1.02 \, \text{W/m}^2\text{K}$$

The overall heat loss coefficient in evacuated solar collector is significantly reduced in the comparison with the value of overall heat loss coefficient of flat plate collector (Example 3.2, $U_L = 7.2 \, \text{W/m}^2\text{K}$).
For Example 4.2 (Fig. 4.1(a))

$$h_r = \varepsilon\sigma(T_p + T_{c1})\left(T_p^2 + T_{c1}^2\right) = 0.05 \times 5.67 \times 10^{-8}(433 + 413)(18.74 \times 10^4 + 17.06 \times 10^4)$$

$$= 0.86 \text{W/m}^2\text{K}$$

and,

$$U_{t1} = \frac{h_1 h_c}{h_1 + h_c} = \frac{5.8 \times 9.5}{5.8 + 9.5} = 3.6 \, \text{W/m}^2\text{K}$$

Then,

$$U_t = \frac{h_r U_{t1}}{h_r + U_{t1}} = \frac{0.86 \times 3.6}{0.86 + 3.6} = 0.694 \, \text{W/m}^2\text{K}$$

The overall heat loss coefficient will be

$$U_L = U_t + U_b = 0.694 + 0.37 = 1.064 \text{W/m}^2\text{K}$$

Hence overall heat loss coefficient in the case of Figure 4.1(a) is slightly higher.

4.3 EVACUATED-TUBULAR COLLECTOR (*Williams, 1983*)

In this section the working principle of various evacuated-tubular collectors will be discussed.

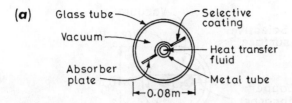

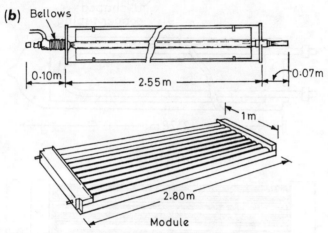

Figure 4.2 (**a**) Cross-sectional and (**b**) frontview of sanyo collector.

4.3.1 Sanyo Evacuated-tube Collector

The cross-sectional view of Sanyo evacuated tube collector has been shown in Figure 4.2(*a*). A cylindrical metal tube is fixed with selectively coated metal absorber like tube in plate configuration. The configuration is then inserted in a glass tube of 0.08 m diameter. The diameter of tube and thickness of absorber depends on material. Generally, copper is used. The effective length of tube is about 2.8 m. The tube is evacuated to reduce top loss coefficient through suppressing convection heat loss between absorber and tube cover. The bottom heat loss is also reduced due to insulating nature of vacuum space. The absorber can also be rotated at the optimum angle to receive maximum solar radiation for a given latitude. This is not possible in evacuated-tube cover collector discussed earlier. There are ten such tubes which are connected in parallel to form one module with an effective area of 1 m × 2.8 m as shown in Figure 4.2(*b*).

4.3.2 Corning Evacuated-tube Collector

Figures 4.3 shows a cross-sectional view, top elevation of corning evacuated-tube collector (*a*) and their module (*b*) with an area of 2.65 m × 0.71 m. Here, a U-shaped copper tube is fitted with selectively coated horizontal copper flat plate in tube below absorber configuration. The assembly is inserted in a pyrex (glass) tube of diameter 0.10 m and length 2.28 m. Further, a U-shaped tubing is also supported by a plate strip. The other end of the tube is sealed after creating vacuum inside the tube. The tubes are connected in series to form one module as shown in Figure 4.3(*b*). There are six tubes connected in series.

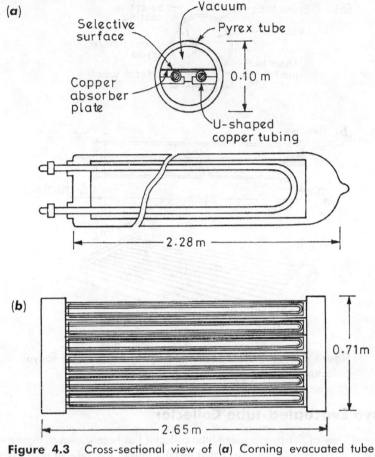

(a)

Vacuum

Selective
surface

Pyrex tube

Copper
absorber
plate

0.10 m

U-shaped
copper tubing

2.28 m

(b)

0.71m

2.65 m

Figure 4.3 Cross-sectional view of (**a**) Corning evacuated tube collector and (**b**) module.

4.3.3 Phillips (Germany) Evacuated-tube Collector

The Phillips evacuated-tube collector is a improved version of evacuated tubular collector with high operating temperature. The top elevation and cross-sectional view is shown in Figure 4.4. In this case too, there is U-shaped copper tube which is fitted in glass tube so that each arm of the tube touches the glass wall. The inner side of the tube is filled with polyurethane foam to avoid heat exchanges between tube unlike corning evacuated-tube collector. The outer side of the tube is selectively coated. This arrangement is further inserted in to another glass tube having bigger diameter. An annular space between inner and outer tube is evacuated to reduce convective heat losses. In this case also, the tubes are connected in series to form the module of required area.

4.3.4 Roberts Evacuated-tube Collector

Figures 4.5 shows a cross-sectional view (*a*) and the top elevation (*b*) of Roberts evacuated-tube collector. In this case, a concentric copper tube is centrally fitted with flat steel absorber plate as shown in

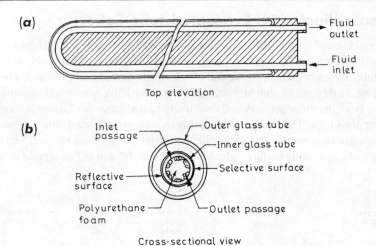

Figure 4.4 Phillips (Germany) evacuated-tube collector.

Figure 4.5(*a*). There is a aluminum reflector below the absorber to reflect back all radiation falling on it towards the absorber. The whole assembly is inserted in a glass tube of diameter of 0.10 m. The glass tube is evacuated to eliminate the convection loss within the tube. The open part of the tube is properly sealed to create vacuum. The working fluid is allowed to pass through the inner tube of a concentric copper tube. The working fluid is heated during its flow through an annular space of a concentric copper tube as shown in Figure 4.5(*b*). The outer portion of a concentric copper tube is perfectly insulated to minimize heat losses at outlet.

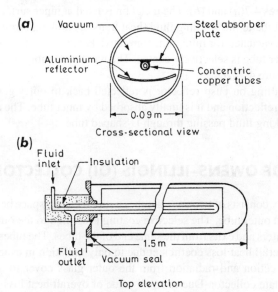

Figure 4.5 Cross-sectional view of Roberts evacuated tube collector.

4.3.5 Owens–Illinois (OI) Evacuated-tube Collector

The cross-sectional view of Owens-Illinois sunpack evacuated-tube collector is shown in Figure 4.6. In this, the working (heat transfer) fluid passes through glass delivery tube which is inserted into selectively coated another tube of bigger diameter. One end of selectively coated tube is sealed. The working fluid is heated while passing through an annular space between glass delivery tube and selectively coated tube in a reverse flow. Both the glass delivery and selectively coated tubes are further inserted into another borosilicate outer glass tube. The annular space between selectively coated tube and borosilicate outer glass tube is evacuated to minimize the convection loss from the selective surface. The collector tubes are often mounted above a white surface, which reflects sunlight onto the underside of the tubes.

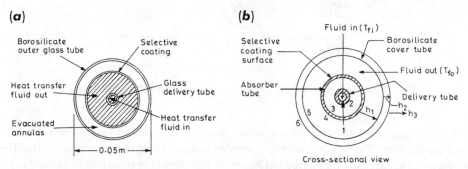

Figure 4.6   Owens–Illinois (OI) sunpack evacuated tube collector.

4.3.6 General Electric (GC) TC-100 Evacuated-tube Collector

The U-shaped copper tube is fitted inside a coaxial glass tube of outer diameter of 0.05 m and length 1.25 m as shown in Figures 4.7(a) and (b). The metal fin is fixed at inner surface of coaxial tube for fast heat transfer from inner tube to the working fluid (heat transfer) flowing through the tubes. An annular space of coaxial tube is evacuated for minimum convective heat loss from inner tube to the outer tube. The outer surface of inner tube is selectively coated. The whole assembly is kept at focal point of cusp reflector as shown in Figure 4.7(a).

The solar radiation falling on cusp reflector is reflected back to outer glass of coaxial tube. It is further transmitted after reflection and it is finally absorbed by inner tube. The absorbed thermal energy is transferred to the working fluid passing through U-shaped tube.

4.4 ANALYSIS OF OWENS–ILLINOIS (OI) COLLECTOR (*Fig. 4.6*)

The OI collector basically consists of two coaxial tubes with evacuated space between a selective coating surface of inner tube and outer tube. The selective coating is applied to the outer surface of inner tube. The heat transfer fluid enters and exits from the same end of the tubes. The tubes are manifolded together at one end. Since the overall heat loss coefficient due to only radiation in evacuated space, conduction in the glass cover, convection and radiation from the outer glass cover to ambient is very small in comparison to the flat plate collector. Due to small value of overall heat loss coefficient, the collector efficiency is relatively insensitive to ambient air temperature, wind velocity and operating temperature up to about 150 °C.

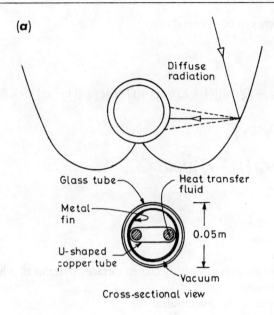

(a)

Diffuse radiation

Glass tube

Heat transfer fluid

Metal fin

0.05 m

U-shaped copper tube

Vacuum

Cross-sectional view

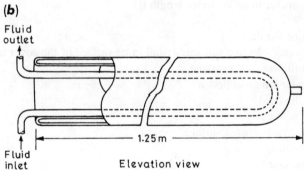

(b)

Fluid outlet

1.25 m

Fluid inlet

Elevation view

Figure 4.7 General electric evacuated tubular collector.

Figure 4.6(b) shows the line sketch of cross-sectional view of a single collector tube. The cover tube is made of low-iron content glass with high transmittance. One end of absorber tube is sealed, while the other is free to move in a flexible support so as to allow for differential thermal expansion between the two tubes.

The rate of useful heat delivered by this collector tube is given by (Beekley and Mather, 1978)

$$\dot{Q} = \dot{m}C_f(T_{fo} - T_{fi}) = F_r A_c[(\alpha\tau)I_{\text{eff}} - U_L(A_L/A_c)(T_{fi} - T_a)] \qquad (4.5a)$$

where

$$F_R = (\sinh\omega_2\xi_1/\{\omega_2\xi_1[\cosh\omega_2\xi_1 + (w_1/w_2)\sinh\omega_2\xi_1]\})F' \qquad (4.5b)$$

$$\omega_1 = U_L P_L F'/2h_3 P_3 \qquad (4.5c)$$

$$\omega_2 = U_L P_L F'[1 + 4h_1 P_1/U_L P_L F']^{1/2}/2h_3 P_3 \qquad (4.5d)$$

$$\xi_1 = h_3 P_3 l/\dot{m}C_f \qquad (4.5e)$$

$$F' = 1/(1 + U_L P_L/h_3 P_3) \qquad (4.5f)$$

The overall heat loss coefficient can be obtained from

$$U_L = \frac{1}{\frac{1}{h_4} + \frac{1}{h_5} + \frac{1}{h_6}} \qquad (4.6a)$$

where $\quad h_4 = \sigma (T_4 + T_5) \left(T_4^2 + T_5^2 \right) / \{[(1 - \varepsilon_4)/\varepsilon_4] + (1/F_{45}) + [(1 - \varepsilon_5)/\varepsilon_5](A_4/A_5)\} \qquad (4.6b)$

$$h_5 = k / \left[\left(\frac{D_4}{2} \ln(D_6/D_5) \right) \right] \qquad (4.6c)$$

$$h_6 = [h + \varepsilon_6 \sigma (T_6 + T_a) \left(T_6^2 + T_a^2 \right)]/(A_6/A_4) \qquad (4.6d)$$

$$h = 5.7 + 3.8V \qquad (4.6e)$$

$1 =$ inner surface of delivery tube
$2 =$ outer surface of delivery tube
$3 =$ inner surface of absorber tube
$4 =$ outer surface of absorber tube (the selectively coated surface, in terms of which U_L is defined)
$5 =$ inner surface of the cover tube
$6 =$ outer surface of the cover tube and $a =$ ambient
$A_c =$ absorber tube diameter times collector length (l)
$A_L = \pi A_c$
$C_f =$ the specific heat of the fluid
$F_{45} =$ shape factor between selective surface (4) and inner surface of the cover tube (5)
$h =$ heat transfer coefficient at ith surface
$I_{\text{eff}} =$ effective solar radiation on collector
$\dot{m} =$ the flow rate of the fluid
$P_i =$ perimeter of ith surface
$P_L =$ perimeter of selectively coated surface
$T_{fi} =$ inlet fluid temperature
$T_{f0} =$ outlet fluid temperature

The variation of an overall heat transfer coefficient (U_L) with absorber temperature and η with $(T_{fi} - T_a)/I_{\text{eff}}$ is shown in Figure 4.8. Figure 4.8(b) also shows the performance of two cover non-selective flat plate collector for comparison. It can be inferred from Figure 4.8(a) that there is not much variation in the value of an overall heat transfer coefficient (U_L) with respect to ambient air temperature (T_a). However, there is sharp increase in the value of an overall heat transfer coefficient (U_L) with respect to absorber plate temperature (T_4). In the Figure 4.8(b), d is the center-to-center distance of two tubes and D_B is the distance of center of tube from the absorber.

EXAMPLE 4.4

Calculate the overall heat transfer coefficient (U_L) for the following OI collector parameters:

- active tube length $(l) = 10.67$ m
- delivery tube i.d. $(D_1) = 0.09$ m

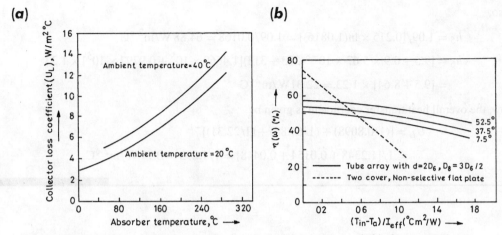

Figure 4.8 (a) Variation of U_L with absorber temperature and (b) characteristic curve.

- delivery tube o.d. $(D_2) = 0.12\,\text{m}$
- absorber tube i.d. $(D_3) = 0.39\,\text{m}$
- absorber tube o.d. $(D_4) = 0.43\,\text{m}$
- cover tube i.d. $(D_5) = 0.49\,\text{m}$
- cover tube o.d.$(D_6) = 0.53\,\text{m}$

$\sigma = 5.67 \times 10^{-8}\,\text{W/m}^2\text{K}$, $T_a = 40\,°\text{C}$, $T_{fi} = 55\,°\text{C}$, $T_4 = 160\,°\text{C}$, $T_5 = 140\,°\text{C}$, $T_6 = 110\,°\text{C}$, $k =$ thermal conductivity of glass cover $= 1.09\,\text{W/m}\,°\text{C}$, $V = 1\,\text{m/s}$, $F_{45} = 0.50$, $\varepsilon_4 = 0.05$ (selective surface) and $\varepsilon_5 = \varepsilon_6 = 0.9$

Solution

Here,

- $A_4 = 3.14 \times 0.43\,\text{m} \times 10.67\,\text{m} = 14.40\,\text{m}^2$
- $A_5 = 3.14 \times 0.49\,\text{m} \times 10.67\,\text{m} = 16.42\,\text{m}^2$
- $A_6 = 3.14 \times 0.53\,\text{m} \times 10.67\,\text{m} = 17.76\,\text{m}^2$
- $A_4/A_5 = 0.8769$ and $A_6/A_4 = 1.23$ and $h = 5.7 + 3.8 = 9.5\,\text{W/m}^2\,°\text{C}$

Further,
The denominator of Equation (4.6b)

$$= (1 - 0.05)/0.05 + (1/0.5) + [(1 - 0.9)/0.9] \times 0.8769$$

$$= 19 + 2 + 0.097 = 21.097$$

and,

$$h_4 = 5.67 \times 10^{-8}(433 + 413)[1.8749 \times 10^5 + 1.70569 \times 10^5]/21.097$$

$$= 0.8098\,\text{W/m}^2\,°\text{C}$$

Also,

$$h_5 = 1.09/[0.215 \times \ln(1.0816)] = 1.09/0.0168 = 64.88 \, \text{W/m}^2 \, ^\circ\text{C}$$

$$h_6 = [9.5 + 0.9 \times 5.67 \times 10^{-8}(383 + 313)[1.4669 \times 10^5 + 0.9797 \times 10^5] \times 1.23$$

$$= [9.5 + 8.64] \times 1.23 = 22.31 \text{W/m}^2 \, ^\circ\text{C}$$

Now, the overall heat transfer coefficient is given by

$$U_L = [(1/0.8098) + (1/64.88) + (1/22.31)]^{-1}$$

$$= 1/[1.2348 + 0.0154 + 0.0448] = 1/1.295 = 0.77 \, \text{W/m}^2 \, ^\circ\text{C}.$$

EXAMPLE 4.5

Calculate convective heat transfer coefficients (h_1 and h_3) at an inner surface of case (i) the delivery and case (ii) an absorber tube for the above example for the case of water flow at the rate of 3 m/s.

Solution

From the above example: an average temperature of water at inner surface of delivery tube (T_1) $< T_{fi} = 55\,^\circ\text{C}$ (say $T_1 = 70\,^\circ\text{C}$) and an average temperature of water at inner surface of absorber tube (T_3) $< T_4 = 160\,^\circ\text{C}$ (say $T_3 = 100\,^\circ\text{C}$).
Since, $L/D_1 = 118.56$ and $L/D_3 = 27.36$, then Equation (2.47d) (Section 2.6.6) will be used to calculate convective heat transfer coefficients. Now,

$$Nu = 0.036 Re^{0.8} Pr^{1/3} \left(\frac{D}{L}\right)^{0.055} \quad \text{for } 10 \le \frac{L}{D} \le 400$$

From Table IIIe, case (i) at $T_1 = 70\,^\circ\text{C}$, $Pr \cong 2.53$, $\mu = 4.01 \times 10^{-4} \, \text{kg/m s}$, $\rho = 977.3 \, \text{kg/m}^3$, $\nu = \mu/\rho = 4.103 \times 10^{-7} \, \text{m}^2/\text{s}$, $K = 0.669 \, \text{W/m}\,^\circ\text{C}$ and $Re = \nu D_1/\nu = (3 \times 0.09/4.103 \times 10^{-7}) = 6.58 \times 10^5$, then,

$$h_1 = \frac{Nu.K}{D_1} = 0.036 \times (6.58 \times 10^5)^{0.8}(2.53)^{1/3} \left(\frac{0.09}{10.67}\right)^{0.055} \times \frac{0.665}{0.09} = 12.583 \text{kW/m}^2 \, ^\circ\text{C}$$

and, case (ii) at $T_1 = 100\,^\circ\text{C}$, $Pr \cong 1.66$, $\mu = 2.67 \times 10^{-4} \, \text{kg/m s}$, $\rho = 955.1 \, \text{kg/m}^3$, $\nu = \mu/\rho = 2.7955 \times 10^{-7} \, \text{m}^2/\text{s}$, $K = 0.684 \, \text{W/m}\,^\circ\text{C}$ and $Re = V D_3/\nu = (3 \times 0.39/2.7955 \times 10^{-7}) = 41.85 \times 10^5$, then

$$h_3 = \frac{Nu.K}{D_3} = 0.036 \times (41.85 \times 10^5)^{0.8}(1.66)^{1/3} \left(\frac{0.39}{10.67}\right)^{0.055} \times \frac{0.684}{0.39} = 12.356 \text{kW/m}^2 \, ^\circ\text{C}.$$

EXAMPLE 4.6

Calculate collector efficiency (F') and flow rate factors (F_R) for the above examples.

Solution

Here, $P_1 = 3.14 \times 0.09 = 0.2826$ m and $P_3 = P_L = 3.14 \times 0.39 = 1.2246$ m.
$h_1 p_1 = 12.583 \times 0.2826 = 3.556$ kW/m °C and $h_3 P_3 = 12.356 \times 1.2246 = 15.131$ kW/m °C.
$U_L P_L = 0.77 \times 1.2246 = 0.943$ W/m °C
The collector efficiency factor is given by

$$F' = 1/(1 + U_L P_L/h_3 P_3) = 1/(1 + 0.943/15131) \approx 1$$

In order to calculate the flow rate factor, one has to evaluate the following parameters:

$$\omega_1 = U_L P_L F'/2h_3 P_3 = 0.943 \times 1/(2 \times 15131) = 3.116 \times 10^{-5}$$

$$\omega_2 = U_L P_L F'[1 + 4h_1 P_1/U_L P_L F']^{1/2}/2h_3 P_3$$

$$= 0.943 \times 1 \left[1 + \frac{4 \times 3556}{0.943 \times 1} \right]^{0.5} /(2 \times 15131) = 3.83 \times 10^{-3}$$

$$\xi_1 = h_3 P_3 l/\dot{m}C_f = \frac{15131 \times 10.67}{3.14 \times (0.045)^2 \times 3 \times 1000 \times 4190} = 2.0199$$

Further, $\omega_2 \xi_1 = 0.0077$, $w_1/w_2 = 8.135 \times 10^{-3}$
After substituting the calculated values in the expression for F_R, one gets

$$F_R = (\sin h\omega_2\xi_1/\{\omega_2\xi_1[\cos h\omega_2\xi_1 + (w_1/w_2)\sin h\omega_2\xi_1]\})F'$$

$$= \frac{0.0077}{0.0077[1.00 + 8.135 \times 10^{-3} \times 0.0077] \times 1} \cong 1.0$$

This example shows that collector efficiency and flow rate factors in the case of evacuated collectors are the same and equals to unity which can't be achieved in the case of flat plate collector.

4.5 EVACUATED-TUBE COLLECTOR WITH HEAT PIPE (*Norton, 1992*)

In this section, the design and working principle of heat pipe and evacuated-tube collector will be discussed.

4.5.1 Heat Pipe

The schematic diagram of heat pipe is shown in Figure 4.9. Heat pipe provides the method of transferring larger amounts of heat from the focal area of a high-concentration solar collector to a fluid with only small temperature difference. It consists of a circular pipe with an annular wick layer situated adjacent to the pipe wall. The circular pipe is perfectly insulated from out side to avoid thermal losses from the circular pipe. Solar energy falls on evaporator and the fluid inside evaporator boils. The vapor migrates to the condenser where heat of vapor is transferred to a circulating fluid loop. The heat available with circulating fluid is further carried away to the end use point. The circulating fluid after releasing its heat is transferred to the boiler by capillary action in the wick or by gravity and cycle repeats. Gravity return heat pipes can operate without wick but cannot be operated horizontally as a result.

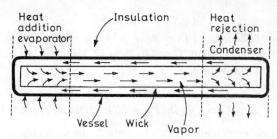

Figure 4.9 Schematic view of heat pipe.

4.5.2 Corning Collector with Internal Reflector

The cross-sectional view of Corning collector with heat pipe and internal cusp reflector has been shown in Figure 4.10(*a*). In this collector, there is a glass tube of diameter 0.10 m and of length 1.4 m. The position of cusp reflector is in half of the tube. The heat pipe is placed at focal point of the reflector. The outer surface of evaporator of heat pipe is a black chrome selective surface to absorb maximum solar radiation and minimum emmitance. The glass tube is also evacuated.

The evaporator of heat pipe is covered with glass tube with cusp reflector and the condenser is exposed to the end point use. The end point use may be immersed with working fluid either directly or through heat exchanger for transferring the heat from condenser to the working fluid.

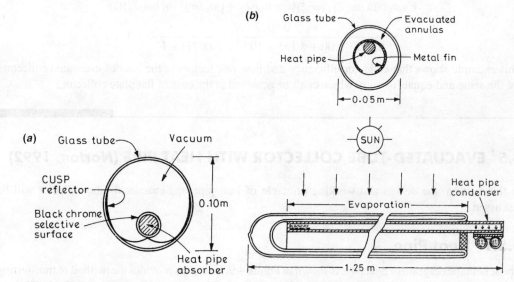

Figure 4.10 (*a*)Cross-sectional view of corning evacuated collector with heat pipe and internal cusp reflector. (*b*) Cross-sectional view of Gumman collector.

4.5.3 Gumman Evacuated-tube Collector

In the Gumman evacuated-tube collector metal fin has been used for maximum heat transfer from absorbed solar energy at the absorber. Figure 4.10(b) shows the cross-sectional view of Gumman collector. The position of heat pipe evaporator and the condenser is also shown in Figure 4.10(b). The evaporator is exposed to solar radiation and condenser is attached with tube in configuration plate. The working fluid flows through tube flat plate of condenser.

4.5.4 Thermal Analysis

The schematic view of a evacuated-tube collector with heat pipe evaporator and condenser is shown in Figure 4.11(a). The fin plate has a selective surface. The condenser of heat pipe transfers the heat to the fluid through a manifold as shown in figure. Each fin plate with heat pipe absorber is enclosed in a separate evacuated cylindrical envelope.

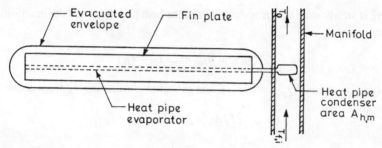

Figure 4.11(a)   Schematic diagram of a evacuated collector with heat pipe.

Figure 4.11(b) shows a electrical network for a system under consideration for the analysis. The figure also shows the thermal resistance (inversely proportional to the heat transfer coefficient) at various points and the rate of energy transferred from one point to another.

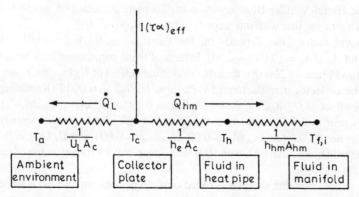

Figure 4.11(b)   Electrical analog for heat transfer for Figure 4.11(a).

In a steady condition, the rate of useful energy available with fin plate can be represented by the equation (Norton, 1992)

$$\dot{Q}_u = A_c \left[\dot{q}_{ab} - U_L(T_p - T_a) \right] \tag{4.7}$$

where $\dot{q}_{ab}$ is defined in Chapter 3.

The rate of thermal energy transferred from fin plate to the fluid in the evaporator of heat pipe can be expressed as follows:

$$\dot{Q}_{p,h} = h_{ph} A_c (T_p - T_h) \tag{4.8}$$

Eliminating T_p from above equations by assuming that there is no heat loss during heat transfer from fin plate to the fluid of evaporator, i.e., steady state condition, one gets

$$\dot{Q}_{p,h} = \frac{h_{ph}}{h_{ph} + U_L} A_c \left[\dot{q}_{ab} - U_L(T_h - T_a) \right] \tag{4.9}$$

Further, there is steady state heat transfer between the heat pipe fluid and the manifold fluid. This can be expressed as

$$\dot{Q}_{h,m} = h_{hm} A_{hm} (T_h - T_{fi}) \tag{4.10}$$

For steady state condition ($\dot{Q}_{p,h} = \dot{Q}_{h,m}$), one gets after eliminating T_h from above equations

$$\dot{Q}_{h,m} = F_r A_c \left[\dot{q}_{ab} - U_L(T_{fi} - T_a) \right] \tag{4.11}$$

where

$$F_r = \frac{1}{\left(\frac{U_L A_c}{h_{hm} A_{hm}} \right) + \left(\frac{U_L}{h_{ph}} + 1 \right)}$$

The instantaneous thermal efficiency can be defined as

$$\eta_i = \frac{\dot{Q}_u}{I A_c} = F_r(\alpha\tau) - F_r U_L \frac{(T_{fi} - T_a)}{I} \tag{4.12}$$

This becomes the Hottel-Whiller-Bliss equation of flat plate collector. A graph between η_i verses $(T_{fi} - T_a)/I$ gives a straight line with intercept $F_r(\alpha\tau)$ and slop $-F_r U_L$.

The heat removal factor (F_r) depends on the three ratios $U_L/h_{ph} = (0.01 - 0.50)$, $U_L/h_{hm} = (0.001 - 0.50)$, and $A_{hm}/A_c = (0.005 - 0.50)$. Effects of these parameters have been shown in Figure 4.12. It is clear from Figure 4.12(a) that there is insignificant effect of U_L/h_{ph} on F_r for $U_L/h_{ph} > 0.01$. For evacuated-tube collector manufactured by Phillips, $U_L/h_{ph} = 0.0038$ (Kamminga 1986). Figure 4.12(b) shows the effect of U_L/h_{hm} on F_r for $U_L/h_{ph} = 0.19$ for different A_{hm}/A_c. This indicates that the value of F_r is near to 0.8 for $A_{hm}/A_c > 0.1$ and $U_L/h_{hm} < 0.04$. Simultaneously, Figure 4.12(c) indicates that F_r is nearly 0.8 for $A_{hm}/A_c > 0.02$, $U_L/h_{hm} = 0.001$ and $U_L/h_{ph} = 0.19$. As indicated in Figure 4.12, the value of F_r can go beyond 0.8 for $U_L/h_{ph} > 0.19$ for any value of ratio of resistances more than 17.

It is important to note that this analysis is based on a single-tube module. The value of F_r increases appreciably when a module of 20 or more tubes are considered.

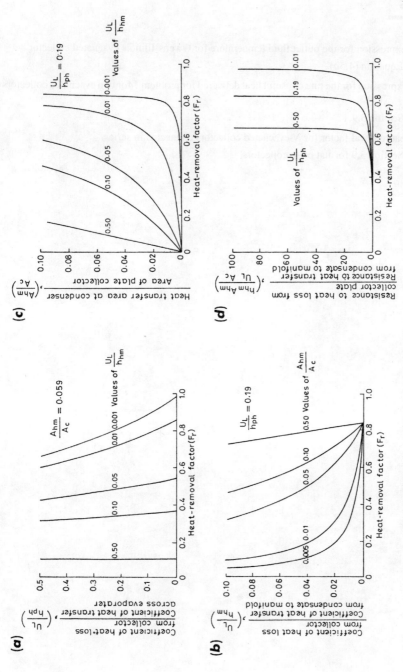

Figure 4.12 Effect of heat-removal factor (F_r) on various parameters.

PROBLEMS

4.1 Derive an expression for the outlet fluid temperature for Owens-Illinois evacuated collector.

Hint Use Equation (4.5*a*).

4.2 Derive an expression for the rate of useful heat delivered for problem 1 for two evacuated collectors connected in series.

Hint See Section 3.9.1.

4.3 Find out heat removal factor for *N*-evacuated collectors connected in series.

Hint See Section 3.9 for flat plate collectors.

CHAPTER 5

Solar Water Heating System

5.1 INTRODUCTION

As we have seen in the third chapter that a flat plate collector can be used for heating of a liquid/air. The flat plate collector is generally used for water heating. The heated water is then stored in an insulated storage tank. The mode of transfer of heated water from the collector to an insulated storage tank can be either natural circulation (thermosiphon) or forced circulation. Further, the transfer of heated water may be carried out either directly or through a heat exchanger. Hence, a solar water heating system shown in Figure 5.1 has the following main components:

  i. a flat plate collector,

  ii. a heat exchanger, and

  iii. an insulated storage tank.

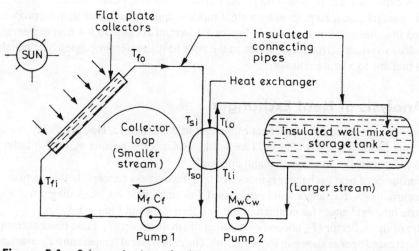

Figure 5.1 Schematic view of solar water heating system showing different components with collector loop.

The use of a heat exchanger mainly depends upon quality of the water and climatic condition of location where a solar water heating system is to be installed. In the third chapter, the flat plate collector has been discussed in detail. In this chapter, we will briefly discuss about a heat exchanger and an insulated storage tank. Here we will also assume that all the connecting pipes used between (a) collector and the heat exchanger and (b) the heat exchanger to an insulated storage tank are perfectly insulated and there is no heat loss. Thus, there is no temperature drop for hot water flowing through it.

The solar water heating system (Fig. 5.1) works under forced circulation mode. A pump 1 has been used in a collector loop to circulate the fluid through the collector and another pump 2 is used to circulate the water in the outer loop between the storage tank and a heat exchanger. The collector loop and outer loop will be referred as smaller stream and larger stream respectively.

5.2 HEAT EXCHANGER (*Lunde, 1980*)

The heat exchanger is essential only to avoid the corrosion due to poor quality of the water and the freezing of water in the collector tubes due to harsh cold climatic condition (Leh, Shimla and Srinagar etc.).

5.2.1 Choice of Fluid

The good quality of water, if available easily, is a ideal heat transfer fluid for a flat plate collector. The good quality of water is a nonflammable, non-toxic, and non-corrosive particularly in plastics and some metals (copper and stainless steel etc.). It has high heat transfer properties, high density, high specific heat and a reasonably high boiling temperature. It has also low viscosity. For good quality of water, no heat exchanger is required. However, the problem, of course, is the freezing point. Hence, a heat exchanger is required for a harsh cold climatic condition where an ambient temperature goes below zero degree centigrade even with good quality of water.

Table 5.1 gives some alternative fluids having low freezing point. The freezing point of the water can be lowered below zero degree by adding some inorganic salts without affecting most of the properties. Most of the fluids in Table 5.1 are non-corrosive, non-flammable, and non-freezing. Except water and air, the rest is expensive and high boiling point. Ethylene and propylene glycol are good choices and indeed are in general use: a 50 percent by weight mixture approaches water in most properties. These are only used in collector loop as shown in Figure 5.1. Air, of course, do not have a freezing problem but the penalty in system performance is due to the poor heat transfer properties. The penalty is about the same as that due to a heat exchanger.

5.2.2 Analysis of Heat Exchanger

There are many kinds of liquid–liquid heat exchanger in the literature. However, only concentric tube and within tank heat exchangers (Fig. 5.2) are used in solar water heating system in India. These are cost effective, inexpensive and easy to maintain locally.

The performance of heat exchanger is measured by effectiveness concept. In this section, an analysis of a countercurrent heat exchanger will be discussed. For simplification, block diagram of Figure 5.2(a) for concentric tube exchanger for countercurrent has been shown in Figure 5.3.

The outlet of the collector (T_{fo}) becomes the inlet of small stream (T_{si}) and passes through an inner tube of the heat exchanger as shown in Figure 5.2(a). The cold water at temperature T_{li} available from the storage tank is allowed to pass in the opposite direction of the small stream for maximum heat transfer

TABLE 5.1 Physical properties of heat transfer fluid

S.No.	Fluid	Temperature (°C)	Density (kg/m³)	Viscosity (g/m.s)	Specific heat (kJ/kg °C)	Thermal conductivity (W/m °C)	Coefficient of thermal expansion °C⁻¹
1	Water	38	993	0.684	4.166	0.628	0.037
		93	963	0.305	4.208	0.661	0.055
2	Ethylene glycol and water, 50% by weight	38	1054	2.3	3.43	0.398	0.057
		93	1016	0.76	3.64	0.433	0.073
3	Propylene glycol and water, 50% by weight	38	1025	3.1	3.64	0.389	0.066
4	Silicone oil	38	935	14.98	1.55	0.144	0.0928
		93	889	6.40	1.63	0.138	0.0928
5	Air (relative humidity=50% and at sea level)	21	1.187	0.018	1.01	0.0260	
		66	1.033	0.021	1.02	0.0292	

$(\dot{Q}_{max})$ from the small stream to the large stream. The inlet temperatures T_{si} and T_{li} to the heat exchanger are considered known, and corresponding outlet temperatures T_{so} and T_{lo} are to be determined.

The maximum quantity of heat transferred from the small stream to the large stream can be expressed as

$$\dot{Q}_{max} = \dot{M}_f C_f (T_{si} - T_{li}) \tag{5.1}$$

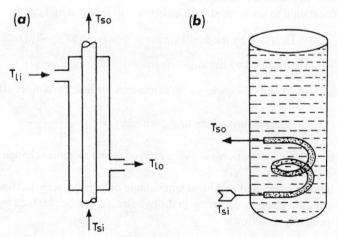

Figure 5.2 Liquid–liquid heat exchanger design (**a**) concentric tube and (**b**) within tank heat exchanger.

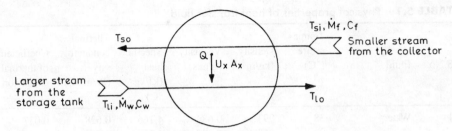

Figure 5.3 Block diagram of countercurrent heat exchanger.

where $\dot{M}_f$ is the mass flow rate of the hot fluid in kg per second available from the collector and C_f is the specific heat of the fluid.

The actual heat exchange ($\dot{Q}$) between the hot fluid of the small stream and the water of the large stream can be expressed as a proportion ε of $\dot{Q}_{max}$, where ε is called as the effectiveness of the heat exchanger, i.e.

$$\dot{Q} = \varepsilon \dot{M}_f C_f (T_{si} - T_{li}) \tag{5.2}$$

The effectiveness of heat exchanger only depends on the exchanger heat transfer properties and the fluid flows and independent of the temperature level. Once these are fixed, the effectiveness can be used to evaluate the performance of heat exchanger in terms of inlet fluid temperature.

In order to express the heat transfer rates and fluid rates for a unit collector area, dividing Equation (5.2) by A_c, we get

$$\dot{q} = \varepsilon \dot{m}_f C_f (T_{si} - T_{li}) \tag{5.3}$$

where $\dot{q} = \dot{Q}/A_c$ and $\dot{m}_f = \dot{M}_f/A_c$.

Referring to Figure 5.3, the hot water available at T_{si} will cool down to T_{so} during its flow in a heat exchanger while cold water available at T_{li} from storage tank will be heated up to T_{lo} during its flow in the opposite direction in the heat exchanger. If there is no heat loss, then the heat loss from the small stream will be the heat gain to the large stream and both will be equal and can be expressed as

$$\text{Heat lost by the small stream} = \dot{q} = \dot{m}_f C_f (T_{si} - T_{so}) \tag{5.4}$$

$$\text{Heat gained by the large stream} = \dot{q} = \dot{m}_w C_w (T_{lo} - T_{li}) \tag{5.5}$$

Equating Equations (5.3) and (5.4), we get the expression for heat exchanger effectiveness, i.e.

$$\text{Effectiveness of the heat exchanger} = \varepsilon = \frac{T_{si} - T_{so}}{T_{si} - T_{li}} \tag{5.6}$$

The above equation defines effectiveness as the ratio of the temperature change actually achieved in the small and large stream respectively.

The flow capacity ratio expressed as ratio of temperature differences required for proper performance or ratio of the temperature changes occurring in the two stream, can be obtained by equating Equations (5.4) and (5.5) respectively as follows:

$$\text{Flow capacity ratio} = \frac{\dot{m}_f C_f}{\dot{m}_w C_w} = \frac{T_{lo} - T_{li}}{T_{si} - T_{so}} \tag{5.7}$$

For countercurrent heat flow, an expression for heat exchanger effectiveness in the terms of the heat transfer properties of the heat exchanger and the flow capacity rates can be expressed as follows:

$$\text{Effectiveness of the heat exchanger} = \varepsilon = \frac{1 - \exp(-B)}{\left[1 - \frac{\dot{m}_f C_f}{\dot{m}_w C_w} \exp(-B)\right]} \tag{5.8}$$

where

$$B = U_x a_x \left[\frac{1}{\dot{m}_f C_f} - \frac{1}{\dot{m}_w C_w}\right] = NTU \left[1 - \frac{\dot{m}_f C_f}{\dot{m}_w C_w}\right]$$

and

$$NTU = \frac{U_x a_x}{\dot{m}_f C_f} = \frac{U_x A_x}{\dot{M}_f C_f}$$

since

$$a_x = \frac{A_x}{A_c} \quad \text{and} \quad \dot{m}_f = \frac{\dot{M}_f}{A_c}.$$

The *NTU* is generally referred as number of transfer unit and for proper heat exchange the value of *NTU* varies between 1 and 10.

EXAMPLE 5.1

Derive an expression for number of transfer unit (*NTU*) in terms of heat exchanger effectiveness.

Solution

Equation (5.8) can be rearranged mathematically as

$$\exp(-B) = \frac{1 - \varepsilon}{1 - \varepsilon \frac{\dot{m}_f C_f}{\dot{m}_w C_w}}$$

Taking log of both sides of this equation and solving for *NTU*, we get

$$NTU = \frac{\ln\left(\frac{1 - \varepsilon \frac{\dot{m}_f C_f}{\dot{m}_w C_w}}{1 - \varepsilon}\right)}{\left(1 - \frac{\dot{m}_f C_f}{\dot{m}_w C_w}\right)}$$

EXAMPLE 5.2

Derive an expression for heat transfer effectiveness and heat transfer unit (*NTU*) for the following cases:

i. if mass flow rate in both the stream are same, and
ii. when one of the flow is very much larger than the other.

Solution

Case (i) If $\dot{m}_f C_f = \dot{m}_w C_w = \dot{m}C$, then Equation (5.8) becomes indeterminate due to $B = 0$. Instead the following equation apply

$$\varepsilon = \frac{NTU}{1 + NTU} \quad \text{and} \quad NTU = \frac{U_x a_x}{\dot{m}c} = \frac{\varepsilon}{1 - \varepsilon}$$

Case (ii) If $\dot{m}_f C_f = \dot{m}C \gg \dot{m}_w C_w$, this means the flow in the collector loop is much higher, then Equation (5.8) becomes

$$\varepsilon = 1 - \exp(-NTU) \quad \text{and} \quad NTU = \frac{U_x a_x}{\dot{m}C} = \ln\frac{1}{1 - \varepsilon}$$

It is important to note that $\dot{m}C$ is the mass flow rate of small stream.

EXAMPLE 5.3

Calculate the temperatures of the return to storage (T_{lo}) and to the collector (T_{so}) for the following configuration of solar water heating system as shown in Figure 5.4. The other parameters are:

i. Collector loop fluid (small stream): 50 percent ethylene glycol with water, $C_f = 3.52$ kJ/kg °C
ii. Outer loop fluid (large stream): water, $C_w = 4.19$ kJ/kg °C
iii. $U_x A_x = 5033 W/°C$
iv. Number of collectors $= 30$

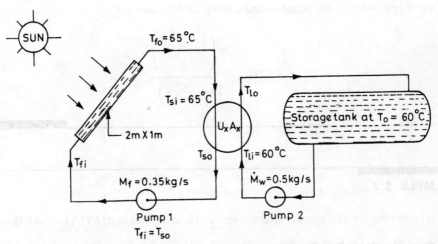

Figure 5.4 Solar water heating system showing the parameters of collector, heat exchanger and storage tank.

Solution

Let us first calculate the following normalized parameters:

Total area of collectors $(A_c) = 30 \times 2 = 60\,\text{m}^2$

$$\dot{m}_w C_w = \frac{\dot{M}_w C_w}{A_c} = \frac{0.5 \times 4190}{60} = 34.92\,\text{W/m}^2\,{}^\circ\text{C}$$

$$\dot{m}_f C_f = \frac{\dot{M}_f C_f}{A_c} = \frac{0.35 \times 3520}{60} = 20.53\,\text{W/m}^2\,{}^\circ\text{C}$$

$$U_x A_x = \frac{U_x A_x}{A_c} = \frac{5033}{60} = 83.88\,\text{W/m}^2\,{}^\circ\text{C}$$

$$NTU = \frac{U_x a_x}{\dot{m}_f C_f} = \frac{83.88}{20.53} = 4.0857$$

$$B = U_x a_x \left[\frac{1}{\dot{m}_f C_f} - \frac{1}{\dot{m}_w C_w} \right] = 83.88 \left[\frac{1}{20.53} - \frac{1}{34.93} \right] = 1.6836$$

After substituting the above values in Equation (5.8), we get

$$\text{Effectiveness of the heat exchanger} = \frac{1 - e^{-1.6836}}{1 - \frac{20.53}{34.93} e^{-1.6836}} = 0.9140$$

The rate of heat transfer from the small stream to large stream can be calculated from Equation (5.3),

$$\dot{q} = 0.9140 \times 20.53(65 - 60) = 93.82\,\text{W/m}^2$$

The outlet temperature from the small stream heat exchanger can be calculated as

$$T_{so} = T_{si} - \frac{\dot{q}}{\dot{m}_f C_f} = 65 - \frac{93.82}{20.53} = 60.43{}^\circ\text{C}$$

The outlet temperature from the large stream heat exchanger can be calculated as

$$T_{lo} = T_{li} + \frac{\dot{q}}{\dot{m}_w C_w} = 60 + \frac{93.82}{34.93} = 62.69{}^\circ\text{C}$$

5.2.3 Heat Exchanger Factor

In this section, a new collector performance equation that directly incorporates the effect of the heat exchanger will be derived.

Referring to Figure 5.1, the basic collector performance equation for $1\,\text{m}^2$ collector area (Equation (3.70)) can be rewritten as

$$\dot{q} = F_R(\alpha\tau)I - F_R U_L(T_{fi} - T_a) \tag{5.9}$$

For small collector flow rate capacity, the outlet temperature from the collector is

$$T_{fo} = T_{fi} + \frac{\dot{q}}{\dot{m}_f C_f} \qquad (5.10)$$

Substitute T_{fi} from Equation (5.10) into Equation (5.9), one gets

$$\dot{q} = F_R(\alpha\tau)I - F_R U_L(T_{fo} - \frac{\dot{q}}{\dot{m}_f C_f} - T_a)$$

or
$$\dot{q} = \frac{F_R(\alpha\tau)I - F_R U_L(T_{fo} - T_a)}{1 - \frac{F_R U_L}{\dot{m}_f C_f}} \qquad (5.11)$$

From Equation (5.3),

$$T_{si} = T_{li} + \frac{\dot{q}}{\varepsilon \dot{m}_f C_f} \qquad (5.12)$$

Since $T_{fo} = T_{si}$ due to negligible heat loss from connecting pipe between outlet of collector and inlet to heat exchanger, then Equation (5.11) can be further written as

$$\dot{q} = \frac{F_R(\alpha\tau)I - F_R U_L(T_{li} + \frac{\dot{q}}{\varepsilon \dot{m}_f C_f} - T_a)}{1 - \frac{F_R U_L}{\dot{m}_f C_f}} \qquad (5.13)$$

The above equation can be solved for $\dot{q}$ as

$$\dot{q} = \frac{F_R(\alpha\tau)I - F_R U_L(T_{li} - T_a)}{1 + \frac{F_R U_L}{\dot{m}_f C_f}\left(\frac{1}{\varepsilon} - 1\right)} \qquad (5.14)$$

Since the heat loss in the connecting pipe between the storage tank and heat exchanger is negligible hence, $T_{li} = T_o$ and Equation (5.14) becomes

$$\dot{q} = \frac{F_R(\alpha\tau)I - F_R U_L(T_o - T_a)}{1 + \frac{F_R U_L}{\dot{m}_f C_f}\left(\frac{1}{\varepsilon} - 1\right)} \qquad (5.15)$$

The above equation can be rewritten as

$$\dot{q} = F_x[F_R(\alpha\tau)I - F_R U_L(T_o - T_a)] \qquad (5.16)$$

where
$$F_x = \left[1 + \frac{F_R U_L}{\dot{m}_f C_f}\left(\frac{1}{\varepsilon} - 1\right)\right]^{-1} \qquad (5.17)$$

The above constant is known as de Winter heat exchanger factor.

Equation (5.16) can be further rewritten as

$$\dot{q} = F_x F_R[(\alpha\tau)I - U_L(T_o - T_a)] \qquad (5.18a)$$

or
$$\dot{q} = F'_x[(\alpha\tau)I - U_L(T_o - T_a)] \qquad (5.18b)$$

where
$$F'_x = F_x F_R$$

EXAMPLE 5.4

Calculate the de Winter heat exchanger factor for $F_R U_L = 5.0\,\text{W/m}^2\,{}^\circ\text{C}$ in Example 5.3.

Solution

From Equation (5.17),

$$F_x = \left[1 + \frac{F_R U_L}{\dot{m}_f C_f}\left(\frac{1}{\varepsilon} - 1\right)\right]^{-1} = \left[1 + \frac{5.0}{20.53}\left(\frac{1}{0.914} - 1\right)\right]^{-1} = 0.9776$$

5.2.4 Natural Convection Heat Exchanger

The temperature difference between the circulating fluid in a heat exchanger and water in the insulated storage tank depends on the flow rate of the fluid in the circulating collector loop as shown in Figure 5.5. This temperature difference determines the heat exchanger parameter $U_x A_x$. For a tank heating water with a given temperature difference, a heat transfer coefficient can be calculated by the formula

$$h = 142\left(\frac{\Delta T_x}{L_0}\right)^{0.25} \tag{5.19}$$

where h is in $\text{W/m}^2\,{}^\circ\text{C}$, ΔT_x is the temperature difference in ${}^\circ\text{C}$ and L_0 is the length of the characteristic path for natural convection in meters. The above equation is valid only for $\Delta T_x = 38{}^\circ\text{C}$. For $\Delta T_x = 65{}^\circ\text{C}$, multiply Equation (5.19) by 1.15 and $\Delta T_x = 21{}^\circ\text{C}$, divide Equation (5.19) by 1.15. For higher ΔT_x, the flow rate in the collector loop should be small.

EXAMPLE 5.5

Calculate the effectiveness of the heat exchanger (ε) and the heat exchanger factor (F_x) for the parameters of the Figure 5.5. Other parameters are:

i. The circulating fluid: glycol solution ($C_f = 3.517\,\text{kJ/kg}\,{}^\circ\text{C}$)
ii. $F_R U_L = 5.069\,\text{W/m}^2\,{}^\circ\text{C}$
iii. An average net heat flux collected by the collector ($\dot{q}$) = 230 W/m^2
iv. Surface area of the heat exchanger (A_x) = 10 m^2.

Solution

Referring to Figure 5.5, we can calculate

$$\dot{m}_f C_f = 84.41\,\text{W/m}^{2\circ}\text{C} \quad \text{and} \quad A_c/A_x = 3.6.$$

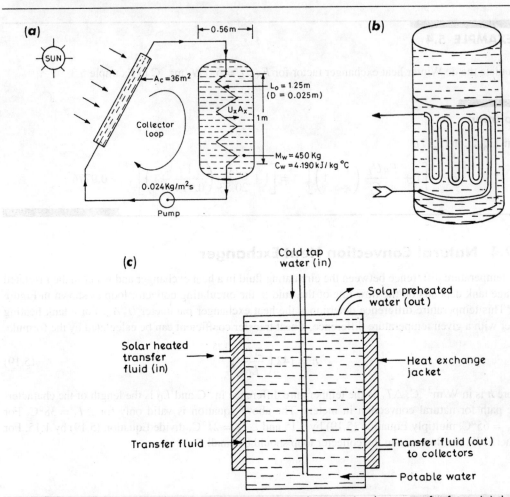

(a)

0.56m

SUN

$A_c = 36m^2$

Collector loop

$L_0 = 1.25m$
(D = 0.025m)

$U_x A_x$

1m

$M_w = 450\,Kg$
$C_w = 4.190\,kJ/kg\,°C$

0.024 Kg/m²s

Pump

(b)

(c)

Cold tap water (in)

Solar preheated water (out)

Solar heated transfer fluid (in)

Heat exchange jacket

Transfer fluid

Transfer fluid (out) to collectors

Potable water

Figure 5.5 Solar water heating system with natural convection heat transfer from, (a) the circulating fluid of heat exchanger to the water in the storage tank, (b) coil type heat exchanger outside tank (Traced tank heat exchanger) and (c) heat exchanger jacket outside tank.

If $\dot{q}$ is the net heat flux collected per unit collector area, then the heat transferred through the heat exchanger to the water of insulated tank is

$$\dot{q}A_c = U_x A_x \Delta T_x$$

In Equation (5.19), $h = Ux$ and substitute ΔT_x from Equation (5.19) in the above equation, we get

$$U_x^5 = \frac{142^4 \dot{q}(A_c/A_x)}{L_0}$$

or,

$$U_x = 52.7 \left[\frac{\dot{q}(A_c/A_x)}{L_0} \right]^{1/5}$$

Substituting the required values in the above equation, we obtain

$$U_x = 52.7 \left[\frac{230 \times 3.6}{125} \right]^{0.20} = 76.92 \, \text{W/m}^2 \, {}^\circ\text{C}$$

Now,

$$NTU = \frac{U_x a_x}{\dot{m}_f C_f} = \frac{76.92 \times (1/3.6)}{84.41} = 0.2531$$

Then from case (ii) of Example 5.2, we have the effectiveness of heat exchanger as

$$\varepsilon = 1 - \exp(-NTU) = 1 - \exp(-0.2531) = 0.2236$$

From Equation (5.17), we have

$$F_x = \left[1 + \frac{F_R U_L}{\dot{m}_f C_f} \left(\frac{1}{\varepsilon} - 1 \right) \right]^{-1} = \left[1 + \frac{5.069}{84.41} \left(\frac{1}{0.2236} - 1 \right) \right]^{-1} = 0.8138$$

5.3 HEAT COLLECTION IN A STORAGE TANK (*Lunde, 1980*)

The hot water from the outlet of collector is fed at the top of storage tank. If the storage tank is in a horizontal position as shown in Figure 5.1, then there is quick mixing between the hot water of the collector and cold water of the storage tank. In the case of vertical position of the storage tank as shown in Figure 5.6 without heat exchanger and Figure 5.5 with heat exchanger, the mixing takes some time due to low density of hot water at the top of the storage tank. The stratification generally took place at very low circulating fluid flow rate. Hence, an analysis of storage tank with and without stratification will be carried out in this section.

5.3.1 Heat Collection with Stratified Storage Tank (Figure 5.6)

From Figure 5.6 and Equation (5.18b), an expression for the total heat collected over a time period of t_T can be expressed as

$$q_T = F_x'[(\alpha\tau)I_T - U_L(T_{si} - T_a)t_T] \tag{5.20}$$

where I_T is the total solar radiation over a time period of t_T.

The hot water fed at the top of storage tank does not affect the temperature of the fluid leaving the tank immediately and hence the collectors operate at a constant efficiency. After time t_s, the first-collected heat at the top reaches the bottom of the tank and hence changes the temperature of the fluid leaving the tank. In this case, the performance of the collector also changes due to change in its inlet temperature. At this time heat collected exactly equals the heat required to raise the temperature of the storage from T_{si} to T_{so}. Mathematically

$$Q_T = \dot{m}_f C_f (T_{fo} - T_{fi})t_s = M_s C_s (T_{so} - T_{si})$$

Since $T_{fo} = T_{so}$ and $T_{fi} = T_{si}$ (Fig. 5.6), the above equation can be solved for t_s as

$$t_s = \frac{M_s C_s}{\dot{M}_f C_f} = \frac{m_s C_s}{\dot{m}_f C_f} \tag{5.21}$$

Here it is important to note that $t_T \leq t_s$.

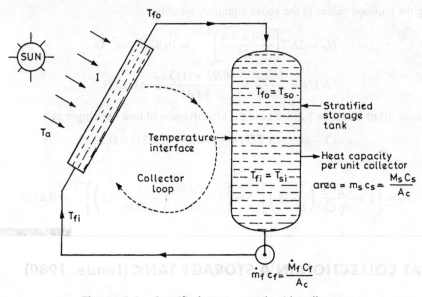

Figure 5.6 Stratified storage tank with collector.

The process can then be repeated with collector operating at new inlet temperature with a new constant efficiency. After one turnover time period t_s, a new storage tank temperature is calculated:

$$T_{so} = T_{si} + \frac{q_T}{m_s C_s}.$$ (5.22)

EXAMPLE 5.6

Calculate the total heat collected and the final storage temperature after six hours operation for the following parameters:

  i. $F_x'(\alpha\tau) = .75$ and $F_x' U_L = 5.0 \, \text{W/m}^2 \,^\circ\text{C}$
  ii. $\dot{m}_f C_f = 45 \, \text{W/m}^2 \,^\circ\text{C}$ and $m_s C_s = 0.32 \, \text{MJ/m}^2 \,^\circ\text{C}$
  iii. Average solar intensity $= 750 \, \text{W/m}^2$, average ambient temperature $= 10^\circ\text{C}$ and $T_{si} = 15^\circ\text{C}$

Solution

From Equation (5.21), the storage turnover time can be calculated as

$$t_s = \frac{m_s C_s}{\dot{m}_f C_f} = \frac{0.32}{45.0} = 0.0071 \, \text{Ms} \cong 2 \, \text{hours}$$

Since the maximum value of t_T is equal to t_s, hence the total heat collected in two hours is therefore calculated from Equation (5.20) as

$$q_T = F_x'[(\alpha\tau)I_T - U_L(T_{si} - T_a)t_T] = F_x'(\alpha\tau)\bar{I} \times t_T - F_x' U_L(T_{si} - T_a) \times t_T$$

$$= 0.75 \times 750.0 \times 0.0071 - 5.0(15.0 - 10.0) \times 0.0071 = 3.8163 \, \text{MJ/m}^2$$

After one turnover time the temperature of entire storage tank can be calculated from Equation (5.22) as

$$T_{so} = T_{si} + \frac{q_T}{m_s C_s} = 15 + \frac{3.8163}{0.32} = 26.93°C$$

During the second turnover time collection will continue with inlet temperature at 26.93°C:

$$q_T = 0.75 \times 750.0 \times 0.0071 - 5.0(26.93 - 10.0) \times 0.0071 = 3.3927 \, MJ/m^2$$

$$T_{so} = T_{si} + \frac{q_T}{m_s C_s} = 26.93 + \frac{3.3927}{0.32} = 37.53°C$$

Similarly, after third turnover time,

$$q_T = 0.75 \times 750.0 \times 0.0071 - 5.0(37.53 - 10.0) \times 0.0071 = 3.0164 \, MJ/m^2$$

$$T_{so} = T_{si} + \frac{q_T}{m_s C_s} = 37.53 + \frac{3.0486}{0.32} = 46.95°C$$

The total heat collected over period of six hour is

$$q_T = 3.8163 + 3.3927 + 3.0164 = 10.2254 \, MJ/m^2$$

5.3.2 Heat Collection with Well-mixed Storage Tank

In this case one turnover time period, t_T, of stratified storage tank is very small and this can be easily achieved in a horizontal storage tank with high circulating fluid flow rate as shown in Figure 5.1. Well-mixed storage might seem to be inherently less efficient due to high temperature of the fluid leaving storage tank. This loss is compensated by the high heat transfer factor attainable with large flow of fluid through the collector. Under this circumstances, T_{si} should be replaced by $\bar{T}_s$ in Equation (5.20) as

$$q_T = F'_x[(\alpha\tau)I_T - U_L(\bar{T}_s - \bar{T}_a)t_T] \tag{5.23}$$

where
$$\bar{T}_s = \frac{T_{so} + T_{si}}{2} \tag{5.24a}$$

Also
$$q_T = m_s C_s(T_{so} - T_{si}) \tag{5.24b}$$

Since $T_{fo} = T_{so}$ and $T_{fi} = T_{si}$, after substituting T_{so} from Equation (5.24a) in the above equation, we obtain

$$\bar{T}_s = T_{si} + \frac{q_T}{2m_s C_s} \tag{5.25}$$

which can be substituted in Equation (5.23) and solved for q_T to give the integrated storage equation

$$q_T = \frac{F'_x[(\alpha\tau)I_T - U_L(T_s - \bar{T}_s)t_T]}{1 + \frac{F'_x U_L t_T}{2m_s C_s}} \tag{5.26}$$

EXAMPLE 5.7

Calculate the hourly variation of temperature for a collector with a well-mixed storage tank for the following parameters:

i. $F'_x = 1.0$, $\alpha\tau = 0.8$, $U_L = 5.0$ W/m² °C, and $T_{si} = 45.0$°C

ii. $m_s C_s = 0.32$ MJ/m² °C

Hourly variation of solar intensity and ambient air are given below:

Time (Hr)	Solar intensity(I) (W/m²)	Ambient air temperature T_a(°C)	q_T (MJ/m²) using (Eq. (5.26))	T_{so}(°C)	q_T (MJ/m²) using (Eq. (5.23))
9	424	11.4	0.5994	46.87	0.5995
10	558	13.5	0.9789	49.93	0.9788
11	641	15.8	1.1980	53.67	1.1981
12	669	18.1	1.2513	57.58	1.2513
13	641	19.8	1.1342	61.12	1.1342
14	558	20.9	0.8589	63.80	0.8589
15	424	21.3	0.4437	65.19	0.4436
Final Average		17.23			
Total	3915		6.4644		

Solution

Here $t_T = 1$ hour$= 3600\,s = 0.0036$ Ms. For first hour, $I = 424$ W/m² and $T_a = 11.4$°C so that from Equation (5.26) one gets

$$q_T = \frac{F'_x[(\alpha\tau)I_T - U_L(T_{si} - \bar{T}_a)t_T]}{1 + \frac{F'_x U_L t_T}{2m_s C_s}}$$

$$= \frac{0.8 \times 424 \times 0.0036 - 5.0(45.0 - 11.4) \times 0.0036}{1 + \frac{5.0 \times 0.0036}{2 \times 0.32}} = \frac{0.6163}{1.0281} = 0.5994\,\text{MJ/m}^2$$

From Equation (5.24b),

$$\bar{T}_{so} = T_{si} + \frac{q_T}{m_s C_s} = 45 + \frac{0.5994}{0.32} = 46.87°C$$

For second hour, $I = 558$ W/m² and $T_a = 13.5$°C so that

$$q_T = \frac{0.8 \times 558 \times 0.0036 - 5.0(46.87 - 13.5) \times 0.0036}{1 + \frac{5.0 \times 0.0036}{2 \times 0.32}} = \frac{1.0064}{1.0281} = 0.9789\,\text{MJ/m}^2$$

and

$$T_{so} = T_{si} + \frac{q_T}{m_s C_s} = 46.87 + \frac{0.9789}{0.32} = 49.93°C$$

Similarly, for other hours q_T and T_{so} can be also calculated and is given in the table. Further, q_T can also be calculated from Equation (5.23) and is given in the same table and it is inferred that the value of q_T calculated from both equation is same for one hour time step.

EXAMPLE 5.8

Estimate q_T and T_{so} for Example 5.7 for five hour time step.

Solution

In this case,

$$t_T = 5 \times 0.0036 = 0.018 Ms$$

$$I_T = (424 + 558 + 641 + 669 + 641) \times 0.0036 = 10.5588 \, \text{MJ/m}^2$$

$$\bar{T}_a = (11.4 + 13.5 + 15.8 + 18.1 + 19.8)/5 = 15.72°C$$

From Equation (5.26),

$$q_T = \frac{0.8 \times 10.5588 - 5.0(45.0 - 15.72) \times 0.018}{1 + \frac{5.0 \times 0.018}{2 \times 0.32}} = \frac{5.81184}{1.140} = 5.0981 \text{MJ/m}^2$$

and given

$$T_{so} = T_{si} + \frac{q_T}{m_s C_s} = 45.0 + \frac{5.0981}{0.32} = 60.93°C$$

The above calculated value of T_{so} is lower by 0.4°C in comparison of one hour time step value of $T_{so} = 61.33$ (from the table in Example 5.7) due to longer time step.

5.3.3 Effect of Heat Load (Lunde, 1980)

In previous section, the integrated storage equation has been derived without considering any heat load. The effect of heating load is shown in Figure 5.7.

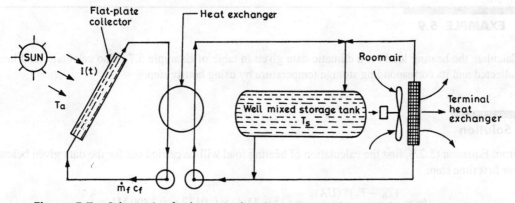

Figure 5.7 Schematic of solar water heating system with terminal heat exchanger.

If l_T is the demand expressed as the head load per unit collector area and it is defined as

$$l_T = \frac{(T_b - T_a)^+ (UA)_b}{A_c} \qquad (5.27)$$

where the positive sign (+) indicates that only positive value in the bracket should be considered for thermal heating of a building and $(UA)_b$ is the overall heat loss coefficient from the room of a building to ambient air through different walls/roof/floor/doors/windows etc. The typical value of $[(UA)_b / A_c] = 0.01386\,\text{MJ/m}^2\,°\text{C hr}$. This can vary depending upon design of a building. The base temperature of room air (T_b) can be considered as $25°\text{C}$ for the present case. Now, the expression for the net heat collected (q_N) can be written as

$$q_N = q_T - l_T. \qquad (5.28)$$

Similar to Equation (5.25), the average temperature of well-mixed storage tank can be written as

$$\bar{T}_s = T_{so} + \frac{q_N}{2m_s C_s}. \qquad (5.29)$$

With the help of Equations (5.23), (5.28) and (5.29), an expression for the net heat collection can be derived as

$$q_N = \frac{F_x'[(\alpha\tau)I_T - U_L(T_{si} - \bar{T}_a)t_T] - l_T}{1 + \frac{F_x' U_L t_T}{2m_s C_s}} \qquad (5.30a)$$

The rise in temperature of storage tank during the day can be obtained from

$$T_{so} = T_{si} + \frac{q_N}{m_s C_s} = T_{si} + \frac{q_T - l_T}{m_s C_s}. \qquad (5.30b)$$

Further, an expression for total heat collected can be obtained by substituting above expression in Equation (5.28) as

$$q_T = \frac{F_x'[(\alpha\tau)I_T - U_L(T_{si} - \bar{T}_a)t_T]}{1 + \frac{F_x' U_L t_T}{2m_s C_s}} + \frac{l_T}{1 + \frac{2m_s C_s}{F_x' U_L t_T}} \qquad (5.31)$$

EXAMPLE 5.9

Calculate the heating load for a climatic data given in table of Example 5.7. Also calculate total heat collected and its corresponding storage temperature by using hourly step.

Solution

From Equation (5.27), first the calculation of heating load will be carried out for the data given below. For first time step:

$$l_T = \frac{(T_b - T_a)^+ (UA)_b}{A_c} = (25 - 11.4) \times 0.0139 = 0.1890\,\text{MJ/m}^2$$

For second time step:

$$l_T = \frac{(T_b - T_a)^+(UA)_b}{A_c} = (25 - 13.5) \times 0.0139 = 0.1599 \, \text{MJ/m}^2$$

Similar calculations can be carried out for other time steps. The results have been shown in the same table.

Time (Hr)	Solar intensity (I) (W/m^2)	Ambient air temperature T_a (°C)	l_T (MJ/m^2)	q_T (MJ/m^2) using Eq. (5.26)	T_{so}(°C)
9	424	11.4	0.1890	0.6046	46.30
10	558	13.5	0.1599	1.0159	48.98
11	641	15.8	0.1279	1.2879	52.61
12	669	18.1	0.0959	1.4057	56.70
13	641	19.8	0.0723	1.3564	60.71
14	558	20.9	0.0570	1.1427	64.10
15	424	21.3	0.0514	0.7742	66.36
Final Average		17.23			
Total	3915		0.7534	7.5874	

After calculating heating load, the total heat collected can be obtained from Equation (5.31) as

$$q_T = \frac{F_x'[(\alpha\tau)I_T - U_L(T_{so} - \bar{T}_a)t_T]}{1 + \frac{F_x' U_L t_T}{2m_s C_s}} + \frac{l_T}{1 + \frac{2m_s C_s}{F_x' U_L t_T}}$$

$$= \frac{1 \times [0.8 \times 424 - 5 \times (45 - 11.4)] \times 0.0036}{1 + \frac{1 \times 5 \times 0.0036}{2 \times 0.32}} + \frac{0.1890}{1 + \frac{2 \times 0.32}{1 \times 5 \times 0.0036}}$$

$$= \frac{0.6163}{1.0281} + \frac{0.1890}{36.555} = 0.6046 \, \text{MJ/m}^2$$

For first time step:
From Equation (5.30b),

$$T_{so} = T_{si} + \frac{q_T - l_T}{m_s C_s} = 45 + \frac{0.6046 - 0.1890}{0.32} = 46.3°C$$

For second time step:

$$q_T = \frac{1 \times [0.8 \times 558 - 5 \times (45 - 13.5)] \times 0.0036}{1 + \frac{1 \times 5 \times 0.0036}{2 \times 0.32}} + \frac{0.1599}{1 + \frac{2 \times 0.32}{1 \times 5 \times 0.0036}}$$

$$= \frac{1.0400}{1.0281} + \frac{0.1599}{36.555} = 1.0159 \, \text{MJ/m}^2$$

$$T_{so} = 46.30 + \frac{1.0159 - 0.1599}{0.32} = 48.98°C$$

Similar calculation can be carried out for other time step. The results have been given in the same table.

EXAMPLE 5.10

Calculate the rise in temperature of storage of Example 5.9 by using entire day time step.

Solution

For entire day time step, use the data of table in Example 5.9

$$q_T = \frac{1 \times [0.8 \times 3915 \times 0.0036 - 5 \times (45 - 17.23) \times 7 \times 0.0036]}{1 + \frac{1 \times 5 \times 0.0036}{2 \times 0.32}} + \frac{0.7534}{1 + \frac{2 \times 0.32}{1 \times 5 \times 0.0036}}$$

$$= \frac{7.7762}{1.0281} + \frac{0.7534}{36.555} = 7.5842 \text{MJ/m}^2$$

$$T_{so} = 45.00 + \frac{7.5842 - 0.534}{0.32} = 67.03°C$$

which is nearly same as obtained by one hour time step in Example 7.9.

PROBLEMS

5.1 Calculate effectiveness of heat exchange (ε) and heat exchanger factor for $\dot{q} = 500$ W/m^2. The other parameters are same as given in Example 5.5(a).

Hint See Example 5.5(a).

5.2 Repeat Example 5.5(a) by assuming $\Delta T_x = 21°C$, $38°C$ and $65°C$.

Hint Find U_x from $\dot{q} A_c = U_x A_x \Delta T_x$ and see Example 5.5(a).

5.3 What will be the total heat collected and the final storage temperature after four hour operation of the system for the same parameters of Example 5.6.

Hint See Example 5.6, only the second turnover time collection calculations are required.

5.4 Calculate the total heat collected after ten hours operation of the system for the following parameters:

  i. $F'_x(\alpha\tau) = 0.5$ and $F'_x U_L = 2.0$ W/m^2 °C
  ii. $\dot{m}_f C_f = 45$ W/m^2 °C and $m_s C_s = 0.32$MJ/m^2 °C
  iii. $\bar{T} = 500$ W/m^2 and $T_a = 5°C$

Hint See Example 5.6.

6.2 DESCRIPTION AND CLASSIFICATION

CHAPTER 6

Solar Air Heaters

6.1 INTRODUCTION

This chapter provides the description and analysis of various types of solar air heaters used in space heating and for drying purposes. The solar air heaters have following advantages over other solar heat collectors:

i. The need to transfer heat from the working fluid to another fluid is eliminated as air is being used directly as the working substance. The system is compact and less complicated.

ii. Corrosion, which can cause serious problems in solar water heater, is completely eliminated.

iii. Leakage of air from the duct does not pose any major problem.

iv. Freezing of working fluid virtually does not exist.

v. The pressure inside the collector does not become very high.

Thus, air heater can be designed using cheaper as well as lesser amount of material and is simpler to use than the solar water heaters.

Air heaters have certain disadvantages also. The first and foremost being the poor heat transfer properties of air. Special care is required to improve the heat transfer. Another disadvantage is the need for handling large volumes of air due to its low density. Also, the thermal capacity of air being low, it cannot be used as a storage fluid. In the absence of proper design the cost of air heater can be very high.

The applicability of a collector, however, depends on various factors such as high efficiency, low fabrication, installation and operational costs and other practical aspects regarding the specific use. Extensive work on solar air heaters has been reported in literature. Various geometries have been proposed and their theoretical investigations carried out. Some of them have been fabricated, tested and their field experiences have been published. Selcuk (1977) gives a comprehensive literature review on solar air heaters. It may be said that the state of the art of solar air heaters has been a technical achievement awaiting commercial exploitation.

6.2 DESCRIPTION AND CLASSIFICATION

A conventional solar air heater is essentially a flat plate collector with an absorber plate, a transparent cover system at the top and insulation at the bottom and on the sides. The whole assembly is encased in a sheet metal container. The working fluid is air, though the passage for its flow varies according to the type of air heater.

Material for construction of air heaters are similar to those of liquid flat plate collectors. The transmission of solar radiation through the cover system and it's subsequent absorption in the absorber plate can be given by expressions identical to those of liquid flat plate collectors. Selective coating on the absorber plate can be used to improve the collection efficiency but cost effectiveness criterion should be kept in mind.

Depending on the type of the absorber plate, the air heater can be non-porous or porous. Figures 6.1 and 6.2 show the basic features of these schematically.

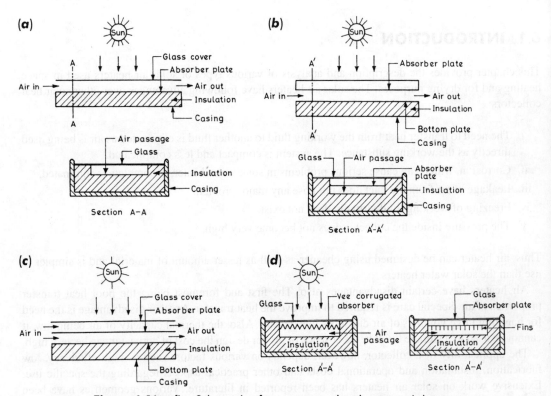

Figure 6.1(a–d)   Schematic of non-porous absorber type air heaters.

6.2.1 Non-porous Type

In non-porous type, air stream does not flow through below the absorber plate but air may flow above and/or behind the plate (Sodha *et al.*, 1982a).

i. In the first type, no separate passage is required and the air flows between the transparent cover system and the absorber plate (Fig. 6.1(a)). In this heater, as the hot air flows above the absorber, the cover receives much of the heat and in turn, looses it to the ambient. Thus, a substantial amount of heat is lost to the ambient and hence this air heater is not recommended.

ii. The non-porous type with air passage below the absorber is most commonly used. A plate parallel to the absorber plate is provided in between the absorber and the insulation, thus forming a passage of high aspect ratio (Fig. 6.1(b)) for the air flow.

iii. In another variety of non-porous type of air heater, the absorber plate is cooled by air stream flowing on both sides of the plate (Fig. 6.1(c)).

It may be noted that the heat transfer between the absorber plate and the flowing air being low, the efficiency of air heaters is less. The performance, however, can be improved by roughening the absorber surface or by using a vee-corrugated plate as the absorber. The heat transfer can also be increased by adding fins to the absorber plate (Fig. 6.1(d)). Turbulence induced to the air flow helps increase the convective heat transfer.

The radiative losses from the absorber plate are significant, unless selective coatings are used, decreasing the collection efficiency. Also, the use of fins may result in a prohibitive pressure drop, thus limiting the applicability of non-porous type.

6.2.2 Porous Type

The second type of air heaters has porous absorber which may include slit and expanded metal, overlapped glass plate absorber and transpired honeycomb.

The air heater with porous type of absorber has the following advantages:

i. solar radiation penetrates to a greater depth and is absorbed along it's path. Thus the radiation loss decreases. Air stream heats up as it passes through the matrix.

ii. the pressure drop is usually lower than the non-porous type.

It may be noted however, that an improper choice of matrix porosity and thickness may cause reduction in efficiencies as beyond an optimum thickness, matrix may not be hot enough to transfer the heat to air stream.

Wire mesh (Fig. 6.2(a)), porous bed formed by broken bottles (Fig. 6.2(b)) and overlapped glass plate (Fig. 6.2(c)) are some examples of porous type of absorbers.

6.3 CONVENTIONAL HEATER

6.3.1 Thermal Analysis

Figure 6.1(b) shows schematics of a conventional air heater. The air to be heated flows in a parallel passage below the absorber. The performance analysis of such a collector does not include the fin effect or the tube-to-plate bond conductance, similar to liquid flat plate collectors.

The thermal performance of such a heater was first investigated analytically by Whillier (1964). The analysis of a steady state model is as follows:

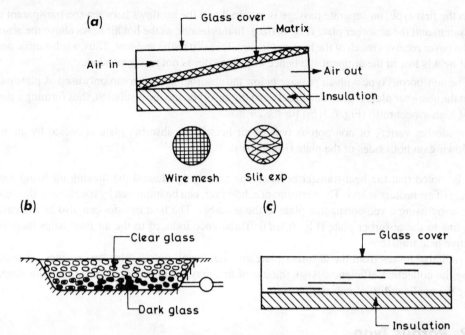

Figure 6.2 (a–c) Schematics of porous absorber type air heaters. (*Selcuk, 1977*)

Let the length and the width of the absorber plate be L_1 and L_2 respectively. Let us consider an element of area $L_2 dx$ at a distance x from the inlet (Fig. 6.3(a)). Energy balance equations for the absorber plate, bottom plate and air stream (Fig. 6.3(b)) can be written as:

$$I(t) = U_t(T_{pm} - T_a) + h_{pf}(T_{pm} - T_f) + h_{rpb}(T_{pm} - T_{bm}) \tag{6.1}$$

$$h_{rpb}(T_{pm} - T_{bm}) = h_{bf}(T_{bm} - T_f) + U_b(T_{bm} - T_a) \tag{6.2}$$

and $$\dot{m}C_{air} d\,T_f = h_{pf}L_2dx(T_{pm} - T_f) + h_{bf}\,L_2dx(T_{bm} - T_f) \tag{6.3}$$

These equations can be solved for the air temperature in a way similar to that of liquid flat plate collector. Then, the rise in the temperature of the air through the duct can be estimated, to write the useful heat gain rate of the collector ($\dot{Q}_u$) in the form:

$$\dot{Q}_u = F_R A_P[I - U_L(T_{fi} - T_a)] \tag{6.4}$$

where F_R is the collector heat removal factor and is given by

$$F_R = \dot{m}C_{air}/(U_L A_P)[1 - \exp[-F'U_L A_P/(\dot{m}C_{air})] \tag{6.5}$$

where $$U_L = U' + (1/F')[U_b h_{bf}/(h_{rpb} + h_{bf} + U_b)] \tag{6.6a}$$

$$U' = U_t + [h_{rpb}U_b/(h_{rpb} + h_{bf} + U_b)] \tag{6.6b}$$

and the collector efficiency factor (F') is given by

$$F' = [1 + U'/h_e]^{-1} \text{ with } h_e = h_{pf} + [h_{bf} h_{rab}/(h_{rpb} + h_{bf} + U_b)] \tag{6.7}$$

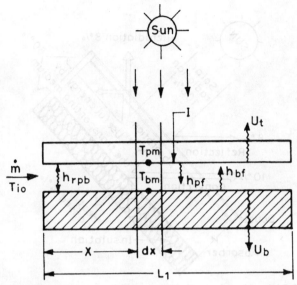

Figure 6.3(a) Heat transfer process in a conventional air heater.

where A_p is the area of the absorber plate (m^2), I the solar intensity (W/m^2), U_b the bottom loss coefficient (W/m^2 °C), U_t the top loss coefficient (W/m^2 °C), h_{bf} the convective heat transfer coefficient between bottom and fluid (W/m^2 °C), h_{pf} the convective heat transfer coefficient between plate and fluid (W/m^2 °C), h_{rpf} the radiative heat transfer coefficient between plate and the bottom (W/m^2 °C), T_a, T_f, T_{fi}, T_{pm} and T_{bm} are respectively the ambient, the fluid, the inlet air, the mean plate and the mean bottom plate temperature (°C), $\dot{m}$ the flow rate (kg/s) and C_{air} the specific heat of air (J/kg °C).

The useful heat gain can be calculated from Equation (6.4) and is in a form similar to that of liquid flat plate collectors. The effect of various parameters on its performance and hence the efficiency can be studied. Heat losses in percentage in air heater has been shown in Figure 6.3(c).

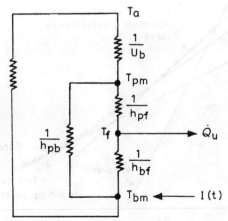

Figure 6.3(b) Thermal resistance representation of heat losses in a conventional air heater.

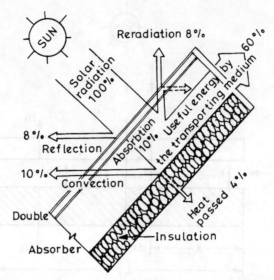

Figure 6.3(c) Heat losses in percentage in a conventional air heater.

Malik and Buelow (1975), after having surveyed the heat transfer process in air heaters have recommended the following correlation for Nusselt number in the case of a smooth absorber plate;

$$Nu = \frac{0.01344 Re^{3/4}}{1 - 1.586 Re^{-1/8}} \qquad (6.8)$$

Selcuk (1977) has presented generalized curves for the performance of non-porous type of air heaters (Fig. 6.4). As seen from the figure, for low temperature rise, transparent covers are not at all required.

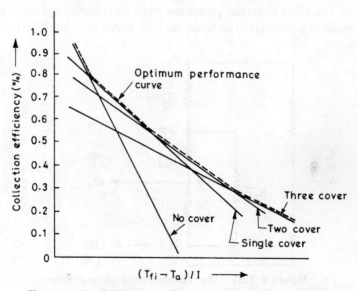

Figure 6.4 Optimum performance curve (*Selcuk, 1977*).

However, beyond a certain temperature, heat loss reduction due to single cover system compensates the transmission losses and becomes advantageous. The number of covers is, however, finally determined from economic considerations.

EXAMPLE 6.1(a)

Air at 28 °C approaches a 0.8 m long and 0.5 m wide flat plate (at 40 °C) with an approach velocity $V_\infty = 4$ m/s. Determine the total rate of heat transfer from the plate to the air.

Solution

The viscosity of air at 32 °C (average temperature) is 1.578×10^{-5} (see Appendix III).
Reynold number can be calculated by using Equation (2.38b) and after the substitution of the parameters it's value is obtained as,

$$Re = \frac{V_\infty L}{\nu} = \frac{(4)(0.8)}{1.578 \times 10^{-5}} = 2.027 \times 10^5$$

The flow being laminar, the heat transfer coefficient is given by Equation (2.47g)

$$\bar{h} = (K/L)\bar{N}u = (K/L)(0.664)Re^{1/2} Pr^{1/3}$$
$$= (0.026/0.8)(0.664)(2.027 \times 10^5)^{1/2}(0.713)^{1/3} = 8.676 \, \text{W/m}^2\text{K}$$

The total rate of heat transfer will be

$$\dot{Q} = \bar{h}A\Delta T = 8.676 \times 0.8 \times 0.5 \times 12 = 41.645 \, \text{W}.$$

EXAMPLE 6.1(b)

Calculate the convective heat transfer coefficient from the plate (1 m × 4 m) to the flowing air at 0.056 kg/s. The air channel depth is 10 mm (Fig. 3.18(e)). The inlet air temperature is 70 °C.

Solution

The convective heat transfer coefficients between the air and two duct walls will be assumed to be same ($h_1 = h_2$). In the present case, the characteristic length is twice the plate spacing ($D_n = 2 \times 0.01$ m), Equation (2.47j). The length-to-characteristic length ratio is

$$\frac{L}{D_n} = \frac{4}{2 \times 0.01} = 200$$

The Reynold number at 70 °C is

$$Re = \frac{\rho V D_n}{\mu} = \frac{\dot{m}D_n}{A_f \mu} = \frac{0.056 \times (2 \times 0.01)}{(0.01 \times 1) \times 2.04 \times 10^{-5}} = 5490$$

Since $Re > 2100$ and L/D_n is large, hence from Equation (2.47j).

$$h_1 = h_2 = \frac{K}{D_n} \cdot Nu = \frac{0.029}{2 \times 0.01} \times 0.0158(5490)^{0.8} = 22.5\,\text{W/m}^2\,{}^\circ\text{C}.$$

EXAMPLE 6.2

The following specifications for a single pass air heater are given: length = 2 m, width = 1 m, spacing = 10 mm, plate temperature = 60 °C, glass temperature = 20 °C, flow rate $(\dot{m}) = 0.0017$ kg/s, ambient air temperature = 9 °C, wind velocity = 3 m/s, bottom insulation thickness = 50 mm.

Calculate the heat transfer coefficient between the absorber and the cover. Also calculate the following parameters:

a. Overall heat transfer coefficient from absorber to ambient.
b. U_L and F'.
c. The flow rate factor.

Solution

The average temperature is 40 °C and the properties of air at this temperature are: $\nu = 1.701 \times 10^{-5}$ m^2/s, $\rho = 0.1130$ kg/m^3, $K = 0.0272$ W/mK, $Pr = 0.7051$, $C_p = 1005.867$ J/kgK (see Appendix III)

The flow rate $\dot{m}$ is given as, $\dot{m} = L\,b\,d\rho/t$.

Thus, $L/t = v = \dot{m}/b\,d\rho = 0.0017/(1 \times 10^{-2} \times 0.1130) = 1.504$ m/s

Reynold number can be calculated by using Equation (2.38b),

$Re = Lv/\nu = 2 \times 1.504/1.701 \times 10^{-5} = 1.768 \times 10^5 < 5 \times 10^5$ (Laminar flow)

From Equation 2.47h, Nusselt number can be calculated as

$$Nu = \frac{0.3387(1.768 \times 10^5)^{1/2}(0.7051)^{1/3}}{\left[1 + \left(\frac{0.0468}{0.7051}\right)^{2/3}\right]^{1/4}} = 122.02$$

The heat transfer coefficient can be calculated by the relation, $Nu = h_{pf}L/K$

The value of h_{pf} obtained is 1.659 W/m^{2}K.

The radiative heat transfer coefficient can be calculated (Equation (3.17)) as

$$h_r = 0.82 \times 5.67 \times 10^{-8}\left[\frac{(60+273)^4 - (20+273)^4}{60-20}\right] = 5.726\,\text{W/m}^2\text{K}$$

Total heat transfer coefficient between the plate and cover is $h_1 = 1.659 + 5.726 = 7.385$ W/m^{2}K.

The heat transfer coefficient from the cover to ambient is given by Equation (2.49a), $h_2 = 5.7 + 3.8\,V$ = 17.1 W/m^{2}K (for $V = 3$ m/s)

a. The overall heat transfer coefficient from absorber to ambient is

$$U_t = \left[\frac{1}{h_1} + \frac{1}{h_2} \right]^{-1} = \left[\frac{1}{7.385} + \frac{1}{17.1} \right]^{-1} = 5.158 \text{ W/m}^2\text{K}.$$

b. The bottom loss coefficient

$$U_b = \left[\frac{L_i}{K_i} + \frac{1}{h_i} \right]^{-1} = 0.699 \text{ W/m}^2\text{K}.$$

From Equation (4.6b), U' can be calculated as

$$U' = 5.158 + \frac{5.726 \times 0.699}{5.726 + 1.659 + 0.699} = 5.653 \text{ W/m}^2\text{K}$$

$$h_e = 1.659 + \left[\frac{1.659 \times 5.726}{5.726 + 1.659 + 0.699} \right] = 2.834 \text{ W/m}^2\text{K}$$

The collector efficiency factor F' calculated from Equation (6.7), will be

$$F' = \left[1 + \frac{U'}{h_e} \right]^{-1} = \left[1 + \frac{5.653}{2.834} \right]^{-1} = 0.334$$

Calculating U' by using Equation (6.6a) and substituting the values, we get $U_L = 6.084 \text{ W/m}^2\text{K}$.

c. The flow rate factor F_R, from Equation (6.5), will be

$$F_R = \frac{0.0017 \times 1005.867}{6.0824 \times 2} \left[1 - \exp \left\{ -\frac{0.334 \times 6.0824 \times 2}{0.0017 \times 1005.867} \right\} \right] = 0.1275$$

There are various types of conventional air heaters which have nonporous type of absorber. The performance analysis of some of them have been discussed here.

6.4 DOUBLE EXPOSURE HEATERS

In the previous section, we discussed the performance of a conventional air heater. It is also called a single exposure air heater, as only one plate is exposed to the solar radiation while the other is insulated from the bottom. If, however, the rear plate is blackened and irradiated (which can be done by removing the insulation and using the reflectors suitably), then the air heater is called a double exposure air heater. Thus, it consists of a flat passage between two parallel metallic plates which are blackened and glazed. As the air stream receives heat from both the plates forming the passage (Fig. 6.5), the efficiency will certainly be increased.

The performance analysis of such a heater can be done in a manner similar to that of a conventional or single exposure air heater.

Energy balance equations for different components of double exposure air heater are:

$$I_1(L_2 dx) = U_t(T_{pm} - T_a)(L_2 dx) + h_{pf}(T_{pm} - T_f)(L_2 dx) + h_{rpb}(T_{pm} - T_{bm})(L_2 dx) \quad (6.9)$$

$$I_b(L_2 dx) + h_{rpb}(T_{pm} - T_{bm})(L_2 dx) = h_{bf}(T_{bm} - T_f)(L_2 dx) + U_b(T_{bm} - T_a)(L_2 dx) \quad (6.10)$$

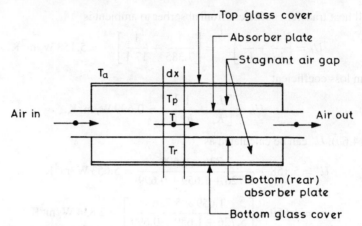

Figure 6.5 Schematic sketch of double exposure solar air heater.

and
$$\dot{m}C_{\text{air}}(dT_f/dx)dx = h_{fb}(T_{pm} - T_f)(L_2 dx) + h_{bf}(T_{bm} - T_f)(L_2 dx) \qquad (6.10a)$$

where I_1 is the solar flux incident on the absorber plate, W/m². These equations can be solved as done in the case of liquid flat plate collectors.

The transient analyses of these air heaters are more appropriate in view of the transient boundary conditions. In the next section, the outlines of transient analysis for a simple case have been presented. The analysis can be extended to other cases as well.

6.5 AIR HEATER WITH FLOW ABOVE THE ABSORBER

Figure 6.1(a) shows the schematic of such a heater. It consists of an absorber plate with a transparent cover at the top and insulation at the bottom. The cover and the plate provides the passage for the air. Solar radiation, after transmission through the cover, is absorbed by the absorber plate. Part of the absorbed energy is convected to air and the heated air moves upwards, and the rest is lost to the ambient through the cover and bottom insulation.

First, we outline the steady state analysis of this air heater and later the transient one.

6.5.1 Steady State Analysis

As was done in the case of conventional air heater (Section 6.3), the steady state performance of air heater with flow above the absorber, can be studied. The difference being that in Section 6.3, the analysis was in forced convection mode whereas in this case it is in the free convection mode. The energy balance conditions can be written as:

Absorber:
$$I(t) = h_{pf}(T_{pm} - T_f) + h_{rpc}(T_{pm} - T_{cm}) + U_b(T_{pm} - T_a) \qquad (6.11)$$

Cover:
$$h_{rpc}(T_{pm} - T_{cm}) + h_{1f}(T_f - T_{cm}) = h_2(T_{cm} - T_a) \qquad (6.12)$$

Fluid:
$$\dot{q}_u = h_{pf}(T_{pm} - T_f) + h_{1f}(T_{cm} - T_f) \qquad (6.13)$$

where $\dot{q}_u$ is the gain rate for useful energy for the fluid per m².

Solving Equations (6.11) and (6.12) for T_{pm} and T_{cm} and substituting the values in (6.13), we can solve for $\dot{q}_u$ to obtain

$$\dot{q}_u = F'[I(t) - U_L(T_f - T_a)] \tag{6.14}$$

where the collector efficiency factor (F') and the heat loss factor (U_L) are given by as follows:

$$F' = B/[h_{1f}(h_{pf} + U_b) + h_{pf}h_2 + h_{rpc}(h_{1f} + h_{pf} + U_b + h_2) + U_b h_2] \tag{6.15}$$

$$U_L = [h_{1f}(h_{pf}U_b + h_{pf\,h_2 + U_b h_2}) + h_{pf}U_b h_2] + h_{rpc}(h_{1f}U_b + h_{1f}h_2 + h_{pf}U_b + h_{pf}h_2)/B \tag{6.16}$$

where

$$B = h_{1f}h_{pf} + h_{pf}h_2 + h_{rpc}(h_{1f} + h_{pf}) \tag{6.17}$$

Once the values of F' and U_L are known, the heat removal factor F_R, in the forced circulation mode, can be calculated for a given flow rate (Equation (6.5)), and the rate of useful energy in terms of inlet air temperature (T_{fi}) can be known from Equation (6.4) (similar to Section 6.3). It may be mentioned here that the collector efficiency factor F' as obtained above and that obtained in Section 6.3 will be different.

EXAMPLE 6.3

Two horizontal surface separated by a 40 mm layer of air have upper and lower temperature of 288 K and 320 K respectively. Determine the rate of heat exchange per square meter between the two surfaces.

Solution

In the given problem $T_{avg} = (288 + 320)/2 = 304$ K. At this average temperature the properties of air are:
$v = 1.61 \times 10^{-5}$ m^2/s, $K = 0.026$ W/mK, $Pr = 0.708$ (see Appendix III).
Grashof number from Equation (2.38d), is

$$Gr = (9.8 \times 32(0.04)^3)/(304(1.61 \times 10^{-5})^2) = 2.547 \times 10^5$$

Nusselt number can be obtained as (Table 2.1)

$$\bar{Nu} = (\bar{h}b/K) = (0.195)(Gr)^{1/4} = 4.38$$

The heat transfer coefficient h can be written as

$$\bar{h} = (\bar{Nu}\,K/b) = \frac{4.38(0.026)}{0.040} = 2.847 \text{ W/m}^2\text{K}$$

The rate of heat transfer per square meter between two surfaces is

$$\dot{q} = \frac{\dot{Q}}{A} = \bar{h}(T_h - T_c) = 2.847 \times 32 = 91.104 \text{ W/m}^2$$

EXAMPLE 6.4

An absorber at 44 °C is placed 5 cm below the glass cover. If the glass cover is at 10 °C, calculate the convective and the radiative heat transfer coefficients between the absorber and the glass cover. Also calculate the total heat transfer coefficient between the absorber and the cover and the overall heat transfer coefficient between the absorber and the ambient.

Solution

The average temperature $T_{avg} = 27\,°C$, $K = 0.026\,W/m.K$
$\nu = 1.568 \times 10^{-5}\,m^2/s$, $Pr = 0.708$, $\beta' = 1/300$ (see Appendix III).
Grashof number is

$$Gr = \frac{g\beta'(T_h - T_c)}{\nu^2}b^3 = \frac{9.8 \times 3.33 \times 10^{-3} \times (44 - 10)(0.05)^3}{(1.568 \times 10^{-5})^2} = 5.64 \times 10^5$$

The value of Nusselt number can be calculated by using Table 2.1

$$\bar{N}u = (\bar{h}\,b/K) = (0.068)(Gr)^{1/3} \quad \text{for,} \quad 4 \times 10^5 < Gr$$

$$= (0.068)(82.623) = 5.618$$

or $\qquad \bar{h} = h_{pf} = (5.618 \times (0.026)/0.05 = 2.921\,W/m^2K$

where, $h_{1f} = h_{pf}$: Then,

$$h_{c,pc} = \left[\frac{1}{h_{pf}} + \frac{1}{h_{1f}}\right]^{-1} = \left[\frac{2}{2.921}\right]^{-1} = 1.46\,W/m^2K$$

The radiative heat transfer coefficient

$$h_{rpc} = \frac{5.67 \times 10^{-8}((44 + 273)^2 + (10 + 273)^2)(44 + 10 + 546)}{1/0.8 + 1/0.8 - 1} = 4.095\,W/m^2K$$

Now, the total heat transfer coefficient

$$h_1 = h_{c,pc} + h_{rpc} = 1.46 + 4.095 = 5.55\,W/m^2K \text{ and, } h_2 = 5.7\,W/m^2K \text{ (for, } V = 0\,m/s).$$

Thus, the overall heat transfer coefficient from the absorber to the ambient

$$U_t = [(1/h_1) + (1/h_2)]^{-1} = [(1/5.55) + (1/5.7)]^{-1} = 2.81\,W/m^2K$$

EXAMPLE 6.5

Calculate F' and U_L for Example 4.4 if a 5 cm thick insulation of thermal conductivity 0.04 W/m K is placed below absorber.

Solution

The bottom loss coefficient

$$U_b = \left[\frac{L_i}{K_i} + \frac{1}{h_i} \right]^{-1} = (1.25 + 0.18)^{-1} = 0.699 \text{ W/m}^2\text{K}$$

where $h_i = h_2$ for $V = 0$ m/s.

Now, $h_{1f} = h_{pf} = 2.921$ W/m^{2}K and $h_2 = 2.81$ W/m^{2}K (from Example 4.4).
Substituting these values in Equation (6.15), we get $F' = 0.689$.
Substituting these values in Equation (6.16), we get $U_L = 3.088$ W/m^2 °C.

6.5.2 Transient Analysis

In view of the time dependent behavior of climatic parameters, the transient analysis of air heater is more appropriate. The analysis followed here (Ranjan *et al.*, 1983) refers to the configuration shown in Figure 6.6.

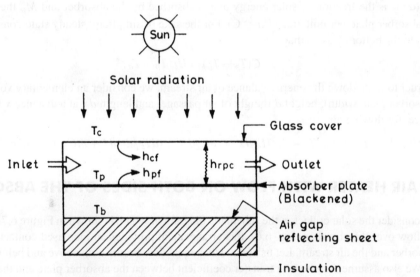

Figure 6.6 Schematic sketch of the system.

Here, an air gap is introduced between the absorber plate and the insulation with the help of a reflecting sheet. This helps in reducing the bottom loss. However, for air heater with no air gap and reflecting sheet, the analysis can be easily done as a special case. The following assumptions have been made in writing the energy balance for the cover and the absorber plate:

i. Temperatures of cover and absorber plate are uniform.

ii. Heat transfer coefficients representing heat transfer from or into air stream are considered constant and their average values are used.

iii. The flow rate is assumed constant.

iv. The side losses are neglected since collector area is taken to be large compared to its thickness.

v. The bulk mean temperature of air changes in the flow direction.

vi. The loss from the plate through the bottom insulation is represented by a steady state expression.

The energy balance for the cover and the absorber plate can be written as:

Cover:

$$\alpha_c I(t) + h_{lf}(T_f - T_c) + h_{rpc}(T_p - T_c) = U_t(T_c - T_a) \tag{6.18}$$

where α_c is the fraction of solar energy that is absorbed by the cover, h_{lf} the heat transfer coefficient between the fluid and the cover, h_{rpc} the radiative heat transfer coefficient between the plate and the cover and $I(t)$ the incident solar radiation.

The equation differs from Equation (6.12) in having a term corresponding to the absorption of solar radiation in the glass cover ($\alpha_c I(t)$) though its value is small, as such it does not contribute much.

Absorber plate:

$$(\alpha\tau)_p I(t) = h_{pf}(T_p - T_f) + h_{rpc}(T_p - T_c) + C(T_p - T_b) + M_p(dT_p/dt) \tag{6.19}$$

where $(\alpha\tau)_p$ is the fraction of solar energy that is absorbed by the absorber and M_p the heat capacity of the absorber plate per unit area (J/m^2 °C). For the sake of simplicity steady state considerations are applied to the bottom loss so that

$$C(T_p - T_b) = U_b(T_b - T_a) \tag{6.20}$$

In order to write down the energy balance of air stream, we consider an elementary volume of width L_2 (absorber plate width), height d (height of air passage) and length dx at a distance x from the inlet. Hence, for flowing air

$$[h_{lf}(T_c - T_f) + h_{pf}(T_p - T_f)]L_2 dx = (\rho L_2 d)C_{\text{air}}(dT_f/dt)dx + \dot{m}C_{\text{air}}(dT_f/dx)dx \tag{6.21}$$

6.6 AIR HEATER WITH FLOW ON BOTH SIDES OF THE ABSORBER

Let us consider the solar collector, the schematic diagram of which is shown in Figure 6.7. As compared to the flow over the absorber (Fig. 6.6), this arrangement provides an increased contact area between the absorber and the air stream. Let us assume that equal flow occurs both above and below the absorber plate. We also assume that the heat transfer coefficient between the absorber plate and the air stream on either side is the same.

The analysis discussed in the previous subsection can easily be extended to study the thermal performance of this heater. Sodha et al. (1982a) have analyzed this heater, the analysis being very approximate as it does not take into account the interaction of cover and bottom plate with air stream; also, the radiation exchanges between the absorber plate and the cover and between the absorber plate and the bottom plate have not been considered.

The steady state heat gain rate for the air stream can be written as (Parker, 1981):

$$\dot{q}_u = h_{pf}(T_{pm} - T_{ft}) + h_{pf}(T_{pm} - T_{fb}) - h_{cf}(T_{ft} - T_c) - h_{cf}(T_{fb} - T_{bm}) \tag{6.22}$$

where T_{ft} is the air temperature in the top passage and T_{fb} that in the lower passage.

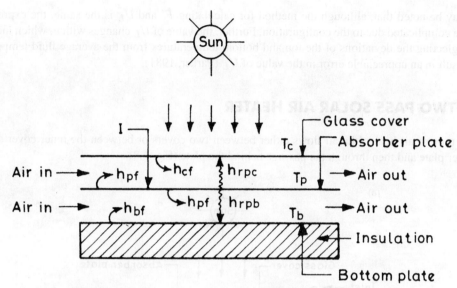

Figure 6.7 Schematic view of a double flow solar air heater.

Energy balance equations for cover, absorber plate and bottom plate can be written and used to eliminate T_p, T_b and T_c from Equation (6.22). The two air temperatures (one above the plate and another below the plate) can be eliminated in the manner given below.

In general, the top heat loss from the plate exceeds the heat loss from the bottom of the plate. As a result, air temperature in the rear passage will be higher than in the top passage. Flow rates are the same on both the sides. If T_f be the average of T_{ft} and T_{fb}, then the deviations of T_{ft} and T_{fb} from T_f must be equal. The deviations between the fluid and the ambient temperatures may also be used to indicate the deviations of T_{ft} and T_{fb} from T_f. Therefore, a fractional deviation n is defined as

$$(T_f - T_a)(1 - n) = (T_{ft} - T_a) \tag{6.23}$$

and

$$(T_f - T_a)(1 + n) = (T_{fb} - T_a) \tag{6.24}$$

These two equations can be used to express T_{ft} and T_{fb} in terms of T_f so that Equation (6.22) can be expressed in terms of mean air temperature T_f.

On solving these equations, we get the expressions for F' and U_L as

$$F' = D/2h_{cf}h_{pf}P + 2h_{pf}U_bh2 + h_{rpc}(h_{cf} + h_{rpb})(P + 2h_{pf})$$

$$+ U_b[2h_{pf} + h2] + (h_{cf}h_{rpb}) + h_{rpb}(h_{1f}[P + 2h_{pf}] + 2h_{pf}h2 + U_bh2) \tag{6.25}$$

$$U_L = 4h_{cf}h_{pf}U_bh2 + 2h_{rpc}([h_{cf} + h_{pf}][h_{rpb}h2 + U_bh2 + h_{rpb}U_b] + h_{cf})$$

$$+ 2h_{rpb}U_b(h_{cf}h_{pf} + h_{cf}h2 + h_{pf}h2) + (1 - n)h_{cf}h2(h_{cf}[2h_{pf} + h_{rpc}$$

$$+ (1 + n)h_{cf}U_b(h_{cf}[h_{rpb} + 2h_{pf}] + h_{rpc}Q)]/D \tag{6.26}$$

where

$$D = [2h_{cf}h_{pf}P + 2h_{pf}U_bh2 + h_{rpc}(Q\{h_{cf} + h_{rpb} + U_b\}$$

$$+ h_{cf}h_r + h_{rpb}Q(h_{1f} + h2)] \tag{6.27}$$

$$P = h_{cf} + U_b + h2 \quad \text{and} \quad Q = h_{cf} + 2h_{pf} \tag{6.28}$$

It may be noted that, although the method for calculating F' and U_L is the same, the expressions become complicated due to the configuration. Further, the value of U_L changes with n, which indicates that neglecting the deviations of the top and bottom temperatures from the average fluid temperature may result in an appreciable error in the value of U_L (Parker, 1981).

6.7 TWO PASS SOLAR AIR HEATER

In this type of air heater, the air flows either between two covers or between the inner cover and the absorber plate and then through the passage behind the plate (Fig. 6.8).

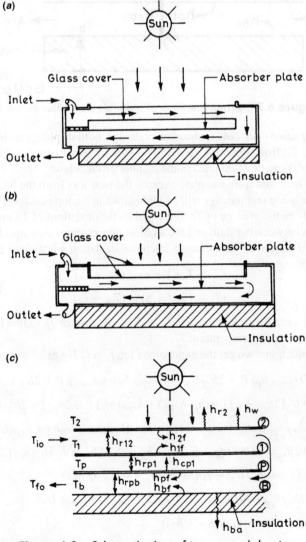

Figure 6.8 Schematic view of two-pass air heaters.

The concept of two pass air heater was introduced by Satcunanathan and Deonarine (1973) and later considered by Caouris *et al.* (1978). Wijeysundera *et al.* (1982) have done thermal performance studies of these air heaters, developed detailed heat transfer model and validated them by the experimental data. They concluded that a two pass mode offers an expensive method of improving the collector efficiency by about 10 to 15 percent.

Figure 6.8(a) shows the configuration of the air heater studied by Satcunanathan and Deonarine (1973). In this type, air enters the flow channel formed by the absorber plate and another plate below it. The inlet air removes heat from the covers so that they are kept cooled and the top heat loss is reduced. Figure 6.8(b) shows the configuration of another two-pass air heater, introduced by Wijeysundera *et al.* (1982). In this type, air flows first through the absorber plate and the inner cover and then through the absorber plate and the bottom plate. In either case, the analysis involves similar considerations. So, we outline the procedure for the configuration shown in Figure 6.8(a). Correspondingly, the linearised heat transfer coefficients signifying various heat transfer processes are shown in Figure 6.8(c).

The analysis corresponds to a quasi-steady state, that is, the collector is assumed to be in a quasi-steady state. Also, the assumptions made in the previous analysis are true in this case as well.

As the system is considered to be in a quasi-steady state, the temperatures of covers, plates and air stream, solar intensity and ambient temperature are not taken to be time dependent. The rise of air temperature along the passage can then easily be calculated. Let us consider, for example, the control volume between the two covers as shown in Figure 6.8(c). Energy balance for flowing air stream can be written as

$$\dot{m}C_{air}\frac{dT_f}{dx}dx = L_2 h_{1f}(T_1 - T_f)dx + L_2 h_{2f}(T_2 - T_f)dx \tag{6.29}$$

The above equation can be integrated and along with the initial condition $T_f|_{x=0} = T_{fi}$, gives,

$$T_f(x) = \frac{(h_{1f}T_1 + h_{2f}T_2)}{h_{1f} + h_{2f}}[1 - \exp(-\nu x)] + T_{fi}\exp(-\nu x) \tag{6.30}$$

where

$$\nu = (h_{1f} + h_{2f})L_2/\dot{m}C_{air}$$

Hence, the average fluid temperature in the passage is,

$$T_{fm} = \frac{1}{L_1}\int_0^{L_1} T_f(x)dx \tag{6.31}$$

or $\quad T_{fm} = \left[\frac{h_{1f}T_1 + h_{2f}T_2}{h_{1f} + h_{2f}}\right][1 - (1/\nu L_1)1 - \exp(-\nu L_1)] + T_{fi}[1 - \exp(-\nu L_1)]/(\nu L_1) \tag{6.31a}$

The energy balance conditions for covers and plates can be written as:

Outer cover : $\quad \alpha_2 I(t) + h_{r12}(T_1 - T_2) = (h_{r2a} + h_w)(T_2 - T_a) + h_{2f}(T_2 - T_{fm}) \tag{6.32}$

Inner cover : $\quad \alpha_1 I(t) + (h_{rp1} + h_{cp1})(T_p - T_1) = h_{r12}(T_1 - T_2) + h_{1f}(T_1 - T_{fm}) \tag{6.33}$

Absorber plate : $\quad \alpha_p I(t) = h_{rpb}(T_p - T_b) + h_{pf}(T_p - T'_{fm}) + (h_{rpm} + h_{cp1})(T_p - T_1) \tag{6.34}$

Bottom plate : $\quad h_{rpb}(T_p - T_b) = h_{ba}(T_b - T_a) + h_{bf}(T_b - T'_{fm}) \tag{6.35}$

where T_{fm} is the mean air temperature in the bottom channel and its expression can be obtained as in Equation (6.31). It may be noted that the outlet temperature of the first channel is the same as the inlet temperature of the second channel. Equations (6.32) to (6.35) can be solved using the expressions for

T_{fm} and T'_{fm} to obtain the outlet air temperature from the collector. The instantaneous thermal efficiency of the collector is given by

$$\eta_i = \frac{\dot{m} C_{\text{air}}(T_{fo} - T_{fi})}{A_p I_1} \tag{6.36}$$

where T_{fo} is the outlet air temperature.

6.8 COMPARISON WITH EXPERIMENTAL RESULTS

Wijeysundera et al. (1982) have validated their heat transfer model by comparing the theoretical predictions with the experimental data of Satcunanathan and Deonarine (1973). The heat transfer coefficients, used, are calculated by using the iterative procedure. Figures 6.9 and 6.10 shows the efficiency and the air temperature rise for both single as well as two-pass air heater respectively. There is a good agreement between the experimental results and theoretical calculations.

It is seen that a two-pass system has 10–15 percent higher efficiency than that of a single pass system. Further, the two pass arrangement shown in Figure 6.8(b), gives a better performance than the one shown in Figure 6.8(a). An extensive parametric study has been carried out by Wijeysundera et al. (1982) to substantiate this fact conclusively.

It may be noted that an improvement in the efficiency of the two-pass system is obtained with very little additional material and construction cost, the main difference being in the design of air passage.

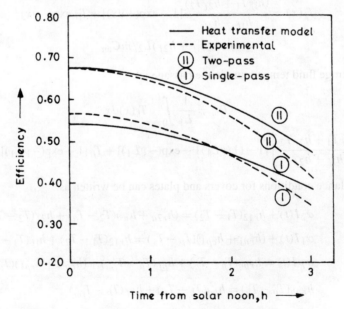

Figure 6.9 Efficiency curve; $\dot{m} = 0.0171$ kg/s; $T_a = 303$ K; $A_c = 1.269$ m^2; gap $= 0.0381$ m; $V = 5.1$ m/s.

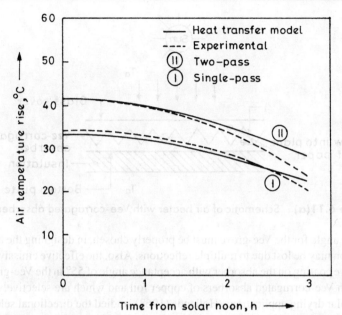

Figure 6.10 Air temperature rise of two pass and single pass air heater. $\dot{m} = 0.0171$ kg/s; $T_a = 303$ K; $A_c = 1.269$ m^2; gap $= 0.0381$ m; $V = 5.1$ m/s.

6.9 HEATER WITH FINNED ABSORBER

In order to improve the heat transfer from the plate to air stream, and hence the efficiency of the heater, fins are added to the rear side of the absorber (Fig. 6.1(d)). This, however, introduces some extra pressure drop. Also the number of fins and their depth cannot be increased beyond a limit because then fan power requirement will also increase.

The heat transfer model of such an air heater can be easily developed since heat transfer from finned surfaces is very common in convective heat transfer problems and is given in detail in books concerning heat transfer. However, the limitation of this model is that there are hardly any appropriate correlations for heat transfer coefficients corresponding to situations encountered in air heaters. Malik and Buelow (1973) and Kuzay *et al.* (1974) have worked on this and Selcuk (1977) has summarized their work. Test results of air heaters with staggered galvanized fins and U-shaped staggered aluminum fins attached to the rear side of the absorber plate have been reported. The efficiencies with fins are substantially higher than conventional air heaters (absorber without fins).

6.10 HEATER WITH VEE-CORRUGATED ABSORBER

The use of a Vee-corrugated absorber (Fig. 6.11(a)) in place of a flat absorber obviously provides a large surface area for heat transfer to the air stream. The convective heat transfer from plate to cover increases in this case but the loss is largely compensated by the increased heat transfer to the flowing air.

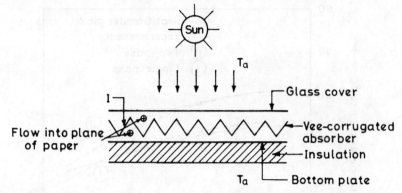

Figure 6.11(a) Schematic of air heater with Vee-corrugated absorber plate.

The acceptance angle for the Vee-grove must be properly chosen, in designing the absorber, as a large amount of radiation may be lost due to multiple reflections. Also, the effective emissivity of the absorber increases. Selective coating on the absorber with acceptance angle of 55° in the Vee-groove is suggested.

Air heaters with Vee-corrugated absorbers of copper foil and which are selectively painted, are used in Australia for solar drying applications. Hollands (1963) studied the directional selectivity, emittance and absorptance properties of Vee-corrugated specular surfaces.

With this type of absorber, flow can be on both sides of the absorber or on one side only. The collector efficiency factor (F') and the heat loss factor (U_L) will be the same as those in case of flat absorber, but the value of h_{pf} increases with the surface area of the absorber (Parker, 1981).

A special case arises when the Vee-groves touch the bottom plate, thus forming triangular ducts. The triangular duct formed restricts the fluid flow only to itself (Fig. 6.11(b)). The ducts are made of solid conductive material (e.g. aluminum) so as to provide sufficient heat conduction around the duct. Consequently, little difference in plate temperature around the triangular duct occurs, as such the radiation heat transfer within the triangle need not be considered. Further, the top and bottom loss coefficients are combined to write the energy balance for the collector as

$$I(t) = \dot{q}_u + U_L(T_{pm} - T_a) \tag{6.37}$$

where, $\qquad U_L = U_b + U_t \quad$ and $\quad \dot{q}_u = h_{pf}(T_{pm} - T_f)$

Figure 6.11(b) Schematic of air heater with Triangular ducts.

Hence, the collector efficiency factor, F' is given as

$$F' = h_{pf}/(h_{pf} + U_L) \tag{6.38}$$

In the next section, air heaters with porous absorber will be discussed.

6.11 REVERSE ABSORBER HEATER

Conventional air heater can provide the hot air at temperature which are 15–30 °C above the ambient temperature. Further, the rise in temperature (say $\cong 100$ °C) can be achieved by reducing the convective and radiative heat losses from the absorber to ambient through the top glass cover. This is possible in a new type of solar air collector known as a reverse flat plate collector (RFP). It can collect solar heat at high temperature ($\cong 200$ °C). The working principle, energy balance and performance study of RFP will be discussed in the next sections.

6.11.1 Working Principle

The schematic view of a reverse flat plate collector has been shown in Figure 6.12(a). The solar radiation is first transmitted by the glass cover $\{\tau I(t)\}$ and then reflected by the polished curved cylindrically surface $\{r^N \tau I(t)\}$ towards the glazed absorber. After absorption, the absorber emits long wavelength radiation which is trapped between the absorber and glazed cover. In this case, the convection and radiation heat losses is suppressed due to hot plate facing downward. There is an insulation of a given thickness above the absorber to reduce the top heat loss due to conduction unlike conventional air heater. There is a gap between absorber and insulation to allow the air to flow above the absorber. The air passing above the absorber is heated by transferring the absorbed thermal energy by convection. The aperture area of RFP collector is the same as that of absorber plate and concentration ratio is 1 : 1.

Other configurations of RFP uses solar radiation from the top surface as shown in Figure 6.12 (b–e). Figure 6.12 (b–e) shows a single and double pass systems respectively.

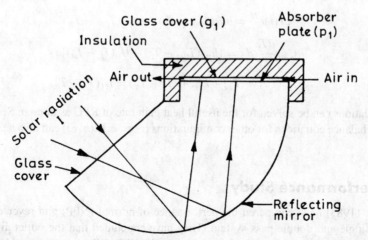

Figure 6.12(a)   Schematic view of a single pass reverse flat-plate (RFP) collector.

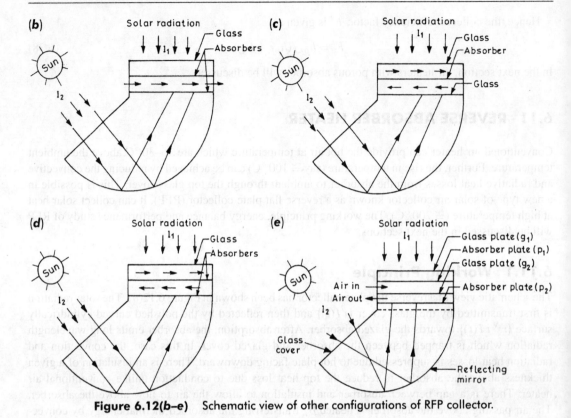

Figure 6.12(b-e) Schematic view of other configurations of RFP collector.

6.11.2 Energy Balance

Referring to Figure 6.12(a), the energy balance equation at absorber plate (T_{pm}), flowing fluid (T_f) and glass cover in the x-direction can be written as:

$$\tau I(t) r^N = h_{pf}(T_{pm} - T_f) + h_{rpg}(T_{pm} - T_g) \tag{6.39a}$$

$$\dot{m} C_{air} \frac{dT_f}{dx} dx = [h_{pf}(T_{pm} - T_f) - U_b(T_f - T_a)] dx \tag{6.39b}$$

and

$$h_{rpg}(T_{pm} - T_g) + h_{fp} + (T_f - T_g) = U_t(T_g - T_a) \tag{6.39c}$$

The above equations can be solved for the useful heat gain rate of RFC as done in Section (6.3).

The energy balance equations for other configurations (Figs. 6.12(b-e)) can also be written in similar ways.

6.11.3 Performance Study

Chandra *et al.* (1983) have compared the performance of normal (NFP) and reverse (RFP) flat plate collector for single and double pass system. They have concluded that the outlet air temperature for RFP collector is higher than the outlet air temperature for NFP collector.

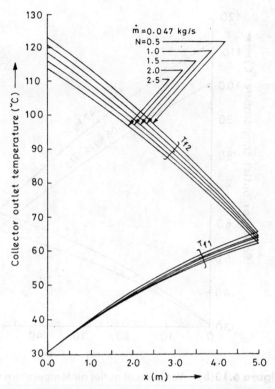

Figure 6.13(a) Variation of outlet air temperature with x for different N.

Effect of number of reflection on the outlet air temperature in two absorber two-pass reverse flat-plate collector (Fig. 6.12(e)) has been shown in Figure 6.13(a). It is clear that the outlet air temperature decreases with an increase of number of reflection due to increase of reflection losses. Further, the outlet air temperature is increased by reducing the mass flow rate (Fig. 6.13(b)). In this case, the convective heat transfer is increased due to longer contact between the air and hot plate.

6.12 AIR HEATERS WITH POROUS ABSORBERS

Porous materials are suitable for application in high temperature heat exchanger. The increased effectiveness of the heat exchange is mainly due to the intimate contact between the gas particles and the porous plate.

The idea of using black porous matrix as an absorber of solar radiation has been used since long. Widely used types of porous absorber are wire mesh, slit and expanded metal and transpired honeycomb absorber.

6.12.1 Matrix Air Heaters

Figure 6.14 shows schematically an air heater with porous matrix as the absorber. The mesh size and porosity depends on the required performance. The top surface behaves as a set of black cavities whose absorptivity greatly exceeds that of a regular solid surface. The matrix surface can also be painted by

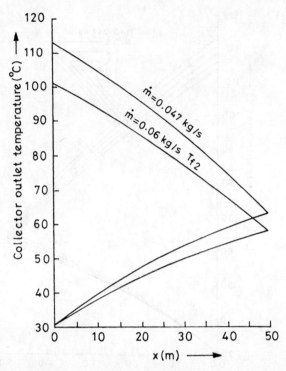

Figure 6.13(b) Variation of outlet air temperature with x for different $\dot{m}$.

selective coatings to reduce the long wave radiation losses. The heat exchanger mechanisms may take any one of the three patterns shown in Figures 6.14(a), (b) and (c). However, pattern in Figure 6.14(b) is preferred, as inlet air remains in contact with the cover reducing the top loss.

If the matrix is a loosely packed semitransparent material, then the solar energy is absorbed along the matrix depth. If it is a closely packed mesh, then most of the absorption takes place at the top surface. A steady state thermal analysis has the following assumptions:

i. Thermal, physical and transport properties of the matrix plate and air are uniform over the entire collector area and do not change with temperature.

ii. Heat conduction and air flow are one dimensional and are perpendicular to the bounding surfaces of the matrix.

iii. Air and matrix pores are at the same temperature at any position within the plate.

iv. Radiation and convection losses take place only from the collector's top surface.

Considering one dimensional flow assumption, the analysis can explain the behavior of air heaters shown in Figures 6.14(a) and (b); whereas Figure 6.14(c) corresponds to a two dimensional study. The steady state temperature distribution can be determined by solving the heat balance equation of the collector.

$$K_f \frac{d^2 T_f}{dy^2} + \frac{\dot{m} C_{\text{air}}}{A_p} \frac{dT_f}{dy} + \dot{Q}_{\text{int}}(y) = 0 \qquad (6.40)$$

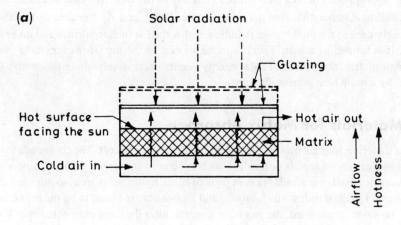

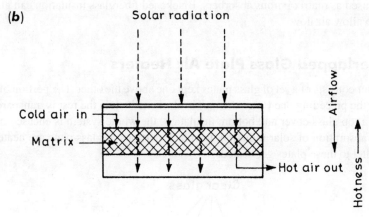

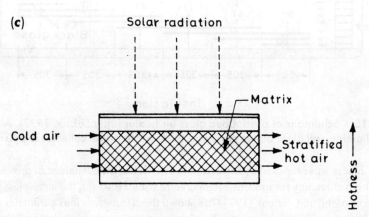

Figure 6.14 (a–c)   Schematic of matrix air heater. (*Lansing et al., 1979*)

subject to the appropriate boundary condition. Here K_f is the effective matrix conductivity, W/mK, $\dot{Q}_{int}$ the internal heat generation rate per unit volume, W/m³ and A_p the area of the absorber plate, m². For a closely packed mesh, the solar radiation is absorbed at the top surface and no heat generation term occurs. It is termed as a matrix with opaque surface at the top. However, for a loosely packed semitransparent matrix, absorption of solar energy occurs successively along the matrix depth, and is characterised by a finite heat generation term.

6.12.2 Materials for Matrix Absorber

The choice of materials for matrix absorber has direct relation to the cost. The cost needs to be low. Thus, the materials like expanded metals or meshes, such as screens for screen doors, strainers, or restaurant grease filters are not used. For similar reason pads of black nylon wool are also not used. Metal wools, available in carbon steel, stainless steel, bronze and aluminum are found to be the best candidates. The carbon steel, however, gets rusted, the wools, in general, filter dust and clog with time. The dust filters can possibly be used as matrix porous absorbers. Blackened fiberglass insulation can also be used if it is thin enough to allow air flow.

6.12.3 Overlapped Glass Plate Air Heaters

This type of heater consists of a set of glass plates kept one above the other. The portion of the glass plate which is behind the preceding one (from the top) is blackened while the rest is transparent (Fig. 6.15). The system has a top glass cover and bottom insulation; the whole system is encased in a sheet metal box. Due to the absorption of solar radiation, the black portions of glass plate are heated and thus the air flowing parallel to these plates get heated.

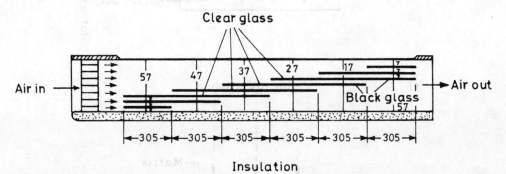

Figure 6.15 Schematic of overlapped glass air heater (after *Selcuk, 1977*). All dimensions are in millimeters; all glass plates are 3 mm thick; width of the collector is 690 mm.

This type of heater was proposed by Löf *et al.* (1961a). It has the advantage of providing low pressure drop and also high efficiency for moderate temperature rises. However, the large glass area requirement increases the cost of the unit. Selcuk (1977) has shown the effect of various parameters. It was observed that the variation of the spacing between the plates and the number of plates slightly influences the collection efficiency. Optimum spacing is about 5 to 7 mm and glass thickness is about 3 mm. The cost of the heat collected decreases with the increase in collector length.

6.12.4 Air Heater With Honeycomb Absorber

The use of honeycomb structures between the cover and the absorber to reduce top convective loss is an established fact. An air heater can also be built up by passing air through honeycomb structure (Fig. 6.16).

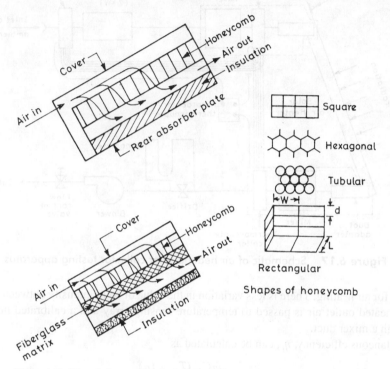

Figure 6.16 Schematic of transpired honeycomb solar air heater.

Honeycomb can be transparent or opaque; and can be of hexagonal or rectangular shape. The honeycomb structures can also be used along with porous matrix materials to reduce convection and radiation to cover glass. Buchberg *et al.* (1971) made extensive study on honeycomb structures and developed an efficient unit having specularly reflecting walls coated with a thin lacquer film. They also made analytical studies to predict the performances of such air heaters.

6.13 TESTING OF SOLAR AIR COLLECTOR

A schematic view of a solar air collector test facility is shown in Figure 6.17. This test facility has been developed at the University of Waterloo. The air flow circuit is an open loop operating. The air at ambient temperature passes through a temperature sensing thermocouple array. The volume flow rate of air is kept at 7.5 and 10 liters/m² s respectively. To keep the inlet air temperature constant up to 80 °C, the inlet air is further allowed to pass through a thermostatically controlled electric heater (2 kW). A pyranometer is fixed at the top of an inclined solar air collector to measure solar intensity [I(t)] during the testing period. The inlet air at constant temperature is then passed through air collector between

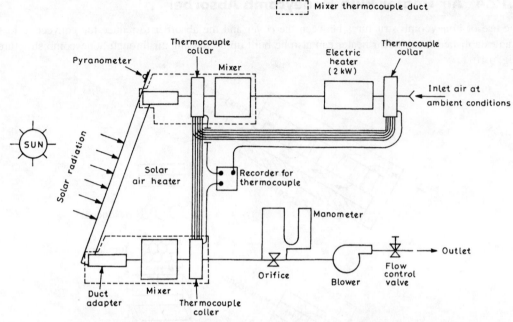

Figure 6.17 Schematic of air heating solar collector testing apparatus.

11AM to 1PM for air heating. There is less variation in the value of solar intensity between 11AM to 1PM. Further, the heated outlet air is passed to temperature sensing array and a calibrated flow measuring orifice through a mixer duct.

The instantaneous efficiency, η_i, can be calculated as

$$\eta_i = \frac{\dot{m} C_{air}(T_{fo} - T_{fi})}{A_p I(t)} \tag{6.41}$$

The plot of η_i with $[(T_{fi} - T_a) \div I(t)]$ for 7.5 and 10 liters/m²s is shown in Figure 6.18 (adapted from Kreith *et al.*, 1980). It is clear from figure that the air collector is more efficient at high flow rate.

6.14 PARAMETRIC STUDIES

6.14.1 Effect of Air Leakage

The performance of solar air collector is mainly affected by air leakage due to poor design and manufacturing of a collector. The air leakages may also be due to the following reasons:

i.   thermal expansion and contraction of hot air
ii. corrosion
iii. vibration settlement
iv. poor workmanship etc.

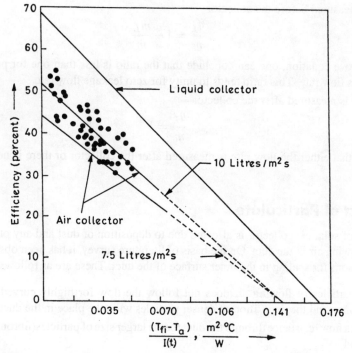

Figure 6.18 Comparison of liquid and air flat plate collectors (characteric curve).

Complete elimination of air leakage is difficult. The detection and correction of small leakage is often ignored.

Air leakage in solar air collector may occur (i) from the inside of the collector to the outside environment and (ii) within the collector between the flow channels at different absolute temperatures.

The measurement of flow rate plays a vital role to evaluate thermal efficiency. If flow rate is measured before collector, the thermal efficiency will be higher because of ignorance of air leakage in the collector. In this case, thermal efficiency is given by

$$\eta_i = \frac{\dot{m} C_{\text{air}} \Delta T}{A_p I(t)} \tag{6.42}$$

The thermal efficiency will be lower if flow rate is measured after the collector due to consideration of air leakage for the same temperature difference.

$$\eta_L = \frac{\dot{m}_o C_{\text{air}} \Delta T}{A_p I(t)} \tag{6.43}$$

The difference in the flow rate before and after the collector gives the air leakage flow rate i.e.

$$\dot{m}_L = \dot{m} - \dot{m}_o \tag{6.44}$$

An efficiency ratio is the ratio of the efficiency with leakage η_{iL} to the efficiency without leakage η_i, and it is given by the following:

a. the flow is measured before the collector

$$\frac{\eta_L}{\eta_i} = 1 - \frac{\dot{m}_L}{\dot{m}}$$
(6.45)

From the above equation, one can conclude that the ratio is less than one for positive value of leakage mass flow rate. This ratio tends to unity for zero leakage flow rate.

b. the flow rate is measured after the collector

$$\frac{\eta_L}{\eta_i} = 1$$
(6.46)

This means that either the flow rate is measured after the collector or there is no leakage in the collector.

6.14.2 Effect of Particulate

The performance of solar air collector is affected due to deposition of dust and any other particulates available in the flowing air in the duct. On the basis of literature survey, it has been observed that there are three main reasons for scaling in the inner surface of the duct. These are as follows:

i. the denser particles of flowing air does not follow the flow for highly curved duct in the air collector if any. And the deposition of denser particles will take place in the duct,

ii. there will be a flow resistance through the duct for the larger size of particles(fibrous dust) available in the air, and

iii. small size of particles available in the air is deposited in a laminar flow condition due to molecular diffusion etc.

The above problems can be minimized as follows:

i. the normal viscous filter should be provided at the inlet of duct,

ii. minimum turns and bends should be used in the design of duct, and

iii. forced mode (higher flow rate) of operation of collector should be used.

6.15 COMPARISON OF PERFORMANCE OF LIQUID AND AIR COLLECTOR

The comparison of a liquid and air flat plate collector is shown in Figure 6.18. The performance of air collector has been plotted at two different flow rate. It can be seen that the performance of a liquid collector is better in comparison to an air collector. It can also be observed that the flow rate play an important and significant role in an air collector. Further, one can observe that the difference in performance of both collector minimizes at higher solar intensities.

6.16 APPLICATIONS OF AIR HEATER

It is technically feasible to use solar-heated air providing energy for almost any application that uses solar-heated liquids.

The important areas of applications are as follows:

i. heating and cooling of buildings.

ii. heating of greenhouse, and

iii. industrial processes such as drying of agricultural crops and timber.

6.16.1 Heating and Cooling

Various types of air heaters have been designed and used in space heating and cooling. Löf *et al.* (1963) had used rock pile storage and overlapped glass plate solar heaters in Denver House. Air heaters are used only in actively heated or cooled buildings. Air heaters are also used with desiccant beds for solar air conditioning. The heat from air heaters can also be used to heat the generator of an absorption air conditioner for cooling purpose.

6.16.2 Drying

Drying is a promising area for the application of air heaters and, for that matter, solar energy. It has been the oldest and the most widely used application of solar energy in the developing countries. The methods have been based on open-air drying. As a result of extensive research, dryers of various designs, have been developed. However, in this chapter the use of drying has been demonstrated as an application of air heaters. Thus, we will discuss only indirect drying, in which solar energy does not come in direct contact with the crops. On the other hand, hot air from a solar air collector is circulated through the crop to reduce it's moisture content. The air can be circulated either by a fan or by natural convection; correspondingly, the heaters are called active or passive dryers. The principle of drying operation is shown schematically in Figure 6.19. In this design, the warm air rises through the air heater due to buoyancy and enters the drying chamber.

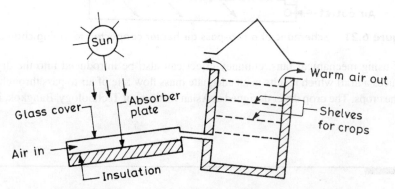

Figure 6.19 Schematic view of an indirect crop dryer.

Depending on the mode of circulating air, a number of designs are possible. Figure 6.20 shows one such design (Headly and Springer, 1973) in which ambient air is heated and is passed into the drying chamber by natural convection. The hot air removes moisture from the crops, becomes cooler and falls to the bottom of the drying chamber. Yam, sweet potato, sorrel and grass can be dried by this system.

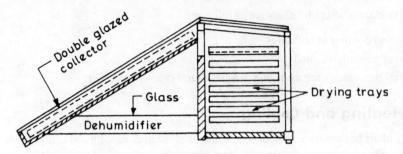

Figure 6.20   Schematic diagram of an indirect crop dryer.

Figure 6.21 shows a two-pass air heater connected to a drying chamber. The hot air passes through the crops immediately after it leaves through the collector. Hence, insulation is not required to be used in the air heater (Satcunanathan, 1973).

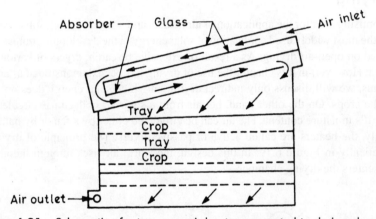

Figure 6.21   Schematic of a two-pass air heater connected to drying chamber.

Instead of using mechanical fans, chimney effect can also be introduced into the dryer. Use of a chimney induces a draft which results in an adequate mass flow rate of air to pass through the collector and then to the crops. The crop dryer designed at Asian Institute of Technology, Bangkok, is an example of this type.

PROBLEMS

6.1   Calculate pressure drop (ΔP) across a single pass air heater for parameters of Figure 6.9.

Hint   Use $\Delta P = F(M_a^2/\rho)(L/D)^3$ where $F = F_0 + \gamma(D/L)$

For laminar flow $(Re < 2550)$; $F_0 = 24/Re$ and $\gamma = 0.9$

For transitional flow $(2550 < Re < 10^4)$; $F_0 = 0.0094$ and $\gamma = 2.92 Re^{-0.15}$

For turbulent flow $(10^4 < Re < 10^5)$; $F_0 = 0.0599 Re^{-0.2}$ and $\gamma = 0.73$.

6.2 What can be the total pressure drop for multi-pass air heater?

Hint Calculate for each pass and then add to get the total pressure drop.

6.3 Calculate the energy consumption for pumping the air through air duct of an air collector.

Hint $E = M(\Delta P/\rho_a)t(M = \text{kg/m}^2\,\text{s})$

Take the value of ΔP, M, ρ_a from Problem 6.1.

6.4 Find out the convective heat transfer coefficient from the absorber to the flowing air over it.

Hint $h_{pf} = (Nuk/D_h)$ where, $D_h = (4A/P)$; $A = \text{cross-sectional area}$, $P = \text{perimeter}$ and $Nu = Nu_0 + \beta_0(W/L)$

i. For $2500 \leq Re$ (Laminar flow) $Nu_0 = 5.385$; $\beta_0 = 0.0148Re$

ii. For $2500 < Re < 10^4$ (Transitional flow) $Nu_0 = 4.4 \times 10^{-4} Re^{1.2}$; $\beta_0 = 9.37 Re^{0.471}$

iii. For $10^4 < Re < 10^5$ (Turbulent flow) $Nu_0 = 0.03 Re^{0.74}$; $\beta_0 = 0.788 Re^{0.74}$.

6.5 Derive an expression for outlet air temperature for a conventional solar air heater.

Hint Use Equations (6.1) to (6.3).

6.6 Calculate the outlet air temperature for different lengths of collector (2 m–10 m) for the following specifications:

$I = 350\,\text{W/m}^2$; $T_a = 15\,°\text{C}$; $W = 1\,\text{m}$; $\dot{m} = 0.02\,\text{kg/s}$;

$U_t = 2.81\,\text{W/m}^2\text{K}$; $2h_{pf} = h_{bf} = 5.842\,\text{W/m}^2\,°\text{C}$; $h_{rpb} = 5.27\,\text{W/m}^2\,°\text{C}$

Also plot the curve between outlet temperature and the length of the collector.

Hint See Problem 6.5.

6.7 Calculate F' for Problem 6.6.

Hint Use Equation (6.7).

6.8 Plot η_i versus $(T_{fi} - T_a)/I$.

Hint $\eta_i = \dot{q}_u/I(t)$; Equation (6.4); $T_{fi} = (T_a + 2)\,°\text{C}$.

6.9 What can be the optimum number of glass cover of a conventional air heater.

Hint See Figure 6.4.

6.10 Derive an expression for fluid temperature for double exposed air heater.

Hint Solve Equation (6.11).

6.11 Repeat Problems 3.6–3.9 of flat plate collector, for air collector in a steady state condition.

Hint Use Equation (6.14).

6.12 Calculate F' for single and double flow air heater and compare the results.

Hint Use Equations (6.15) and (6.25).

6.13 Write down the energy balance equation of absorber, considering the effect of conduction through the plate.

Hint Use Equation (2.4).

6.14 Derive an expression for air temperature (T_f) for double exposure air heater.

Hint Use Equations (6.9) to (6.11).

6.15 Write down an expression for the ratio of useful energy for configuration of various air heaters shown in Figure 3.18.

Hint Replace F' and U_L of Figure 3.18 in Equation (6.4).

CHAPTER 7

Solar Crop Drying

7.1 INTRODUCTION

It is well known that food is a basic need of a human being after air and water. Food holds a key position in the development of a country. This chapter deals with drying of food to avoid food losses between harvesting and consumption. High moisture content is one of the reasons for its spoilage during the course of storage at time of harvesting. High moisture crops are prone to fungus infection, attacks by insects, pests and the increased respiration of agricultural produce. To solve this problem many drying technique have been developed. Drying has the following advantages:

    i. facilitate early harvest
    ii. permits planning the harvest season
    iii. helps in long term storage
    iv. helps farmers to fetch better returns
    v. helps farmers to sell a better quality product
    vi. reduce the requirement of storage space
    vii. helps in handling, transport and distribution of crops
    viii. permits maintaining viability of seeds

Drying helps in reducing the moisture content of a product to a level below which deterioration does not occur and the product can be stored for a definite period. Different crops have different level of safe moisture content as given in Table 7.1.

7.2 WORKING PRINCIPLE

The three modes of drying are: (i) open sun, (ii) direct and (iii) indirect in the presence of solar energy. The working principle of these modes mainly depends upon the method of solar energy collection and its conversion to useful thermal energy.

TABLE 7.1 Initial and final moisture content and maximum allowable temperature for drying for some crops (Brooker *et al.*, 1992; Sharma *et al.*, 1993)

Sl. No.	Crop	Initial moisture content (%, w.b.)	Final moisture content (%, w.b.)	Max. allowable temp. (°C)
1	Paddy, raw	22–24	11	50
2	Paddy, Parboiled	30–35	13	50
3	Maize	35	15	60
4	Wheat	20	16	45
5	Corn	24	14	50
6	Rice	24	11	50
7	Pulses	20–22	9–10	40–60
8	Oil seed	20–25	7–9	40–60
9	Green peas	80	5	65
10	Cauliflower	80	6	65
11	Carrots	70	5	75
12	Green beans	70	5	75
13	Onions	80	4	55
14	Garlic	80	4	55
15	Cabbage	80	4	55
16	Sweet potato	75	7	75
17	Potatoes	75	13	75
18	Chillies	80	5	65
19	Apples	80	24	70
20	Apricot	85	18	65
21	Grapes	80	15–20	70
22	Bananas	80	15	70
23	Guavas	80	7	65
24	Okra	80	20	65
25	Pineapples	80	10	65
26	Tomatoes	96	10	60
27	Brinjal	95	6	60

7.2.1 Open Sun Drying (OSD)

Figure 7.1 shows the working principle of open sun drying (OSD) by using solar energy. The short wavelength solar energy $[I(t)]$ falls on the uneven crop surface having an area of A_t. A part of this energy is reflected back and the remaining part is absorbed $[\alpha_c I(t)]$ by the surface depending upon the colour of crops. The absorbed radiation is converted into thermal energy and the temperature of crop $[T_c]$ starts increasing. This results in long wavelength radiation loss from the surface of crop to ambient air through moist air. In addition to long wavelength radiation loss there is convective heat loss too due to the blowing wind through moist air over the crop surface. Evaporation of moisture takes place in the form of evaporative losses and so the crop is dried. Further a part of absorbed thermal energy is conducted into the interior of the product. This causes a rise in temperature and formation of water vapor inside the crop and then diffuses towards the surface of the crop and finally losses the thermal energy in the form of evaporation. In the initial stages, the moisture removal is rapid since the excess moisture on the surface of the product presents a wet surface to the drying air. Subsequently, drying depends upon the rate at which the moisture within the product moves to the surface by a diffusion process depending upon the type of the product (Sodha *et al.*, 1985).

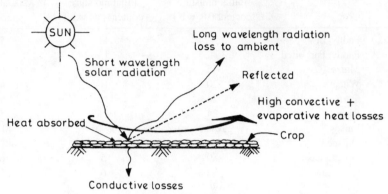

Figure 7.1 Working principle of open sun drying.

In open sun drying, there is a considerable loss due to various reasons such as rodents, birds, insects and micro-organisms. The unexpected rain or storm further worsens the situation. Further, over drying, insufficient drying, contamination by foreign materials like dust, dirt, insects, and micro-organisms as well as discolouring by UV radiation are characteristic for open sun drying. In general, open sun drying do not fulfil the international quality standards and therefore it can not be sold in the international market.

With the awareness of inadequacies involved in open sun drying, a more scientific method of solar energy utilisation for crop drying has emerged termed as controlled drying or solar drying. These will be discussed in the forthcoming sections.

7.2.2 Direct Solar Drying (DSD)

The principle of direct solar crop drying is shown in Figure 7.2. This is also called cabinet dryer. A part of incidence solar radiation on the glass cover is reflected back to atmosphere and remaining is transmitted inside cabinet dryer. Further, a part of transmitted radiation is reflected back from the surface of the crop. The remaining part is absorbed by the surface of the crop. Due to the absorption of solar radiation, crop temperature increases and the crop starts emitting long wavelength radiation which is not allowed to escape to atmosphere due to presence of glass cover unlike open sun drying. Thus the temperature above the crop inside chamber becomes higher. The glass cover serves one more purpose of reducing direct convective losses to the ambient which further becomes beneficial for rise in crop and chamber temperature respectively. However, convective and evaporative losses occur inside the chamber from heated crop. The moisture (the vapor formed due to evaporation) is taken away by the air entering into the chamber from below and escaping through another opening provided at the top as shown in Figure 7.2.

A cabinet dryer has the following limitations:

  i. due to its small capacity its use is limited to small scale applications
 ii. discolouration of crop due to direct exposure to solar radiation
iii. moisture condensation inside glass cover reducing its transmitivity

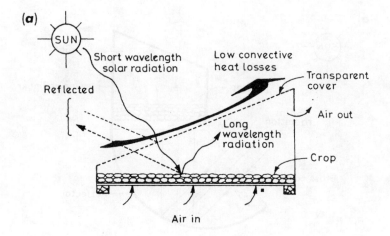

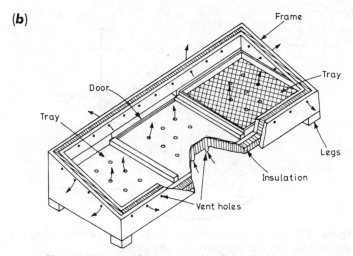

Figure 7.2 Working principle of direct solar drying.

    iv. sometimes the insufficient rise in crop temperature affecting moisture removal

    v. limited use of selective coatings on the absorber plate.

In order to solve some of the above problems, indirect solar drying is preferred and it has been discussed in the next section.

7.2.3 Indirect Solar Drying (ISD)

In this case, the crop is not directly exposed to solar radiation to minimise discolouration and cracking on the surface of the crop. Goyal and Tiwari(1999a) have proposed and analysed reverse absorber cabinet dryer (RACD). The schematic view of RACD without and with glass is shown in Figure 7.3. The drying chamber is used for keeping the crop in wire mesh tray. A downward facing absorber (reverse absorber) is fixed below the drying chamber at a sufficient distance ($\cong 0.05$m) from the bottom of the

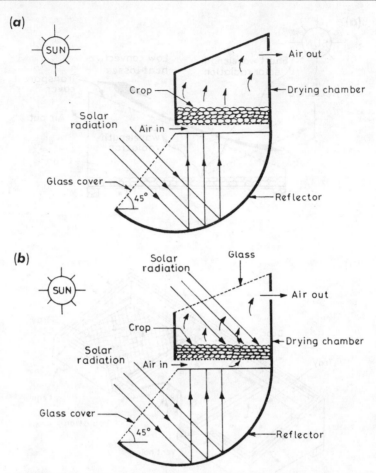

Figure 7.3 Schematic view of single tray reverse absorber cabinet dryer (*a*) without glass and (*b*) with glass.

drying chamber. A cylindrical reflector is placed under the absorber fitted with the glass cover on its aperture to minimise convective heat losses from the absorber. The absorber can be selectively coated. The inclination of the glass cover is taken as 45° from the horizontal to receive maximum radiation. The area of absorber and glass cover are taken equal to the area of bottom of the drying chamber. Solar radiation after passing through the glass cover is reflected by cylindrical reflector towards the absorber. After absorption, a part of this is lost to ambient through the glass cover and the remaining is transferred to the flowing air above it by convection. The flowing air is thus heated and passes through the crop placed in the drying chamber. The crop is heated and moisture is removed through a vent provided at the top of drying chamber.

Figure 7.4 describes another principle of indirect solar drying which is generally known as conventional dryer. In this case, a separate unit termed as solar air heater is used for solar energy collection for heating of entering air into this unit. The air heater is connected to a separate drying chamber where the crop is kept. The heated air is allowed to flow through wet crop. Here, the heat from moisture evaporation is provided by convective heat transfer between the hot air and the wet crop. The drying is

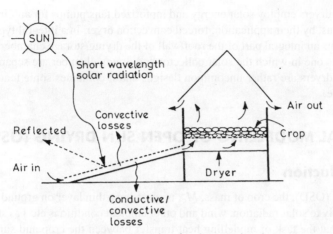

Figure 7.4 Working principle of indirect solar drying system.

basically achieved by the difference in moisture concentration between the drying air and the air in the vicinity of crop surface. A better control over drying is achieved in indirect type of solar drying systems and the product obtained is of good quality.

The classification of solar crop dryer is given in Table 7.2.

TABLE 7.2 Classification of crop drying using solar energy

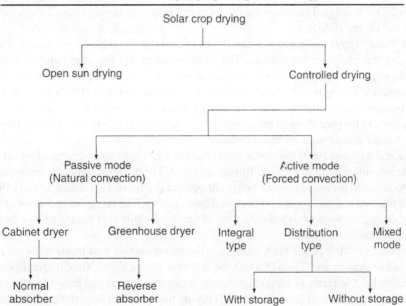

In a passive solar dryer, air is heated and circulated naturally by buoyancy force or as a result of wind pressure or in combination of both. Normal and reverse absorber cabinet dryer and greenhouse dryer operates in passive mode.

The active solar dryers employ solar energy and motorized fans/pumps for air circulation. All active solar dryers are, thus, by their application, forced convection dryer. In a integral type active dryers, the solar collector forms an integral part of the roof/wall of the drying/storage chamber. A distributed type active solar dryer is one in which the solar collector and drying chamber are separate units (Fig. 7.4). Mixed-mode type dryers are rather uncommon designs and it combines some features of the integral and distributed type.

7.3 THERMAL MODELLING OF OPEN SUN DRYING (OSD)

7.3.1 Introduction

In open sun drying (OSD), the crop of mass, M_c, is spread in a thin layer on ground (horizontal surface and exposed directly to solar radiation, wind and other ambient conditions etc.) as shown in Figure 7.1. It is well known that the task of modelling heat transfer between the crop and surrounding air in the presence of solar energy is a complex phenomenon. Recently, an attempt has been made by Anwar and Tiwari (2001) for open sun drying for different crops by using linear analysis. It has been observed that convective heat transfer vary significantly from crop to crop due to presence of different level of moisture content. Their results are within 16 percent internal uncertainty.

Sodha *et al.* (1985) presented an analytical model based on the principle of heat and mass transfer at the product and included the effect of wind speed (v), relative humidity (γ), product thickness and heat conducted to the ground. The analytical models are helpful in predicting the hourly variation of product temperature (T_c) and moisture content under constant rate (Q_e) and falling rate period of drying. Garg and Kumar (2000) developed an analytical model to determine the drying characteristics of any product under open sun drying (OSD).

Patil and Ward (1989) evaluated the feasibility of solar and combined solar-natural air drying of rapeseed using the computer simulations. The study revealed that the solar-natural air-drying regime appears to be superior of that of solar drying alone in respect of the required total drying time and in achieving complete drying within the safe storage time. Zaman and Bala (1989) presents a set of simple empirical equations for natural air flow solar drying of rough rice in mixed-mode type dryer and open floor drying system for prediction of moisture content. Komilov *et al.* (1992) optimized the temperature conditions in solar drying units by mathematical modelling.

Diamante and Munro (1993) derived a mathematical model for solar drying of sweet potato slices based on the simplified form of Fick's diffusion equation. The model could describe the solar drying of sweet potato slices to moisture content below 20 percent dry basis. Goyal and Tiwari (1999b) studied the effect of thermal energy storage on deep bed grain drying. The rate of evaporation of moisture and humidity of drying air were analysed with time of the day for different depths of bed, heat capacity of storage material and flow rate.

In this section, an attempt has been made to estimate convective heat transfer for the same crop by using the data of Anwar and Tiwari (2001) for different drying time. Also, thermal predictions have also been carried out in terms of crop temperature, moisture removal and equivalent solar temperature for experimental validation. It is inferred that (i) drying time plays an important role on convective heat transfer and (ii) there is fair agreement between theoretical and experimental results.

7.3.2 Computational Procedure for Convective Heat Transfer

The convective heat transfer coefficient (h_c) in open sun drying (Fig. 7.1) can be determined using the following (Section 2.6.3)

$$Nu = \frac{h_c X}{K_v} = C(Gr\,Pr)^n,$$

or

$$h_c = \frac{K_v}{X} C(Gr\,Pr)^n \tag{7.1}$$

where Nu, Gr and Pr are Nusselt, Grashof and Prandtl numbers, respectively, defined in Chapter 2 and, C and n are constants to be determined. K_v and X are thermal conductivity of humid air and characteristic dimension. The rate of heat utilized to evaporate moisture ($\dot{q}_{ev}$) is given as,

$$\dot{q}_{ev} = 0.016\,h_c[P(T_c) - \gamma P(T_e)] \tag{7.2}$$

where an expression for partial vapor pressure $P(T)$ has been given in Example 2.9. Substituting h_c in Equation (7.2) from Equation (7.1), we get

$$\dot{q}_{ev} = 0.016\frac{K_v}{X} C(GrPr)^n[P(T_c) - \gamma P(T_e)] \tag{7.3}$$

The moisture evaporated (m_{ev}) is determined by dividing Equation (7.3) by the latent heat of vaporisation (λ) and multiplying the area of the tray (A_t) and time interval (t).

$$m_{ev} = \frac{\dot{q}_{ev}}{\lambda} A_t t = 0.016\frac{K_v}{X\lambda} C(Gr\,Pr)^n[P(T_c) - \gamma P(T_e)]A_t t \tag{7.4}$$

Let

$$0.016\frac{K_v}{X\lambda}[P(T_c) - \gamma P(T_e)]t A_t = Z,$$

$$\frac{m_{ev}}{Z} = C(Gr\,Pr)^n \tag{7.5}$$

Taking logarithm of both sides of Equation (7.5), we obtain

$$\ln\left[\frac{m_{ev}}{Z}\right] = n \ln(Gr\,Pr) + \ln C \tag{7.6}$$

This is the form of a linear equation, $Y = mX_0 + C_0$, where

$$Y = \ln\left[\frac{m_{ev}}{Z}\right], m = n, X_0 = \ln[Gr\,Pr] \text{ and } C_0 = \ln C$$

Thus, $C = e^{C_0}$.

The convective heat transfer coefficient, h_c, can be evaluated from Equation (7.1) after getting C and n.

Here, it is important to note that the values of m_{ev}, T_c, T_e, and γ are experimentally determined at time interval of t. The procedure has been discussed in 7.3.5. The Gr and Pr have been computed by using the temperature dependent physical properties of humid air given in Appendix III g.

EXAMPLE 7.1

Calculate C and n for green chillies by considering six consecutive data (N) of Table 7.4a.

Solution

The data upto 90 minutes for chillies and other calculations are produced below:

Data for green chillies

Time (min)	T_c	T_e	T_i	m_{ev}	γ	Gr $\times 10^5$	Pr	$X_0 =$ $\ln(Gr.Pr)$	Z	$Y =$ $\ln\left(\frac{m_{ev}}{Z}\right)$	h_c
15	39.34	40.80	40.07	5.9	0.365	3.7452	0.7055	12.48	0.0009	1.8929	3.325
30	42.66	42.67	42.66	8.7	0.360	0.0234	0.7056	7.409	0.0011	2.0474	1.602
45	45.54	44.43	44.98	10.9	0.360	2.4008	0.7058	12.04	0.0013	2.0892	3.155
60	47.39	45.06	46.22	13.2	0.355	4.8376	0.7058	12.74	0.0015	2.1478	3.504
75	48.79	45.29	47.04	12.7	0.355	7.0770	0.7059	13.12	0.0017	2.0117	3.710
90	49.49	45.60	47.54	15.3	0.355	7.7388	0.7059	13.21	0.0018	2.1541	3.763

By using the values of X_0, Y, and $N(= N + 1 = 6 + 1 = 7)$, we have

$$m = \frac{N \sum X_0 Y - \sum X_0 \sum Y}{N \sum X_0^2 - \left(\sum X_0\right)^2} = n = 0.145$$

and

$$C_0 = \frac{\sum X_0^2 \sum Y - \sum X_0 \sum X_0 Y}{N \sum X^2 - \left(\sum X_0\right)^2} = 0.2914$$

Now,

$$C = \exp(C_0) = 1.3378$$

EXAMPLE 7.2

Calculate convective and evaporative heat loss coefficient from the green chillies for Example 7.1.

Solution

From Example 7.1, $C = 1.3378$ and $n = 0.145$. The value of convective heat transfer coefficient can be calculated from the following formula [Equation (7.1)]

$$h_c = \frac{K_u}{L} 1.3378 (Gr\, Pr)^{0.145}$$

After knowing h_c, an evaporative heat loss coefficient can be calculated as,

$$h_{ew} = 0.016\, h_c \{P(T_c) - \gamma P(T_e)\}/(T_c - T_e)$$

The results for h_c and h_{ew} are given below:

Time (min)	T_i	$K_u \times 10^5$	$Gr\,Pr$	h_c (W/m^2 °C)	h_{ew} (W/m^2 °C)
15	40.07	0.0271	2.6422	3.325	−133
30	42.66	0.0273	0.0165	1.602	−26366
45	44.98	0.0274	1.6945	3.155	296
60	46.22	0.0275	3.4144	3.504	166
75	47.04	0.0276	4.9956	3.710	124
90	47.54	0.0276	5.4628	3.763	118

From the above table, it is clear that the value of h_{ew} is very high and unrealistic with negative value. It is due to small temperature difference between the value of T_c and T_e.

7.3.3 Prediction of Crop Temperature and Moisture Evaporation

In order to write the energy balance for different component of OSD (Fig. 7.1), the following assumptions have been made:

  i. the heat capacity of crop tray, and moist air have been neglected,
 ii. particle to particle conduction is negligible,
iii. the heat flow is unidirectional, and
 iv. there is no stratification along the depth of the crop due to small depth of crop.

The energy balance equation on crop surface for moisture evaporation can be written as

$$\alpha_c I(t)A_t - h_{rc}(T_c - T_e)A_t - 0.016h_c[P(T_c) - \gamma P(T_e)]A_t - h_i(T_c - T_a)A_t = M_c C_c \frac{dT_c}{dt} \qquad (7.7)$$

where T_c, T_e and T_a are crop temperature, temperature just above the crop surface (2–3 mm) and ambient air temperature. h_{rc} and h_i are radiative and convective heat transfer coefficient from crop surface to environment (T_e) and h_i is overall bottom heat loss coefficient. α_c, C_c and I(t) are absorptivity of the crop surface, specific heat of the crop and solar radiation falling on it.

Energy balance equation of moist air above the crop

$$h_{rc}(T_c - T_e)A_t + 0.016h_c[P(T_c) - \gamma P(T_e)]A_t = h_2(T_e - T_a)A_t \qquad (7.8)$$

where $h_2 = 5.7 + 3.8\,V$ is convective and radiative heat transfer coefficient from environment to ambient air and V is the wind velocity. Moisture evaporated (m_{ev}) can be evaluated as

$$m_{ev} = 0.016\frac{h_c}{\lambda}[P(T_c) - \gamma P(T_e)]A_t t \qquad (7.9)$$

Since operating temperature range for open sun drying lies between 32 and 50 °C, hence the partial vapor pressure can be linearised as

$$P(T) = R_1 T + R_2 \qquad (7.10)$$

where, R_1 and R_2 are constant and it can be obtained by regression method by using the data of steam table (Appendix Vb). By substituting the linear expression of partial vapor pressure in Equation (7.7) and (7.8), the new equations becomes,

$$\alpha_c I(t) A_t - h_{rc}(T_c - T_e) A_t - 0.016 h_c [(R_1 T_c + R_2)$$

$$- \gamma(R_1 T_e + R_2)] A_t - h_i (T_c - T_a) A_t = M_c C_c \frac{dT_c}{dt} \tag{7.11}$$

$$h_{rc}(T_c - T_e) + 0.016 h_c (R_1 T_c + R_2) - 0.016 h_c \gamma (R_1 T_e + R_2) = h_2 (T_e - T_a) \tag{7.12}$$

From Equation (7.12)

$$T_e = \frac{(rc + 0.016 R_1) T_c + R_2 [0.016 h_c (1 - \gamma)] + h_2 T_a}{h_{rc} + 0.016 h_c \gamma + h_2} \tag{7.13}$$

and Equation (7.11) can be rewritten as

$$\alpha_c I(t) A_t + h_i A_t T_a - [(h_{rc} + h_i) A_t + 0.016 h_c R_1 A_t] T_c + (h_{rc} + 0.016 h_c R_1 \gamma) A_t T_e$$

$$- 0.016 h_c R_2 A_t (1 - \gamma) = M_c C_c \frac{dT_c}{dt} \tag{7.14}$$

Substituting Equation (7.13) in (7.14)

$$\{[\alpha I(t) A_t + h_i A_t T_a] - [(h_{rc} + h_i) A_t + 0.016 h_c R_1 A_t]\} T_c + [(h_{rc} + 0.016 h_c R_1 \gamma) A_t]$$

$$\times \left\{ \frac{(h_{rc} + 0.016 h_c R_1) T_c + R_2 [0.016 h_c (1 - \gamma)] + h_2 T_a}{h_{rc} + 0.016 h_c \gamma R_1 + h_2} \right\}$$

$$- 0.016 h_c R_2 A_t (1 - \gamma) = M_c C_c \frac{dT_c}{dt} \tag{7.15}$$

This is the form of the first order differential equation

$$\frac{dT_c}{dt} + a T_c = f(t) \tag{7.16}$$

where

$$a = \frac{[(h_{rc} + h_i) A_t + 0.016 h_c R_1 A_t] - \frac{A_t (h_{rc} + 0.016 h_c R_1 \gamma)(h_{rc} + 0.016 h_c R_1)}{h_{rc} + 0.016 h_c \gamma R_1 + h_2} - 0.016 h_c R_2 A_t (1 - \gamma)}{M_c C_c} \tag{7.17a}$$

and

$$f(t) = \frac{\alpha I(t) A_t + h_i A_t T_a + \left[\frac{h_2 (h_{rc} + 0.016 h_c R_1 \gamma) A_t}{h_{rc} + 0.016 h_c \gamma R_1 + h_2} \right] T_a + \frac{(h_{rc} + 0.016 h_c R_1 \gamma) A_t R_2 . 0.016 h_c (1 - \gamma)}{h_{rc} + 0.016 h_c \gamma R_1 + h_2} - 0.016 h_c R_2 A_t (1 - \gamma)}{M_c C_c} \tag{7.17b}$$

The analytical solution of the Equation (7.17) (see Section 9.9.4) is

$$T_c = \frac{\overline{f(t)}}{a} (1 - e^{-a\Delta t}) + T_{co} . e^{-a\Delta t} \tag{7.18}$$

After knowing the value of T_c from above equation for a given design and climatic parameters, the value of T_e can be obtained from Equation (7.13) and m_{ev} can be determined by Equation (7.9) which can be rewritten as

$$m_{ev} = 0.016\frac{h_c}{\lambda}[(R_1 T_c + R_2) - \gamma(R_1 T_e + R_2)]A_t t \qquad (7.19)$$

EXAMPLE 7.3

Derive an expression for time interval to raise the temperature of crop from T_{co} to T_c for a given climatical parameter and crop mass.

Solution

Equation 7.18 can be written as $\quad e^{-a\Delta t} = \dfrac{T_c - \dfrac{\overline{f(t)}}{a}}{T_{co} - \dfrac{f(t)}{a}}$

Taking log of both side and solving for Δt, we get

$$\Delta t = -\frac{1}{a}\ln\left[\frac{T_c - \dfrac{\overline{f(t)}}{a}}{T_{co} - \dfrac{\overline{f(t)}}{a}}\right]$$

EXAMPLE 7.4

Using the data of Tables 7.5 and 7.6, compute Δt to raise the temperature of green chillies from 39°C to 42°C for the following parameters:

$$\overline{I(t)} = 1005\ \text{W/m}^2$$
$$\overline{T_a} = 37\ °\text{C}.$$

Solution

The values of a and $f(t)$ can be evaluated for Equations (7.17 a and b), respectively and is given as

$$a = 5.3832 \times 10^{-4} \quad \text{and} \quad \overline{f(t)} = 0.037.$$

Substituting these value in Δt in equation of Example 7.3, we get

$$\Delta t = -\frac{1}{5.3832 \times 10^{-4}}\ln\left[\frac{42 - \frac{0.037}{5.3832\times 10^{-4}}}{39 - \frac{0.037}{5.3832\times 10^{-4}}}\right] = -\frac{1}{5.3832 \times 10^{-4}}\ln\left[\frac{42 - 68.73}{39 - 68.73}\right]$$

$$= -\frac{1}{5.3832 \times 10^{-4}} \times -0.1064 = 198\ \text{sec}$$

7.3.4 Analysis for Steady State Condition

In a steady state condition, $\frac{dT_c}{dt} = 0$ in Equation (7.11), then Equations (7.11) and (7.12) can be rewritten as

$$\alpha I(t)A_t - h_{rc}(T_c - T_e)A_t - 0.016h_c[(R_1T_c + R_2) - \gamma(R_1T_eR_2)]A_t - h_i(T_c - T_a)A_t = 0 \quad (7.20)$$

$$(h_{rc} + 0.016h_cR_1)T_c + R_20.016h_c(1 - \gamma) + h_2T_a = (h_{rc} + 0.016h_cR_1\gamma + h_2)T_e \quad (7.21)$$

From Equation (7.20), expression of T_c can be written as

$$T_c = \frac{\alpha I(t) + h_iT_a + (h_{rc} + 0.016h_cR_1\gamma)T_e - 0.016h_cR_2(1 - \gamma)}{h_{rc} + 0.016h_cR_1 + h_i} \quad (7.22)$$

Substituting Equation (7.22) in (7.21), we can get the expression of T_e as

$$T_e = \frac{h\alpha}{ht}I(t) + \frac{Hh_i + h_2}{h_T}T_a + \frac{H_i}{h_T}0.016h_c(1 - \gamma)R_2 \quad (7.23)$$

where

$$h_T = \frac{(h_{rc} + 0.016h_cR_1)(h_{rc} + 0.016h_cR_1\gamma)}{h_{rc} + 0.016h_cR_1 + h_i} + h_{rc} + 0.016h_cR_1\gamma + h_2$$

$$H = \frac{h_{rc} + 0.016h_cR_1}{h_{rc} + 0.016h_cR_1 + h_i}$$

$$H_i = \frac{h_i}{h_{rc} + 0.016h_cR_1 + h_i}$$

T_e in Equation (7.23) is equivalent to solair temperature (T_{sae}) for open sun crop drying on the crop surface in steady state condition, i.e., $T_e = T_{\text{sae}}$.

7.3.5 Experimental Setup for Open Sun Drying (OSD)

Experiment were performed during the months of summer in the year 1999 as shown in Figure 7.5(*a*). The natural cooling of Kabuli chana were performed for indoor simulation as shown in Figure 7.5(*b*).

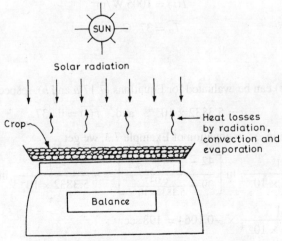

Figure 7.5(a)   Open sun drying.

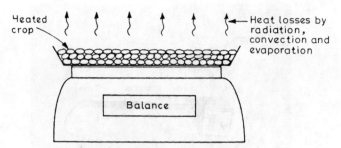

Figure 7.5(b) Natural drying by natural cooling.

The crops were given some treatments such as size reduction, peeling and soaking with water. The treatments given and the resulting bulk densities are summarized in Table 7.3.

A wire mesh tray of 0.45 m × 0.19 m size was used for drying the crop. A six-channel digital temperature indicator up to the range of 125°C with least count of 0.1°C with copper constantan thermocouple was used to measure the crop and air temperatures at different points. A dial type hygrometer with least count of 1 was used for relative humidity measurement just above the crop. For gravimetric analysis, an electronic balance of 1 kg capacity with a least count of 0.1 kg was used. The crop was kept on the electronic balance using the tray. Thermocouple were placed for measuring crop temperature and temperature of air just above the crop surface. Precautions were taken so that placement of the thermocouple does not affect the reading being shown by the balance. The dial type hygrometer was kept on the crop surface with its sensing part facing downwards towards the crop surface. Every time, it was kept two minutes before recording observations. The weight of the hygrometer was finally deducted from the electronic balance reading. All observation were recorded at 15 min time intervals. The whole unit was kept in open sun at a place with negligible wind velocity. The difference in weight gave the quantity of water evaporated during that time interval. The whole set up for cauliflower has been shown in Figure 7.5(c). The experiment was repeated five times for obtaining more accurate results for each crop.

In case of natural cooling of *Kabuli chana*, a known quantity of water soaked crop was heated up to 50°C and then allowed to cool naturally in the wire mesh tray kept directly on the electronic balance. Observations were recorded at every one minute interval. The datas recorded have been given in Table 7.4.

The number of data of Table 7.4 have been obtained by using the initial observation. Hence the value of N for calculating C and n should be always taken one more than actual observation i.e., $N = N + 1$. This will be valid only for transient condition.

TABLE 7.3 Treatment given to crops before drying and the bulk density

No.	Crop	Treatment given	Bulk density (kg/m³)
1	Green chillies	No treatment	280
2	Green peas	No treatment	575
3	Kabuli chana	Soaked in water for 6 hr to raise the moisture content up to 30 (w.b.)	550
4	Onion	Peeled and cut with the help of slicer in the form of flakes of 2 mm thickness	450
5	Potato	Peeled and cut with the help of slicer in the form of slices of 2 mm thickness (average diameter 35 mm)	500
6	Cauliflower	The flower was cut into small pieces of 2 cm size	415

Figure 7.5(c) Photograph of experimental set-up.

7.3.6 Methodology and Input Parameters for Computation

The value of C, n and h_c were computed at each stage of drying time by developing a new computer program. The input data taken from Anwar and Tiwari (2001) for determination of C, n and h_c with respect to time for various crops are given in Table 7.4.

TABLE 7.4a Input data for green chillies

Time (min)	$\overline{T}_c(°C)$	$\overline{T}_e(°C)$	$T_i(°C)$	$m_{ev}(g)$	$\overline{\gamma}$	$Gr \times 10^5$	Pr
15	39.34	40.80	40.070	5.9	0.365	3.74	0.7055
30	42.66	42.67	42.665	8.7	0.360	0.02	0.7056
45	45.54	44.43	44.985	10.9	0.360	2.40	0.7058
60	47.39	45.06	46.225	13.2	0.355	4.83	0.7058
75	48.79	45.29	47.040	12.7	0.355	7.07	0.7059
90	49.49	45.60	47.545	15.3	0.355	7.73	0.7059
105	49.35	45.99	47.670	18.9	0.345	6.65	0.7059
120	48.07	46.03	47.050	17.6	0.340	4.12	0.7059
135	47.85	45.98	46.915	19.5	0.335	3.79	0.7059
150	47.65	45.97	46.810	16.2	0.335	3.42	0.7059

TABLE 7.4b Input data for green pea

Time (min)	$\overline{T}_c(°C)$	$\overline{T}_e(°C)$	$T_i(°C)$	$m_{ev}(g)$	$\overline{\gamma}$	$Gr \times 10^5$	Pr
15	39.39	46.95	43.170	12.4	0.380	17.38	0.7057
30	41.14	47.97	44.555	15.1	0.395	14.98	0.7058
45	42.22	47.65	44.935	22.2	0.380	11.76	0.7058
60	42.34	47.25	44.795	18.4	0.355	10.68	0.7058
75	42.76	47.55	45.155	32.7	0.350	10.30	0.7058
90	43.07	47.38	45.225	21.2	0.350	9.24	0.7058
105	43.21	47.30	45.255	19.6	0.350	8.76	0.7058
120	43.71	47.24	45.475	15.1	0.350	7.51	0.7058
135	44.67	47.98	46.325	14.1	0.350	6.85	0.7059
150	45.40	49.03	47.215	14.3	0.350	7.29	0.7059

TABLE 7.4c Input data for kabuli chana

Time (min)	$\bar{T}_c(°C)$	$\bar{T}_e(°C)^*$	$T_i(°C)$	$m_{ev}(g)$	$\bar{\gamma}$	$Gr \times 10^5$	Pr
15	38.11	39.76	38.935	10.4	0.380	4.41	0.7054
30	41.27	42.29	41.780	13.8	0.400	2.46	0.7056
45	43.46	43.38	43.420	20.1	0.400	0.18	0.7057
60	43.86	43.66	43.760	22.7	0.385	0.45	0.7057
75	43.92	43.16	43.540	25.6	0.370	1.72	0.7057
90	43.39	41.90	42.645	23.7	0.365	3.49	0.7056
105	42.86	41.13	41.995	23.4	0.360	4.14	0.7056
120	43.76	41.99	42.875	19.5	0.355	4.11	0.7057
135	44.49	42.84	.43.665	16.2	0.355	3.73	0.7057
150	44.07	42.35	43.210	15.8	0.350	3.95	0.7057

$^*\bar{T}_e$; The average values of T_e for kabuli chana taken by Anwar and Tiwari (2001) in Table 4 were incorrect, which is being corrected in this table

TABLE 7.4d Input data for onion flakes

Time (min)	$\bar{T}_c(°C)$	$\bar{T}_e(°C)$	$T_i(°C)$	$m_{ev}(g)$	$\bar{\gamma}$	$Gr \times 10^5$	Pr
15	34.09	39.11	36.600	13.8	0.395	14.66	0.7053
30	34.60	39.62	37.110	16.9	0.400	14.37	0.7053
45	35.27	40.12	37.695	18.5	0.395	13.58	0.7054
60	35.79	40.47	38.130	20.2	0.380	12.89	0.7054
75	36.38	40.99	38.685	27.3	0.375	12.44	0.7054
90	36.67	41.26	38.965	32.1	0.380	12.26	0.7054
105	36.83	41.54	39.185	24.3	0.385	12.47	0.7054
120	37.06	41.65	39.355	23.4	0.375	12.08	0.7055
135	37.07	41.78	39.425	23.0	0.370	12.36	0.7055
150	37.50	42.52	40.010	22.8	0.370	12.90	0.7055

TABLE 7.4e Input data for potato slices

Time (min)	$\bar{T}_c(°C)$	$\bar{T}_e(°C)$	$T_i(°C)$	$m_{ev}(g)$	$\bar{\gamma}$	$Gr \times 10^5$	Pr
15	37.66	38.63	38.145	38.4	0.370	2.67	0.7054
30	32.85	39.16	36.005	41.7	0.375	18.86	0.7053
45	34.35	40.06	37.205	28.3	0.380	16.29	0.7053
60	35.39	40.82	38.105	38.0	0.380	14.97	0.7054
75	36.70	41.22	38.960	37.0	0.380	12.07	0.7054
90	37.63	41.64	39.635	35.5	0.380	10.45	0.7055
105	37.83	42.35	40.090	39.0	0.370	11.58	0.7055
120	37.88	42.98	40.430	37.0	0.360	12.91	0.7055
135	37.52	43.07	40.295	46.4	0.360	14.12	0.7055
150	37.18	43.02	40.100	40.0	0.360	14.96	0.7055

TABLE 7.4f   Input data for cauliflower

Time (min)	$\bar{T}_c(°C)$	$\bar{T}_e(°C)$	$T_i(°C)$	$m_{ev}(g)$	$\bar{\gamma}$	$Gr \times 10^5$	Pr
15	33.30	39.17	36.235	39.17	0.365	17.38	0.7053
30	34.09	39.75	36.920	39.75	0.355	16.32	0.7053
45	35.15	40.68	37.915	40.68	0.345	15.36	0.7054
60	36.52	41.26	38.890	41.26	0.340	12.69	0.7054
75	37.35	41.64	39.495	41.64	0.340	11.23	0.7055
90	38.49	42.03	40.260	42.03	0.335	9.01	0.7055
105	39.48	42.51	40.995	42.51	0.325	7.51	0.7055
120	40.07	43.12	41.595	43.12	0.320	7.40	0.7056
135	39.79	42.70	41.245	42.70	0.320	7.15	0.7056
150	38.38	42.08	40.230	42.08	0.320	9.43	0.7055

Numerical computation has been carried out for the solar intensity and ambient temperature of 15 June between 11 and 13.30 hours for the climatic conditions of Delhi. The solar intensity during this time has little variation from 990 to 1020 W/m^2. Equations (13), (18), (19) and (23) have been computed by using MATLAB software for predicting crop temperature, temperature above the crop surface, the rate of moisture removal and equivalent solar temperature. The design parameters for different crops used for computations have been given Tables 7.5 and 7.6, respectively.

TABLE 7.5   Constant used in modelling

Parameters	Values
A_t	0.45×0.19
g	9.81
X	$0.45 \times 0.19/[2 \times (0.45 + 0.19] = 0.0668$
λ	2.26×10^6
ε	0.9
σ	5.67×10^{-8}
t	15×60
h_i	5.7

TABLE 7.6   Input initial values of various crops used for modelling

Parameter	Green chillies	Green pea	Kabuli chana	Onion flakes	Potato slices	Cauli-flower
α	0.65	0.8	0.8	0.7	0.8	0.65
C	1.3158	0.9769	1.3105	1.0064	1.0200	0.9784
n	0.1556	0.2196	0.2098	0.2579	0.2965	0.2323
C_c	3950	3060	3060	3810	3520	3900
γ	0.35	0.362	0.371	0.3825	0.371	0.337
T_c	38.21	38.83	36.89	33.91	43.61	33.03
T_e	40.23	46.40	38.23	38.23	38.33	39.00
M_c	0.607	0.6157	0.6135	0.6029	0.6246	0.6246

7.3.7 Results and Discussion

7.3.7.1 Heat and mass transfer

The convective heat transfer coefficient at different drying time for various crops (green chillies, green pea, kabuli chana, onion, potato and cauliflower) taken for study is presented in Table 7.7. These reveal that the convective heat transfer coefficient vary significantly with increase of drying time. This is mainly because of porosity, shape, size and moisture content in the crops. This effect has not been considered by Anwar and Tiwari (2001). The h_c for green chillies and onion flakes increase with increase in drying time (Tables 7.7a and 7.7d), whereas for cauliflower h_c decrease during drying (Table 7.7f). For green pea and kabuli chana (Tables 7.7b and 7.7c) h_c increase in first 90 minutes of drying time then decrease steadily with increase of drying time. For potato slices h_c increase to be highest in 30 minutes of drying and then decrease steadily with increase in drying time (Table 7.7e). These results are expected.

However, the present values of convective heat transfer coefficient (h_c) for different crop after 150 minutes drying are very much closed to the results reported by the Anwar and Tiwari (2001). The h_c

TABLE 7.7a C, n and h_c for green chillies

Time (min)	C	n	h_c
15	1.0000	0.1516	2.69
30	1.2809	0.1607	2.79
45	1.2894	0.1570	2.97
60	1.3013	0.1536	3.11
75	1.3311	0.1466	3.10
90	1.3378	0.1451	3.17
105	1.3251	0.1482	3.34
120	1.3200	0.1511	3.47
135	1.3161	0.1544	3.62
150	1.3158	0.1556	3.68

TABLE 7.7b C, n and h_c for green pea

Time (min)	C	n	h_c
15	1.0000	0.2102	7.78
30	1.0007	0.2148	8.19
45	1.0057	0.2220	8.92
60	1.0050	0.2214	8.68
75	1.0131	0.2281	9.50
90	1.0125	0.2278	9.34
105	1.0102	0.2268	9.10
120	0.9999	0.2243	8.61
135	0.9866	0.2219	8.13
150	0.9769	0.2196	7.75

TABLE 7.7c C, n and h_c kabuli chana

Time (min)	C	n	h_c
15	1.0000	0.2026	5.24
30	1.0058	0.2116	5.59
45	1.1752	0.2196	6.08
60	1.2249	0.2270	6.51
75	1.2183	0.2310	6.91
90	1.2257	0.2291	7.11
105	1.2347	0.2271	7.24
120	1.2585	0.2218	7.11
135	1.2855	0.2153	6.84
150	1.3105	0.2098	6.62

TABLE 7.7d C, n and h_c for onion flakes

Time (min)	C	n	h_c
15	1.0000	0.2354	10.47
30	1.0001	0.2422	11.50
45	1.0006	0.2455	11.98
60	1.0012	0.2475	12.27
75	1.0034	0.2524	13.08
90	1.0061	0.2575	14.02
105	1.0065	0.2584	14.18
120	1.0065	0.2585	14.15
135	1.0064	0.2583	14.10
150	1.0064	0.2579	14.03

TABLE 7.7e C, n and h_c for potato slices

Time (min)	C	n	h_c
15	1.0000	0.3172	19.03
30	1.0012	0.3164	26.84
45	1.0179	0.3033	24.70
60	1.0190	0.3024	25.12
75	1.0196	0.3006	24.62
90	1.0189	0.2986	23.83
105	1.0191	0.2976	23.56
120	1.0207	0.2960	23.20
135	1.0193	0.2969	23.63
150	1.0200	0.2965	23.70

TABLE 7.7f C, n and h_c for cauli flower

Time (min)	C	n	h_c
15	1.0000	0.2451	12.49
30	1.0002	0.2487	13.04
45	0.9990	0.2421	11.81
60	0.9923	0.2339	10.32
75	0.9964	0.2368	10.64
90	0.9887	0.2342	10.02
105	0.9785	0.2321	9.46
120	0.9725	0.2309	9.13
135	0.9775	0.2318	9.19
150	0.9784	0.2323	9.23

for kabuli chana comes to 6.61 instead of 8.45 in case of Anwar and Tiwari (2001) due to the incorrect value taken for their calculation. Whereas h_c in present model for green chillies and onion flakes are 3.68 and 14.03 respectively, which are same as the model of Anwar and Tiwari (2001).

7.3.7.2 Thermal modelling

Thermal prediction of various crops, namely green chillies, green pea, kabuli chana, onion flakes, potato slices and cauliflower is shown in Figures 7.6–7.11 respectively. These figures show that there are fair

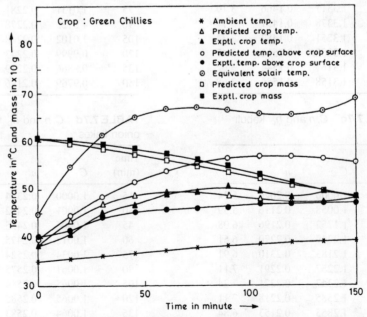

Figure 7.6 Prediction of crop temperature (T_c) and mass (M_c) under open sun drying for green chillies.

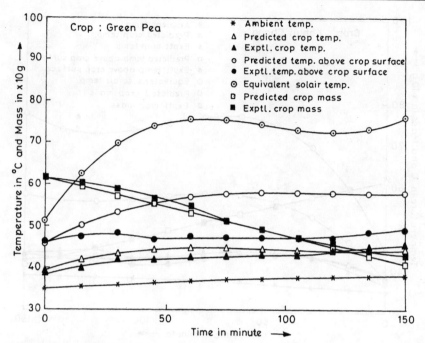

Figure 7.7 Prediction of crop temperature (T_c) and mass (M_c) under open sun drying for green pea.

agreements between predicted and measured crops temperature and moisture removal. The predicted value of crop temperature (T_c) for the kabuli chana, onion flakes, potato slices and cauliflower are very close to the experimental values (Figures 7.7, 7.9, 7.10 and 7.11). The deviation between predicted and measured crop temperature for green chillies and green pea (Figures 7.6 and 7.7) may be due to change in velocity of air during experiment and the day of observation, which has been considered constant for modeling.

Predicted temperature above the crop surface (T_e) for all the cases is much higher than the experimental temperature above the crop surface due to difference in the height of measurement from the crop surface. The prediction of temperature above the crop surface is made near to the crop surface (Figures 7.6 to 7.11). It is further noted that the rate of moisture removal remains the same during the experiments. This may be referred as first stage drying. Due to this there is linear variation in the crop mass as expected. The predicted values of mass of the crop during drying are very closed to the observed values in all the crops except potato slices. The experimental values of mass in potato slices are lower (higher moisture evaporated) than the predicted value because of during experiment the potato slices were turned at regular interval and exposing more area (Figure 7.10). It is also important to mention that there is a strong needs to further extend this work up to final drying of the crop up to optimum moisture (safe storage moisture) of each crop.

7.3.7.3 Equivalent solair temperature

Equivalent solair temperature is the combine effect of solar radiation, ambient air temperature and long wave radiant heat exchange on the crop surface. From Figures 7.6 to 7.11, it is further noted that the

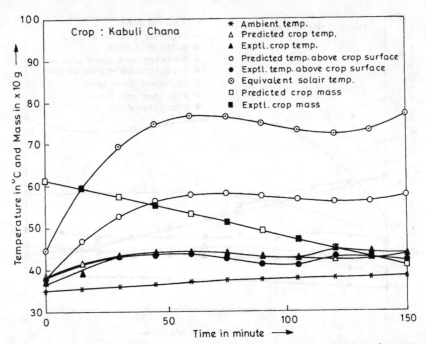

Figure 7.8 Prediction of crop temperature (T_c) and mass (M_c) under open sun drying for kabuli chana.

equivalent solar temperature (T_{sae}) is significantly much higher than temperature above the crop surface as expected. However, the measured surrounding temperature (T_e) is significantly lower than equivalent solar temperature in all the cases. It is due to fact that this temperature is measured at higher distance from crop surface while equivalent solar temperature is just above the crop surface.

7.4 THERMAL ANALYSIS OF CABINET DRYER

In order to write the energy balance for different components of cabinet dryer (Fig. 7.2), the following additional assumptions have been made:

i. the heat capacity of glass, crop tray, drying chamber wall and air have been neglected,

ii. volume shrinkage is negligible during drying process,

iii. particle to particle conduction is negligible,

iv. the heat flow is negligible,

v. there is no condensation of water vapor in drying chamber, and

vi. there is no stratification along the depth of the crop due to small depth of crop.

The energy balance equations are:

Crop surface:

$$[\alpha_c \tau I(t)]A_c = M_c C_c \frac{dT_c}{dt} + h(T_c - T_{ch})A_c \tag{7.24}$$

where $h = h_{rc} + h_{cc} + h_{ec}$ (the sum of radiative, convective and evaporative heat transfer coefficient)

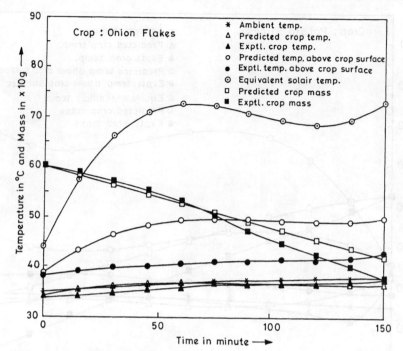

Figure 7.9 Prediction of crop temperature (T_c) and mass (M_c) under open sun drying for onion flakes.

Drying chamber:

$$h(T_c - T_{ch})A_c = V_1(T_{ch} - T_a) + h_s A_s(T_{ch} - T_a)$$

where

$$V_1 = \frac{NV}{3} \tag{7.25}$$

The performance of conventional cabinet dryer will be discussed in the next Section 7.5.1.

7.5 ENERGY BALANCE FOR REVERSE ABSORBER CABINET DRYER (RACD)

7.5.1 Thin Layer Drying

Referring to Figure 7.3, the energy balance equation for different components are as follows:

At absorber plate: $\quad \tau\rho'\alpha_p I' = h_{pf}(T_p - T_f) + h_{rpc}(T_p - T_c) + U_t(T_p - T_a) \tag{7.26}$

where, ρ', α_p and τ are the reflectivity of reflector, the absorptivity of the plate and the transmittivity of glass cover respectively.

At working fluid (Air): $\qquad h_{pf}(T_p - T_f) = h_{fc}(T_f - T_c) \tag{7.27}$

At crop surface: $\quad [h_{fc}(T_f - T_c) + h_{rpc}(T_p - T_c)]A_c = M_c C_c \dfrac{dT_c}{dt} + h(T_c - T_{ch})A_c \tag{7.28}$

where, $h = h_{rc} + h_{cc} + h_{ec}$ (the sum of radiative, convective and evaporative heat transfer coefficient)

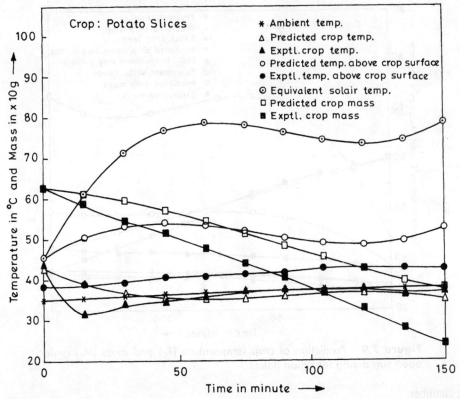

Figure 7.10 Prediction of crop temperature (T_c) and mass (M_c) under open sun drying for potato slices.

At chamber:
$$h(T_c - T_{ch})A_c = V_1(T_{ch} - T_a) + h_s A_s(T_{ch} - T_a)$$

where
$$V_1 = \frac{NV}{3}. \tag{7.29}$$

The overall thermal efficiency of conventional and reverse absorber cabinet dryer in Figure 7.12 (Goyal and Tiwari, 1997) shows that there is about 10 percent improvement in the performance of a reverse absorber cabinet dryer (RACD) over conventional dryer. The RACD has following advantages:

i. the overall heat loss in RACD is significantly reduced

ii. the absorptivity of absorber is unaffected due to placement of crop in RACD

The hourly variation of the crop temperature and moisture removal from the crop for reverse absorber cabinet dryer (RACD) with and without glass cover have been shown in Figures 7.13. Figure 7.13(a) shows that there is significant increase in crop temperature ($\cong 10°C$) due to additional solar energy available from glass cover of the dryer. Similar effect has been observed for moisture removal in Figure 7.13(b). The comparison of overall thermal efficiency of RACD dryer with and without glass cover has been also shown in Figure 7.14. This shows that RACD with glass cover gives better results.

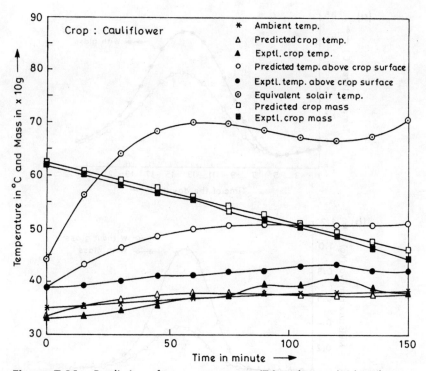

Figure 7.11 Prediction of crop temperature (T_c) and mass (M_c) under open sun drying for cauliflower.

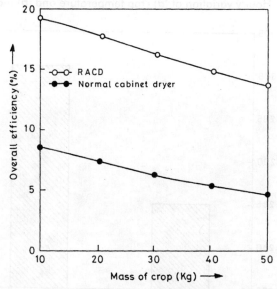

Figure 7.12 Comparison of overall thermal efficiency of RACD and conventional cabinet dryer.

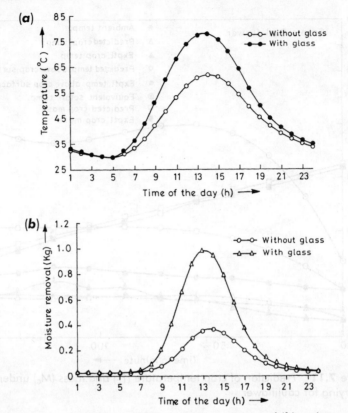

Figure 7.13 Hourly variation of (*a*) crop temperature and (*b*) moisture removal.

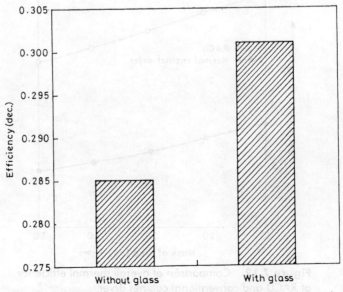

Figure 7.14 Overall thermal efficiency of RACD with and without glass cover.

EXAMPLE 7.5

Derive an expression for the rate of heat transferred from the fluid to the crop for RACD.

Solution

From Equation (7.26), we get $\quad T_p = \dfrac{\tau \rho' \alpha_p I' + h_{pf} T_f + h_{rpc} T_c + U_t T_a}{h_{pf} + h_{rpc} + U_t}$

Substitute $h_{pf}(T_p - T_f)$ from above equation into Equation (7.27), we get

$$h \tau \rho' \alpha_p I' - U_{fc}(T_f - T_c) - U_{fa}(T_f - T_a) = h_{fc}(T_f - T_c)$$

This gives, $\qquad T_f = \dfrac{h \tau \rho' \alpha_p I' + U_{fc} T_c + U_{fa} T_a + h_{fc} T_c}{U_{fc} + U_{fa} + h_{fc}}$

Also, $\qquad T_p - T_f = \dfrac{\tau \rho' \alpha_p I' - h_{rpc}(T_f - T_c) - U_t(T_f - T_a)}{h_{pf} + h_{rpc} + U_t}$.

Further, $\qquad h_{pf}(T_p - T_f) = h \tau \rho' \alpha_p I' - U_{fc}(T_f - T_c) - U_{fa}(T_f - T_a)$

where, $\quad h = \dfrac{h_{pf}}{h_{pf} + h_{rpc} + U_t} ; U_{fc} = \dfrac{h_{pf} h_{rpc}}{h_{pf} + h_{rpc} + U_t} \quad$ and $\quad U_{fa} = \dfrac{h_{pf} U_t}{h_{pf} + h_{rpc} + U_t}$

Now, the rate of heat transferred from the fluid to the crop is given by,

$$\dot{q}_{fc} = h_{fc}(T_f - T_c) = H \tau \rho' \alpha_p I' - U_{ca}(T_c - T_a)$$

where $\qquad H = \dfrac{h_{fc} h}{U_{fc} + U_{fa} + h_{fc}} \quad$ and $\quad U_{ca} = \dfrac{h_{fc} U_{fa}}{U_{fc} + U_{fa} + h_{fc}}$

7.5.2 Deep Bed Grain Drying

In this case, the energy balance equation for absorber will be same as Equation (7.26). In addition to this, there will be four sets of equations for general deep bed drying condition. These are as follows:

Drying rate equation (Brooker et al., 1992)

If M and M_e are moisture content and equilibrium moisture content (decimal dry basis), then

$$\frac{\partial M}{\partial t} = -K_d(M - M_e) \tag{7.30}$$

where $\quad k_d = a \exp(-b/T_c) \quad$ and $\quad M = 0.01[\ln(-\gamma)/\{2.31 \times 10^{-5}(T_f + 55.815)\}]^{\frac{1}{2.99}}$

For rough rice, $a = 13.88 \text{ s}^{-1}$ and $b = 3818.2 \text{ K}$ (Verma et al., 1985).

Mass balance equation

$$\dot{M} \frac{\partial H}{\partial Y} = -\rho_c \frac{\partial M}{\partial t} \tag{7.31}$$

Energy balance equation

The exchange of thermal energy between air and grain is written as

$$\dot{M}(C_f + C_v H)\frac{\partial T_f}{\partial Y} = \rho_c C_v(T_f - T_c)\frac{\partial M}{\partial t} - \rho_c(C_c + C_1 M)\frac{\partial T_c}{\partial t} + \rho_c \lambda \frac{\partial M}{\partial t} \qquad (7.32)$$

Heat transfer rate equation

The rate of heat transfer between air and grain is given by

$$\rho_c(C_c + C_1 M)\frac{\partial T_c}{\partial t} = h_v(T_f - T_c) + \rho_c \lambda \frac{\partial M}{\partial t} \qquad (7.33)$$

Heat utilization factor

The heat utilization factor (*H.U.F.*) may be defined as the ratio of temperature decrease due to cooling of the air during drying and the temperature increase due to heating of air. The expression for *H.U.F.* is given by

$$H.U.F. = \frac{\text{Heat Utilised}}{\text{Heat Supplied}} = \frac{T_f - T_c}{T_f - T_a} \qquad (7.34)$$

Coefficient of performance

The coefficient of performance (*C.O.P.*) of a grain dryer is expressed as

$$C.O.P. = \frac{T_c - T_a}{T_f - T_a} \qquad (7.35)$$

The variation of heat utilization factor (*H.U.F.*) and coefficient of performance with time of the day has been shown in Figure 7.15(*a*). It is clear that the *H.U.F.* increases with time of the day. It may be due to fact that the temperature decrease $(T_f - T_c)$ is faster in the beginning due to more moisture content in the crop and the temperature increase $(T_f - T_a)$ is slow at later stage due to decrease on insolation level

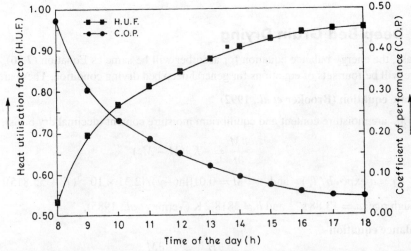

Figure 7.15(a)   Hourly variation of HUF and COP with time of the day.

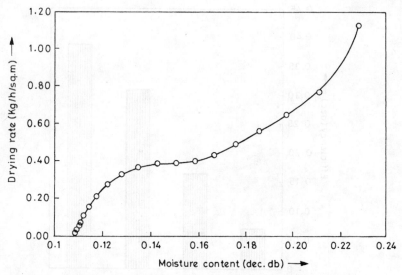

Figure 7.15(b) Variation of drying rate with moisture content.

as expected. Similarly, the coefficient of performance ($C.O.P.$) decreases with time of the day similar nature of the crop ($T_c - T_a$) and the fluid ($T_f - T_a$) temperature difference with respect to ambient air temperature.

Thermal efficiency

Instantaneous thermal drying efficiency (η_i) of the system can be defined as

$$\eta_i = \frac{\dot{M}_{ev} \times \lambda}{3600 \times I'(t)} \times 100 \tag{7.36}$$

Overall daily thermal efficiency of the system (η_o) is defined by the ratio of heat energy utilised in vaporising the moisture to that of heat collected but the reverse absorber collector can be determined using the following expression,

$$\eta_o \frac{\lambda \sum\limits_{t=1}^{t=24} \dot{M}_{ev}(t)}{(3600) \sum\limits_{t=1}^{t=24} I'(t)} \tag{7.37}$$

The overall thermal efficiency depends on the amount of moisture removal from the crop for a given insolation. This also determines the drying rate. The variation of drying rate with moisture content has been shown in Figure 7.15(b). This indicates that drying rate is more for more moisture content as expected.

7.6 ENERGY BALANCE FOR INDIRECT SOLAR DRYING (ISD) SYSTEM

The energy balance for ISD system will be as follows (Fig. 7.4):

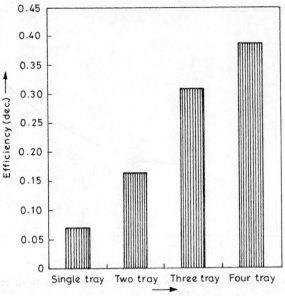

Figure 7.16 Efficiency curve of different trays system.

7.6.1 Solar Air Heater

Absorber Plate:
$$\tau\alpha_p I W dx = U_t(T_p - T_a)W dx + h_{pf}(T_p - T_f)W dx \qquad (7.38)$$

Working Fluid:
$$h_{pf}(T_p - T_f)W dx = \dot{m}C_f \frac{dT_f}{dx}dx + U_b(T_f - T_a)W dx \qquad (7.39)$$

where
$$U_t = \left[\frac{1}{h_{pg}} + \frac{1}{h_{ga}}\right]^{-1}$$

7.6.2 Drying Chamber

and,
$$\dot{m}C_f(T_{fo} - T_{fi}) = M_c C_c \frac{dT_c}{dt} + h(T_c - T_{ch})A_{ch} \qquad (7.40)$$

$$h(T_c - T_{ch})A_{ch} = \dot{m}C_f(T_{ch} - T_a) + h_s A_s(T_{ch} - T_a) \qquad (7.41)$$

Energy balance for multi-trays in drying chamber can also be written with help of Equation (7.40). In multi-trays, the thermal energy of first tray is used for drying for second tray and so on. The results for different trays has been summarised in Figure 7.16. It is seen that the overall efficiency increases with increase of number of trays.

EXAMPLE 7.6

Derive an expression for the rate of useful energy available from air heater to the drying chamber for ISD.

Solution

From Equation (7.38), we get

$$T_p = \frac{\tau \alpha_p I + U_t T_a + h_{pf} T_f}{U_t + h_{pf}}$$

Substitute T_p from above equation into Equation (7.39) and after rearranging one has,

$$\frac{dT_f}{dx} + aT_f = f(t)$$

where $\qquad a = \dfrac{(U_b + U_{fa})W}{\dot{m}_f C_f}$ and $\quad f(t) = \dfrac{h \tau \alpha_p I + (U_p + U_{fa})W T_c}{\dot{m}_f C_f}$

$$h = \frac{h_{pf}}{U_f + h_{pf}} \quad \text{and} \quad U_{fa} = \frac{h_{pf} U_t}{h_{pf} + U_t}$$

The solution of above equation is

$$T_f = \frac{f(t)}{a}(1 - e^{-ax}) + T_{fi} e^{-ax}$$

with, $T_{fi} = T_f |_{x=0}$.

The rate of useful energy available to the drying chamber is

$$\dot{Q}_u = \dot{m}_f C_f (T_f |_{x=L} - T_{fi}) = \dot{m}_f C_f \left[\frac{f(t)}{a} - T_{fi} \right] (1 - e^{-aL}) = \dot{m}_f C_f (1 - e^{-aL}) \left[\frac{f(t)}{a} - T_{fi} \right]$$

PROBLEMS

7.1 Calculate C and n for all crops of Table 7.3 by using the different number of observations of Table 7.4.
Hint See Example 7.1.

7.2 Calculate the rate of convective and evaporation heat loss for the Problem 7.1.
Hint See Example 7.2.

7.3 Derive an expression for the crop temperature in the case of normal and reverse absorber cabinet dryer.
Hint
   a. For cabinet dryer, solve Equations (7.24) and (7.25) by eliminating T_{ch}.
   b. For reverse absorber cabinet dryer, eliminate T_p, T_f and T_{ch} by using Equations (7.26), (7.27) and (7.29) from Equation (7.28) and then solve. For detail, see Section 9.9.4.

7.4 Derive an expression for the rate of solar energy required to raise the temperature of crop from T_{co} to T_c as a function of design parameters.

Hint Rewrite Equation (7.18) as,

$$\frac{f(t)}{a} = \frac{T_c - T_{co}e^{-a\Delta t}}{1 - e^{-a\Delta t}}$$

and solve above equation for $I(t)$.

7.5 Calculate the time interval to raise its temperature from 40°C to 50°C of the crop of Tables 7.5 and 7.6 for the following dates: Mass= 500 kg, $I(t) = 500$ W/m² and $A_t = 1$ m².

Hint See Example 7.4.

7.6 Compute instantaneous thermal efficiency (η_i) for Figure 7.6 and plot η_i versus $\frac{T_{co} - T_a}{I(t)}$.

7.7 Plot η_i versus $\frac{T_c - T_a}{I(t)}$ for other crops.

7.8 Calculate the C and n and corresponding the rate of convective and evaporative heat transfer for different number of consecutive observations of Table 7.4.

Hint Consider first eight observations and repeat the calculation of Examples 7.1 and 7.2. Similarly, continue same computation for other sets of observations.

7.9 Derive an expression for the rate of heat transfer from plate to crop for RACD.

Hint In Example 7.5, eliminate T_f from expression of T_p and follow similar procedure for Example 7.6.

CHAPTER *8*

Solar Concentrators

8.1 INTRODUCTION

Solar concentrator is a device which concentrates the solar energy incident over a large surface onto a smaller surface. The concentration is achieved by the use of suitable reflecting or refracting elements, which results in an increased flux density on the absorber surface as compared to that existing on the concentrator aperture. In order to get a maximum concentration, an arrangement for tracking the sun's virtual motion is required. An accurate focussing device is also required. Thus, a solar concentrator consists of a focussing device, a receiver system and a tracking arrangement. Temperature as high as 3000°C can be achieved using solar concentrators, and hence they have potential applications in both thermal and photovoltaic utilization of solar energy at high delivery temperatures.

Solar concentrating devices have been used since long. In Florence, as early as 1695, a diamond could be melted by solar energy. Lavoisier carried out a number of experiments with his double-lens concentrator. The knowledge of concentrator dates back even to the time of Archimedes, whose book "On Burning Mirrors" is an evidence of this fact. Many uses of concentrators were reported in the eighteenth and nineteenth centuries, particularly in heat engines and steam production. The advantages of concentrator are as follows:

  i. It increases the intensity by concentrating the energy available over a large surface onto a smaller surface (absorber).
  ii. Due to the concentration on a smaller area, the heat-loss area is reduced. Further, the thermal mass is much smaller than that of a flat plate collector and hence transient effects are small.
  iii. The delivery temperatures being high, a thermodynamic match between the temperature level and the task occurs.
  iv. It helps in reducing the cost by replacing an expensive large receiver by a less expensive reflecting or refracting area.

However, concentrator is an optical system and hence the optical loss terms become significant. Further, it operates only on the beam component of solar radiation, resulting in the loss of diffuse component. Although the basic concepts of flat plate collectors are applicable to concentrating systems, a number of complications arise because of nonuniform flux on absorbers, wide variations in shape, temperature and heat-loss behavior of absorbers and finally the optical considerations in the energy balance conditions. It may be noted that the higher the concentration of a collector, higher is the precision of optics and more is the cost of the unit. In addition to the complexity of the systems, the maintenance requirements are also increased.

8.2 CHARACTERISTIC PARAMETERS

Definitions of the several terms that characterize concentrating collectors have been given here:

8.2.1 Aperture Area (A_a)

It is the plane opening of the concentrator through which the incident solar flux is accepted. It may be defined by the physical extremities of the concentrator (Fig. 8.1).

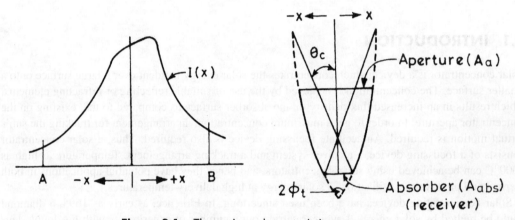

Figure 8.1   The various characteristic parameters.

8.2.2 Acceptance Angle ($2\theta_c$)

It is the limiting angle over which incident ray path may deviate from normal to the aperture plane and still reach the absorber. Concentrators with large acceptance angle need to be moved only seasonally while concentrators with small acceptance angles must be moved continuously to track the sun (Fig. 8.1).

8.2.3 Absorber Area (A_{abs})

This is the total area receiving the concentrated radiation. It is also from this area that useful energy is delivered to the system (Fig. 8.1).

8.2.4 Geometric Concentration Ratio (C)

It is defined as the ratio of the collecting aperture area to the area of the absorber. It's value varies from unity (flat plate collector) to several thousand (parabolic dish).

$$C = A_a / A_{abs} \qquad (8.1a)$$

8.2.5 Local Concentration Ratio or Brightness Concentration Ratio

It may so happen that the absorber in some systems may not be fully or uniformly illuminated, thus, in order to characterize this local concentration this term is defined. It is defined as the ratio of flux arriving at any point on the absorber to the incident flux at the entrance aperture of the concentrating system.

8.2.6 Intercept Factor (γ)

It is the fraction of focussed energy intercepted by the absorber of a given size (Fig. 8.1).

$$\gamma = \frac{\int_A^B I(x)dx}{\int_{-\infty}^{\infty} I(x)dx} \tag{8.1b}$$

For a typical concentrator-receiver design, it's value depends on the size of the absorber, γ usually has a value greater than 0.9. If the radiation is normal to the aperture, the value of γ is 1.

8.2.7 Optical Efficiency (η_0)

It is defined as the ratio of the energy absorbed by the absorber to that incident on the collector. It includes the effect of mirror surface shape and reflection, transmission losses, tracking accuracy, shading by the receiver, cover transmission absorptance of the absorber and solar beam incident angle.

$$\eta_0 = \frac{S}{I_b} \tag{8.1c}$$

where S is defined in Section 8.7.

8.2.8 Thermal Efficiency (η_c)

It is the ratio of the useful energy delivered to the energy incident on the aperture.

$$\eta_c = \dot{q}_u / I_b \tag{8.1d}$$

8.2.9 Concentration Ratio (C)

The area concentration ratio is $C = A_a/A_r$.

Let us consider the circular concentrator with aperture area A_a and receiver area A_r viewing the sun of radius r at a distance R as shown in Figure 8.2. The half angle subtended by the sun is θ_s. If the concentrator is perfect, the radiation from the sun on the aperture (and consequently on receiver) is the fraction of the radiation emitted by the sun which is intercepted by the aperture.

$$Q_{s \to r} = A_a \frac{r^2}{R^2} \sigma T_s^4$$

A perfect receiver (i.e., blackbody) radiates energy equal to $A_r T_r^4$, and a fraction of this, E_{r-s}, reaches the sun.

$$Q_{r \to s} = A_r \sigma T_r^4 E_{r-s}$$

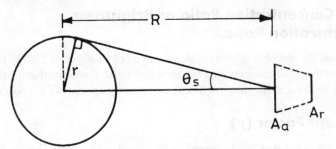

Figure 8.2 Schematic of the sun at T_s at a distance R from a concentrator with aperture area A_a and receiver area A_r.

When T_r and T_s are the same, $Q_{s \to r}$ will be equal to $Q_{r \to s}$ by the second law of thermodynamics. Thus,

$$A_a \frac{r^2}{R^2} \sigma T_s^4 = A_r \sigma T_r^4 E_{r-s}$$

The maximum possible value of E_{r-s} is unity, the maximum concentration ratio for circular concentrators is,

$$\left(\frac{A_a}{A_r} \right)_{max} = \frac{R^2}{r^2} = \frac{1}{\sin^2 \theta_s}. \tag{8.1e}$$

Similar relation can be obtained for linear concentrators as well.

8.3 CLASSIFICATION

Solar concentrators may be classified as (i) tracking type and (ii) non-tracking type. Tracking may be continuous or intermittent and may be one-axis or two-axes. As the sun may be followed by moving either the focussing part or the receiver or both; concentrators can be classified accordingly. Further, the system may have distributed receiver or central receiver.

The concentrators may also be classified on the basis of optical components. They may be (i) reflecting or refracting type, (ii) imaging or non-imaging type, and (iii) line focussing or point focussing type. The reflecting or refracting surface may be one piece or a composite surface, it may be a single stage or two stage type system and may be symmetric or asymmetric. In practice, however, hybrid and multistage systems, incorporating various levels of the features, occur frequently.

8.4 TYPES OF CONCENTRATORS

There are a number of methods by which the flux of radiation on receivers can be increased. Some of them have been discussed in this section.

8.4.1 Tracking Concentrators

8.4.1.1 One-axis tracking concentrators

Concentrators with one-axis tracking are used to achieve moderate concentration. A few of them have been described below:

i. Fixed mirror solar concentrator (FMSC)

This concentrator consists of a fixed mirror with a tracking receiver system (Fig. 8.3). The fixed mirror is composed of long, narrow flat strips of mirror arranged on a reference circular cylinder of a chosen radius R.

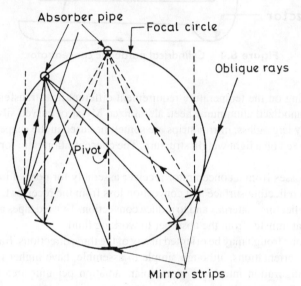

Figure 8.3   Fixed mirror solar concentrator.

The width of mirror strip is compatible with the diameter of the absorber pipe to be used. The angle of each element is so chosen that the focal distance of the array is twice the radius of the reference cylinder. The array produces a narrow focal line that lies along the same circular path with the diurnal motion of the sun. The focal line can be easily tracked by the movable receiver pipe that is made to rotate about the centre of curvature of the reflector module. Thus the delicate part (the mirrors) of the system can be rigidly fixed.

The image width at the absorber is ideally the same as the projected width of the mirror element. Thus, the concentration ratio is approximately the same as the number of mirror elements, ignoring the solar beam spread. As the aperture is fixed and concave in shape, the mirror strips result in shading with very high or very low sun altitude angles. Also, due to the strips, edge losses occur during reflection. However, mirrors can be suitably designed to have less than 10% of the total energy lost over a years time. Some (Fixed Mirror Solar Concentrator) models have shown overall efficiencies in the range of 40–50%.

ii. Cylindrical parabolic concentrator

A cylindrical parabolic trough is a conventional optical imaging device used as a solar concentrator. It consists of a cylindrical parabolic reflector and a metal tube receiver at its focal plane (Fig. 8.4). The receiver is blackened at the outside surface and is covered by concentrator and rotated about one axis to track the sun's diurnal motion. The heat transfer fluid flows through the absorber tube, gets heated and thus carries heat.

Such concentrators have been in use for many years. The aperture diameter, rim angle and absorber size and shape may be used to define the concentrator. The absorber tube may be made of mild steel or copper and is coated with a heat resistant black paint. Selective coatings may be used for better

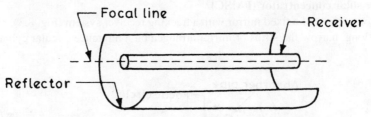

Figure 8.4 Cylindrical parabolic concentrator.

performance. Depending on the temperature requirement different heat transfer liquids may be used. Reflectors may be of anodized aluminum sheet, aluminized Mylar or curved silvered glass. Since it is difficult to curve a very large glass, mirror strips are sometimes used in the shape of parabolic cylinder. The reflecting part is fixed on a light weight structure. The concentration ratio for a cylindrical absorber varies from 5 to 30.

The major energy losses from a concentrator-receiver assembly for normal incidence are the losses during reflection from reflecting surface and convection loss from the receiver to surroundings. Efforts are made to use high reflecting materials and to reduce convection. Twisted tapes are used in the absorber tube to cause large heat transfer from the absorber to working fluid.

A cylindrical parabolic trough may be oriented in any of the three directions: East–West, North–South or polar. The first two orientations, although simple to assemble, have higher incidence angle cosine losses. The polar configuration intercepts more solar radiation per unit area as compared to other modes and thus gives the best performance.

iii. Linear Fresnel lens/reflector

A linear Fresnel lens solar concentrator is shown in Figure 8.5. It consists of linear grooves on one surface of the refracting material. The groove angles are chosen with reference to a particular wavelength of incident beam so that the lens acts as a converging one for the light which is incident normally.

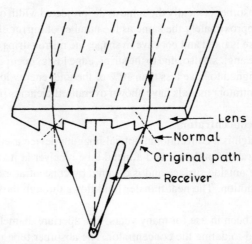

Figure 8.5 Schematic view of Fresnel lens concentrator.

Although, both glass and plastic can be used as refracting materials for fabricating Fresnel lenses, glass is seldom used because it is difficult to mould, and has large surface tension. Plastic lenses, on the other hand, are economical and the mould last for an appreciable time. Plastic Fresnel lenses with 20 grooves per mm have been molded.

The Fresnel lens may be installed with either the grooves facing the sun or the grooves facing downward. In the first case, the ineffective facts of the grooves prevent a part of the input light from being transmitted to the focus (According to Snell's law the refracted light is deviated away from the normal on moving from a denser to a rarer medium). Also dust is accumulated in these grooves resulting in a reduced performance. In the second case, the concentrator has a high surface reflection loss and large off-axis aberrations. While reflection loss causes low efficiency, the aberrations result in a low concentration ratio. The fact that the beam is not incident normally also affects. The focal length of the lens varies rapidly with the change of angle of incidence. So, for a better performance, the optical system needs to track the path of the sun.

Fresnel reflectors can also be used as concentrating devices. Figure 8.6 shows such a configuration, which is made up of smaller flat or curved components. It consists of a number of mirror elements mounted suitably, so that all incident parallel rays of light, after reflection, are focussed at a common point. Ideally, mirror elements must be parabolic in shape, but to simplify the manufacturing and assembling problems, flat mirrors are generally used.

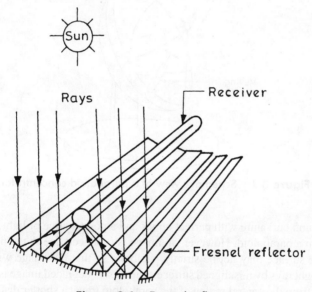

Figure 8.6   Fresnel reflectors.

8.4.1.2 *Two-axes tracking concentrators*

In order to achieve a high concentration for high temperature solar processes, concentrators with double curvatures are used. These require two axes tracking of the sun. Some of these have been described below:

i. Paraboloidal dish concentrator
A paraboloid is produced when a parabola rotates about its optic axis (Fig. 8.7). When it is used to concentrate solar radiation, a high concentration ratio is achieved.

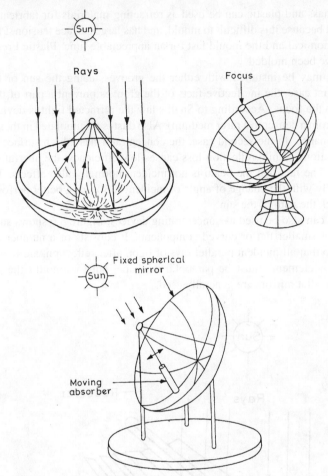

Figure 8.7 Schematic view of a paraboloid concentrator.

Due to the compound curvature with perfect optics and a point source of light, theoretically, light is focussed at a point in a paraboloid. However, an image is produced due to the finite angular subtense of the sun. Also, the surface may not be strictly parabolic so that the image will be enlarged due to misdirection of the light rays by misaligned surface elements. A degraded image is obtained if the object is off-axis. The rays from the central region of the paraboloid travel a shorter distance in arriving at the focus, whereas, the rays from the edges travel a larger distance, resulting in a spread of the image. Thus, a three dimensional image of the sun in the shape of an ellipsoid is formed (Fig. 8.8).

The thermal losses from a paraboloid are primarily radiative and can be reduced by decreasing the absorber aperture area. This, however, results in a smaller intercept factor. The optimum intercept factor is about 0.95–0.98. The larger the surface errors, the larger must be the absorber size to achieve the optimum beam intercept.

High collection efficiency and high quality thermal energy are the features of a paraboloid or parabolic dish type of concentrator. The delivery temperatures being very high, these devices can be used as sources for a variety of purposes.

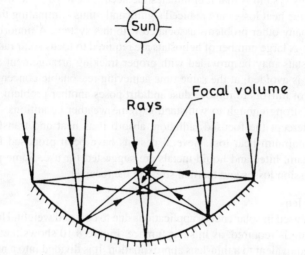

Figure 8.8   Illustration of formation of ellipsoid image in paraboloid.

ii. Central tower receiver

The system consists of a central stationary receiver to which the solar radiation is reflected by heliostats (Fig. 8.9). A heliostat is composed of a large array of mirrors fixed to a supporting frame. This frame can be used to track the sun, as desired.

The heliostats are installed in the open space and together they act like a dilute paraboloid. They focus solar radiation on a central receiver, which is stationary. Concentration ratios as high as 3000 can be achieved. The absorbed energy can be extracted from the receiver and delivered at a temperature and pressure suitable for driving turbines for power generation.

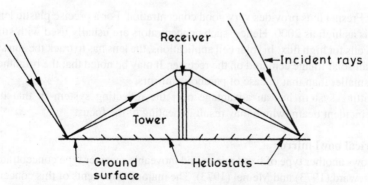

Figure 8.9   Schematic view of central receiver-heliostat system.

The advantage of this system is that it eliminates the need for carrying the working fluid over large distances. As a result, the heat losses are reduced significantly thus eliminating the need for insulation. There are, however, many other problems associated with this system. A majority of them are due to heliostats and receiver. A large number of heliostats are required to focus solar radiation requiring large free space. The heliostats may be provided with proper tracking arrangement and may be arranged such that self shading is avoided; at the same time achieving reasonable concentration. In addition to the cost, the cleaning of mirrors to remove dust and dirt poses another problem. Further, the heliostat arrangement must be strong enough to withstand extreme weather conditions. The receiver must be able to effectively intercept the focused radiation, absorb it as heat and transfer this energy to the working fluid with minimum heat loss. Several designs have been proposed for the receiver. Heat transfer fluids like steam, hitec and liquid metals are suggested for use. Some problems arise due to reflection and transmission losses and thermal stress in the receiver.

iii. Circular Fresnel lens

Lenses are usually not used in solar energy applications due to cost and weight. However, these are used where high temperature is required, as in solar furnace. Figure 8.10 shows the principle of this lens. Optically, the lens is equivalent to a thin-lens approximation. It is divided into a number of zones which are spaced at a few tenths of a millimeter; the space can also be few centimeters. Within each aperture zone, the tilt of the lens surface is so adjusted that its optical behavior resembles that of a conventional spherical lens of the same focal length. The focus of the annulus need not be curved, but is required to have the correct tilt so as to refract the light to the focus. This is because, the absorbing surface is usually much larger than the width of a Fresnel zone on the lens. Meinel (1977) has given the equation for the tilt of the facet as a function of the aperture zone and focal length.

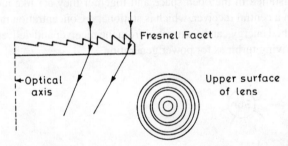

Figure 8.10 Schematic diagram of circular Fresnel lens.

The circular Fresnel lens provides very good concentration. For a precise plastic lens, the brightness concentration is as high as 2000. Hence, such concentrators are usually used with silicon and gallium arsenide solar cells for high flux. In solar cell applications, the lens has to track the sun since it is required to keep the small solar image centered on the receiver. It may be noted that the brightness concentration in this case is smaller than that in case of parabolic mirrors.

The transmitting system has an advantage over the reflecting system in that it absorbs certain wavelength of incident beam, which may result in heating of the focus.

iv. Hemispherical bowl mirror

Figure 8.11 shows another type of fixed mirror and moveable receiver type concentrator, independently proposed by Steward (1973) and Meinel (1973). The major components of this concentrator are a fixed hemispherical mirror, a tracking linear absorber and a supporting structure. The hemisphere produces a highly aberrated optical image. However, because of its symmetry all rays entering into the hemisphere

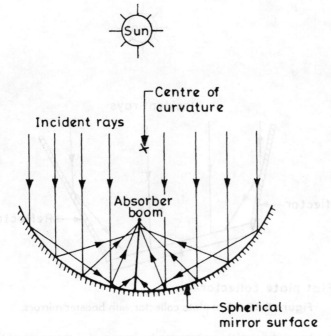

Figure 8.11   Cross-sectional view of hemispherical mirror concentrator.

after reflection cross the paraxial line at some point between the focus and the mirror surface. Therefore, an absorber pivoted about the center of curvature of the hemisphere intercepts all reflected rays. The absorber is to be moved so that its axis is always aligned with solar rays passing through the center of the sphere. This requires two axes tracking. Though this motion can be set in a number of ways, the simplest one is to adopt equatorial mount in which the absorber is driven around a polar axis, during the day, at a constant angular speed of 15 degree/hour. Through a slow continuous motion or a periodic adjustment about an axis normal to polar axis, correct declination can be maintained. It may be noted that this type of concentrator gives a lesser concentration, owing to spherical aberration, than that in paraboloids.

8.4.2 Non-tracking Concentrators

Tracking concentrators provide high delivery temperatures but require accurate tracking device and fine surface accuracies and hence are expensive. However, for medium temperature operation, less expensive concentrators have been designed, without the tracking requirement. The description of some of these concentrators is given below:

i. Flat receiver with booster mirror
Figure 8.12 shows a flat receiver with plane reflectors at the edges to reflect additional radiation into the receiver. Mirrors are also called booster mirrors. The concentration ratio of these concentrators is relatively low, with a maximum value less than four. As the solar incidence angle increases, the mirrors become less effective. For a single collector, booster mirrors can be used on all the four sides. When the sun angle exceeds the semiangle of booster mirrors, the mirror actually starts casting shadow on the absorber. In case of an array of collectors, booster mirrors can be used only on two sides.

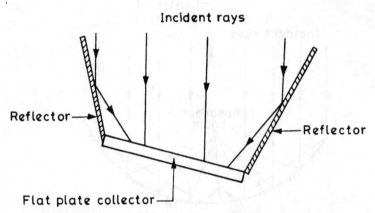

Figure 8.12 Flat plate collector with booster mirrors.

The efficiency of a boosted flat plate system can be increased if the angle of the flat mirrors can be changed several times during the year. The advantage of such a system is that it makes use of the diffuse radiation in addition to the beam radiation. The attainable temperature and collection efficiency will be higher than that of a flat plate collector of the same collection area.

ii. Tabor-Zeimer circular cylinder

Figure 8.13 shows such a concentrator. It is a very simple cylindrical optical system which consists of an inflated plastic cylinder with a triangular pipe receiver. The cylinder has a clear portion on the top to

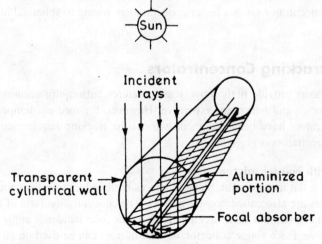

Figure 8.13 Schematic of Tabor-Zeimer cylindrical concentrator.

permit radiation to enter and fall on its rear portion which is aluminized to act as a mirror. The incident radiation is reflected by the mirror and is focussed on the absorber near the bottom of the cylinder.

A concentration of about 3 can be achieved without tracking. It can be placed along East–West axis and requires only seasonal tracking. The concentrator uses, in addition to beam component, some diffuse radiation. The delivery temperatures and collection efficiencies are higher than that possible with an ordinary flat plate collector.

iii. Compound parabolic concentrator

This concentrator is a nonimaging one and belongs to a family of concentrator which has highest possible concentration permissible by thermodynamic limit for a given acceptance angle. Further, it has a large acceptance angle and needs to be intermittently turned towards the sun.

The first design of a compound parabolic concentrator (CPC) was found independently by Winston (1965) and Baranov (1966). It consists of two parabolic segments, oriented such that focus of one is located at the bottom end point of the other and vice versa (Fig. 8.14). The axes of the parabolic segments subtend an angle, equal to acceptance angle, with the CPC axis, and the slope of the reflector surfaces at the aperture plane are parallel to the CPC axis. The receiver is a flat surface parallel to the aperture joining two foci of the reflecting surfaces.

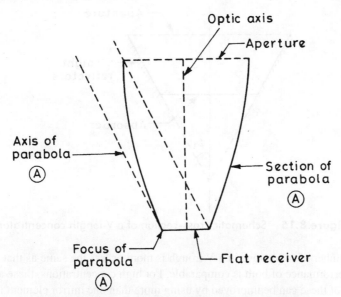

Figure 8.14 Schematic cross-section of a CPC.

Rays incident in the central region of the aperture undergo no reflection whereas those near the edges undergo one or more reflections. The number of reflections depend on the incident angle, collector depth and concentration ratio (Rabl, 1976). To reduce cost of the unit, the CPC can be truncated in height to half, without any significant change in concentration.

Extensive investigations on this concentrator have led to several modified designs of the ideal CPC. The salient modifications can be listed as follows:

i. The use of receiver shapes such as fins, circular pipes for better optical and thermal performance.

ii. Truncation of CPC height to reduce the physical size and cost.

iii. Asymmetric orientation of source and aperture to deliver seasonal varying outputs.

iv. Design of CPC as a second stage concentrator.

In view of the above modification, the reflecting surface of all resulting concentrators may not be parabolic, but these still belong to nonimaging group of concentrators.

The CPC can be used in a nontracking mode for concentration ratios of about 6. However, for higher ratios, the reflector surface area becomes very large and hence cannot be used.

iv. V-trough

Figure 8.15 shows schematically such a concentrator. It consists of highly reflecting side walls which reflect solar rays to a receiver plate placed at the base of the trough. The trough is aligned in East–West direction, so as to avoid diurnal tracking. These concentrators provide higher concentration (of the order of 3 in straight wall case) than flat plate collectors with booster mirrors, because in the latter case the acceptance angle is very large and so the concentration is low. Different combinations of depth to base-width ratios and cone angle are possible for optimum performance depending on the frequency of seasonal tilt adjustments.

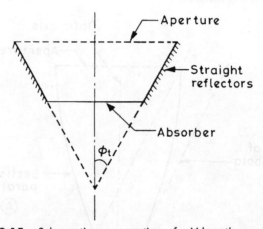

Figure 8.15 Schematic cross-section of a V-length concentrator.

The average number of reflections in a V-trough is more or less the same as that in a CPC. For low concentrations, performance of both is comparable. For high concentrations, these appear impractical. The performance of these can be improved by using more than one mirror element in each side wall at suitable angle thus resulting in polygonal troughs.

8.5 GEOMETRICAL OPTICS IN CONCENTRATORS

Geometrical optics plays an important role in designing almost any solar concentrator. A major part of the design involves following the paths of rays through a system of reflecting surfaces. This is known as *ray tracing method*. It is based on two fundamental laws, namely, the laws of reflection and laws of refraction (Snell's law).

8.5.1 Ray Tracing in a Reflecting Surface

It is primarily governed by the law of reflection which states that the incident ray, the reflected ray and the normal to the surface drawn at the point of incidence lie in one plane and the angle of incidence is equal to the angle of reflection. In a vector notation, we can write for the direction of reflected ray as (Rabl, 1985)

$$\hat{r} = -\hat{i} + 2(\hat{i}.\hat{n}) - \hat{n} \tag{8.2a}$$

where the vectors, shown in Figure 8.16, $\hat{r}$, $\hat{i}$ and $\hat{n}$ are respectively the unit vectors along the reflected, incident and normal to the reflecting surface at the point of incidence. Given an incident ray and the shape of the surface, geometry tells the point of incidence and the normal to the surface at the point of incidence. Then Equation (8.1) gives the direction of reflected ray.

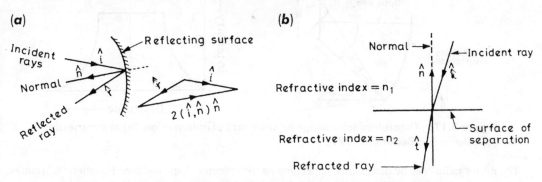

Figure 8.16 Definition of unit vector (**a**) for reflection (**b**) for refraction.

8.5.2 Ray Tracing in a Refracting Surface

The law of refraction governs ray tracing in refracting surfaces. It states the ratio of sine of angle of incidence to sine of angle of refraction is a constant; further, the incident ray, the refracted ray and the normal to the surface at the point of incidence all lie in one plane. We can write a vector equation for the direction of refracted ray. If n_1 and n_2 are the refractive indices of media 1 and 2 respectively, $\hat{i}$ and $\hat{n}$ are respectively the unit vectors along the incident ray and normal drawn at the point of incidence and $\hat{t}$ is the unit vector along the refracted ray, then in vector notation the refracted ray is given as

$$\hat{t} = (1/n)[\hat{i} - \hat{n}(\hat{i}.\hat{n}) + \sqrt{(\hat{i}.\hat{n})^2 + (n^2 - 1)}] \tag{8.2b}$$

where $n = n_1/n_2$ is the relative index of refraction of two media. The direction of reflected and refracted rays can be determined by using Eqs. (8.1) and (8.2), respectively.

8.6 THEORETICAL SOLAR IMAGE

In imaging type concentrators, image of sun is formed on the receiver, and, is in general not, a distinct one. Even a perfect optical system produces an image of finite size because the sun is not a point source. Sun–earth distance causes solar disc to subtend an angle of 32′ at a point on earth surface; which is responsible for the finite size of image in any optical system.

Figure 8.17 illustrates the above fact. R_f is a reflector and R_v is a plane receiver kept normal to the axis of the concentrator. Rays from solar disc are shown to be incident at the edge and after reflection, form the image on the receiver. If the distance between the point of incidence and the focus (mirror radius) is r, then the image size W is given by,

$$W = \frac{2r(\tan\ 16')}{\cos\phi} \tag{8.3}$$

where ϕ is the angle between the optical axis of the concentrator and the normal at the point of incidence (Fig. 8.17(a)).

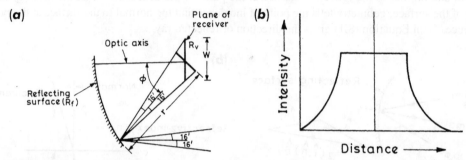

Figure 8.17 Theoretical solar image (**a**) schematic of its formation, (**b**) its cross section (Ref. Duffie and Beckman, 1991).

The mirror radius can be determined depending on the reflector shape and then Equation (8.3) can be used to calculate the theoretical image spread. Figure 8.17(b) shows the spread of an ideal solar image on plane receiver kept normal to the axis of a parabolic reflector.

It may be noted that this theoretical image is subjected to modifications due to two facts. First, the solar disc is not uniformly bright, the center being the brightest. Second, the actual concentrating systems are not precise optical systems. Thus, the image is significantly distorted from the theoretical size, forming larger images.

8.7 THERMAL ANALYSIS

The energy balance considerations, similar to flat plate collectors, are applied to describe the performance of concentrators. The complications occur in the calculation of thermal losses due to the following reasons:

    i. receiver shapes are widely variable and the radiation intensity at the receiver is not uniform.

    ii. the temperature being high, edge losses and conduction effects are significant.

Thus, it is not possible to give a general analysis for the estimation of thermal losses of concentrators. Each receiver has to be analyzed separately. However, from a basic knowledge of flat plate collector, we can derive the expression for collection efficiency or thermal efficiency in terms of inlet fluid temperature, fluid flow rate, ambient temperature and solar intensity.

As defined in Section 8.2, the thermal efficiency, η_c, is given by,

$$\eta_c = (\dot{q}_u / I_b) \tag{8.4}$$

where I_b is the incident solar energy and $\dot{q}_u$ is the rate of useful energy per unit aperture area.

With no losses, $\dot{q}_{loss} = 0$ and $\dot{q}_u = \dot{q}_{ab} = S$; $\eta_c = \eta_0 = \dfrac{S}{I_b}$

where S is the absorbed radiation per unit area of unshaded aperture and is given as, $I_b\rho(\gamma\tau\alpha)_n K_{\gamma\tau\alpha}$, I_b is the beam component of the incident radiation, ρ is the specular reflectance of the concentrator, γ, τ and α are functions of the angle of incidence of radiation on the aperture and $K_{\gamma\tau\alpha}$ is an incidence angle modifier.

With losses, $\dot{q}_u = \dot{q}_{abs} - \dot{q}_{loss} = S - U_L(T_r - T_a)$ (8.5)

where $\dot{q}_{abs}$ and $\dot{q}_{loss}$ are respectively the rate of energy absorbed and lost per unit aperture area.

In some of the concentrators, the acceptance angle may be large so that in addition to beam component, some part of diffuse radiation also comes in. In the case of high concentration systems, even the direct component is curtailed. I_b is, thus, to be corrected accordingly.

As has been mentioned earlier, the losses depend on the geometry of the system. Let us consider linear concentrating systems with cylindrical receivers (Fig. 8.4), the overall heat transfer coefficient from the surroundings to the fluid in the tube (Eq. (3.29)) is

$$U_0 = \left[\frac{1}{U_L} + \frac{D_0}{h_{fi} D_i} + \frac{D_0 \ln(D_0/D_i)}{2K} \right]^{-1}$$ (8.6)

where U_L is the heat transfer coefficient from the receiver to the ambient; D_i and D_0 are the inside and outside tube diameters; h_{fi} is the heat transfer coefficient inside the tube, and K the thermal conductivity of the tube.

The useful energy gain, $\dot{q}_{u'}$, in terms of the absorbed solar radiation per unit length is

$$\dot{q}_{u'} = \frac{\dot{Q}_u}{L} = \frac{A_a S}{L} - \frac{A_r U_L}{L}(T_r - T_a)$$ (8.7)

where $\dot{Q}_u = A_a \dot{q}_u$, A_a is the aperture area and A_r is the area of the receiver.

In a steady state condition, the useful energy gain per unit collector length, in terms of the energy transfer to the fluid at local fluid temperature T_f is

$$\dot{q}_{u'} = \frac{(A_r/L)(T_r - T_f)}{\frac{D_0}{h_{fi} D_i} + \left(\frac{D_0}{2K} \ln \frac{D_0}{D_i} \right)}$$ (8.8)

Elimination of T_r from Equations (8.7) and (8.8) gives the rate of useful energy per unit-length as

$$\dot{q}_{u'} = F' \frac{A_a}{L} \left[S - \frac{A_r}{A_a} U_L(T_f - T_a) \right]$$ (8.9)

Equation (8.9) can be used to get the expression for $\dot{Q}_u$ and $\dot{q}_u$

$$\dot{Q}_u(W) = F' A_a \left[S - \frac{A_r}{A_a} U_L(T_f - T_a) \right]$$ (8.9a)

and,

$$\dot{q}_u(W/m^2) = F' \left[S - \frac{A_r}{A_a} U_L(T_f - T_a) \right]$$ (8.9b)

In terms of inlet fluid temperature, T_{fi}, for forced circulation mode (similar to Equation (3.70)), Equation (8.9b) becomes

$$\dot{q}_u(\text{W/m}^2) = F_R \left[S - \frac{A_r}{A_a} U_L(T_{fi} - T_a) \right] \tag{8.9c}$$

or

$$\dot{Q}_u(\text{W}) = F_R A_a \left[S - \frac{A_r}{A_a} U_L(T_{fi} - T_a) \right] \tag{8.9d}$$

where the collector efficiency factor, F', is given as

$$F' = \frac{1/U_L}{\dfrac{1}{U_L} + \dfrac{D_0}{h_{fi} D_i} + \left(\dfrac{D_0}{2K} \ln \dfrac{D_0}{D_i} \right)} \tag{8.10}$$

or

$$F' = U_0/U_L \tag{8.11}$$

Using Equations (8.4) and (8.5), we get

$$\eta_c = (\dot{q}_{\text{abs}}/I_b) - (\dot{q}_{\text{loss}}/I_b) = \eta_0 - (\dot{q}_{\text{loss}}/I_b) \tag{8.12}$$

The collector heat loss factor (U_L) with respect to collector aperture area is defined as,

$$U_L = \dot{q}_{\text{loss}}/(T_r - T_a) \tag{8.13}$$

where T_r and T_a are respectively the receiver plate and ambient temperature. Hence, from Equation (8.9b)

$$\eta_c = \frac{\dot{q}_u}{I_b} = F'[\eta_0 - (A_r/A_a)\{U_L(T_f - T_a)/I_b\}] \tag{8.14}$$

As in the case of flat plate collector, the collection efficiency in terms of average fluid temperature can be written as

$$\eta_c = F'[\eta_0 - (A_r/A_a)\{U_L(T_f - T_a)/I_b\}] \tag{8.15}$$

where the collector efficiency factor, F', is given by the ratio between thermal resistance from the receiver surface to ambient and the thermal resistance from fluid to ambient. In terms of inlet fluid temperature (T_{fi}), Equation (8.15) becomes

$$\eta_c = F_R[\eta_0 - (A_r/A_a)\{U_L(T_{fi} - T_a)/I_b\}] \tag{8.16}$$

where the heat removal factor, F_R is

$$F_R = \frac{\dot{m}C_p}{A_r U_L} \left[1 - \exp\left(-\frac{A_r U_L F'}{\dot{m}C_p} \right) \right] \tag{8.17}$$

here $\dot{m}$ is the flow rate in kg/sec and C_p is the fluid heat capacity.

EXAMPLE 8.1

A linear parabolic concentrator with aperture, $a = 2.00$ m and focal length, $f = 1.00$ m is continuously adjusted about a horizontal east–west axis. It is to be fitted with a liquid-heating receiver unit whose length is 10.0 m. A strip of the reflector 0.21 m wide is shaded by the receiver. The receiver is designed to be just large enough to intercept all of the specularly reflected beam radiation when the incident beam radiation is normal to the aperture. The normal beam radiation is 950 W/m^2. Here $\alpha\tau$ for the receiver is 0.78 with radiation normal to the aperture, ρ is 0.84. The inlet fluid temperature is 180°C, the ambient temperature is 20°C and F_R is 0.85.

Solution

Since the radiation is normal to the aperture $\gamma = 1$. A fraction of the reflector 0.21/2.00, or 0.10, is shaded by the receiver, so (1−0.10) or 0.90 of the reflector is effective.
The product $\rho\gamma\tau\alpha$ is, $\rho\gamma\tau\alpha = 0.84 \times 1 \times 0.78 = 0.655$. Based on the area of the unshaded aperture,

$$\dot{q}_{ab} = S = 950 \times 0.655 = 622.25 \text{ W/m}^2$$

At an estimated mean receiver surface temperature of 200°C, the overall heat transfer coefficient, U_L is 13.7 W/m^2°C.
At a concentration ratio A_a/A_r of $(2.00 - 0.21) \times 10/(0.21 \times 10) = 8.5$
From Equation (8.9d),

$$\dot{Q}_u = 0.85 \times 10.0(2.00 - 0.21)[622 - 13.7/8.5(180 - 20)]$$
$$= 0.85 \times 10.0 \times 1.79[364] = 5540 \text{ W} = 5.54 \text{ kW}.$$

EXAMPLE 8.2

A cylindrical parabolic concentrator with width 2.0 m and length 8 m has an absorbed radiation, per unit area of aperture, of 400 W/m^2. The receiver is a cylinder painted flat black and surrounded by an evacuated glass cylindrical envelope. The absorber has a diameter of 55 mm, and the transparent envelope has a diameter of 85 mm. The collector is designed to heat a fluid entering the absorber at 220°C at a flow rate of 0.04 kg/s. The value of C_p for the fluid is 3.26 kJ/kg°C. The heat transfer coefficient inside the tube is 280 W/m^2°C and the overall loss coefficient is 12 W/m^2°C. The tube is made of stainless steel ($K = 16$ W/m°C) with a wall thickness of 5 mm. If the ambient temperature is 22°C, calculate the useful gain and exit fluid temperature.

Solution

The area of the receiver $A_r = \pi D.L = \pi \times 0.055 \times 8 = 1.382$ m^2
Taking into account shading of the central part of the collector by receiver,
$A_a = (2.0 - 0.085)8 = 15.32$ m^2
From Equation (8.10),

$$F' = \cfrac{1/U_L}{\cfrac{1}{U_L} + \cfrac{D_0}{h_{fi}D_i} + \left(\cfrac{D_0}{2K}\ln\cfrac{D_0}{D_i}\right)} = \cfrac{1/12}{\cfrac{1}{12} + \cfrac{0.055}{280\times0.045} + \cfrac{0.055}{2\times16}\ln\left(\cfrac{0.055}{0.045}\right)} = 0.946$$

F_R is given by Equation (8.17), and its value is

$$F_R = \frac{\dot{m}C_p}{U_L}\left[1 - \exp\left(-\frac{U_L F'}{\dot{m}C_p}\right)\right]$$

Here,

$$\frac{\dot{m}C_p}{A_r U_L F'} = \frac{0.04 \times 3.26 \times 10^3}{1.382 \times 12 \times 0.946} = 8.31$$

Substituting the above values in expression for F_R, we get

$$F_R = \frac{\dot{m}C_p}{A_r U_L}\left[1 - \exp\left(-\frac{A_r U_L F'}{\dot{m}C_p}\right)\right] = 7.87\left[1 - \exp\left(-\frac{1}{8.31}\right)\right] = 0.89$$

Now, the useful gain is (Equation (8.9d)),

$$\dot{Q}_u = A_a \times F_R\left[S - \frac{A_r \times U_L}{A_a}(220 - 22)\right] = 15.32 \times 0.98\left[400 - \frac{1.382 \times 12}{15.32}(220 - 22)\right] = 2787.5 \text{ W}$$

and the exit fluid temperature

$$T_0 = T_i + \frac{\dot{Q}_u}{\dot{m}C_p} = 220 + \frac{2787.5}{0.04 \times 3260} = 241.4°\text{C}$$

In order to improve the performance of a concentrator, the thermal losses must be minimized. This implies that the absorber should be small. But, smaller size of absorber implies a lower intercept factor. Thus, it leads to poor optical performance. Hence, a compromise has to be made between optical and thermal performance, of any concentrator to choose suitable concentration ratio. From the thermodynamic consideration, upper limit of concentration of any concentrating collector has been derived (Rabl, 1975, Rabl and Winston, 1976). It states that the maximum possible concentration is the reciprocal of the radiation shape factor between the sun and the concentrator (both considered black surfaces). This limit is valid for a given acceptance angle. Practical values of acceptance angle vary from a minimum value subtended by the disc of sun (about 1/2°) to 180°, which corresponds to a flat plate collector receiving radiation from a full hemisphere.

For two dimensional geometries, such as line focus or trough like concentrators, the radiation shape factor F_{A-S} is

$$F_{A-S} = \sin \theta_c \tag{8.18}$$

so, thermodynamic concentration limit

$$C = \frac{1}{\sin \theta_c} \tag{8.19}$$

For three dimension geometries such as point-focus of dish like concentrators, the shape factor and concentration limits are:

$$F_{A-S} = \sin^2\theta_c \tag{8.20}$$

$$C = 1/\sin^2\theta_c \tag{8.21}$$

EXAMPLE 8.3

Calculate the concentration ratio for a two-dimensional geometry if the acceptance angle is 30° and compare the result with that of a three-dimensional geometry.

Solution

For two dimensional geometry, from Equation (8.19), we have, $C = 1/\sin\theta_c$.
In the given problem, $\theta_c = 15°$, $C = 1/\sin 15 = 3.86$.
For three dimensional geometry, from Equation (8.21), we have, $C = 1/\sin^2\theta_c$.
Substituting the value of θ_c, we get, $C = 1/\sin^2 15 = 14.93$.
Thus, for the same acceptance angle the concentration ratio is more in three dimensional geometry.

If the optical considerations are applied, then the maximum limit is determined from the minimum acceptance angle. That is, if the concentrator is always facing the sun, the acceptance half angle (θ_c) is equal to half angle of solar disc (about $1/4°$). Equations (8.19) and (8.21) show that the concentration limit of a two dimensional concentrator is of the order of 200 and that of a three dimensional concentrator is of the order of 40,000.

However, in practice, due to the finite acceptance angles, the concentration are different. Further, in the nontracking collectors, the cosine effect reduces the energy gain at hours far from solar noon. The concentrations of some concentrators for different types of absorbers is given below:

8.7.1 Cylindrical Parabolic Concentrator

The concentrator for a flat absorber (one-sided) is

$$C = \frac{\sin\phi\cos(\phi + \theta_c)}{\sin\theta_c} - 1 \tag{8.22}$$

For two-sided flat absorber, concentration can be calculated by using the relation,

$$C_{2\text{-sided}} = \frac{1}{2}[C_{1\text{-sided}} + 1] \tag{8.23}$$

For a tube absorber, the concentration

$$C = (\sin\phi)/(\pi \sin\theta_c) \tag{8.24}$$

The optimum value of rim angle (ϕ) for maximum geometrical concentration can be determined by the condition

$$\left.\frac{dt}{d\phi}\right|_{(\phi)\text{opt}} = 0 \tag{8.25}$$

For a flat absorber (one-sided), the $(\phi)_{\text{opt}}$ and the corresponding concentration are given by

$$(\phi)_{\text{opt}} = (\pi/4) - (\theta_c/2) \tag{8.26}$$

and
$$C_{(\phi)_{opt}} = (1/2)[(1/\sin\theta_c) - 3] \qquad (8.27)$$

For a cylindrical absorber,
$$(\phi)_{opt} = (\pi/2) \qquad (8.28)$$

and
$$C_{(\pi)_{opt}} = (1/\pi)(1/\sin\theta_c) \qquad (8.29)$$

We can see that if the optical limit is considered, i.e. $\theta_c = 1/4°$, then for a cylindrical absorber, concentration becomes 73 and for flat absorber (one-sided), it is 113. Thus, the highest possible concentrations and the practical values lie below these limits.

8.7.2 Three Dimensional Concentrator

For a flat absorber (one-sided), the concentration

$$C = \frac{\sin^2\phi \cos^2(\phi + \theta_c)}{\sin^2\theta_c} - 1 \qquad (8.30)$$

and for a spherical absorber
$$C = \frac{\sin^2\phi}{4\sin^2\theta_c} \qquad (8.31)$$

The optimum rim angle can be determined, as done earlier. We see that the condition of flat plate absorber (one-sided) remains the same while for a spherical absorber its value is $(\phi)_{opt} = \pi/2$.

For $\theta_c = 1/4°$, maximum possible concentration is 1300 in both the absorbers.

8.7.3 Hemispherical Bowl Mirror

For a circular cylinder absorber, the concentration

$$C = \frac{\sin^2\phi}{\sin\theta_c} \qquad (8.32)$$

The optimum value of ϕ is $(\pi/2)$ and for $\theta_c = 1/4°$, the maximum possible concentration is 229. Thus this concentrator cannot achieve the concentration ratios of paraboloid type concentrator.

It may be seen that due to the fixed position, its aperture is not always normal to the sun rays; as such, the concentration is decreased by the cosine factor, i.e.

$$C_{eff} = \frac{\sin^2\phi}{\sin\theta_c}\cos\theta \qquad (8.33)$$

where θ is the angle of incidence of sun rays on the reflector.

EXAMPLE 8.4

Calculate the concentration ratio for a hemispherical bowl mirror if the acceptance half angle is 60°. Compare the result with cylindrical parabolic concentrator and a spherical absorber.

Solution

From Equation (8.32), we have, $C = \sin^2 \phi / \sin \theta_c$,
Substituting the values, $\theta_c = 30°$ and $\phi = \pi/2$ in the above equation, $C = 2.0$
Now, from Equation (8.24), we have, $C = \sin \phi / \pi \sin \theta_c$, thus, $C = 0.637$.
Further, for a spherical absorber, from Equation (8.31), we have,

$$C = \sin^2 \phi / 4 \sin^2 \theta_c = 1$$

8.7.4 V-trough

The concentration for a flat absorber is,

$$C = 1/ \sin(\theta_c + \phi_t) \tag{8.34}$$

where ϕ_t is the trough half angle (Fig. 8.1(a)). For smaller values of ϕ_t, C becomes large. Though C is maximum for $\phi_t = 0$, it cannot be used as a concentrator in that case, since the condition implies a point receiver.

It is very interesting to compare the concentration of the above concentrators with the corresponding ideal values.

8.8 TRACKING METHODS

8.8.1 Three Dimensional Concentrators

Point focus concentrating systems or three dimensional concentrators are required to continuously track the sun's motion, for maximum brightness concentration and radiation balance. This is because their acceptance angles are small and considering the fact that the absorber must be kept small. Hence, the collectors must be tracked along two axes to constantly keep the sun on the receivers.

There are several geometries of fully tracking mountings such as polar, azimuth, altitude-altitude and horizontal yoke mountings (Vasilevskies, 1966). The simplest is the polar mounting in which one axis is placed parallel to the earth's axis and the second axis orthogonal. If the collector, at a location, is oriented such that one axis is inclined with the horizontal at an angle equal to the latitude of that place, then that axis is parallel to the earth's axis. Other axis is kept normal to it. This mounting is called the polar mounting. The diurnal solar motion requires to be tracked about the polar axis only.

Generally, in all other configurations, two-axes tracking is required. A number of complications arise as a result, since the motions become nonlinear. There are other mountings in which motion is mechanically convenient. One such mounting is the attitude-azimuth mounting in which one axis is horizontal and the other vertical. In this case, the obvious advantage is that gravitational attraction being in the vertical plane of the concentrator, engineering problems are minimized.

The equation of motion for different mountings can be easily written, as has been dealt with in the works of Vasilevskies (1966) and Meinel (1977). The mounting for heliostats result in the generalized case of nonlinear motion in two coordinates. The kinematics for the generalized heliostat are given by Vartanyan *et al.* (1974) and Meinel (1977).

8.8.2 Two Dimensional Concentrators

Two dimensional or line focussing concentrators are moved only in one coordinate to track the sun's motion. There are a number of mounting configurations possible. The commonly used mountings are with the long axis, facing either north–south or east–west. When the troughs are oriented north–south, the collectors are moved to track the diurnal motion of the sun. In this case, the long axis can be horizontal or with an inclination, equal to the latitude of that place, to the horizontal. The latter mounting is the polar mounting, discussed earlier. If in addition to the east–west tracking, the north–south tracking is introduced then the sun always stays normal to the aperture so that availability of solar radiation on the aperture plane is maximum. But two-axis tracking results in a larger engineering problems.

It may be noted that in the north–south orientation, with east-west tracking, collectors must be separated from each other so as to eliminate self shading. This problem is acute when the sun's angle is shallow during morning and afternoon. In this case, more land area is required compared to east–west orientation.

The east–west orientation is the most common for line focussing systems. The sun's motion then coincides with the long axis. Hence, the performance is similar to flat plate collector, i.e. it follows the pattern of solar intensity, maximum at noon and falls off on either side. Except for the north–south corrections, which can be taken care of manually, there is no further tracking in the north–south direction.

As these concentrators are used to give moderate concentration, we can afford to have broad acceptance angle. As a result, these concentrators collect some diffuse energy from the sun and also the tracking does not pose any complicated engineering problem.

8.9 MATERIALS FOR CONCENTRATORS

As can be seen, the availability of suitable materials determines the effective use of solar energy. This is the case for concentrators as well. We outline here the major material problems associated with different components of solar concentrators.

8.9.1 Reflecting and Refracting Surfaces

The reflectors should have high reflectivity and good specular reflectance. No surface, however, is perfectly specular because of micro-roughness or surface undulation. For low concentration ratio, the reflector need not be highly specular. Glass silvered on the rear or second surface is usually used as mirror materials. Front surface mirrors can also be used. Aluminum can also be used for the purpose. It is protected by a coating of aluminum oxide, magnesium fluoride or cerium oxide. Other materials such as, metallized plastic films and thin metal sheets are also useful. Aluminum with a total reflectivity of about 85–90 percent and silver with total reflectance around 95 percent are very good reflecting surfaces for solar energy applications. Silver is not adequate for front-surface mirror since its film is tarnished.

Glass is the most durable material and with low iron content, it can be used as a transmitting material. Plastics are, now a days, used as refracting substances. Acrylic is found to be an excellent material for Fresnel lenses. Polymethyl methacrylate is generally used as it is weather resistant. Besides, acrylic can be easily molded or extruded.

The reflectors and refractors should be light weight so that they can be oriented easily. They should be able to withstand wind and other weather extremes, as dust, sand and other contamination strongly affect the performance, although, the effect varies with the type of material used, collector configuration, exposure time and site. The surfaces must be cleaned properly.

8.9.2 Receiver Cover and Surface Coatings

A variety of receiver shapes are used. A simple flat receiver can be used in parabolic concentrators or simply nonimaging type such as the V-trough, especially in low-temperature applications. Glass and transparent plastic films are generally used as cover materials for the receiver. Glass should have low iron content to reduce absorption, and its surface should be etched to reduce reflection losses. A double etching process can give better results, but the cost is increased. Plastics are not suitable as cover materials because of low service temperature limit and ultraviolet degradation. Evacuated tube receivers have also been used to eliminate convection losses.

Coatings are required to have strong solar absorptivity weather resistance, stability at high temperatures and cost effectiveness. Black paints are usually good non-selective paints. Among the selective coatings, black chrome appears to be the best. It can be electroplated on steel, copper, aluminum and other metals. Other metal oxide coatings such as black copper oxide and black nickel are also used; selective paints like leadsulphide particles in a silicon binder are also used.

8.9.3 Working Fluids

It is the working fluid that finally takes away the heat from the receiver for further use. Therefore, for effective heat transfer, the fluid should be stable at high temperatures, be noncorrosive and safe, besides, it also needs to be cost effective. Air is attractive for heating and cooling applications, but the heat transfer is very poor. The commonly used materials are pressurized water, liquid metals, thermirol 55 and Mobile therm 603.

8.9.4 Insulation

Insulation is required to reduce heat losses from the unirradiated portions of the receiver and the pipes carrying working fluid. Besides being cost effective, the insulation should be strong enough to withstand high temperature fluctuations. Fibre glass with and without binder, urethane foams and mineral fibre blankets are commonly used.

PROBLEMS

8.1 Derive an expression for concentration ratio for a circular concentrator.

Hint Since $Q_{s-r} = Q_{r-s}$ for $E_{r-s} = 1$ (see Section 7.2, Duffie and Beckman 1991).

8.2 What is the maximum concentration ratio for a circular concentrator?

Hint Use $\theta_s = 0.27°$ in Problem 8.1.

8.3 A cylindrical receiver of 70 mm diameter has an emittance of 0.9 and temperature at 250°C. The absorber is covered with a tubular glass cover of diameter 100 mm at 50°C and the annular space is evacuated. Find out
   a. convective heat transfer coefficient from cover tube to ambient, $h_{c,c-a}$
   b. radiative heat transfer coefficient from cover tube to ambient, $h_{r,c-a}$
   c. radiative heat transfer coefficient from absorber (receiver) to cover, $h_{r,r-c}$
   d. the overall heat loss coefficient, U_L

The ambient temperature air temperature is 10°C.

Hint a. $Nu = 0.40 + 0.54 \, Re^{0.52}$ for $0.1 < Re < 1000$
$\qquad = 0.30 \, Re^{0.6}$ for $1000 < Re < 50,000$

b. $h_{r,c-a} = 4\varepsilon\sigma\bar{T}^3 \, (\bar{T} = (50 + 10)/2)$

c. $h_{r,r-c} = \frac{\sigma(T_2^2 + T_1^2)(T_2 + T_1)}{\frac{1-\varepsilon_1}{\varepsilon_1} + \frac{1}{F_{12}} + \frac{(1-\varepsilon_2)A_1}{\varepsilon_2 A_2}}$, $F_{12} = 1$ for present case

d. $U_L = \left[\frac{A_r}{(h_{c,c-a} + h_{r,c-a})A_a} + \frac{1}{h_{r,r-c}} \right]^{-1}$.

8.4 Prove that the overall heat transfer coefficient from outer surface of cylindrical receiver to flowing fluid through it is given by

$$U = \left[\frac{D_0 \ln(D_0/D_i)}{2K} + \frac{D_0}{D_i h_{fi}} \right]^{-1}$$

in a steady state condition.

Hint See Section 8.7 and Chapter 2.

8.5 Prove that the rate of useful energy per unit length $\dot{q}_{u'}$ for natural circulation is given by

$$\dot{q}_{u'} = F' \frac{A_a}{L} \left[S - \frac{A_r}{A_a} U_L(T_f - T_a) \right]$$

where

$$F' = \frac{1/U_L}{\left[\frac{1}{U_L} + \frac{D_0}{h_{fi}D_i} + \frac{D_0}{2K} \ln \frac{D_0}{D_i} \right]} = \frac{U_0}{U_L}$$

and,

$$U_0 = \left[\frac{1}{U_L} + \frac{D_0}{h_{fi}D_i} + \frac{D_0}{2K} \ln \frac{D_0}{D_i} \right]^{-1}$$

Hint In a steady state condition, from Equation (8.7),

$$\dot{q}_{u'} = \frac{1}{L}[A_a S - A_r U_L(T_r - T_f)]$$

and also,

$$\dot{q}_{u'} = \frac{U}{L} A_r(T_r - T_f)$$

eliminate T_r to get the results.

8.6 Find out an expression for rate of heat carried away from concentrator in a forced circulation mode which is given by

$$\dot{Q}_u = F_R A_a \left[S - \frac{A_r}{A_a} U_L(T_{fi} - T_a) \right]$$

with

$$F'' = \frac{F_R}{F'} = \frac{\dot{m}C_f}{A_r U_L F'} \left[1 - \exp\left(-\frac{A_r U_L F'}{\dot{m}C_f} \right) \right]$$

Hint Follow the same procedure given in Section 3.7.4 for flat plate collector.

8.7 Plot the curve between η_i and $(T_{fi} - T_a)/I_b(t)$.

Hint $\eta_i = \dot{q}_u/I_b(t)$

8.8 Repeat Problems 3.7 and 3.8 of the flat plate collector, to the case of concentrator.

8.9 Concentrator of Problem 8.3 has the specifications as given below:
$S = 500 \, \text{W/m}^2$, $K = 16 \, \text{W/m}^\circ\text{C}$, $C_f = 3.26 \, \text{kJ/kg}^\circ\text{C}$, $\dot{m} = 0.054 \, \text{kg/s}$,
$h_{fi} = 200 \, \text{W/m}^2{}^\circ\text{C}$. Calculate the following parameters:

a. U_0 and F_R.

b. $\dot{q}_u$ and T_{f0}.

Hint See Problems 8.5 and 8.6, and $T_{fo} = T_{fi} + \dot{Q}_u/\dot{m}C_f$

8.10 Calculate the shape factor for a two dimensional concentrator with acceptance half angle equal to 40° and compare the result with that of a three dimensional concentrator.

Hint Use Equations (8.18) and (8.20) respectively.

8.11 Calculate the maximum concentration for the V-trough concentrator if the acceptance half angle is 45°.

Hint Use Equation (8.31).

8.12 Calculate the concentration limit in a concentrator with two dimensional geometry and compare the result with that of a concentrator with three dimensional geometry.

CHAPTER 9

Solar Distillation

9.1 INTRODUCTION

Supply of drinking water is a major problem in underdeveloped as well as in some developing countries. Along with food and air, water is a basic necessity for man. Man has been dependent on rivers, lakes and underground water reservoirs for fresh water but the pollution of rivers and lakes by industrial effluent and sewage has caused scarcity of fresh water in many towns and villages near lakes and rivers. Surveys show that about 79 percent of water available on the earth is salty, only one percent is fresh and the rest 20 percent is brackish. Many developing countries (e.g., India) have given utmost priority to rural water supply in their development plans. Major UN organizations, through Rajiv Gandhi National Drinking Water Mission, have been actively involved in promoting projects aimed at supplying drinking water in Indian villages, where people travel upto 30 km for fresh water.

Distillation of brackish or saline water, wherever it is available, is a good method to obtain fresh water. However, the conventional distillation processes such as multi-effect evaporation, multi-stage fresh evaporation, thin film distillation, reverse osmosis and electrodialysis are energy intensive techniques, and are not feasible for large fresh water demands. Therefore, solar distillation is an attractive alternative because of its simple technology; non-requirement of highly skilled labour for maintenance work and low energy consumption. As such, it can be used at any place without much problem.

A classification of published literature on solar distillation is given in Table 9.1. (Sodha *et al.*, 1980; Sodha *et al.*, 1981c, Tiwari and Rao, 1983, Tiwari and Yadav, 1985, Tiwari *et al.*, 1989; Malik *et al.*, 1982). The details of some of the designs along with the performance are discussed in subsequent sections.

9.2 WORKING PRINCIPLE

Figure 9.1 shows various components of energy balance and thermal energy loss in a conventional solar distiller unit. It is an airtight basin, usually constructed out of concrete/cement, galvanised iron sheet (GI) or fibre reinforced plastic (FRP) with a top cover of transparent material like glass, plastic etc. The inner surface of the rectangular base is blackened to efficiently absorb the solar radiation, incident at the surface. There is a provision to collect the distillate at lower end of the glass cover. The brackish or saline water is fed into the basin for purification. The working principle of the distiller unit is described here in.

TABLE 9.1 Classification of solar distillation system

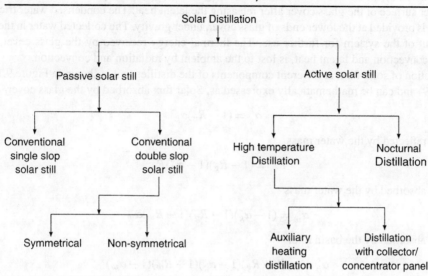

The solar radiation, after reflection and absorption by the glass cover is transmitted inside an enclosure of the distiller unit. This transmitted radiation $[\tau_g I(t)]$ is further partially reflected $[R'_w I(t)]$ and absorbed $[\alpha'_w I(t)]$ by the water mass. The attenuation of solar flux in water mass depends on it's absorptivity and depth. The solar radiation finally reaches the blackened surface where it is mostly absorbed. After absorption of solar radiation at the blackened surface, generally known as the basin liner, most of the thermal energy is convected to water mass and a small quantity is lost to the atmosphere, by conduction. Consequently, the water gets heated, leading to an increased difference of water and glass cover temperatures. There are basically three modes of heat transfer, radiation $(\dot{q}_{rw})$, convection $(\dot{q}_{cw})$

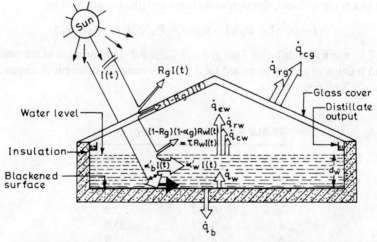

Figure 9.1 Energy flow diagram in a conventional solar still.

and evaporation ($\dot{q}_{ew}$) from the water surface to the glass cover. The evaporated water gets condensed on the inner surface of the glass cover after releasing the latent heat. The condensed water trickles into the channels provided at the lower ends of glass cover, under gravity. The collected water in the channel is taken out of the system for further use. The thermal energy received by the glass cover, through radiation, convection and latent heat, is lost to the ambient by radiation and convection.

The fraction of solar flux, at different components of the distiller unit is shown in Figure 9.1 (Tiwari *et al.*, 1989) and can be mathematically expressed as, Solar flux absorbed by the glass cover,

$$\alpha'_g = (1 - R_g)\alpha_g \tag{9.1a}$$

Solar flux reflected by the water mass

$$R'_w = (1 - R_g)(1 - \alpha_g)R_w \tag{9.1b}$$

Solar flux absorbed by the water mass

$$\alpha'_w = (1 - \alpha_g)(1 - R_g)(1 - R_w)\alpha_w \tag{9.1c}$$

Solar flux absorbed by the basin liner

$$\alpha'_b = \alpha_b(1 - R_g)(1 - \alpha_g)(1 - R_w)(1 - \alpha_w) \tag{9.1d}$$

Solar flux lost to the ambient, through water and glass cover, will be

$$(1 - \alpha_b)(1 - R_g)(1 - \alpha_g)(1 - R_w)(1 - \alpha_w) \tag{9.1e}$$

If, however, attenuation of solar flux within the water mass is considered, then the Equations (9.1c) to (9.1e) become,
Solar flux absorbed by the water mass

$$\alpha'_w = (1 - R_g)(1 - \alpha_g)(1 - R_w)[(1 - \Sigma\mu_j\exp(-\eta_jd_w)] \tag{9.1f}$$

Solar flux absorbed by the basin liner

$$\alpha'_b = \alpha_b(1 - R_g)(1 - \alpha_g)(1 - R_w)\Sigma\mu_j\exp(-\eta_jd_w) \tag{9.1g}$$

and the energy lost to the ambient, through water mass and glass cover, will be

$$(1 - \alpha_b)(1 - R_g)(1 - \alpha_g)(1 - R_w)\Sigma\mu_j\exp(-\eta_jd_w) \tag{9.1h}$$

The values of μ_j and η_j for different j are given in Table 9.2. The variation of attenuation factor with water depth and fraction of energy absorbed (α'_w) in water mass for different absorptivity are given in Tables 9.3(a) and 9.3(b), respectively.

TABLE 9.2 The values of μ_j and η_j

j	μ_j	$\eta_j(m^{-1})$
1	0.237	0.032
2	0.193	0.45
3	0.167	3.0
4	0.179	35.0
5	0.224	255.0

TABLE 9.3(a) Variation of attenuation factor with water depth

d_w	$\Sigma \mu_j \exp(-\eta_j d_w)$
0.20	0.51
0.10	0.5492
0.08	0.5648
0.06	0.5858
0.04	0.6185
0.02	0.6756
0.01	0.7344
0.008	0.7565
0.004	0.831
0.0	1.0

TABLE 9.3(b) Variation of α_b with α_w (with dye) for $d_w = 0.10$, $R_g = R_w = 0.05$, $\alpha_g = 0.0$ and $\alpha_b = 0.8$

α_w	α_b' without attenuation
0.0	0.17
0.2	0.31
0.4	0.45
0.6	0.58
0.8	0.72
1.0	0.86

EXAMPLE 9.1

Calculate the attenuation factor for 0.15 m water depth.

Solution

Attenuation factor $= \Sigma \mu_j \exp(-\eta_j d_w)$

Substituting the different values of j, μ_j, η_j from Table 9.2 in the above expression; calculating the values for $d_w = 0.15$ and summing up gives,

For $j = 1; \mu_j = 0.237; \exp(-\eta_j d_w) = 0.9952$; Attenuation factor $= 0.2359$

For $j = 2; \mu_j = 0.193; \exp(-\eta_j d_w) = 0.9347$; Attenuation factor $= 0.1804$

Similarly, for $j = 3$; Attenuation factor $= 0.106$

For $j = 4$; Attenuation factor $= 0.939 \times 10^{-3}$

For $j = 5$; Attenuation factor $= 3.03 \times 10^{-18}$

Hence, effective attenuation factor is, $\Sigma \mu_j \exp(-\eta_j d_w) = 0.523$

EXAMPLE 9.2

Calculate the value of α_w' at a depth of 0.10 m for a given $\alpha_w = 0.2$. The value of R_g and R_w is 0.05 and that of $\alpha_b = 0.075$ and $\alpha_g = 0.0$.

Solution

Substituting the values in Equations (9.1c) and (9.1f) we have,

$$\alpha_w' = (1 - 0.0)(1 - 0.05)(1 - 0.05) \times 0.2 = 0.1805, \quad \text{without attenuation}$$

$$\alpha_w' = (1 - 0.05)(1 - 0.0)(1 - 0.05)[1 - 0.5492], \quad \text{with attenuation}$$

$$= 0.95 \times 0.95 \times 0.4508 = 0.4068$$

9.3 THERMAL EFFICIENCY

The thermal efficiency of distiller unit can be defined as the ratio of the amount of thermal energy utilised to get a certain amount of distilled water to the incident solar energy within a given time interval.

9.3.1 Instantaneous Efficiency

If the evaporation process inside the distiller unit can be considered as an isobaric atmospheric process at thermal equilibrium, then all the absorbed solar radiation is utilised for evaporation and thermal losses. An energy balance for steady state around the water basin can be written as [Tamini, 1987]:

$$[\alpha_w' + \alpha_b']I(t)A_S = \dot{Q}_{ew} + \dot{Q}_{\text{losses}} \tag{9.2}$$

$$[\text{Rate of energy in}] = [\text{Rate of energy out}]$$

where, $\dot{Q}_{ew} = \dot{m}_w L$ and $\dot{Q}_{\text{losses}} = U_L'(T_w - T_a)A_s$, U_L' is the overall heat transfer coefficient from water to the ambient through top, bottom and sides of the distiller unit and it is assumed that,

$$(\alpha_w' + \alpha_b') = (\alpha\tau)_w$$

Here in the Tamini model, $\dot{Q}_{\text{losses}}$ does not include the evaporative heat loss. The analysis has been compared with that of flat plate collector. However, it may be noted that whereas in a flat plate collector, the upward losses should be minimum, in a solar distiller the upward losses should be maximum in order to get a higher yield. In the conventional solar distiller, to be discussed later, the radiative, convective and evaporative losses from the water to the condensing cover (glass) are grouped together and taken as the total internal heat transfer coefficient from water to glass.

Equation (9.2) can be rewritten as,

$$\dot{Q}_{ew} = \dot{m}_w L = (\alpha\tau)_w I(t) A_s - U_L'(T_w - T_a) A_s \tag{9.3}$$

The expression for instantaneous efficiency (η_i), can be given as,

$$\eta_i = \frac{\dot{m}_w L}{I(t)A_s} = (\alpha\tau)_w - U_L' \frac{(T_w - T_a)}{I(t)} \tag{9.4}$$

The plot of η_i versus $(T_w - T_a)/I$ will represent a straight line with $(\alpha\tau)_w$ and $-U_L'$ as the intercept and the slope respectively, where U_L' can be taken as a constant (Nayak *et al.*, 1980). The expression for η_i is similar to that for a conventional flat plate collector except for the heat removal factor. Thus, according to Tamini (1987), the distiller unit can be considered as a special type of flat plate collector that collects the solar energy and produces distilled water. An actual variation of η_i versus $\frac{T_w - T_a}{I(t)}$ for a solar still will be discussed in detail in section 9.9.4.

9.3.2 Overall Thermal Efficiency

The overall thermal efficiency of the distiller unit in the passive and active modes of operation can be mathematically expressed as,

$$\eta_{\text{passive}} = \frac{\sum \dot{m}_w L}{A_s \int I(t)dt} \times 100 \tag{9.5}$$

and

$$\eta_{\text{active}} = \frac{\sum \dot{m}_w L}{\left[A_s \int I(t)dt + N A_c \int I'(t)dt\right]} \times 100 \tag{9.6}$$

where N is number of collectors connected either is series or parallel.

Here, the latent heat of vaporisation (L) in Joule/kg can be considered temperature dependent, and can be given as: (Fernandez and Chargoy, 1990 and Toyama, 1972).

$$L = 3.1615 \times 10^6 [1 - 7.6160 \times 10^{-4} T] \qquad (9.7)$$

for temperature higher than 70°C, and

$$L = 2.4935 \times 10^6 [1 - 9.4779 \times 10^{-4} T + 1.3132 \times 10^{-7} T^2 - 4.7974 \times 10^{-9} T^3] \qquad (9.8)$$

for operating temperatures less than 70°C.

EXAMPLE 9.3

Calculate the value of latent heat of vaporisation (L) at 40°C.

Solution

Using Equation (9.8), we have

$$L = 2.4935 \times 10^6 [1 - 9.4779 \times 10^{-4} \times 40 + 1.3132 \times 10^{-7} \times 40^2 - 4.7974 \times 10^{-9} \times 40^3]$$

$$= 2398.5 \times 10^3 \, \text{J/kg} \, °\text{C}.$$

9.4 HEAT TRANSFER

The heat transfer in solar distillation systems can be classified in terms of external and internal modes. The external heat transfer mode is primarily governed by conduction, convection, and radiation processes, which are independent of each other. These heat transfer occur outside the solar distiller, from the glass cover and the bottom and side insulation. Heat transfer within the solar distiller is referred to as internal heat transfer mode which consists of radiation, convection, and evaporation (Fig. 9.1). In this case, convective heat transfer occurs simultaneously with evaporative heat transfer and these two heat transfer processes are independent of radiative heat transfer. The classification of element of heat transfer in a solar distiller is given in Table 9.4.

9.5 EXTERNAL HEAT TRANSFER

9.5.1 Top Loss Coefficient

Due to the small thickness of the glass cover, the temperature in the glass may be assumed to be uniform. The external heat transfer, radiation, and convection losses from the glass cover to the outside atmosphere $\dot{q}_g$ can be expressed as

$$\dot{q}_g = \dot{q}_{rg} + \dot{q}_{cg} \qquad (9.9)$$

$$\dot{q}_{rg} = \epsilon_g \sigma \left[(T_g + 273)^4 - (T_{sky} + 273)^4 \right] \qquad (9.10)$$

$$\dot{q}_{rg} = h_{rg}(T_g - T_a) \qquad (9.10a)$$

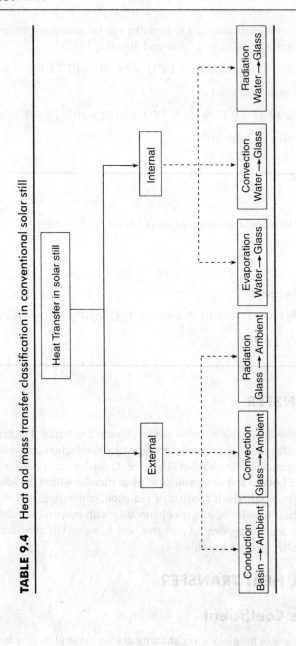

TABLE 9.4 Heat and mass transfer classification in conventional solar still

and

$$\dot{q}_{cg} = h_{cg}[T_g - T_a] \qquad (9.11a)$$

with

$$h_{rg} = \frac{\epsilon_g \sigma[(T_g + 273)^4 - (T_{sky} + 273)^4]}{(T_g - T_a)} \qquad (9.11b)$$

On substituting the expressions for $\dot{q}_{rg}$ and $\dot{q}_{cg}$ in Equation (9.9), we get

$$\dot{q}_g = h_{1g}(T_g - T_a) \qquad (9.12)$$

where

$$h_{1g} = h_{rg} + h_{cg} \qquad (9.13)$$

The empirical relation for h_{1g} can be discussed for the following conditions:

Case (i): The expression for h_{1g} is given by:

$$h_{1g} = 5.7 + 3.8\,V \qquad (9.14a)$$

where V is the wind velocity in m/s. This expression includes the effect of free convection and radiation from the glass cover as discussed by Watmuff *et al.* (1977).

Case (ii): In case the radiation and convection losses are to be evaluated separately, the radiative heat transfer coefficient (h_{rg}) can be obtained from Equation (9.11b) and the convective heat transfer coefficient, h_{cg}, can be obtained from the relation,

$$h_{cg} = 2.8 + 3.0\,V \qquad (9.14b)$$

There is, however, no significant change in the performance of the distillation system by considering h_{1g} as represented either by case (i) or case (ii).

EXAMPLE 9.4

Calculate the radiative heat transfer coefficient from glass at 32°C to an ambient at 20°C, and the convective heat transfer coefficient when a wind is blowing at a velocity of 3 m/s. Compare the result with the combined radiative and convective heat transfer coefficient.

Solution

From Equation (9.11b), the value of radiative heat transfer coefficient is

$$h_{rg} = \frac{\epsilon_g \sigma\left[(T_g + 273)^4 - (T_{sky} + 273)^4\right]}{T_g - T_a}$$

with $T_{sky} = T_a - 6.0 = 20 - 6 = 14°C$ (from Equation (3.16)). Now,

$$h_{rg} = 0.9 \times 5.67 \times 10^{-8} \frac{[8.65 \times 10^9 - 6.78 \times 10^9]}{12} = 7.95 \text{ W/m}^2$$

and, $h_{cg} = 2.8 + 3.0V = 11.8 \text{ W/m}^2 \text{ °C}$ (from Equation (9.14b))

The total heat transfer coefficient from glass to ambient (h_{1g}) = 19.75 W/m^2

The combined radiative and convective heat transfer coefficient from Equation (9.14a) is

$$h_{1g} = 5.7 + 3.8V = 17.1 \, \text{W/m}^2 \, {}^\circ\text{C}.$$

There is not much difference in the value of h_{1g}.

9.5.2 Bottom and Side Loss Coefficient

Heat is also lost from the water in the basin to the ambient through the insulation and subsequently by convection and radiation from the bottom or side surface of the basin. The bottom loss coefficient (U_b) can be written as

$$U_b = \left[\frac{1}{h_w} + \frac{1}{h_b}\right]^{-1} = \left[\frac{1}{h_w} + \frac{1}{K_i/L_i} + \frac{1}{h_{cb} + h_{rb}}\right]^{-1} \tag{9.15}$$

The side heat loss coefficient (U_e) can be approximated as

$$U_e = U_b A_{ss}/A_s \tag{9.16}$$

If A_{ss} is very small in comparison to A_s, for small water depth, U_e can be neglected. Here, A_{ss} is the surface area in contact with water and A_s is the area of the basin of the distiller.

The rate of heat loss per m^2 from basin liner to ambient can be written as,

$$\dot{q}_b = h_b(T_b - T_a) \tag{9.16a}$$

where

$$h_b = \left[\frac{L_i}{K_i} + \frac{1}{h_{cb} + h_{rb}}\right]^{-1} \tag{9.16b}$$

EXAMPLE 9.5

Calculate the bottom loss coefficient from a solar distiller with 5 mm thick insulation ($K_i = 0.04 \, \text{W/m}^\circ\text{C}$). Given $h_w = 100 \, \text{W/m}^2 \, {}^\circ\text{C}$.

Solution

From Equation (9.15), the bottom loss coefficient is given as,

$$U_b = \left[\frac{1}{h_w} + \frac{1}{K_i/L_i} + \frac{1}{h_{ch} + h_{rb}}\right]^{-1}$$

Here, $h_{cb} + h_{rb} = 5.7 \, \text{W/m}^2 \, {}^\circ\text{C}$ (Equation (9.14a)), $h_w = 100 \, \text{W/m}^2 \, {}^\circ\text{C}$, $K_i = 0.04 \, \text{W/m}\,{}^\circ\text{C}$ and $L_i = 5 \, \text{mm} = 0.5 \times 10^{-2}$ m (given). Then,

$$U_b = \left[\frac{1}{100} + \frac{1}{0.04/0.5 \times 10^{-2}} + \frac{1}{5.7}\right]^{-1} = 3.22 \, \text{W/m}^2 \, {}^\circ\text{C}.$$

9.6 INTERNAL HEAT TRANSFER

The internal heat transfer mode, that is, the heat exchange from the water surface to the glass cover inside the distillation unit is governed by radiation, convection, and evaporation and hence these heat transfer modes are discussed separately.

9.6.1 Radiative Loss Coefficient (h_{rw})

In this case, the water surface and the glass cover are considered as infinite parallel planes (approximation is valid for small inclination of the glass cover and for large width of the distiller unit). The rate of radiative heat transfer ($\dot{q}_{rw}$) from the water surface to the glass cover for these infinite parallel planes is given by

$$\dot{q}_{rw} = \epsilon_{eff} \sigma \left[(T_w + 273)^4 - (T_g + 273)^4 \right] \tag{9.17}$$

$$\dot{q}_{rw} = h_{rw}(T_w - T_g) \tag{9.18}$$

where h_{rw} is the radiative heat transfer coefficient from the water surface to the glass cover and is given by

$$h_{rw} = \epsilon_{eff} \sigma \left[(T_w + 273)^2 + (T_g + 273)^2 \right] [T_w + T_g + 546] \tag{9.19}$$

It is to be noted that water and glass are considered to be parallel surfaces and thus the radiation shape factor is 1 in this case.

9.6.2 Convective Loss Coefficient (h_{cw})

Heat transfer occurs across humid air in the distillation unit by free convection, which is caused by the effect of buoyancy, due to density variation in the humid fluid, which occurs due to the temperature gradient in the fluid. Hence, the rate of heat transfer from the water surface to the glass cover ($\dot{q}_{cw}$) by convection in the upward direction through humid fluid can be estimated by

$$\dot{q}_{cw} = h_{cw}(T_w - T_g) \tag{9.20}$$

The coefficient h_{cw} can be determined by the relation,

$$Nu = \frac{h_{cw} d_f}{K_f} = C(Gr \, Pr)^n \tag{9.21}$$

$$Gr = \frac{d_f^3 \rho_f^2 g \beta'}{\mu_f^2} \Delta T' \tag{9.22}$$

$$Pr = \mu_f C_f / K_f \tag{9.23}$$

$$\Delta T' = \left[\Delta T + \frac{(P_{w0} - P_{g0})(T_{w0} + 273)}{268.9 \times 10^3 - P_{w0}} \right] \tag{9.24}$$

For a normal operating temperature range, say 50°C and $\Delta T' = 17°C$, the expression for Grashof number given above is reduced to

$$Gr = 2.81 \times 10^7 \, d_f^3 \tag{9.25}$$

TABLE 9.5(a) Value of Grashof number
(Gr) for different average spacing (d_f)

d_f (m)	Gr	C	n
0.15	0.948×10^5	0.21	1/4
0.20	2.248×10^5	0.21	1/4
0.25	4.39×10^5	0.075	1/3

As can be seen from Equation (9.22), the Grashof number depends on the average spacing between the water and the glass cover for a normal operating temperature range (Table 9.5(a)). Table 9.5(a) gives the value of Grashof number for different spacing. The values of C and n appearing in Equation (9.21) for different Grashof numbers are given in Table 9.5(b). The value of Prandtl number remains constant and is given by Equation (9.23). For the normal operating temperature range and $d_f = 0.25$ m ($C = 0.075$ and $n = 1/3$) (Table 9.5(a)). Dunkle (1961) derived the following expression for h_{cw}.

$$h_{cw} = 0.884 \left[T_w - T_g + \frac{(P_w - P_g)(T_w + 273)}{268.9 \times 10^3 - P_w} \right]^{1/3} \tag{9.26}$$

TABLE 9.5(b) The values of C and n for different Grashof number ranges

Case	C	n	Grashof number	Source
	1.00	0	$Gr < 10^3$	Mull &
I	0.21	1/4	$10^4 < Gr < 3.25 \times 10^5$	Reiher (1930)
	0.075	1/3	$3.2 \times 10^5 < Gr < 10^7$	
	0.07477	0.36	$2.5 \times 10^3 < Gr < 6 \times 10^4$	
II	0.05238	0.36	$2.5 \times 10^5 < Gr < 10^7$	Held(1931)
	0.05814	0.4	$2 \times 10^3 < Gr < 5 \times 10^4$	De Graf &
III	3.8	0.0	$5 \times 10^4 < Gr < 2 \times 10^5$	Vander (1953)
	0.04836	0.37	$2 \times 10^5 < Gr$	
IV	0.3	1/4	$2.8 \times 10^3 < Gr < 2.1 \times 10^5$	Jakob & Gupta
	0.1255	1/3	$4.2 \times 10^5 < Gr < 4.2 \times 10^9$	(1954)

9.6.3 Evaporative Loss Coefficient (h_{ew})

The mass transfer coefficient h_e in terms of convective heat transfer coefficient h_{cw} (Equation (9.26)) is given by

$$\frac{h_e}{h_{cw}} = \frac{L}{C_{pa}} \frac{M_w}{M_a} \frac{1}{P_T} \tag{9.27}$$

where P_T is total gas pressure; M_w the mass of water vapor; M_a mass of air and C_{pa} the specific heat per unit volume at constant pressure of the mixture.

The above equation has been derived by Baum (1964), making the following assumptions:

a. negligible exchange of water vapor with the boundary layers at the water and glass surface, and

b. P_w and P_g are considerably smaller than P_T.

The above equation can also be derived by making use of the Lewis relation (Malik *et al.*, 1982).

The rate of heat transfer per unit area from the water surface to the glass cover can be obtained by substituting the appropriate values for the parameters in Equation (9.27). Thus,

$$\dot{q}_{ew} = 0.013 h_{cw}(P_w - P_g) \tag{9.28}$$

Similar equation has also been derived by Cooper, and is given as

$$\dot{q}_{ew} = 0.0162 h_{cw}(P_w - P_g) \tag{9.28a}$$

Equation (9.28) can be rearranged as $\dot{q}_{ew} = h_{ew}(T_w - T_g)$ (9.29)

$$h_{ew} = 16.273 \times 10^{-3} h_{cw} \frac{P_w - P_g}{T_w - T_g} \quad \text{(Cooper, 1973)} \tag{9.30}$$

It is important to mention here that the value of h_{ew} can be more realistic for larger value of $(T_w - T_g)$.

The values of P_w and P_g (for the range of temperature $10°C - 90°C$) can be obtained from the expression. (Ref. Fernandez and Chargoy 1990),

$$P(T) = \exp\left(25.317 - \frac{5144}{T + 273}\right) \tag{9.31}$$

Table 5(a) in Appendix V gives some other expressions for the saturated vapor pressure. Table 5(b) gives the variation of saturated vapor pressure with temperature. The variation of saturated vapor pressure (N/m^2) with temperature ($°C$) is shown in Figure 9.2.

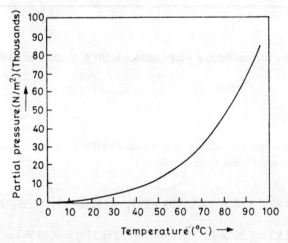

Figure 9.2   Variation of saturated vapor pressure with temperature.

The total internal heat transfer coefficient can be obtained by combining Equations (9.19), (9.26) and (9.30), and is given as

$$h_{1w} = h_{rw} + h_{cw} + h_{ew} \tag{9.32}$$

EXAMPLE 9.6

Calculate the saturated vapor pressure at $30°C$.

Solution

Substituting the value of T in Equation (9.31), we have

$$P(30°C) = \exp\left(25.317 - \frac{5144}{303}\right) = 4188.52 \, N/m^2$$

9.7 OVERALL HEAT TRANSFER

9.7.1 Top Loss Coefficient

The top loss coefficient (U_t) from the water surface to the ambient air can be written as

$$U_t = \left[\frac{1}{h_{1g}} + \frac{1}{h_{1w}}\right]^{-1} \tag{9.33}$$

Hence the rate of heat lost in upward direction of distillation system is

$$\dot{q}_t = U_t(T_w - T_a) \tag{9.34}$$

EXAMPLE 9.7

Calculate the top loss coefficient from a water surface at 20°C to the ambient air at 8.5°C. Given the temperature of glass = 12°C.

Solution

The radiative, convective and evaporative heat transfer coefficients from water to glass are given by Equations (9.19), (9.26) and (9.30), respectively.
Thus,

$$h_{rw} = 0.82 \times 5.67 \times 10^{-8}\left[(20+273)^2 + (12+273)^2\right][20+12+546]$$

$$= 0.82 \times 5.67 \times 10^{-8}[85849 + 81225][578] = 4.49 \, W/m^2\,°C$$

Here, in the given problem,

$$\epsilon_{eff} = \left[\frac{1}{0.9} + \frac{1}{0.9} - 1\right]^{-1} = 0.82 \quad \text{for} \quad \varepsilon_g = \varepsilon_w = 0.9$$

Also, $P_w = 2367.69 \, N/m^2$ and $P_g = 1447.18 \, N/m^2$ (Equation (9.31)).
Further,

$$h_{cw} = 0.884\left[(20-12) + \frac{(2367.69 - 1447.18)(293)}{268.9 \times 10^3 - 2367.69}\right]^{1/3} = 1.84 \, W/m^2\,°C$$

$$h_{ew} = 16.273 \times 10^{-3} \times 1.84 \times (920.51/8) = 3.445 \, W/m^2\,°C$$

Now, $h_{1w} = 3.445 + 1.84 + 4.49 = 9.775\,\text{W/m}^2\,°\text{C}$ (Equation (9.32))
$h_{1g} = 17.1\,\text{W/m}^2\,°\text{C}$ (from Example 9.4)
The top loss coefficient from water surface to ambient air can be given (from Equation (9.33)) as

$$U_t = [(1/17.1) + (1/9.775)]^{-1} = (0.058 + 0.102)^{-1} = 6.25\,\text{W/m}^2\,°\text{C}$$

9.7.2 Bottom Loss Coefficient

The rate of heat lost through the bottom of the insulation from water to ambient air is

$$\dot{q}_{bg} = U_b(T_w - T_a) \tag{9.35}$$

where U_b is the bottom heat loss coefficient.

The rate of total heat lost per m² from the water surface to the ambient through the top and the bottom of the system can be obtained by adding Equations (9.34) and (9.35) and is given as

$$\dot{q}_{\text{loss}} = U_L(T_w - T_a) \tag{9.36}$$

where $U_L = U_t + U_b$, is the overall heat transfer coefficient.

9.8 DETERMINATION OF DISTILLATE OUTPUT

The hourly distillate output per m² from a distiller unit can be obtained as

$$\dot{m}_{ew} = \frac{\dot{q}_{ew}}{L} \times 3600 = \frac{h_{ew}(T_w - T_g)}{L} \times 3600 \tag{9.37}$$

9.9 PASSIVE SOLAR STILLS

9.9.1 Conventional Solar Still

The conventional single basin solar still is the most practical design for an installation to provide distilled drinking water for daily requirement. The conventional solar stills can be classified as

  i. Symmetrical double-sloped (Fig. 9.1).
 ii. Non symmetrical double-sloped.
iii. Single-sloped (Fig. 9.3(a)).

The symmetrical double-sloped solar still is positioned in the east–west direction and is free of orientation requirement. In other two cases, the low wall faces equator to receive maximum solar radiation for higher yield.

9.9.2 Basin Construction

The basin of the distillation system is made water tight to avoid any water leakage and the surface inside is blackened so as to absorb maximum solar radiation. The bottom and the sides of the basin are insulated to reduce the heat loss through conduction. The basin may be constructed of (i) concrete/bricks/cement (ii) galvanized iron sheet (G.I. sheet) or (iii) fibre reinforced plastic (FRP) (Fig. 9.3(c)).

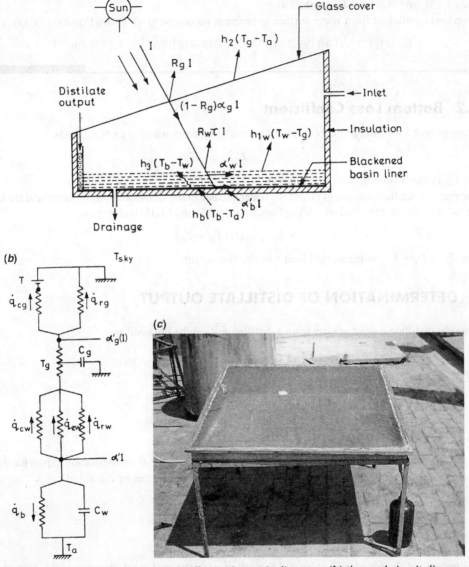

Figure 9.3 A single slope solar still (**a**) schematic diagram, (**b**) thermal circuit diagram and (**c**) photograph.

9.9.3 Thermal Analysis of Conventional Solar Still

The thermal circuit of a single slope basin solar still is shown in Figure 9.3(b).

Energy Balances: The basin heat flux components at various points are shown in Figure 9.3(a). The following assumptions have been made in writing the energy balance in terms of Joules per sec per m²:

i. inclination of the glass cover is very small,

ii. the heat capacity of the glass cover, the absorbing material and the insulation (bottom and sides) is negligible, and

iii. the solar distiller unit is vapor-leakage proof.

The energy balance for different components of the still are as follows:

Glass Cover

$$\alpha'_g I(t) + [\dot{q}_{rw} + \dot{q}_{cw} + \dot{q}_{ew}] = \dot{q}_{rg} + \dot{q}_{cg}$$

Rate of energy absorbed   Rate of energy received from water surface by radiation convection and evaporation   Rate of energy lost to air

(9.38)

Water Mass

$$\alpha'_w I(t) + \dot{q}_w = (MC)_w \frac{dT_w}{dt} + \dot{q}_{rw} + \dot{q}_{cw} + \dot{q}_{ew}$$

Rate of energy absorbed   Rate of energy convected from basin liner   Rate of energy stored   Rate of energy transferred to glass cover

(9.39)

Basin Liner

$$\alpha'_b I(t) = \dot{q}_w + [\dot{q}_b + \dot{q}_b(A_{ss}/A_s)]$$

Rate of energy absorbed   Rate of energy transferred to water   Rate of energy lost by conduction through bottom/sides

(9.40)

where α'_g, α'_w and α'_b are given by Equations (9.1, a, c, d). The various q's in terms of respective heat transfer coefficients, are given in sections (9.5 and 9.6). The side area A_{ss} being very small in comparison to the basin liner area of the solar distiller, the term A_{ss}/A_s can be neglected.

On substitution of the expressions for all q's Equations (9.38) to (9.40) can be rewritten as

$$\alpha'_g I(t) + h_{1w}(T_w - T_g) = h_{1g}(T_g - T_a) \tag{9.41}$$

$$\alpha'_w I(t) + h_w(T_b - T_w) = (MC)_w \frac{dT}{dt} + h_{1w}(T_w - T_g) \tag{9.42}$$

$$\alpha'_b I(t) = h_w(T_b - T_w) + h_b(T_b - T_a) \tag{9.43}$$

Substituting the values of T_g and T_b from Equations (9.41) and (9.43) in Equation (9.42) and simplifying, we get

$$\frac{dT_w}{dt} + aT_w = f(t) \tag{9.44}$$

where

$$a = \frac{U_L}{(MC)_w} \tag{9.45}$$

$$f(t) = \frac{(\alpha\tau)_{eff} I(t) + U_L T_a}{(MC)_w} \tag{9.46}$$

$$(\alpha\tau)_{eff} = \alpha_b' \frac{h_w}{h_w + h_b} + \alpha_w' + \alpha_g' \frac{h_{1w}}{h_{1w} + h_{1g}} \tag{9.47}$$

and

$$U_L = U_b + U_t; U_b = \frac{h_w h_b}{h_w + h_b}, U_t = \frac{h_{1w} h_{1g}}{h_{1w} + h_{1g}} \tag{9.48}$$

9.9.4 Approximate Solution For (T_w)

In order to obtain an approximate solution of Equation (9.44), the following assumptions have been made:

   i. The time interval $\Delta t (0 < t < \Delta t)$ is small.
  ii. The function $f(t)$ is constant, i.e. $f(t) = \overline{f(t)}$ for the time interval Δt.
 iii. a is constant during the time, interval Δt.

The value of h_{1w} can be determined by considering known values of water and glass temperatures at

$$t = 0, \quad \text{i.e.} \quad T_w|_{t=0} = T_{w0} \quad \text{and} \quad T_g|_{t=0} = T_{g0}$$

The solution of Equation (9.44) can be written as

$$T_w = \frac{\overline{f(t)}}{a}[1 - \exp(-a\Delta t)] + T_{w0} \exp(-a\Delta t) \tag{9.49}$$

where T_{w0} is the temperature of basin water at $t = 0$ and $\overline{f(t)}$ is the average value of $f(t)$ for the time interval between 0 and t.

EXAMPLE 9.8

Calculate the time taken for the water temperature to rise to 40°C from an initial value of 20°C. Given $(\alpha\tau)_{eff} = 0.65$, $\overline{I(t)} = 900 \text{ W/m}^2$; $\overline{T}_a = 8.5°C.(MC)_w = 10 \times 4190 = 41900 \text{ J/°C}$; $U_L = 9.48 \text{ W/m}^2 \text{ °C}$.

Solution

From Equation (9.49), we have

$$\Delta t = -\frac{1}{a} \ln \left[\frac{T_w - \frac{\overline{f(t)}}{a}}{T_{w0} - \frac{\overline{f(t)}}{a}} \right]$$

From Equations (9.45) and (9.46), we get

$$a = \frac{U_L}{(MC)_w} = \frac{9.48}{41900} = 2.26 \times 10^{-4}$$

$$\overline{f(t)} = \frac{0.65 \times 900 + 9.48 \times 8.5}{41900} = 0.0159$$

Hence, $\quad \Delta t = -\dfrac{1}{2.26 \times 10^{-4}} \ln \left[\dfrac{40 - \dfrac{0.0159}{2.26 \times 10^{-4}}}{20 - \dfrac{0.0159}{2.26 \times 10^{-4}}} \right] = 2239 \text{ sec} = 37 \text{ minutes}$

The average temperature of water, T_w is given by

$$\bar{T}_w = \frac{1}{t} \int_0^t T_w \, dt = \frac{\overline{f(t)}}{a} \left[1 - \frac{(1 - e^{-a\Delta t})}{a} \right] + T_{w0} \frac{(1 - e^{-a\Delta t})}{a} \tag{9.50}$$

The average glass temperature in terms of the water temperature can be obtained from Equation (9.41) and is given as,

$$\bar{T}_g = \frac{\alpha_g' I(t) + h_{1w} \bar{T}_w + h_{1w} \bar{T}_a}{h_{1w} + h_{1g}} \tag{9.51}$$

The calculated values of T_w and T_g are used to evaluate the internal heat transfer coefficient h_{1w}. The value of T_w at the end of an interval becomes the initial condition for the next interval and so on. The accuracy of the computation depends on the selected time interval, being high, for a smaller time interval. Usually the time interval taken is one hour.

The rate of evaporative heat loss, the hourly yield and the daily efficiency of the solar still are given as follows:

$$\dot{q}_{ew} = h_{ew}(T_w - T_g) \tag{9.52}$$

and $\qquad \dot{m}_{ew} = \dfrac{h_{ew}(T_w - T_g)}{L} \times 3600 \text{ kg/m}^2\text{h} \tag{9.53}$

After obtaining the hourly yield, $\dot{m}_{ew}$, from Equation (9.53), the value of $(MC)_w$ for the next interval becomes

$$(MC)_w - (\dot{m}_{ew} C) \tag{9.54}$$

Further, the instantaneous efficiency of distillation unit can be determined as follows:

$$\eta_i = \frac{\dot{q}_{ew}}{I(t)} = \frac{h_{ew}(T - T_g)}{I(t)} \tag{9.55a}$$

and $\qquad \eta_i = \dfrac{h_{ew} h_{1g}}{h_{1w} + h_{1g}} (T_w - T_a) \quad \text{(from Equation (9.51))} \tag{9.55b}$

The above equation can be rewritten by substituting the value of T_w from Equation (9.49)

$$\eta_i = \frac{h_{ew} h_{1g}}{h_{1w} + h_{1g}} \frac{1}{U_L} \times \left[(\alpha\tau)_{eff}(1 - e^{-a\Delta t}) + U_L \frac{(T_{w0} - T_a)}{I(t)} e^{-a\Delta t} \right] \tag{9.56}$$

The variation of η_i with $\frac{T_{w0} - T_a}{I}$ has been shown in Figure 9.4 for a typical climatic and design, parameters. The slope of Figure 9.4 is positive (Equation (9.56)) unlike the slope of Figure 3.8(b) for flat plate collector. It is due to the fact that η_i in the case of solar still is higher for maximum top loss coefficient (U_t) which includes h_{ew}. In the case of flat plate collector it should be minimum for higher η_i. The non-linear behavior of Figure 9.4 is due to temperature dependence of h_{ew}.

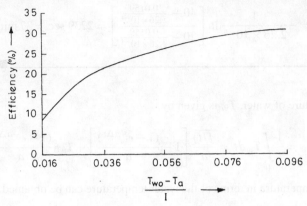

Figure 9.4 Characteristic curve of a single slope solar still (Fig. 9.3).

EXAMPLE 9.9

Calculate the rate of evaporative heat loss from the solar still given in Example 9.7.

Solution

From Example 9.7, $h_{ew} = 3.445\,\text{W/m}^2\,°\text{C}$
The rate of evaporative heat loss, from Equation (9.52), is

$$\dot{q}_{ew} = 3.445(20 - 12) = 27.56\,\text{W/m}^2.$$

EXAMPLE 9.10

Calculate the hourly output from the still given in Example 9.7.

Solution

Given $L = 2390 \times 10^3\,\text{J/kg}$, the hourly yield is

$$\dot{m}_{ew} = \frac{h_{ew}(T_w - T_g)}{L} \times 3600\,\text{kg/m}^2\text{h} \quad \text{(From Equation (6.53))}$$

and hence, $\qquad \dot{m}_{ew} = \dfrac{3.445 \times (20 - 12)}{2390 \times 10^3}\,3600 = 0.0415\,\text{kg/m}^2\text{h}.$

EXAMPLE 9.11

Calculate instantaneous efficiency of distillation unit given in Example 9.9.

Solution

From Equation (9.55a),

$$\eta_i = \frac{\dot{q}_{ew}}{I(t)} = \frac{27.56}{900} = 0.0306 = 3.06\%.$$

9.10 EFFECT OF VARIOUS PARAMETERS

The yield from a solar still depends on the temperature difference between the water in the basin and the condensing cover. As seen from Equation (9.53), the higher the value of $T_w - T_c$, the greater is the yield.

9.10.1 Effect of Wind Velocity

As concluded by Cooper (1969 a), the output increases by 11.5 percent for average wind velocities from 0–2.15 m/s, while the increase is only 1.5 percent for average wind velocities from 2.15 m/s to 8.81 m/s. Thus, wind at higher velocities has a lesser influence on the distillation rate.

The wind blowing over the glass cover causes faster evaporation from it resulting in a fall in the temperature; thus the yield from the solar still increases for larger water depth in the still. However, for smaller water depth the wind has no effect on the output. Even for larger water depth, wind velocity above a particular value (around 5 m/s) has not much effect on the yield.

However, as the wind velocity increases, the convective heat loss from the glass cover to ambient increases hence the glass cover temperature decreases which increases the water-glass cover temperature difference and hence the overall yield.

9.10.2 Effect of Ambient Temperature

With the decrease in the ambient temperature, the glass temperatures decrease and the difference $(T_w - T_g)$ increases, but there is a general fall in the overall temperature of the system, hence the output increases.

9.10.3 Effect of Solar Radiation and Loss Coefficient

Solar insolation is an important parameter in the determination of yield from a solar still. The output will, to an extent, depend upon the distribution of the radiation throughout the day, however, this is a second-order effect and it is usually sufficient to take into account the total radiation received on each day. The yield is maximum when the daily insolation and the mean ambient temperature are consistently high, with low loss coefficient. The loss coefficient has less influence at higher ambient temperatures.

9.10.4 Effect of Double-glass Cover and Cover Inclination

The effective thermal barrier between the two glass covers impedes the rejection of heat through the condenser, as a result the output is reduced. [This is beneficial in the case of flat plate collector, as the heat loss is reduced]. Even with very high water temperatures the governing factor is the low water-glass temperature difference. Although a high water temperature leads to a higher evaporation, the low temperature difference results in a considerably reduced total energy transfer. The high water temperature causes greater base and side losses. Increasing the value of the thermal conductance of air in between glass, will result in a slightly improved performance, but the combined effects of radiative and convective heat transfer across the air gap lowers the output. From economical and constructional point of view, doubling the amount of glass, which constitutes a substantial part of the total cost of the still, is not desirable.

The variation of distillate production with glass cover inclination has been studied by Cooper (1969a). It was found that the evaporation rate decreases with cover slope variation from 0–45°, rises at about 60° and falls again beyond 75°. Although, the output changes very little with change of inclination of glass cover.

9.10.5 Effect of Salt Concentration on Output

Experiments have shown that as the salt concentration of the water to be distilled increases right up to the saturation point, the output of the still falls linearly. However, as the salt concentration of the water increases, there is an increase in the corrosion damage to the components of the still and thus it becomes necessary to use materials which are not readily oxidized.

9.10.6 Effect of Thermal Capacity on Output

Cooper (1969 a) has studied the effect of water depth on the distillate output, as shown in Figure 9.5. As can be seen from the figure, without insulation, the gains from decreasing the water depth are only marginal, but with insulation, the difference is more marked, particularly, at lower water depths.

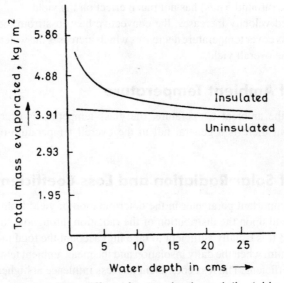

Figure 9.5 Effect of water depth on daily yield.

9.10.7 Effect of Charcoal Pieces on the Performance of a Still

Charcoal pieces affect the performance of a still because of their wettability, large absorption coefficient for solar radiation and their property to scatter, rather than reflect, the solar radiation. As concluded from studies done by Akinsete and Duru (1979), the effect of charcoal is most pronounced in the mornings and on cloudy days when the values of direct radiation are low. The presence of charcoal pieces utilises the diffuse radiation much better than the conventional unlined still. It was also seen that the charcoal lined still is relatively insensitive to basin-water depth as long as a good amount of the charcoal remains uncovered.

9.10.8 Effect of the Formation of Algae and Mineral Layers on Water and Basin Liner Surface on the Performance of a Still

As concluded by Cooper (1972), the presence of deposits on the surface of the basin water and basin liner have a detrimental effect on the output, assuming that no other factor becomes significant. He showed that surface reflection appears to be more detrimental than basin liner reflection because of the absorbing properties of the basin liner, except at normal incidence of insolation.

9.11 OTHER DESIGN OF SOLAR STILL

From the previous sections, it has been observed that the daily yield per m^2/day in single basin solar still (Fig. 9.3) mainly depends on the temperature difference between the evaporative and condensing surfaces. Various scientists had made an attempt to maximise the daily yield per m^2/day in a single basin solar still in a passive mode by changing its design to get maximum temperature difference between the evaporative and condensing surfaces. Some of the developed design will be discussed in the following sections:

9.11.1 Single Slope Solar Still with Condenser (Fig. 9.6) (*Faith, 1998*)

In a conventional solar still, the glass cover is used for transmission of solar energy as well as for condensation of water vapor evaporated from water surface. During the condensation, the latent heat is given to the glass cover which raised the glass cover temperature and hence reduces the overall temperature difference between the evaporative and condensing surface. In order to increase this difference, the condensing surface is separated from the solar still chamber as shown in Figure 9.6. There is little

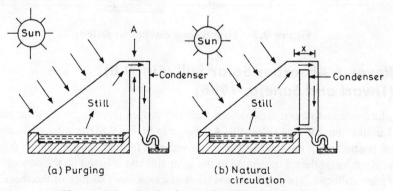

Figure 9.6 Single slope solar still with condenser.

condensation on the slopped surface, mainly condensation takes place in the attached condenser due to transfer (purging) of vapor from solar still chamber to condensing chamber. Since most of condensation is taking place in the condensing chamber, the temperature difference between glass cover and water is more which causes faster evaporation and distillate output is more. In this case, the still efficiency is increased by 45 percent. Further, the distillate output can be increased by natural circulation of vapor as shown in Figure 9.6(b).

9.11.2 Hybrid Single Slope Solar Still (Fig.9.7) (Abu-Qudair et al., 1996)

The condensing chamber of Figure 9.6 can be further improved for faster condensation as shown in Figure 9.7. The proposed change in design uses electrically operated fan and condensing chamber to increase the distillate output. This is referred as hybrid solar distillation system. The active components i.e., electrical blower and condenser can be attached with distillation unit having collector panel too (Fig. 9.11).

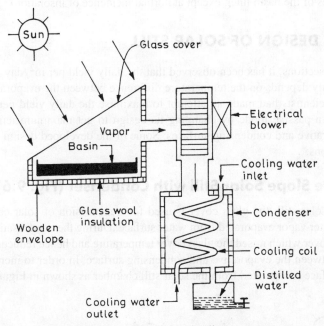

Figure 9.7 Hybrid solar distillation system.

9.11.3 Reverse Absorber Solar Still (Fig. 9.8) (Tiwari and Suneja, 1998)

In this case, the condensing cover is separated from the surface receiving (G_1) solar energy as shown in Figure 9.8 unlike conventional solar still. As shown in Figure 9.8, the solar radiation is allowed to be absorbed at the bottom of the solar still after transmission through the glass cover (G_1). This design consists of a cylindrical reflector integrated to the solar still and is based on the concept of an inverted absorber flat plate collector. The condensation takes place on inner surface of the metallic condensing

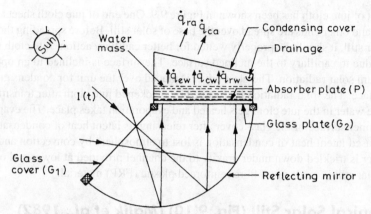

Figure 9.8 Schematic view of a reverse absorber solar still.

cover. The absorbed solar energy is transferred to the water mass by convection. The water gets heated. There is heat loss from the water surface to inner surface of metallic condenser by radiation, convection and evaporation. The evaporated water is condensed on inner surface of the condensing cover after realising its latent heat of condensation. The condensed water is trickled down under gravity to the drainage provided at the lower end of the condensing cover. Due to separation of condensing cover (cold surface) and receiving of solar energy surface, the temperature difference between condensing cover and water surface is increased for higher yield.

9.11.4 Multi-Wick Solar Still (Fig. 9.9) (*Sodha et al., 1981b*)

In this design, the maximum temperature difference between the condensing cover and water surface can be achieved by reducing the heat capacity of the water mass in the basin. In other word, a water film is maintained on the absorber for fast heating and quick evaporation. The water film is achieved by using a porous multi-wick (jute cloth) as shown in Figure 9.9. Each jute cloth layer is separated from other by providing a black polythene sheet between them so that each jute cloth can act independently.

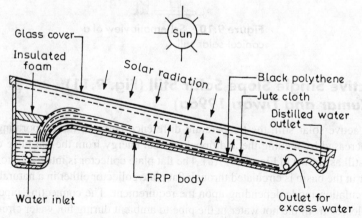

Figure 9.9 Cross-sectional view of Multi-wick solar still.

An arrangement of jute cloth has been shown in Figure 9.9. One end of jute cloth sheets is dipped in the water reservoir and other ends are spread over the base of solar still. Before spreading the jute cloth over the base of solar still, it should be properly wetted for better capillary action. Jute cloth sucks the water from reservoir due to capillary to the inclined surface. The surface is inclined to an optimum angle to receive maximum solar radiation. The glass cover is placed over the unit for condensation of the vapor on its inner surface. The solar radiation is absorbed by blackened jute cloth after transmission from the glass cover. The water in the jute cloth gets heated and evaporation takes place. The evaporated water is condensed on inner surface of the glass cover after releasing its latent heat of condensation to the glass cover. The released latent heat of condensation is lost to atmosphere by convection and radiation. The condensed water is trickled down under gravity to the channel provided at lower end of the solar still. The body of solar still is made up of fibre reinforced plastic (FRP) material.

9.11.5 Conical Solar Still (Fig. 9.10) (*Malik et al., 1982*)

In the above design of solar stills, the area of evaporating and condensing surfaces are equal. In the present design, the temperature difference between evaporating and condensing surfaces can be increased by fast cooling the condensing surface. This can be achieved by increasing heat transfer coefficient from the condensing surface to atmosphere. This can be obtained by increasing the surface area as shown in Figure 9.10. In this, impure water is enclosed in a transparent twin-one arrangement. Solar energy trapped within the enclosure heats up the water which causes evaporation and then condensation of the inner surface of the transparent upper cone. Condensed water droplets slide down in the water pan and are collected in the bottom cone.

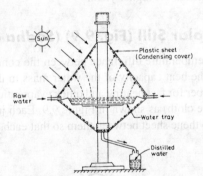

Figure 9.10 Schematic view of a conical solar still.

9.11.6 Active Single Slope Solar Still (Fig. 9.11) (*Kumar and Tiwari 1996a*)

In the case of active solar still, the temperature difference between the evaporating and condensing surfaces are increased by feeding the additional thermal energy from the flat plate collector in to the basin of solar still as shown in Figure 9.11(*a*). The flat plate collector is integrated to the basin of solar still. The water in the basin is circulated through flat plate collector either in a natural circulation mode or a forced circulation mode depending upon the requirement. The connecting pipes are insulated to avoid thermal losses from the hot water in the pipe to ambient during hot water circulation through it. In an active solar still, the water in the basin is heated directly as well as indirectly through a flat plate

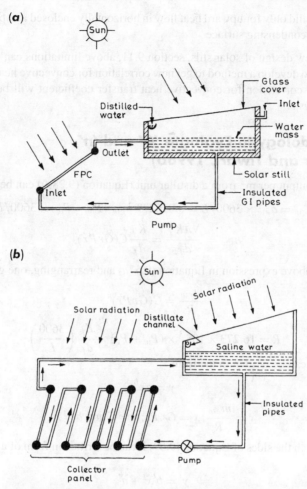

Figure 9.11 A schematic view of active solar still. (**a**) with single collector and (**b**) with number of collectors connected in series.

collector. The collector should be operated only during sunshine hour. The rise in the temperature of water in the basin mainly depends on number of collectors connected in series (Fig. 9.11(*b*)).

9.12 MODIFIED INTERNAL HEAT TRANSFER

Equation (9.26) has been derived by Dunkle (1961) to determine the convective heat transfer coefficient from water surface to the inner glass cover of a conventional solar still (Fig. 9.3(*a*)). This expression has the following limitations as described herein.

i. Firstly, this equation is valid only for normal operating temperature $\cong 50°C$ in a solar still and equivalent temperature difference of $\Delta T = 17°C$.

ii. Secondly, this is independent of cavity volume due to $n = 1/3$ and,

iii. Thirdly, this is valid only for upward heat flow in horizontally enclosed air space, i.e. for a parallel evaporative and condensing surface.

As we have seen in new design of solar stils, section 9.11, above limitations can't be fullfiled. Hence, there is a strong need to develop a method to get new correlation for convective heat transfer coefficient. In this section, a new correlation for convective heat transfer coefficient will be discussed by using linear regression analysis.

9.12.1 Methodology to Evaluate 'C' and 'n' (*Kumar and Tiwari, 1996b*)

The hourly distillate output per m^2 from a distiller unit (Equation (9.28a)) can be written as

$$\dot{m}_{ew} = \dot{q}_{ew} \times 3600/L = 0.016273 \, h_{cw} \, (P_w - P_w) \times 3600/L \tag{9.57a}$$

where

$$h_{cw} = \frac{Nu K_f}{d_f} = \frac{K_f}{d_f} C(Gr\,Pr)^n \tag{9.57b}$$

After substitution of above expression in Equation (9.57a) and rearranging, one gets

$$\frac{\dot{m}_{ew}}{R} = C(Gr\,Pr)^n \tag{9.58}$$

where

$$R = 16.273 \times 10^{-3} \times (P_w - P_g) \times \frac{K_f}{d_f} \left(\frac{3600}{L}\right)$$

Now Equation (9.58) can be written in the form

$$y = ax^b \tag{9.59}$$

where

$$y = \frac{\dot{m}_{ew}}{R}; x = Gr.Pr; a = C \text{ and } b = n$$

By taking log on both the sides of Equation (9.59) it can be written in form of a straight line equation given by

$$y' = b'x' + a' \tag{9.60}$$

where

$$y' = \ln y; a' = \ln n; b' = b \text{ and } x' = \ln x.$$

By using the linear regression analysis the coefficients a' and b' in Equation (9.60) can be obtained as

$$a' = \left(\frac{\sum y'}{N}\right) - b'\left(\frac{\sum x'}{N}\right) \tag{9.61a}$$

and

$$b' = \frac{N\left(\sum x'y'\right) - \left(\sum x'\right)\left(\sum y'\right)}{N\left(\sum x'^2\right) - \left(\sum x'\right)^2} \tag{9.61b}$$

where N is the number of experimental observations under steady state condition. In a quasi-steady state condition N becomes $N + 1$ (Example 7.1).

After knowing a' and b' from Equations (9.61), the 'C'' and 'n'' can be obtained by the following expressions

$$C = \exp(a') \text{ and } n = b'$$

Once the value of C and n are known, convective heat transfer coefficient can be obtained from Equation (9.57).

9.12.2 Experimental Setup

A schematic diagram of an experimental setup is shown in Figure 9.12. The constant temperature both has 40 litre capacity and effective evaporating surface area 36 cm × 26 cm. It consist of heating coil to heat water mass, a stirrer to maintain the uniform temperature and a proportional control to fix a desired temperature (steady state condition). The calibrated thermocouples were used to measure the various temperature namely outer and inner surface temperatures of condensing cover, water and vapor temperature etc. All thermocouples were attached to the digital temperature indicator having least count of 0.1°C. The distillate out-put was collected in graduated measuring cylinder of least count 1 ml at interval of 10 minutes. Specification of condensing cover has been given in Table 9.6

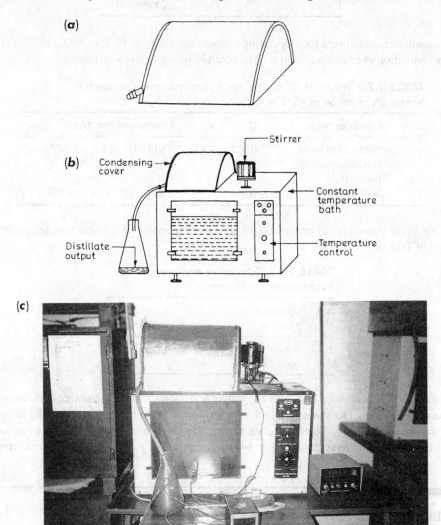

Figure 9.12   (*a*) View of condensing cover (*b*) view of experimental setup and (*c*) photograph of experimental setup.

TABLE 9.6 Specification of condensing cover

Dimension	Size
Shape	Semi-cylinder
Material	Aluminium
Thickness	20 gauge
Length	38 cm
Breadth	26 cm
Heagtit	13 cm
Rivet	2 mm (dia)

Experiments were conducted for a operating temperature range of 45°C to 80°C. In such condition, there is no limitation to evaluate C' and n'. The result obtained is given in Table 9.7.

TABLE 9.7 Values of 'C' and 'n' for a semi-cylindrical shape for temperature range of 45°C to 80°C

S.No.	Operating mode	C	n	Grashof number (Gr)
1	Without fan above condensing cover (natural)	0.023	0.438	$2.614 \times 10^6 < Gr < 5.527 \times 10^6$
2	With fan (forced mode)	0.058	0.379	$2.891 \times 10^6 < Gr < 5.987 \times 10^6$

Table 9.8 given the values of convecture (h_{cw}) and evaporative (h_{ew}) heat transfer coefficient for both cases of Table 9.7.

TABLE 9.8 Convective and evaporative heat transfer coefficient

S. No.	Operating mode	h_{cw} (W/m^2 °C)	h_{ew} (W/m^2 °C)
1	Natural mode	3.88	110
2	Forced mode	4.05	115

From Table 9.7, it is inferred that both convective and evaporative heat transfer coefficient depends on the value of C and n. The value of C and n for semi-cylindrical shape is significantly different then these values recommended by Dunkle. Hence, there is a strong need to evaluate first C and n for a given shape, size and operating temperature range before designing a condensing cover for maximum yield.

PROBLEMS

9.1 Plot the curve between fraction of energy absorbed by basin liner (α'_b) and the depth of water in the basin.

Hint Refer Equation (9.1g) and Table 9.3a and $R_g = \alpha_g = R_w = \alpha_w = 0.05$ and $\alpha_b = 0.8$.

9.2 Plot the curve between the rate of radiative ($\dot{q}_{rw}$), convective ($\dot{q}_{cw}$) and evaporative ($\dot{q}_{ew}$) heat transfer with water temperature T_w, for $T_w - T_g = 10°C$ and $30°C$.

   Hint Consider $T_w = 20, 30, 40, 50, 60, 70$ and $80°C$.

9.3 What is an ideal solar still?

   Hint Solar still with water film ($d_w \approx 0.0$).

9.4 Draw thermal circuit diagram of a solar still of Figure 9.10.

   Hint Refer Figure 9.3b.

9.5 Plot internal heat transfer coefficients for Problem 9.2.

   Hint See Problem 9.2.

9.6 Calculate bottom heat transfer coefficient for different thickness of bottom insulation and draw curve between h_b and L_i.

   Hint $L_i = 0, 2\,cm, 4\,cm, 6\,cm, 8\,cm$ and $10\,cm$; $K_i = 0.04\,W/m°C$ and Equation (2.49a) with $V = 0.0$
   ($h_i = h_c$).

9.7 Write down the energy balance equation for double slope solar still.

   Hint Assume the water temperature is the same for east (T_{gE}) and west (T_{gW}) glass temperatures and h_{1E} and h_{2W} will be different. See figure given below.

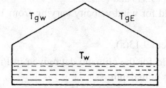

Double slope solar still

9.8 Determine side loss coefficient for water depth ranging from 2 to 15 cm for basin area of $1\,m \times 1\,m (A_s)$.

   Hint See Equation 9.16.

9.9 Discuss Equation 9.50 for the two limiting cases: $\Delta t \to 0$ and ∞.

   Hint See Equation 9.50.

9.10 Derive an expression for h_{cw} for cases II, III and IV in Table 9.5.

   Hint See Table 9.5.

9.11 Write down energy balance equation for each component of active solar still.

   Hint Add $\dot{q}_u$ from Equation (3.70) on the left hand side of Equation (9.42). See figure given below.

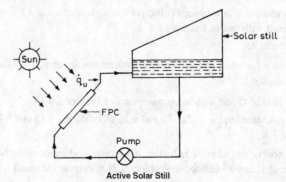

Active Solar Still

9.12 How is the energy balance for non-symmetrical double slope solar still different from that of symmetrical double slope solar still?

Hint Through glass cover area ($A_{g1} \neq A_{g2}$).

9.13 a. Calculate the hourly variation of water (T_w), glass cover (T_g) and yield $\dot{m}_{ew}$ for 5 cm water depth, for the following parameters: Design parameters: $\alpha'_b = 0.6$; $A_s = 1\,m^2$; $C_w = 4190\,J/kg°C$; $L = 2.26 \times 10^6 J/kg$; $V = 0.0\,m/s$.

Climatic parameters:

Time (hrs)	1	2	3	4	5	6	7	8	9	10	11	12
$I(t)W/m^2$	0	0	0	0	0	98	102	402	662	845	954	990
$T_a(°C)$	11.3	10	9.0	8.3	8.0	8.5	9.8	12.1	15.4	19.2	23.5	27.6

Time (hrs)	13	14	15	16	17	18	19	20	21	22	23	24
$I(t)W/m^2$	954	845	662	402	102	98	0	0	0	0	0	0
$T_a(°C)$	30.6	32.6	33.4	32.6	31	28	25	21.5	18.7	16	14	12.6

Hint Use Equations (9.49), (9.51) and (9.53).

b. Calculate the hourly yield for wind velocity varying from 1 m/s to 5 m/s for 1 cm and 10 cm depth of water.

Hint Use the data of Problem 9.13(a).

c. Calculate the hourly yield for the climatic data given in Problem 9.13(a) and compare the results with Problem 9.13(a) for 1 cm depth.

Hint See Problem 9.13(a).

9.14 Draw the hourly variation of radiative, convective and evaporative heat transfer coefficients of Problem 9.13(a).

Hint Determine h_{rw}, h_{cw} and h_{ew} for each initial condition, i.e., T_{w0} and T_{g0} for each set of calculation.

9.15 Plot η_i versus $(T_{w0} - T_a)/I(t)$ for Problem 9.13(a).

9.16 Derive an expression for the water and the glass temperatures, hourly yield for Problems 9.7 and 9.11.

Hint See section 9.9.3.

9.17 Plot F' and U with time of the day of Problem 9.13(a).

9.18 Derive an expression for Δt from Equation 9.49.

Hint See Example 9.8.

9.19 Calculate the radiative heat transfer coefficient from glass cover at 12°C to an ambient at 8.5°C and compare the result with that of Example 9.4.

Hint See Example 9.4.

9.20 Derive expressions for water and glass cover temperatures for an active solar still.

Hint See section 9.9.3.

9.21 Calculate the value of C and n by using the results of Problems 9.13.

Hint Use the calculated value of T_w, T_g and $\dot{m}_{ew}$ of Problems 9.13 and follow the procedure given in Example 7.1.

9.22 Calculate the hourly variation of radiative, convective and evaporative heat transfer coefficients by using the value of C and n obtained in Problem 9.21. Compare the result with Problem 9.14.

9.23 Derive an expression for Δt for active solar still (Problem 9.11).

Hint See Problem 9.18.

9.24 Calculate hourly variation of T_w, T_g and $\dot{m}_{ew}$ for active solar still for date of Problem 9.13(a).

Hint Use $F_R = 0.65$, $A_c = 2\,\text{m}^2$ and $I'(t) = I(t)$ and $U_L = 8\,\text{W/m}^2\,{}^\circ\text{C}$.

9.25 Calculate C and n for the results of active solar still (Problem 9.24).

Hint See Problem 9.21.

9.26 Repeat Problem 9.22 for active solar still.

9.27 Write down energy balance equation for reverse absorber solar still (Figure 9.8).

Hint Energy balance for reverse absorber is

$$\alpha_b \rho^N \tau I(t) = h_3(T_b - T_w) + U'_t(T_b - T_a)$$

ρ reflectivity and N number of reflection.

9.28 Write down energy balance equation for multi-wick solar still, (Figure 9.9).

Hint Use $\dot{m}_w C_w \dfrac{dT_w}{dx}$ in place of $(MC)_w \dfrac{dT_w}{dt}$.

9.29 Derive an expression for T_w and $\bar{T}_w$ as a function of x for Problem 9.28.

Hint Follow similar procedure as done is Section 9.9.4 by considering x as a variable.

9.30 Calculate $\bar{T}_w$ for Problem 9.29 by using the date of Problem 9.13(a). Also calculate $\bar{T}_g$ and $\dot{m}_{ew}$.

Hint After knowing $\bar{T}_w$, $\bar{T}_g$ can be calculated by using Equation 9.41 for $\alpha'_g = 0$ and $\dot{m}_{ew}$ from Equation (9.53).

CHAPTER 10

Solar House

10.1 INTRODUCTION

Passive solar house is a new concept in architecture and describes a way to design a building keeping under consideration the site, climate, local building materials and the sun. It implies a special relationship of building design and natural processes that offer the potential for an inexhaustible source of energy. This concept, although, is not entirely new, as can be seen from it's description in vernacular architecture. The statement of Socrates (400 B.C.): "Now in houses with south aspect, the sun's rays penetrate into porticos in winter, but in summer the path of the sun is right over our heads and above the roof. If then, this is the best arrangement one should build the south side loftier to keep out the cold wind", reflects this. Ancient Iranian architecture used the concepts of clustering (decreasing surface to volume ratio, to reduce thermal load) of thick walls (large thermal storage capacity to smoothen out the temperature fluctuations), of plantations for shade and living in the basements (during extreme cold and heat). Iranians also introduced the concept of wind towers, which along with cooling by earth and water evaporation, made the buildings comfortable in summer. The concept of clustering and mass walls were also used in medieval India. American Indians used the passive solar techniques as early as 1100 A.D.

However, scientific application of solar energy for passive heating started around 1881 when Prof. E.L. Morse (1881) was granted a patent on glazed south facing dark wall for keeping the house warm. This idea was not used for a long time until 1972, when Trombe (1972, 1974) repatented the concept and successfully applied it.

In recent years, however, people have chosen to abandon these long standing considerations, getting increasingly dependent on the mechanical control of the indoor environment, rather than exploiting the climatic and other natural processes to meet their requirements. As a result, even a minor power or equipment failure can make these buildings uninhabitable. Further, with the advent of energy crisis, effective utilization of energy resources have become the basis of building planning. Now, there is a renewed interest in passive solar heating and cooling systems, because these systems which operate on energy available in the immediate environment are simple in concept and require little or no maintenance. Also, these systems do not generate thermal pollution, as they require no external energy input and produce no physical by-products or waste. As solar energy is extensively distributed over the world, expensive transportation and distribution networks are eliminated.

The passive solar concepts need to be included in every step of a building's design. Although, the conventional or active solar heating systems can be somewhat independent of the conceptual

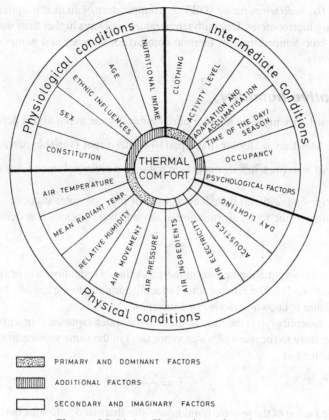

PRIMARY AND DOMINANT FACTORS

ADDITIONAL FACTORS

SECONDARY AND IMAGINARY FACTORS

Figure 10.1(a) Thermal comfort chart.

organization of a building, it is difficult to add a passive system to a building once it has been designed. Thus, the information regarding various passive concepts/systems must lead to necessary degree of accuracy at each stage of building design. A judicious application of the passive solar concepts provides thermal comfort in an economical way. The thermal comfort chart (Figure 10.1(a), gives various factors influencing the thermal comfort. The factors can be divided into three conditions, viz. (a) physical, (b) physiological and (c) intermediate.

These factors can further be classified as primary and dominant; additional; and secondary and imaginary. For example, in physical conditions, the primary factors are the air temperature, mean radiant temperature, the relative humidity and the air movement whereas in intermediate conditions, clothing and activity level are the dominant factors. Similarly, in physiological conditions all the factors are the additional factors, and in intermediate conditions, occupancy and adaptation and acclimatization are the additional factors.

10.1.1 Physical Parameters

10.1.1.1 Air temperature

There are basically two temperature generally used for thermal comfort. These are *wet bulb* and *dry bulb temperatures*. *Wet bulb temperature* (WBT) is a temperature at which vapor pressure of air is equal to saturation pressure. *Dry bulb temperature* (DBT) is a temperature of air at atmospheric pressure and it is measured by ordinary thermometer. Dry bulb temperature is always higher than wet bulb temperature. Recommended dry bulb temperature for thermal comfort for summer and winter are $(25 \pm 1)°C$ and $(20 \pm 1)°C$ respectively.

10.1.1.2 *Relative humidity*

This is related to moist air which is a mixture of dry air and water vapor and is defined as

$$Moist\ air = Dry\ air\ (fixed\ part) + Water\ vapor\ (variable\ part)$$

If the absorption of water reaches its maximum value in a given volume of dry air, then it is known as *saturated air*.

Humidity or *specific humidity* (ω) is defined as ratio of mass of water vapor (m_v) to mass of dry air (m_a) in a given volume of the moist air (mixture of dry and water vapor). It is defined as

$$\omega = \frac{m_v}{m_a}$$

It is clear from the above equation that the numerical value of humidity ratio or specific humidity (ω) is always greater than one (1). The ratio of actual specific humidity to the specific humidity of saturated air at given temperature is known as *degree of saturation*.

Now, the *relative humidity* (γ) is the ratio of mass of the water vapor (m_v) in a given volume of moist air at a given temperature to the mass of water vapor (m_s) in the same volume of saturated air at same temperature. It is defined as

$$\gamma = \frac{m_v}{m_s}$$

The recommended value of the relative humidity (γ) for thermal comfort for both summer and winter condition is (50 ± 5) percent $[0.50 \pm 0.05]$. Its numerical value is always less than 100 percent [1].

The variation of humidity ratio (specific humidity) with dry bulb temperature for different relative humidity at atmospheric pressure has been shown in Figure 10.1(*b*). It is also referred as *psychometric chart*.

10.1.1.3 *Air movement*

The temperature of the body remains constant between 36.9°C to 37.2°C. The temperature above 40°C may be fatal. As mentioned earlier that thermal comfort air temperature for human being is around 20°C to 25°C for both winter and summer respectively.

If the air surrounding the body of human being is circulated with certain movement then there will be a heat transfer (respiration) from the body to an environment. This reduces the thickness of air film near body. As the air movement near the body increases, the level of discomfort reduces. This is true only when the air temperature of the surrounding is lower than the body temperature otherwise the level of discomfort increases.

An increase of air movement helps in heat transfer (respiration). It will create discomfort for higher air temperature and relative humidity. The recommended value of air movement is around 0.2 m/s and 0.4 m/s in winter and summer respectively inside the room. This condition is achieved inside the room by operating the ceiling/table fan in the summer condition.

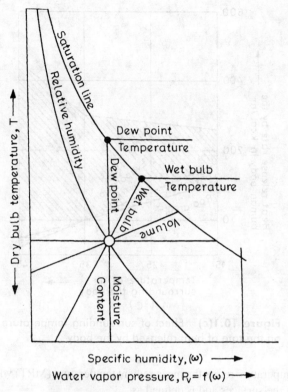

Figure 10.1(b) Psychometric chart.

10.1.1.4 Mean radiant temperature

The human body loses heat to the surrounding by radiation (R), convection (C) and evaporation (E) (respiration). The total heat loss (Q) can be written as

$$Q = (R + C) + E$$

where ($R + C$) are sensible heat loss (Q_S) and E is latent heat loss (Q_L).

The sensible heat loss mainly depends on temperature difference between the surface of the body and surroundings while latent heat loss depends on difference in water vapor pressure.

The heat exchange between man and environment can be expressed as follows:

$$M - W = Q + S$$

where M is metabolic rate, W the work done by man and S the rate of heat storage.

In summer condition, the body temperature rises due to storage of energy (S being +ve). In such situation, blood flow rate increases through extremities.

In winter condition, the temperature of body tends to fall (S being −ve) and hence the blood flow decreases. This gives rise to shivering.

The rate of heat released by human body with surrounding temperature is shown in Figure 10.1(c).

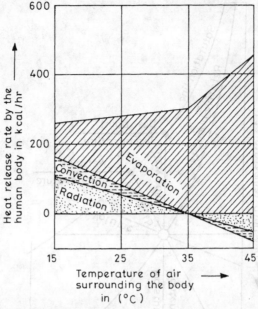

Figure 10.1(c) Effect of surrounding temperature on the rate of heat released by the body.

Here surrounding temperature refers the mean radiant temperature (MRT) which is the area weighted average temperature of the surfaces and is defined as

$$MRT = \frac{\sum_i T_i A_i}{\sum_i A_i}$$

where T_i and A_i are the temperature and area of different surfaces of the living enclosure, respectively.

10.1.1.5 Air pressure

A person feels more comfortable at atmospheric pressure. At higher altitude pressure is reduced and human being feels more discomfort due high pressure within the body.

10.1.1.6 Air ingredients

Each person requires $0.65\,m^3$/hr of oxygen (O_2) under normal condition and produces $0.2\,m^3$/hr of corbon-dioxide (CO_2). Human body requires at least this rate of oxygen for survival and to sustain its action inside as well outside of the building. In the open condition this requirement is met by the plants which releases oxygen to maintain the level of oxygen in the environment. The carbon dioxide, released by human being, is consumed by the plants to balance the eco-system. The optimum level of oxygen and carbon-dioxide in the atmosphere are (20.94×10^4 ppm) and 320 ppm.

However, the optimum level of oxygen and carbon dioxide should also be maintained inside the building either by passive or active modes. Some of conventional methods to maintain this level is to have a provision roof vent (roshandan)/ventilator/exhaust fan etc. in the building. In such way, there is direct contact between environment and living space of the building.

10.1.1.7 Air electricity

A person feels shocks at skin when a thin layer of air passes over the skin. This can be observed while rubbing the skin in the presence of thin layer of air. This is due to friction between thin air layer and skin.

10.1.1.8 Acoustics

A person feels more comfortable in a living space for effective working condition up to maximum sound level of 120db (0.01 W/m^2). An average person can hear frequencies from 20 to 16,000 Hz. Unwanted sound creates a sense of irritation in a living space and it is termed as *noise pollution*.

10.1.1.9 Day lighting

In the absence of natural lighting in a building, an artificial lighting is created by using electrical appliances e.g. electric bulb/lamp/tube light etc. By using electric bulb, the thermal heat gain in a building is significant because in a lamp, only a fraction of the electrical energy is converted into light energy. The heating effect of day lighting is 1 Watt per lumen.

10.1.2 Physiological Parameters

10.1.2.1 Nutritional intake

Human comfort is influenced by physiological factors determined by the rate of heat generated inside body and the rate of heat dissipated to the environment.

The rate, at which the body produces heat, is known as *metabolic rate*. A healthy person produces heat at the rate of 60 W ($\approx$35 W/m^2) during the sleeping and it can go up to 600 W ($\approx$350 W/m^2) during the hard work.

It is well known that a person requires optimum nutritional intake. It depends upon composition of the body. It varies from person to person and from season to season. For example, nutritional intake (healthy food) for a person becomes more during winter period and vice versa during summer. Due to this reason person feels more thermal comfort after taking more caloric value food during the winter period in comparison to the summer period.

10.1.2.2 Age

With the increase in age, a person feels more cold in winter and warm in summer because of the

  i. decrease of metabolic rate (the rate of generating heat within the body)
 ii. change in food habit
iii. decrease in nutritional intake
 iv. change in activity level etc.

10.1.2.3 Ethnic influences

This deals with the way the people live in a particular region including their food habits, clothing etc.

10.1.2.4 Sex

A female produces more heat due to the energy generated within body in comparison to man and feels more comfortable in cold climate.

10.1.2.5 Constitution

This refers to the constitution of body of individuals. It is well known that a weak person feels more cold in winter than a healthy one and vice versa in summer.

10.1.3 Intermediate Parameters

10.1.3.1 Clothing

The clothing intervenes with air movement across the skin and as a result, it decreases the potential of heat transfer [conduction, convection and evaporation (respiration)] from the body to room air. The color of the clothing is also important. For example, in

 i. desert, the thin, loose fitting and light color clothing is suitable. This gives lower absorption of solar radiation and heat transfer from outside to skin and provide air movement across skin due to loose fitting
 ii. hot and humid climate, the loose, light color and porous clothing is advised. This provides fast heat transfer from the skin to environment air due to porosity of cloth
 iii. cold climate, dark color, thick and tight fitting is recommended for higher absorption of solar radiation and minimum heat transfer from skin to environment air

10.1.3.2 Activity level

As it is mentioned earlier that the rate of heat dissipation by a person depends on activity of the concerned person. It varies from 60 W ($\approx$35 W/m^2) to 600 W ($\approx$350 W/m^2). Hence, the human body can be considered as a heat engine converting thermal energy into mechanical energy with efficiency of 20 percent. In a space of 6 m^3, the room air temperature may raise to 48°C/hr if man's body dissipate energy at rate of 335 KJ/hr. In order to provide thermal comfort, the heat generated should be removed by providing ventilator to room.

10.1.3.3 Adaption and acclimatisation

The adaption and acclimatisation means that a person should be first become immune with artificial similar climate before starts living in a actual harsh climatic condition.

10.1.3.4 Time of the day/season

The recommended thermal comfort air temperature is 20°C. It is known that ambient air temperature changes with the

 i. time of the day due to change in the level of solar intensity (insolation)
 ii. month of year from season to season

For example, there is variation of ambient air temperature from 5°C to 15°C in the winter months and from 30°C to 45°C in the summer months for northern climatic condition.

Therefore, the requirement of heating/cooling load depends on the time of day/season.

10.1.3.5 Occupancy

As we have seen that a person can produce heat due to energy generated within body. If number of persons is increased for a given volume of an enclosed space, then total amount of heat produced will also be increased depending upon numbers. This leads to increase in room air temperature. This results in affecting living as well working condition of person.

10.1.3.6 Psychological factors

Sometimes by expressing the level of peak winter and summer conditions in a group people feels comfortable by sharing their thought with others. This gives the pleasure of individual talking in a group. This is due to psychological factor.

India has been divided into a number of climatic zones for the simplification of analysis in the solar energy systems. The criteria for classification of these climatic zones in India is given in Table 10.1.

The heating/cooling requirement of a building can be determined if the solar temperature, defined below, is known.

10.2 SOLAIR TEMPERATURE AND HEAT FLUX

Solair temperature combines the effect of solar radiation, ambient air temperature and long wave radiant heat exchange with the environment. Physically solair temperature can be interpreted as the temperature

TABLE 10.1 Criteria for the classification of climates

Climate	Mean monthly temperature	Relative humidity (%)	Precipitation (mm) (rain/snow fall)	Number of clear days	Example
Hot and dry (HD)	>30	<55	<5	>20	Jodhpur
Warm and humid (WH)	>30	>55	>5	<20	Bombay
Moderate (MO)	25–30	<75	<5	<20	Bangalore
Cold and cloudy (CC)	<25	>55	>5	<20	Srinagar
Cold and sunny (CS)	<25	<55	<5	>20	Leh
Composite (CO)	This applies, when six months or more do not fall within and of the above categories				New Delhi

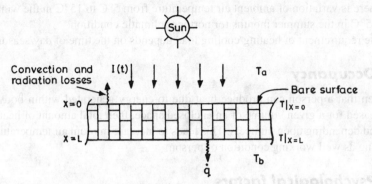

Figure 10.2(a i)   A bare surface exposed to solar radiation.

of the surroundings that will produce the same heating effect as the incident radiation in conjunction with the actual external air temperature.

Let us consider a surface, exposed to solar radiation, at temperature $T|_{x=0}$, as shown in Figure 10.2. The energy balance can be described as follows:

10.2.1 For Bare Surface (Fig. 10.2(a))

The energy balance at $x = 0$ can be written as

$$\alpha I(t) = -K\frac{\partial T}{\partial x}\bigg|_{x=0} + h_{ra}(T|_{x=0} - T_a) + \varepsilon\Delta R + h_{ca}(T_{x=0} - T_a) \qquad (10.1a)$$

where α is the absorptivity of the surface, K the thermal conductivity, $\varepsilon\Delta R$ is as given in Equation (2.58d) and its value is 60 W/m^2.

Equation (10.1a) can be rewritten as

$$-K\frac{\partial T}{\partial x}\bigg|_{x=0} = \alpha I(t) - h_1(T_{x=0} - T_a) - \varepsilon\Delta R$$

where $h_1 = h_{ra} + h_{ca}$, $h_{ca}(= 2.8 + 3V)$ and h_{ra} are given by Equations (2.49b) and (2.58e), respectively.

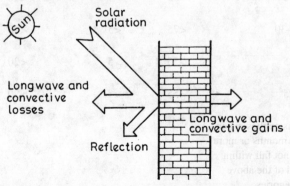

Figure 10.2(a ii)   Energy flow diagram for bare surface (Trombe wall).

Further, the above equation can be rewritten as

$$\dot{q} = -K \frac{\partial T}{\partial x}\bigg|_{x=0} = h_1 \left[\frac{\alpha}{h_1} I(t) + T_a - \frac{\varepsilon \Delta R}{h_1} - T|_{x=0} \right]$$

$$= h_1 [T_{sa} - T|_{x=0}] \tag{10.1b}$$

where $\qquad T_{sa} = \dfrac{\alpha}{h_1} I(t) + T_a - \dfrac{\varepsilon \Delta R}{h_1}$ is the solair temperature for bare surface. $\qquad$ (10.1c)

In a steady state condition, the rate of heat conducted from $x = 0$ to $x = L$ can also be written as

$$\dot{q} = \frac{K}{L} (T|_{x=0} - T|_{x=L}) \tag{10.1d}$$

This rate of heat is further transferred from $x = L$ to living space by convection and radiation and can be expressed as

$$\dot{q} = h_{si}(T|_{x=L} - T_b) \tag{10.1e}$$

Equations (10.1b), (10.1d) and (10.1e) can be rearranged as

$$\frac{\dot{q}}{h_1} = [T_{sa} - T|_{x=0}]$$

$$\frac{\dot{q}}{\frac{K}{L}} = T|_{x=0} - T|_{x=L}$$

and

$$\frac{\dot{q}}{h_{si}} = [T|_{x=L} - T_b]$$

Adding the above three equations, we get an expression for $\dot{q}$ as

$$\dot{q} = U(T_{sa} - T_b) \tag{10.1f}$$

where

$$U = \left[\frac{1}{h_1} + \frac{L}{K} + \frac{1}{h_{si}} \right]^{-1}$$

This gives the rate at which the heat is transferred during the day time from the exposed surface to the living space maintained at temperature T_b. The total energy per m^2 in terms of Joule during sunshine hour t_T, can be obtained as

$$Q_{ud} = U(T_{sa} - T_b) \times t_T \times 3600 \tag{10.1g}$$

If the bare surface is covered with insulation of thickness L_i, having thermal conductivity of K_i, during the night time, then

$$\dot{q} = U(T_a - T_b) \tag{10.1h}$$

where

$$U = \left[\frac{1}{h_1} + \frac{L_i}{K_i} + \frac{L}{K} + \frac{1}{h_{si}} \right]^{-1}$$

The total energy per m^2 in terms of Joule during off-sunshine hours t'_T, can be obtained as

$$Q_{un} = U(T_{sa} - T_b) \times t'_T \times 3600 \tag{10.1i}$$

where $t = t_T + t'_T = 24$ hours.

Figure 10.2(a iii) A man working under bare roof exposed to solar radiation.

The net total energy per m^2 available in terms of Joule during the day/night will be sum of the value obtained from Equations (10.1g) and (10.1i) respectively. This can be written as

$$Q_T = Q_{ud} + Q_{un} \qquad (10.1j)$$

When a person is working inside room with bare roof exposed to solar radiation, the heat flux Q_T enters the room through roof by conduction. This flux provides thermal uncomfort inside the room as shown in Figure 10.2(a iii).

10.2.2 For Wetted Surface (Fig. 10.2(b))

In this case, in addition to convection and radiation, there will be evaporation also (Fig. 2(b i)).

Equation (10.2a) can be rewritten as:

$$\alpha I(t) = -K \frac{\partial T}{\partial x}\bigg|_{x=0} + h_{ra}(T|_{x=0} - T_a)$$

$$+ \varepsilon \Delta R + h_{ca}(T|_{x=0} - T_a) + h_{ea}(T|_{x=0} - T_a) \qquad (10.2a)$$

or

$$\dot{q} = -K \frac{\partial T}{\partial x}\bigg|_{x=0} = h_1(T_{sa} - T|_{x=0}) \qquad (10.2b)$$

$$T_{sa} = \frac{\alpha}{h_1} I(t) + T_a - \frac{\varepsilon \Delta R}{h_1} \qquad (10.2c)$$

where $h_1 = h_{ra} + h_{ea} + h_{ca}$ and $h_{ea}(= h_{ew})$ is given by Equation (2.77).

In wetted surface, an expression for the rate of heat conducted from $x = 0$ to $x = L$ and the rate of heat convected and radiated from $x = L$ to the living space in a steady state condition will be similar to Equations (7.1d) and (7.1e), respectively.

With the help of Equations (10.1d), (10.1e) and (10.2b), an expression for the rate of heat transfer from an exposed wetted surface to the living space will be same as Equation (10.1f), except the change in numerical value of h_1 for the wetted surface.

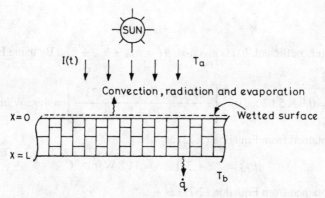

Figure 10.2(b i) A wetted surface exposed to solar radiation.

Similarly, the expression for the net energy gain/loss during day/night in this case can also be obtained.

Figure 10.2(b ii) shows a man working inside room with wetted roof. He feels more comfort due to reduction of heat flux by evaporation from the roof. It is found that there is 25 percent reduction in thermal flux inside room as shown in Figure 11.2(b iii).

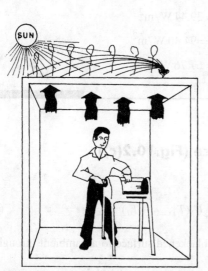

Figure 10.2(bii) A man working under wetted roof exposed to solar radiation.

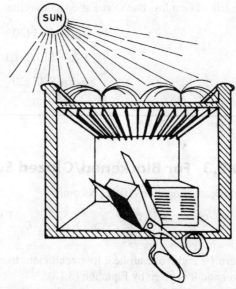

Figure 10.2(biii) Showing 25% reduction in thermal flux.

EXAMPLE 10.1

Calculate the total heat transfer coefficient (h_1) and the rate of heat loss due to radiation, convection and evaporation for wetted surface.

Given: Temperature of wetted surface = 20°C, ambient temperature = 12°C, relative humidity = 0.6, wind velocity = 3 m/s, emissivity = 0.9.

Solution

The total heat transfer coefficient, h_1, is given as, $h_1 = h_{ra} + h_{ca} + h_{ea}$. By using Equation (3.17), h_{ra} can be determined as

$$h_{ra} = 0.9 \times 5.67 \times 10^{-8} \left[\frac{(20+273)^4 - (12+273)^4}{20-12} \right] = 4.93 \, \text{W/m}^2 \, ^\circ\text{C}$$

Also, h_{ca} can be obtained from Equation (3.12) as

$$h_{ca} = 2.8 + 3.0 \times 3 = 11.8 \, \text{W/m}^2 \, ^\circ\text{C}$$

Now, h_{ea} can be obtained from Equation (2.77) as

$$h_{ea} = 16.273 \times 10^{-3} \times 2.8 \left[\frac{(2346.5 - 0.6 \times 1433.5)}{20-12} \right] = 8.47 \, \text{W/m}^2 \, ^\circ\text{C}$$

The total heat transfer coefficient is

$$h_1 = 4.93 + 11.8 + 8.47 = 25.2 \, \text{W/m}^2 \, ^\circ\text{C}$$

The rate of heat loss due to radiation, convection and evaporation are

$$\dot{q}_{ra} = 4.93 \, (20 - 12) = 39.44 \, \text{W/m}^2$$

$$\dot{q}_{ca} = 11.80 \, (20 - 12) = 94.40 \, \text{W/m}^2$$

$$\dot{q}_{ea} = 8.47 \, (20 - 12) = 67.76 \, \text{W/m}^2$$

10.2.3 For Blackened/Glazed Surface (Fig.10.2(c))

In this case, Equation (10.1a) becomes

$$\alpha \tau I(t) = -K \left. \frac{\partial T}{\partial x} \right|_{x=0} + U_t (T|_{x=0} - T_a) \tag{10.3a}$$

where U_t is the overall heat loss coefficient from the blackened surface to an ambient through glass cover and it is given by Equation (3.18).

Equation (10.3a) can be rearranged as follows:

$$\dot{q} = -K \left. \frac{\partial T}{\partial x} \right|_{x=0} = U_t \left(\frac{\alpha \tau}{U_t} I(t) + T_a - T|_{x=0} \right) \tag{10.3b}$$

or

$$\dot{q} = -K \left. \frac{\partial T}{\partial x} \right|_{x=0} + U_t (T_{sa} - T|_{x=0}) \tag{10.3c}$$

In the case of blackened/glazed surface too, an expressions for the rate of heat conducted from $x = 0$ to $x = L$ and the rate of heat convected from $x = L$ to the living space will be similar to Equations (10.1d) and (10.1e) respectively.

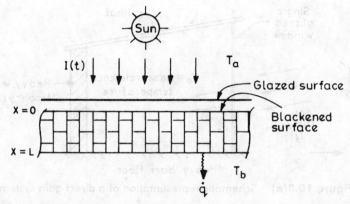

Figure 10.2(c) Schematic view of blackened and glazed surface.

Similarly, an expression for the rate of heat transfer from an exposed blackened/glazed surface to the living space can be derived in a steady state condition by using Equations (10.1d), (10.1e) and (10.3b) as

$$\dot{q} = U_L(T_{sa} - T_b) \qquad (10.3d)$$

where

$$U_L = \left[\frac{1}{U_t} + \frac{L}{K} + \frac{1}{h_{si}}\right]^{-1} \qquad (10.3e)$$

$$T_{sa} = \frac{\alpha\tau}{U_t} \cdot I(t) + T_a \qquad (10.3f)$$

The above derived equation can be used to determine rate of thermal energy gain/loss for heating/cooling of a building during day and night, respectively.

10.3 THERMAL GAIN

In order to understand and analyze the passive systems, three basic concepts, viz. direct gain, indirect gain and isolated gain, need to be studied.

10.3.1 Direct Gain

For direct gain, the sunlight is admitted through a window or wall of glass, facing south (north, for southern hemisphere) to heat up the walls, floors and objects (consequently also the air) in the room (Fig. 10.3(a))

In this approach, the space becomes a live-in solar collector, heat storage and distribution system. Direct gain systems are continuously working, as they collect and use every bit of energy that passes through the glazing—direct or diffuse (Fig. 10.3(b)). In order to avoid overheating during day and cooling at night (i.e. large variations of air-temperature), the floor and walls must be constructed of materials capable of storing heat for use at night. The commonly used materials for heat storage are masonry and water. Double glazing Fig. 10.3(c) is used to reduce the heat loss from the room to outside air and the windows are covered by insulation at night for the same purpose.

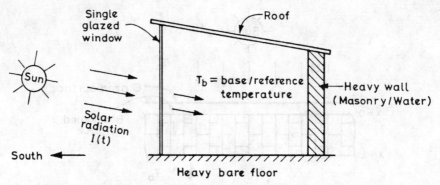

Figure 10.3(a) Schematic representation of a direct gain system.

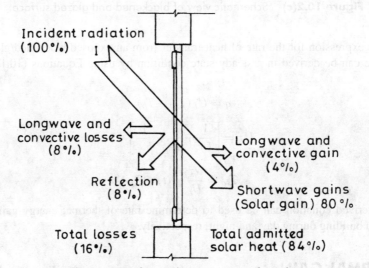

Figure 10.3(b) Schematic view of window.

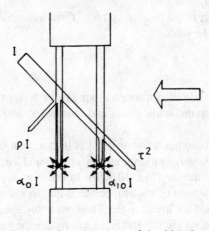

Figure 10.3(c) Schematic view of double glazed window.

The rate of useful energy gain (Fig. 10.3(b)) is given by

$$\dot{q} = \tau I(t) - U_t(T_b - T_a)$$ (10.4)

where τ is the transmissivity, U_t the overall heat transfer coefficient from living room to ambient through the glass cover, T_b the base or reference temperature in the building and T_a the ambient air temperature. From Equation (10.4), we have

$$\dot{q} = U_t \left[\frac{\tau}{U_t} I(t) + T_a - T_b \right] = U_t[T_{sa} - T_b]$$ (10.4a)

where $T_{sa} = (\tau / U_t)I(t) + T_a$, which is similar to Equation (10.3c).

Further, Equation (10.4a) can also be used to find out total net energy gain during sunshine hour (t_T is the total sunshine hour). This can be obtained as follows:

$$Q_T = \int_0^{t_T} \dot{q} \, dt = \left[\tau \int_0^{t_T} I(t)dt - U_t \left(T_b t_T - \int_0^{t_T} T_a dt \right) \right]$$

If

$$\bar{I}(t) = \frac{1}{t_T} \int_0^{t_T} I(t)dt \quad \text{and} \quad \bar{T}_a = \frac{1}{t_T} \int_0^{t_T} T_a \, dt$$

Then

$$Q_T = \tau \bar{I}(t) t_T - U_t(T_b - \bar{T}_a)t_T$$ (10.5)

EXAMPLE 10.2(a)

Calculate the total net energy gain in terms of kWh a direct gain system at temperature 20°C, irradiated with solar radiation of intensity 800 W/m^2 for 4 hours. Given ambient temperature is 12°C, transmissivity is 0.9 and overall heat loss coefficient is 6 W/m^2 °C.

Solution

From Equation (10.5), we have

$$Q_T = 0.9 \times 800 \times 4 \times 3600 - 6 \times (20 - 12) \times 4 \times 3600$$

$$= (2880 - 192) \times 3600 = 2688 \times 3600 = 9.67 \times 10^6 \text{ Joules}$$

Since $1 \, J = 2.778 \times 10^{-7}$ kWh, then,

$$Q_T = 2.69 \, \text{kWh}$$

For double glazed window (Fig. 10.3(c)), an expression for Q_T remains the same as given by Equation (10.5) except τ is replaced by τ^2 and the value of U_t should be used for double glazed window.

EXAMPLE 10.2(b)

Calculate U_t for double glazed window with air cavity of 0.05 m $(C = 4.75 \, \text{W/m}^2 \, ^\circ\text{C}$, Fig. 2.5(c)). Given: $K_g = 0.78 \, \text{W/m} \, ^\circ\text{C}$ and $L_g = 0.003$ m.

Solution

Following Example 2.1(a)

$$U_t = \left[\frac{1}{9.5} + \frac{0.003}{0.05} + \frac{1}{4.75} + \frac{0.003}{0.05} + \frac{1}{5.8} \right]^{-1}$$

$$= [0.1053 + 0.06 + 0.2105 + 0.06 + 0.1724]^{-1}$$

$$= [0.608]^{-1} = 1.64 \, \text{W/m}^2 \, ^\circ\text{C}$$

It is clear that the value of U_t for double glazed window is reduced.

EXAMPLE 10.2(c)

Calculate Q_T for double glazed window for Example 10.2(a). The value of U_t for double glazed is 1.99 W/m² °C.

Solution

From Equation (10.5), we get

$$Q_T = 0.80 \times 800 \times 4 \times 3600 - 1.64(20 - 12) \times 4 \times 3600$$

$$= 9.216 \times 10^6 - 0.189 \times 10^6 = 9.03 \times 10^6 \, \text{J} = 2.51 \, \text{kWh}$$

10.3.2 Indirect Gain

In indirect gain, the sun's radiation first strike a thermal mass located between the sun and the living space (Fig. 10.2(a ii)). The radiation, absorbed by the mass is converted into thermal energy and then transferred into the living space. The advantage of indirect gain over the direct one is that more uniform room-air temperature is achieved. The two basic indirect gain systems are (i) thermal storage walls/roof and (ii) roof ponds. The details of thermal storage walls/roof and roof ponds are discussed below:

Both walls and roof can be used but a south facing wall is preferred to the roof because of the abundance of solar radiation on the south wall in winter and the difficulty in supporting a heavy roof. A thermal storage wall absorbs sun's radiation on it's outer surface and then transfers this heat into the building through conduction. Thermal storage wall may be either masonry or water.

10.3.2.1 *Simple glazed wall*

A Trombe wall is a thick wall made of concrete, adobe, stone or composites of brick blocks and sand. In order to increase the absorption, the outer surface is painted black and are glazed.

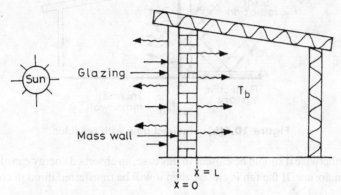

Figure 10.4(a) Schematic representation of an indirect gain system with single glazing.

For a simple glazed wall of thickness L and thermal conductivity K (Fig. 10.4(a)), the heat flux entering the room, maintained at a temperature, T_r, can be calculated by

$$\dot{q} = \alpha\tau I(t) - U_t(T|_{x=0} - T_a) \qquad (10.6)$$

where U_t is the overall heat loss coefficient from the outer surface of the wall to the ambient through glazing. In the steady state, the amount of heat flux conducted and subsequently convected and radiated into the space remains and $\dot{q}$ can be written as

$$\dot{q} = \frac{K}{L}(T|_{x=0} - T|_{x=L}) \qquad (10.7)$$

$$= h_{si}(T|_{x=L} - T_b) \qquad (10.8)$$

Solving Equations (10.6) to (10.8) gives

$$\dot{q} = U_L\left[\frac{(\alpha\tau)I(t)}{U_t} + T_a - T_b\right]$$

or

$$\dot{q} = U_L[T_{sa} - T_b] \qquad (10.9)$$

where

$$U_L = \left[\frac{1}{U_t} + \frac{L}{K} + \frac{1}{h_{si}}\right]^{-1} \qquad (10.10)$$

Figure 10.4(b) shows the description of another indirect gain system with reflector. In this case some of the reflected radiation from the horizontal surface will be also available for thermal heating. Another advantages of this reflector is to use it as a movable insulation during the night/low insulation period. Hence, the heat loss during these period from an enclosure is reduced. Mathematically the value of intensity will be increased in Equation (10.6).

The combination of direct and indirect gain has been shown in Figure 10.4(c). In this case, a small fan is provided at the top and a vent at bottom. For faster transfer of the heat, the fan should be on alongwith

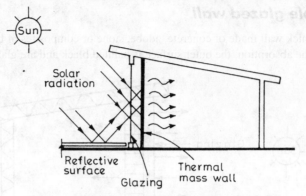

Figure 10.4(b) Indirect gain with reflector.

bottom opening. Otherwise it should be closed. In this case, an absorbed energy can also be immediately transferred to an enclosure. If the fan is closed, then it will be transferred through conduction.

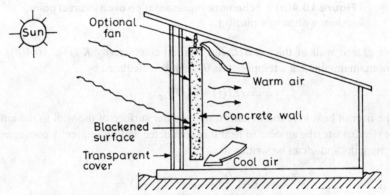

Figure 10.4(c) Combination of direct and indirect gain.

EXAMPLE 10.3

Calculate the rate of heat flow through a south-facing concrete wall with mean incident solar radiation of $250\,\text{W/m}^2$, ambient air temperature $13°\text{C}$, wall thickness 30 cm, wall conductivity $0.72\,\text{W/m}\,°\text{C}$, mean room temperature, $20°\text{C}$, $h_c = 8.7\,\text{W/m}^2\,°\text{C}$, $h_r = 3.8\,\text{W/m}^2\,°\text{C}$, $\alpha = 0.6$ and $h_{si} = 8\,\text{W/m}^2\,°\text{C}$.

Solution

First U value should be calculated by using the expression given in Equation (10.1f) and considering $h_1 = h_c + h_r = (8.7 + 3.8) = 12.5\,\text{W/m}^2\,°\text{C}$. Now,

$$\frac{1}{U} = \frac{1}{8.7 + 3.8} + \frac{0.30}{0.72} + \frac{1}{8} = 0.622$$

or

$$U = 1.609\,\text{W/m}^2\,°\text{C}.$$

From Equation (10.1c), we have

$$T_{sa} = (0.6)(250)/(12.5) + 13 = 25°C$$

Here, $\varepsilon \Delta R = 0$ for the south wall. The rate of heat flow can be obtained from Equation (10.1f) as

$$\dot{q} = 1.609 \times (25 - 20) = +8.045 \, W/m^2$$

The positive sign indicates that the heat is gained from the environment to the living space and heat has to be removed at the above rate from the living space to maintain the space at 20°C.

The same problem, if done with an ambient air temperature of 5°C gives

$$T_{sa} = (0.6)(250)/(12.5) + 5 = 17°C$$

and,

$$\dot{q} = 1.609 \times (17 - 20) = -4.827 \, W/m^2$$

This indicates that the heat has to be added at this rate to the living space to maintain the space at 20°C.

EXAMPLE 10.4

If the south wall in Example 10.3 is covered with a 4 cm thick movable night insulation (NI) with $K = 0.025 \, W/m°C$, then calculate the net heat flux into the room.

Solution

With night insulation, the overall heat loss coefficient from room to the ambient air will be

$$U = \left(\frac{1}{12.5} + \frac{0.04}{0.025} + \frac{0.30}{0.72} + \frac{1}{8} \right)^{-1} = 0.45 \, W/m^2 \, °C$$

The rate of heat loss from the room to ambient during the night will be

$$\dot{q} = 0.45(13 - 20) = 0.45 \times -7 = -3.15 \, W/m^2$$

Comparing from the result of Example 10.3, we observe the loss during the night time. Hence,

$$\text{Net rate of heat flux} = 8.045 - 3.15 = 4.895 \, W/m^2$$

This is true only when the duration of day and night are the same otherwise the gain and loss should be multiplied by the duration of day and night respectively before adding.

EXAMPLE 10.5

Consider a south facing brick wall, with $\alpha = 0.8$ and $\tau = 0.71$. Mean solar radiation on south face is $310 \, W/m^2$. If the average ambient temperature is 14°C, calculate the mean heat flux into a room maintained at 18°C for Trombe wall of 305 mm thick concrete without vents. The external wall heat transfer coefficient from glazed surface is 5 W/m² °C and $h_{si} = 8$ W/m² °C.

Solution

Using Equations (10.9) and (10.10), we have

$$\dot{q} = \left(\frac{1}{5.0} + \frac{0.305}{0.62} + \frac{1}{8} \right)^{-1} \left(\frac{0.8 \times 0.71 \times 310}{5} + 14 - 18 \right)$$

$$= 1.224(35.22 + 14 - 18) = 38.21 \, \text{W/m}^2$$

EXAMPLE 10.6(a)

For Example 10.5 and for an average ambient temperature of 8°C, calculate the night loss for the above system if the inside temperature (T_b) is maintained at 18°C and also calculate the net heat flux into the room for the same duration of day/night.

Solution

For the Trombe wall described in Example 10.5,

$$\text{Rate of night losses} = 1.22(8 - 18) = -1.22 \times 10 = -12.2 \, \text{W/m}^2$$

Since the day and night durations are same, i.e., 12 hours each, then

$$\text{Net heat flux} = \text{Rate of heat flux gain} + \text{Rate of night losses}$$

$$= 38.21 - 12.2 = 26.01 \, \text{W/m}^2$$

and   Net energy gain (in Joule) $= 26.01 \times 12 \times 3600 = 1.12 \times 10^6 \, \text{J} = 0.31 \, \text{kWh}$

EXAMPLE 10.6(b)

Calculate the net energy in terms of kWh for Example 10.6 (a) when the Trombe wall is exposed solar radiation to eight and four hours respectively.

Solution

a. For eight hour exposure:

$$\text{Net energy gain} = (38.21 \times 8 - 12.2 \times 16) \times 3600 = 3.98 \times 10^5 \, \text{J}$$

$$= 3.98 \times 10^5 \times 2.778 \times 10^{-7} = 0.11 \, \text{kWh}$$

b. For four hour exposure:

$$\text{Net energy gain} = (38.21 \times 4 - 12.2 \times 20) \times 3600 = -3.28 \times 10^5 \text{ J}$$

$$= -3.28 \times 10^5 \times 2.778 \times 10^{-7} = -0.091 \text{ kWh}$$

This indicates that there is a net gain for eight hour exposure and vice versa for four hour exposure. This means that some external thermal energy is to be fed for four hour exposure to maintain the living space temperature.

10.3.2.2 Water wall

Water wall works on the principle similar to that of a Trombe wall. Here, drums of water stacked one above the other and behind the collector glazing, are used as a storage medium (Fig. 10.5).

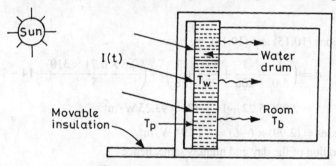

Figure 10.5 Water Wall.

From the experiments conducted using different materials, like water, concrete, bricks and hollow bricks, it was inferred by Mellogh *et al.* (1979) that the solar heating factor is a function of thickness of the storage wall and any increase in thermal conductivity results in a proportional increase in the efficiency of solar wall.

If T_p is the temperature of the metallic surface of the drums, the energy balance equations, for the steady state, can be written as:

$$\dot{q} = (\alpha\tau)I(t) - U_t(T_p - T_a) \qquad (10.11)$$

$$\dot{q} = h'_1(T_p - T_w) \qquad (10.12)$$

$$\dot{q} = h'_2(T_w - T_{si}) \qquad (10.13)$$

$$\dot{q} = h_{si}(T_{si} - T_b) \qquad (10.14)$$

here, h'_1 and h'_2 are convective heat transfer coefficient from absorber to water and water to other metallic surface.

From Equations (10.11) to (10.14), we get

$$\dot{q} = U_L(T_{sa} - T_b) \qquad (10.15)$$

where

$$\frac{1}{U_L} = \left(\frac{1}{U_t} + \frac{1}{h'_1} + \frac{1}{h'_2} + \frac{1}{h_{si}} \right) \qquad (10.16a)$$

and

$$T_{sa} = \frac{\alpha\tau}{U_t}I(t) + T_a \qquad (10.16b)$$

EXAMPLE 10.7

Consider a south-facing water wall with the parameters given in Example 10.5 with $h'_1 = 206\ \text{W/m}^2\ {}^\circ\text{C}$ and $h'_2 = 206\ \text{W/m}^2\ {}^\circ\text{C}$. Now calculate

   i. rate of mean heat flux into a room maintained at 18°C
  ii. rate of night losses from the system if the inside temperature is maintained at 18°C and the ambient temperature is 12°C
 iii. rate of net heat flux into the room, for the above system for equal duration of day and night and
  iv. rate of night losses if the glazing is covered with a 3 cm thick movable night insulation.

Solution

   i. Using Equations (10.15) and (10.16), we have

$$\dot{q} = \left(\frac{1}{5.0} + \frac{1}{206} + \frac{1}{206} + \frac{1}{8} \right)^{-1} \left(\frac{0.8 \times 0.71 \times 310}{5} + 14 - 18 \right)$$

$$= 2.99(35.22 + 14 - 18) = 93.35\ \text{W/m}^2$$

  ii. The night losses $= (2.99) \times (-6) = -17.94\ \text{W/m}^2$

 iii. Since the duration of the day and night is same, hence

Net heat flux = Heat flux gain during the day + Heat loss during the night (Equation 10.1j)

$$= 93.35 - 17.94 = 75.41\ \text{W/m}^2$$

  iv. Losses with insulation $= \left(\dfrac{1}{5} + \dfrac{0.03}{0.025} + \dfrac{1}{206} + \dfrac{1}{206} + \dfrac{1}{8} \right)^{-1} (12 - 18)$

$$= 0.65 \times -6 = -3.90\ \text{W/m}^2$$

This is clear from the above calculation that there is a reduction in night losses due to movable insulation from $-17.94\ \text{W/m}^2$ to $-3.90\ \text{W/m}^2$.

$$\text{Net heat flux} = 93.35 - 3.90 = 89.45\ \text{W/m}^2$$

If the wall exposure period is reduced from 12 hr to 8 hr, then the net energy available with movable insulation is

$$= (93.35 \times 8 - 3.9 \times 16) \times 3600 = 2.46 \times 10^6\ \text{J} = 0.68\ \text{kWh}$$

In this case it is assumed that U_t for the glazed south wall with and without movable insulation is same.

10.3.2.3 *Transwall*

A all *admitting translucent* wall (or transwall) (Fig. 10.6) is a transparent thermal storage wall admitting solar energy partially. Introduction of a layer of concrete wall behind the water mass decreases the average absorption of solar energy but it also reduces the temperature fluctuations. Thus transwall

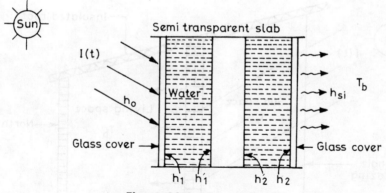

Figure 10.6 Transwall.

absorbs marginally less energy as compared to the pure Trombe wall. Transwall is more useful for heat transfer when day-time heating load is significant.

The rate of useful energy gain for a transwall can be written as

$$\dot{q} = \tau I(t) - U_t(T_b - T_a) \tag{10.17}$$

where

$$U_t = \left[\frac{1}{h_0} + \frac{L_g}{K_g} + \frac{1}{h_1} + \frac{1}{h_1'} + \frac{L_t}{K_t} + \frac{1}{h_2} + \frac{1}{h_2'} + \frac{L_g}{K_g} + \frac{1}{h_{si}}\right]^{-1}$$

and

$$\tau = (1 - R_g)(1 - \alpha_g)(1 - \alpha_w)(1 - \alpha_t)(1 - \alpha_w)(1 - \alpha_g)$$

where α_g, α_w, α_t, are respectively, the absorptivity of the glass cover, water and semitransparent material and R_g is the reflectivity of the glass cover.

Equations (10.17) can be rewritten as

$$\dot{q} = U_t \left[\frac{\tau}{U_t} I(t) + T_a - T_b\right] \tag{10.18}$$

or

$$\dot{q} = U_t[T_{sa} - T_b] \tag{10.19}$$

10.3.2.4 Solarium

A solarium is an integration of direct gain and thermal storage concepts. It consists of sun space (with thick mass wall on the south side), linking living and sun space (Fig. 10.7).

The thermal-link between the collector space and the storage mass, helps in heat retention and distribution and also enhances the efficiency of the system. The overheating of sunspace during the summer can be avoided by the use of shading. The movable insulation/shutters minimize the heat losses.

There is no direct heat transfer between the absorbing surface and ambient. The net heat flux gain for zone I, partition wall and zone II can be expressed as

$$\dot{q} = (\alpha\tau)I(t) - h_{TS}(T|_{x=0} - T_{ss}) \tag{10.20}$$

$$\dot{q} = \frac{K}{L}(T|_{x=0} - T|_{x=L}) \tag{10.21}$$

and

$$\dot{q} = h_{si}(T|_{x=L} - T_b) \tag{10.22}$$

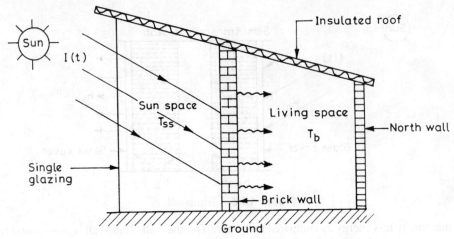

Figure 10.7 Schematic design of a solarium.

The unknown air temperature of zone 1, T_{ss}, is related to the ambient air temperature through the following relation

$$h_0(T_{ss} - T_a) = h_{TS}(T|_{x=0} - T_{ss}) \tag{10.23}$$

where h_0 is the overall heat loss coefficient from the air temperature of zone 1 to the ambient through the glazing and h_{TS} is a convective and radiative heat transfer coefficient from the wall's outside surface to sun space.

Simplification of the above equations gives

$$T_{ss} = h_m \left[\frac{1}{h_0} T|_{x=0} + \frac{1}{h_{TS}} T_a \right]$$

where

$$h_m = \left(\frac{1}{h_0} + \frac{1}{h_{TS}} \right)^{-1} \tag{10.24}$$

After substituting the value of T_{ss} in Equation (10.20), we get

$$\dot{q} = U \left[\frac{\alpha \tau I(t)}{h_m} + \frac{h_0}{h_m} \left(1 - \frac{h_m}{h_{TS}} \right) T_a - T_b \right] \tag{10.25}$$

where

$$\frac{1}{U} = \left(\frac{1}{h_m} + \frac{L}{K} + \frac{1}{h_{si}} \right) \tag{10.26}$$

Substitution of the value of h_m from Equation (10.24) in to Equation (10.25) and some algebraic simplification gives

$$\dot{q} = U \left[\frac{\alpha \tau I(t)}{h_m} + T_a - T_b \right] \tag{10.27}$$

The above equation can be written as

$$\dot{q} = U[T_{sa} - T_b] \tag{10.28}$$

where
$$T_{sa} = \frac{(\alpha\tau)I(t)}{h_m} + T_a \qquad (10.29)$$

It may be noted that the expression for T_{sa} is similar to that in Equation (10.3f).

10.3.3 Isolated Gain

Isolating the building, the collector of solar energy and the storage results in a greater flexibility in the design and operation. The most common application of this concept is the natural convective loop of which the thermosyphoning water heater (Fig. 10.8(a)) represents the simplest version.

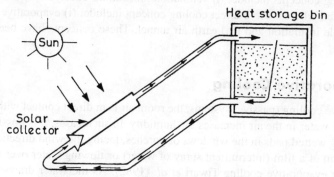

Figure 10.8(a) Isolated gain-convective loop.

Figure 10.8(b) shows the working principle of collector cum rock bed storage system integrated with apartment. Hot air from collector is allowed to enter the apartment as shows in the figure and the cooled air is returned to the collector through basement and rock bed. In the case of hot air not allowed to the apartment, the available thermal energy is stored in the rock bed for later use, preferably during the night time.

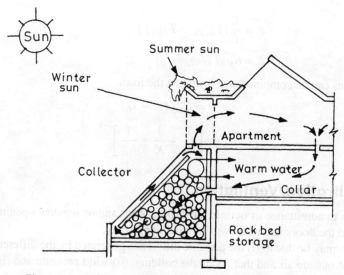

Figure 10.8(b) View of rock bed cum collector loop.

10.4 VARIOUS THERMAL COOLING CONCEPTS

Thermal cooling, for reduction of peak cooling power demand of a building, utilize a number of natural heat rejection mechanisms, viz. ventilation, evaporation, infrared radiations to the sky and earth contact cooling.

The first step in passive cooling is the reduction of unnecessary thermal loads, viz. exterior load due to climate and interior loads due to people, on the buildings. The climate dependent loads involve conduction of heat through the building skin, infiltration of outside air and penetration of short wavelength radiations directly. The cooling concepts may be direct and indirect.

The direct cooling concepts include: (i) ventilation/infiltration/ courtyard, (ii) wind tower/air vents and (iii) earth shelter, whereas the indirect cooling concept include: (i) evaporative cooling/roof pond (ii) shading/movable insulation and (iii) earth air tunnel. These concepts have been described in the subsequent sections.

10.4.1 Evaporative Cooling

In direct evaporative cooling (passive) systems, the room air is in direct contact with the water surface. The evaporation of water in the air increases the humidity. In such cases, it is possible to cool a small building by placing wetted pads in the windows or porches, facing the wind direction.

Water in the form of a film (intermittent spray of water) or flowing water over the roof, is another method of indirect evaporative cooling Tiwari *et al.* (1982). As the water draws heat from the roof surface, it leaves a cooler ceiling surface below, which acts as a radiative cooling panel for the space. The indoor temperature is lowered without elevating the humidity level. The indirect evaporative cooling by roof pond is more effective if the roof is covered with plants and movable insulation.

In the case of heating $\dot{q}$ is enhanced while it is reduced for cooling. In the case of evaporative cooling of walls/roof, the rate of heat flux reduced can be determined by using the expression

$$\dot{q} = -K \frac{\partial T}{\partial x}\bigg|_{x=0} = h_1(T_{sa} - T|_{x=0}) \tag{10.30}$$

$$\dot{q} = \frac{K}{L}(T|_{x=0} - T|_{x=L}) \tag{10.31}$$

$$\dot{q} = h_{si}(T|_{x=L} - T_b) \tag{10.32}$$

Above equations can be combined and written in the form

$$\dot{q} = U_w(T_{sa} - T_b)$$

where
$$U_w = \left[\frac{1}{h_1} + \frac{1}{K} + \frac{1}{h_{si}}\right]^{-1} \tag{10.33}$$

10.4.2 Infiltration/Ventilation

Infiltration refers to admittance of outside air through door and/or window openings and cracks, and interstices around the doors and windows, into the living space.

The infiltration may be due to (i) the pressure difference generated by the difference in temperatures and humidities of outside air and that inside the building, (ii) wind pressure and (iii) entry and exit of occupants.

Convective heat losses due to ventilation are attributable to the air exchange rate, temperature difference between inside and outside the building and heat capacity of air.

The ventilation losses can be given by,

$$Q_v = 0.33 \ N \ V(T_i - T_a) \tag{10.34}$$

where N is the number of air exchanges per hour, V the volume of the building in m^3, T_i the temperature of air inside the building in °C.

Local topography and surface texture affect the wind conditions considerably. Spacing of buildings at six times their height in a grid iron pattern results in proper wind movement with a uniform flow and removal of stagnant zones.

Windows play a dominant role in inducing ventilation. The ventilation rate is affected by climate, wind direction, size of inlet and outlet, volume of the room, shading devices and the internal partitions.

10.4.3 Wind Tower

Wind towers are designed to harness the wind, cool the air and circulate it through the building. A wind tower operates according to the time of the day and the presence or absence of wind. During the day, hot ambient air enters the tower through the openings in the sides and is cooled as it comes in contact with the cool tower. The cool air being denser than the warm air, sinks down through the tower, creating a downdraft. The draft is faster in the presence of wind. At night, the tower operates like a chimney. Heat that has been stored in walls during the day warms the cool night air in the tower, the pressure at the top of the tower being reduced, an updraft is created.

The concept of wind tower works well in individual units and not in multi-storeyed apartments.

10.4.4 Earth-air Tunnel

An earth-air tunnel exploits the constant ground temperature few meters below the surface, where it remains constant throughout the year (Table 10.2). The air passing through a tunnel or a buried pipe gets cooled in summers and heated in winters. Parameters like surface area of pipe, length and depth of the tunnel below ground, dampness of the earth, humidity of inlet air and it's velocity affect the exchange of heat between air and the surrounding soil.

TABLE 10.2 Ground temperatures for various surface conditions at a depth of 4 meters

Surface condition	Ground temperature
Dry Sunlit	27.5°C
Wet Sunlit	21.5°C
Wet Shaded	21.0°C

Figure 10.9(a) shows the cross sectional view of an earth tunnel below the ground at 4 m depth and an air available from atmosphere is allowed to pass through it. The shape of the tunnel is cylindrical with radius r and length L. As the air passes through the tunnel there is a heat transfer from inner surface of the tunnel to a flowing air by forced convection. Depending upon the air temperature, the air is either heated or cooled. If the temperature of air is below temperature of inner surface of the tunnel, then heat is transferred from surface to air for heating. This occurs in the winter season. For summer it is vice versa.

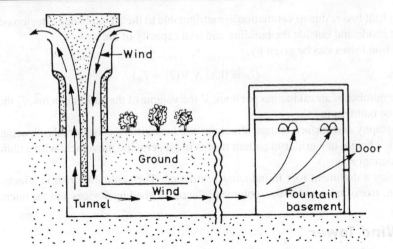

Figure 10.9(a) A combination of underground stream and a wind tower.

Referring to Figure 10.9(b), the energy balance for an elemental length dx can be written as

$$\dot{m}_a C_a \frac{dT(x)}{dx} dx = 2\pi \, rh_c(T_0 - T(x))dx \qquad (10.35)$$

where $h_c = 2.8 + 3V$, V is the speed of air flowing through the tunnel, $\dot{m}_a = \pi r^2 \rho V$ and C_a are mass flow rate and specific heat of air and $T(x)$ is the temperature of air as a function of x.

The solution of above equation with $T(x = 0) = T_{fi}$ can be written as

$$T(x) = T_0 \left(1 - e^{-\frac{2\pi rh_c}{\dot{m}_a C_a}x} \right) + T_{fi} e^{-\frac{2\pi rh_c}{\dot{m}_a C_a}x} \qquad (10.36)$$

Now
$$T_{fo} = T(x = L)$$

The rate of thermal energy carried away by the flowing air is

$$\dot{Q}_u = \dot{m}_a C_a (T_{fo} - T_{fi})$$

or
$$\dot{Q}_u = \dot{m}_a C_a (T_0 - T_{fi}) \left[1 - e^{-\frac{2\pi rh_c}{\dot{m}_a C_a}L} \right] \qquad (10.37)$$

Figure 10.9(b) Flow direction through an earth-air tunnel.

The energy available in one hour $= \dot{Q}_u \times 3600$ J.

The volume of hot air in one hour $= \pi r^2 V \times 3600$.

If V_0 is the volume of a room to be heated then, the number of air change (N) can be obtained as

$$N = (\pi r^2 V \times 3600)/V_0, \text{ number of air change per hour}$$

For $L \Rightarrow \infty$ means very large tunnel length, then

$$T_{fo} = T_0 \text{ and } \dot{Q}_u = \dot{m}_a C_a (T_0 - T_{fi})$$

This indicates the withdrawal of maximum thermal energy to heat living space.

For $L \Rightarrow 0$ means small tunnel length, then

$$T_{fo} = T_{fi} \text{ and } \dot{Q}_u = 0$$

This indicates no withdrawal of heat from tunnel and hence there is a need to optimize the length, radius and velocity of air for thermal heating/cooling of a living space. The variation of $\dot{Q}_u$ with L for winter and summer conditions for a typical set of parameters has been shown in Figure 10.9(c).

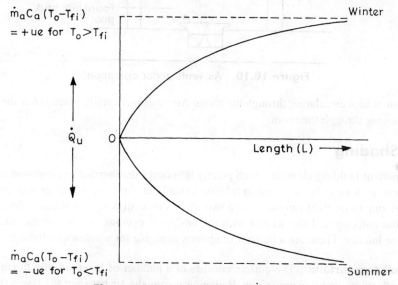

Figure 10.9(c) Variation of $\dot{Q}_u$ with L.

Figure 10.9(c) indicates that the value of $\dot{Q}_u$ will be positive for winter condition due to lower value of ambient air temperature ($T_a = T_{fi}$) than underground surface temperature (T_o). Further, the value of $\dot{Q}_u$ becomes negative due to higher value of ambient air. It is also important that the length of tunnel should be optimised for a given volume of living space.

A combination of wind-tower and earth air tunnel is effective in increasing the draft of air in the earth air tunnel for the cooling rate of building air (Fig. 10.9(a)).

10.4.5 Air Vent

Air vents are used in the areas where dusty winds make the working of wind tower impossible. These are suited for single units and work well in hot and dry and warm and humid climates. A typical vent

(Fig. 10.10) is a hole in the apex of domed or cylindrical roof with a protective cap over it. Openings in the protective cap over the vent direct the wind across it. As air flows over the curved surface, its velocity increases thus decreasing its pressure at the apex of the curved roof, thereby inducing the hot air under the roof to flow out through the vent.

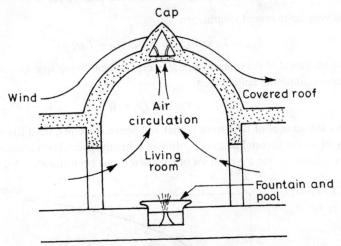

Figure 10.10   Air vents under operation.

Thus the air is kept circulating through the room. Air vents are usually placed over the living rooms to cool air moving through the room.

10.4.6 Shading

Another important building element which greatly influences the thermal environment is the window. Window openings have to be designed in relation to sunlight, ventilation and air-movement. Window designing relating to sunlight can be done in two ways: one with design of shading devices to prevent radiation from entering and the second with the design of openings to permit the adequate natural lighting of the interior. There are a number of options available for windows including:

i. **Self inflating curtain:**   The curtain consists of a number of layers of thin flexible material of high reflectivity and low emissivity. Radiation warms the air between the layers, increasing the pressure and decreasing the density in the upper part of the system. The pressure pushes the layers apart and causes fresh air intake of air from the bottom. Thus the system of reflecting layers separated by air gaps provides good insulation and when the insulation is not required the air is evacuated from the sides. The rate of energy transfer through a self inflating curtain is given as

$$\dot{q} = (1 - f)\tau I(t) - U(T_b - T_a) \tag{10.38}$$

where f is the shading factor having a value 1 for complete shading and less than 1 for partial shading.

ii. **Window quilt shade:**   The quilt consists of a sandwich of fine layers, assembled with an ultra-sonic fibre welder. The quilt is enclosed in decorative polyester fibre.

iii. **Venetian blind between the glasses:**   This is an effective system to reduce the heat loss through a double glazed window. In this case, the characteristic dimension of the unit is small, hence the convective heat transfer is stopped. In this case, $\dot{q}$ is the same as that in Equation (10.4a) with lower value of τ and U.

iv. **Transparent heat mirrors:** A method to reduce the heat loss from the glazed surface is to coat the glass by a film which to a large extent reflects the infra-red radiation from the surface. However, this coating also reduces the transmissivity of the glass for solar radiation, and a suitable compromise has to be made. The coating may consist of single or multiple layers of different substances, deposited by vacuum evaporation or spray technique. The heat mirror gives much less heat loss and a higher transmission than multipane system.

v. **Heat trap:** To reduce the heat transfer, a reasonable thickness of insulating material with good transmissivity may be introduced.

vi. **Optical shutter:** An optical shutter consists of three layers of transparent sheets and one layer of cloud gel. It is opaque at high temperatures and highlight intensities. It can be used for reducing air-conditioning loads and preventing overheating in green houses and solar collector systems.

Till now, we have discussed the design of windows in relation to solar radiation. Another function, ventilation is also performed by them. All the three basic functions of ventilation, viz. (i) inflow of fresh air to replace the used internal air and removal of the products of combustion of fuel-less cooking and gas cooking, (ii) to cool the body by increasing evaporation of moisture from the skin, and (iii) heating or cooling the interior of the building, may be achieved in three different ways, the stack effect (temperature difference), wind pressure and mechanical means respectively.

10.4.7 Rock Bed Regenerative Cooler (*Sodha et al., 1986*)

A cross sectional view of a regenerative cooler is shown in Figure 10.11. It uses two rock beds set side by side as shown in the figure. It also acts as a heat exchanger. It is separated by an air space. A damper has been used between two rock beds to divert the incoming air from house towards cooling rock bed by water spray. The rock beds are cooled alternatively. It also absorbs water during cooling. The air passing through a dry rock bed (already cooled in the first cycle) is cooled by transferring its heat to the rock bed and the cooled air is allowed to pass into the room. The humid air produced during its evaporation cycle from rock bed is vented to the outside. After getting the rock bed warm, the damper is reversed for further cooling. In the meantime other cooled rock bed is used for cooling similarly.

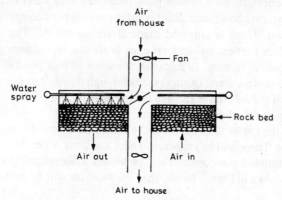

Figure 10.11 A regeneration cooler.

10.4.8 Radiative Cooling

As we have seen that the sky temperature is always lower than ambient air temperature by 12°C (Equations (2.56) and (2.57)) with clear night sky. The northern sky (in the northern hemisphere) is often cool enough even during the day. It acts as a cooler heat sink with respect to ambient air the day/night. A horizontal surface is the most effective radiative configuration.

Exposed horizontal surface losses heat to an ambient air by convection and radiation till its temperature reaches equals to dry bulb air temperature. Also, there is heat losses from ambient air/surface to the sky by long wave length radiation exchange ($\varepsilon \Delta R$, Chapter 2). If the net heat exchange between surface and sky reduces the roof surface temperature to the wet bulb temperature of surrounding air, condensation of moisture of air starts taking place on the roof. The condensation will further cool the roof due to fast heat loss from enclosed room to surface of the roof.

If surface is inclined as shown in Figure 10.12, then the cooled air from the inclined surface will trickle down towards an internal courtyard due to its high density. Then the trickled cooled air enters the room through the openings at lower level as shown in the figure. However, this does not work in the windy condition because the winds becomes the carrier of cooled air. To avoid this, the roof should be covered with transparent polyethylene sheet. The polyethylene sheet also has limited life. This problem can be solved by covering the roof by corrugated metallic sheet with the provision of an openings at lower and upper end from inside the roof as shown in Figure 10.13.

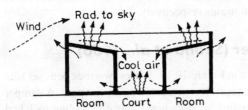

Figure 10.12 Cooling by radiation through open loop.

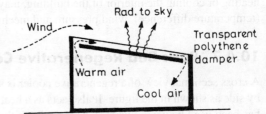

Figure 10.13 Cooling by radiation through closed loop.

10.4.9 Heating and Cooling

Figure 10.14 shows a schematic view of roof pond integrated with water circulating column working under natural (thermosyphan) flow mode. Roof pond is covered permanently with thin heat exchanger membrane (copper plate). There is louvered shade above membrane. The louvered shade is closed during night time to reduce the heat loss and it is open to allow solar radiation to fall on roof pond. This is precisely done for thermal heating. In this case the water of roof pond is heated. The hot water is allowed to pass through down-comer. During passes through down-comer, it releases heat to the water outside and cooled down towards lower end due to high density. Since hot water is in contact with room air, therefore thermal enegy is released to the room.

For thermal cooling, the louvered shade should be closed during the day to cut off solar radiation and opened during the night. There will be evaporative heat loss from water. The water will be cooled. The cold water will pass through down-comer. It will be heated while returning to roof pond by absorbing heat from room air as shown in Figure 10.14. Thus the room air will be cooled.

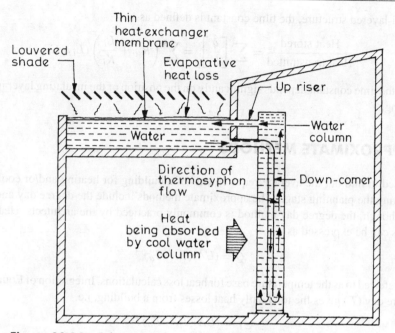

Figure 10.14   Schematic view of roof pond with thermosyphon flow.

10.5   TIME CONSTANT

For most of the commonly encountered thermal structures the boundary conditions change constantly. As such, the steady state methods give only the approximate solutions. The exact analyses of the heat transmission in transient conditions is a complex one. An approximate estimation of the time dependent response of thermal structures can be done by the assumption that the temperature, at any instant, is constant throughout the structure. If in time dt, there is a small change dT in the temperature, the change in the internal energy is equal to the net heat flow rate across the boundary, i.e.

$$\rho C V \, dT = h' A (T - T_a) dt \tag{10.39a}$$

where h' is the heat transfer coefficient from the building to ambient. The solution of Equation (10.39a) gives

$$\left[\frac{T(t) - T_a}{T_0 - T_a} \right] = \exp \left\{ -t \left(\frac{h' A}{\rho C V} \right) \right\} \tag{10.39b}$$

where $(\rho C V / h' A)$ is called the *time constant* of the system. The time constant gives the time taken for the difference between the system temperature and that of the surroundings to change to 36 percent of the initial value.

If the quantity $(h' L / K)$, Biot number, is less than 0.1, the error introduced by assuming that the temperature is uniform throughout the structure, at any instant, is less than 5 percent.

For a multi-layered structure, the time constant is defined as,

$$\frac{\text{Heat stored}}{\text{Heat transmitted}} = \sum_i \left[\frac{\dot{q}}{U}\right]_i = \sum_i \left(R_{so} + \frac{L_i}{K_i}\right)(L_i \rho_i C_i) \tag{10.39c}$$

The value of the time constant depends significantly on the position of the insulating layer in the element of the building.

10.6 APPROXIMATE METHODS

These are used to find out the average requirements of a building for heating and/or cooling purposes and used during the planning stage. The approximate methods include the degree day and steady state methods, although, the degree day method is commonly practiced by the architects. Heat losses from the buildings can be expressed as

$$\dot{Q} = (UA)(T_b - T_a) \tag{10.40a}$$

where T_b is referred to as the temperature base for heat loss calculations. Integration of Equation (10.40a) over an entire day (T) gives the total daily heat losses from a building, i.e.

$$Q_T = (UA)(T_b - \bar{T}_a)T \tag{10.40b}$$

where $\bar{T}_a$ is the average daily temperature.

The term over the entire day period is evaluated only when it is positive, i.e. when $T_a < T_b$. The number of degree days per month (°C hr) are defined as

$$DD = (T_b - \bar{T}_a)T n_{d'} \tag{10.41}$$

where $n_{d'}$ is the number of days in a month for which the heating is required.

Monthly space heating load Q_m is calculated from the formula,

$$Q_m = Q_T \times n_{d'} = (UA)DD \tag{10.41a}$$

The product of building overall energy loss coefficient and the area, UA, can be determined in several ways.

For structures whose conventional fuel requirements are known, UA can be calculated as the amount of energy required to heat the building for a given period (considering heating value of fuel H_F and the furnace efficiency (η_F)) divided by the total number of degree days during that period, thus,

$$UA = (M_F \times H_F \times \eta_F)/DD \tag{10.42}$$

where M_F is the mass of fuel.

The value of H_F and η_F for different fuels is given in Appendix IV.

EXAMPLE 10.8

Find the overall heat loss coefficient of a building if the number of degree days in the month of January are 750×24°C-hr and 90 liters of oil has to be used to keep the space comfortable.

Solution

Using Equation (10.42), $UA = M_F \times H_F \times \eta_F / DD$.

After substituting the known parameters from Appendix IV in the above equation, we have

$$UA = 90 \times 43000 \times 80/(750 \times 100 \times 24)$$

$$= 172 \, \text{kJ/h°C} = 47.8 \, \text{W/°C}$$

EXAMPLE 10.9

A resident at Leh has $(UA) = 500 \, \text{W/°C}$. If the degree days for January are $750 \times 24°\text{C-hr}$, calculate the total space heating load for January.

Solution

From Equation (10.41a), we get

$$\text{Space heat load} = 500 \times 750 \times 24 \times 3600 \, \text{J} \quad (1 \, \text{hr} = 3600 \, \text{sec})$$

$$= 32.4 \, \text{GJ}$$

10.7 SOLAR-LOAD RATIO METHOD

The solar-load ratio (SLR) method, widely used for designing direct-gain, collector-storage wall and sunspace systems was developed by Balcomb *et al.* (1980, 1983a, b). It is a method of calculating annual requirements of auxiliary energy, based on the simulation studies of passive heating systems performance. The SLR method is based on a set of defined terms that must be properly used for the correlations to work. The important definitions are as follows:

Solar wall: This is the glazed building wall providing the solar gains that are to be estimated.

Solar aperture: It is the glazed portion of the wall that admits solar radiation.

Net glazing area (A_r): It is the area of the solar aperture after mullions, framing, etc., have been subtracted.

Projected area (A_p): The projected area is the projection of the net glazing area on a vertical plane normal to the azimuth of the glazing. It is to be noted that for direct-gain and collector storage wall, A_r and A_{rp} are the same; whereas, for sunspaces, A_r will be substantially larger than A_{rp}.

The SLR correlations do not include the losses through the solar aperture.

Net reference load (L_{ns}): It is the monthly heat loss from the nonsolar portions of the building, i.e. from all of the building envelope except the solar aperture.

Net load coefficient: It is the net reference load per degree of temperature difference between indoors and outdoors.

The net reference load (in Wh) for the month is,

$$L_{ns} = 24(UA)_{ns}(DD) \tag{10.43}$$

where $(UA)_{ns}$ is in W/°C and DD is the number of degree-days in the month (°C-day).

Gross reference load L: It is the heat loss from the building, including both solar and nonsolar portions of the buildings.

Total load coefficient: It is the gross reference load per degree of difference between indoor and outdoor temperature, i.e., a UA for the whole building

$$L = 24(UA)(DD) \tag{10.44}$$

Solar savings fraction (f_{ns}): It is the fraction of the net reference load that is met by solar energy.

$$f_{ns} = \frac{L_{ns} - L_A}{L_{ns}} = 1 - \frac{L_A}{L_{ns}} \tag{10.45}$$

where L_A is the auxiliary energy required for each month and is given by

$$L_A = L_{ns}(1 - f_{ns}) \tag{10.46}$$

Load collector ratio (LCR): It is the ratio of the net load coefficient to the projected area of the solar aperture.

$$LCR = 24(UA)_{ns}/A_{rp} \tag{10.47}$$

Load collector ratio for the solar aperture (LCR_s): It is the ratio of the loss coefficient of the solar aperture (based on the projected area) to the projected area.

The unutilizability method (the "double-U method") for design of direct-gain and collector-storage wall systems is based on the concept that a passively heated building can be considered as a collector with finite heat capacity. The analyses are done for estimation of the auxiliary energy requirements in two limiting cases. The first, is for an infinite capacitance structure which can store all energy, in excess of loads. The second is for a zero capacitance structure that can store no energy. Correlations are then used for the auxiliary energy requirements of a real structure, that will lie somewhere in between these two limits.

Figure 10.15 shows the monthly energy streams entering and leaving a direct-gain structure.

Solar energy absorbed in the room is,

$$\bar{I}_T N(\overline{\alpha\tau})A_r = N\bar{S}A_r \tag{10.48}$$

where N is the number of days of the month and $\bar{S} = \overline{\alpha\tau} \cdot \bar{I}_T$.

The energy lost through the building shell by conduction, infiltration etc. is given as the load. When the solar energy is insufficient to meet loads, auxiliary energy L_A must be supplied. When there is excess absorption of solar energy, as compared to load and this energy cannot be stored; then this Q_D must be vented or "dumped". The diagram does not show the energy stored; at any time during a month sensible heat may be stored in or removed from the structure if it has a thermal capacitance.

Let us consider a hypothetical building with infinite storage capacity. During a month, all absorbed energy in excess of the load is stored in the structure. The temperature of the conditioned space remains

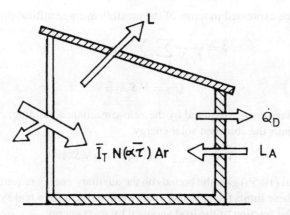

Figure 10.15 Monthly energy streams for a direct-gain building.

constant, as infinite storage capacitance means zero temperature change. This stored energy can be used at any time to offset auxiliary energy needs.

A monthly energy balance gives the auxiliary energy requirement $L_{A,i}$ for the infinite capacity buildings, is given as,

$$L_{A,i} = (L - N\bar{S}\,A_r)^+ \tag{10.49}$$

where L is the load for the entire building, i.e. the auxiliary energy required if the transmittance of the glazing is zero. The '+' sign indicates that only positive values of the difference of monthly energy quantities in the bracket are considered.

Now, let us consider a hypothetical zero storage capacity building. As there is no storage capacity, the energy deficits must be met with by the use of auxiliary energy and excess solar energy is dumped as it is absorbed.

The temperature in the building is again fixed, but by addition or removal of energy rather than storage.

An instantaneous energy balance on the structure gives the rate at which energy must be removed from the structure, $\dot{Q}_D$

$$\dot{Q}_D = [I(t)(\alpha\tau)A_r - (UA)_h(T_b - T_a)]^+ \tag{10.50}$$

A critical radiation level may be defined as that at which the gains just offset the losses:

$$I_{T_c} = (UA)_h(T_b - T_a)/(\alpha\tau)A_r \tag{10.51}$$

The absorbed radiation above this level must be dumped and is "unutilizable". The dumped energy for the month, Q_D can be written as

$$Q_D = A_r(\overline{\alpha\tau}) \int_{month} (I_T - I_{TC})^+ dt \tag{10.52}$$

Over a month, I_{TC} can be considered to be constant at its monthly average value given by,

$$I_{TC} = (UA)_h(T_b - \bar{T}_a)/(\alpha\tau)A_r \tag{10.53}$$

Energy below I_{TC} is useful and above it must be dumped.

Equation (7.52) can be expressed in terms of the monthly average utilizability Φ, defined as

$$\bar{\phi} = \frac{1}{\bar{I}_T N} \sum_{\text{days}} \sum_{\text{hours}} (I_T - I_{TC})^+ \tag{10.54}$$

and

$$Q_D = N \bar{S} A_r \bar{\phi} \tag{10.55}$$

The amount of auxiliary energy required by the zero-capacitance building, can be given as the load plus dumped energy minus the absorbed solar energy.

Thus,

$$\bar{L}_{A,Z} = L - (1 - \bar{\phi}) \bar{N} \bar{S} A_r \tag{10.56}$$

Equations (10.49) and (10.56) give the bounds on the auxiliary energy requirements. Correlations, to indicate where within these limits the auxiliary energy requirement for a real system will be, have been developed in terms of the fraction of the load supplied by solar energy

$$f = 1 - L_A/L \tag{10.57}$$

A solar-load ratio is defined as,

$$X = \frac{\bar{I}_T N (\overline{\alpha\tau}) A_r}{L} = \frac{N \bar{S} A_r}{L} \tag{10.58}$$

For infinite-capacitance system, X is the solar fraction f_i

$$f_i = (1 - L_{A,i}/L) = X \tag{10.59}$$

For zero-capacitance case,

$$f_z = (1 - \bar{\phi}) X \tag{10.60}$$

The monthly solar fraction, f, was found to correlate with the solar-load ratio X and a storage-dump ratio, is given by,

$$Y = \frac{C_b \Delta T_b}{\bar{I}_T (\overline{\alpha\tau}) A_r \bar{\phi}} = \frac{C_b \Delta T_b}{\bar{S} A_r \bar{\phi}} = \frac{N C_b \Delta T_b}{Q_D} \tag{10.61}$$

where C_b is the effective thermal capacitance, ΔT_b the difference in upper and lower temperatures (i.e. the range over which the building temperature is allowed to vary) and Y is dimensionless.

PROBLEMS

10.1 Name the three most important parameters that determine thermal comfort.
  Hint Refer Figure 10.1(a).

10.2 What is the unit of illumination?
  Hint Lux, 1 Lux $= 10^{-2}$ W/m^2.

10.3 Calculate U value for a multilayered exposed wall (shown below) for the following given specifications:

$$L_1 = L_2 = L_3 = 5\,\text{cm}, \; K_1 = K_2 = K_3 = 0.67\,\text{W/m}^\circ\text{C}, \; h_i = 5.7\,\text{W/m}^2\,{}^\circ\text{C}, \; h_0 = 9.5\,\text{W/m}^2\,{}^\circ\text{C}$$

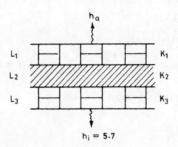

10.4 Calculate U value for the following wall configurations:
   a. glass wall, $L_g = 5\,\text{mm}, \; K_g = 0.78\,\text{W/m}\,{}^\circ\text{C}$
   b. brick wall, $L = 15\,\text{cm}, \; K = 0.67\,\text{W/m}\,{}^\circ\text{C}$
   c. glazed brick wall, $L = 15\,\text{cm}, \; K = 0.67\,\text{W/m}\,{}^\circ\text{C}$
   d. water wall, $h_1' = h_2' = 100\,\text{W/m}^2\,{}^\circ\text{C}$
   e. trans wall, $h_1 = h_1' = h_2 = h_2' = 100\,\text{W/m}^2\,{}^\circ\text{C}$
      $h_0 = 15\,\text{W/m}^2\,{}^\circ\text{C}$ for bare wall and $7\,\text{W/m}^2\,{}^\circ\text{C}$ for glazed wall and $h_{si} = 5.7\,\text{W/m}^2\,{}^\circ\text{C}$.
   Hint Use Equations (10.10), (10.16), (10.17) and (10.26).

10.5 Calculate the total heat transfer coefficient for wetted surface, having temperature 35°C, exposed to ambient air temperature at 32°C and relative humidity of 50 percent.

$$P(T) = \exp\left(25.317 - \frac{5144}{T + 273}\right)$$

   Hint Given $\varepsilon_1 = \varepsilon_2 = 0.9, \; \sigma = 5.67 \times 10^{-8}\ \text{W/m}^2\,\text{K}^4, \; h_{ca} = 2.8\,\text{W/m}^2\,{}^\circ\text{C}$. See Example 10.1.

10.6 Calculate solair temperature for wetted and glazed roof for $I(t) = 300\,\text{W/m}^2$. Given $\alpha\,(\text{wetted}) = 0.9, \alpha\tau\,(\text{glazed}) = 0.8$.
   Hint Use Equations (10.2c) and (10.3f). Take h_1 for wetted surface from Problem 10.5 and U_t for glazed surface from Example 3.1b.

10.7 Derive an expression for energy saved during sunshine hour for glazed Trombe and water wall.
   Hint See Section 10.3.1.

10.8 Calculate the rate of heat lost from a room maintained at 20°C through different walls of Problems 10.3 and 10.4 exposed to an ambient air temperature of 10°C.
   Hint Use $\dot{q} = U(20 - 10)$.

10.9 Find out the net rate of energy saved through different walls of problems 10.3 and 10.4 exposed to an average solar intensity of $350\,\text{W/m}^2$, for a duration of 8 hours. Given $\alpha = \tau = 0.9, \; T_b = 20°\text{C}, T_a = 10°\text{C}$.
   Hint See Example 10.2 and use Equation (10.5).

10.10 Write down an expression for overall heat transfer coefficient, for double glazed wall with an air gap of air conductance C.
   Hint

$$U = \left[\frac{1}{h_0} + \frac{L_{g1}}{K_{g1}} + \frac{1}{C} + \frac{L_{g2}}{K_{g2}} + \frac{1}{h_i}\right]^{-1}$$

10.11 Derive an expression for the rate of net heat transferred into the room having south wall as water wall or brick wall.

Hint See Sections 7.3.2(i) and (ii).

10.12 Derive Equation 10.34.

Hint $\dot{Q}_v = \dot{m}_a \, C_a (T_i - T_a)$.

$$\text{Consider,} \quad \dot{m}_a = \frac{V\rho}{t} = \frac{V \times 1.2}{(3600/N)}, \, Ca = 1000 \, \text{J/kg}^\circ\text{C}, \rho = 1.2 \, \text{Kg/m}^3$$

and use the definition of number of air change.

CHAPTER 11

Other Applications

11.1 COLLECTION-CUM-STORAGE WATER HEATER

The collection-cum-storage water heater which can further be classified as: (i) built-in storage water heater and (ii) shallow solar pond water heater is described in detail as follows:

11.1.1 Built-in Storage Water Heater

A collection-cum storage water heater combines both collection and storage in the same unit, thus eliminating the need of a separate insulated tank for storage of hot water. These water heaters employ water as a transport fluid for energy transfer from the collection unit to the storage one. Water, apart from being a low cost and easily available fluid, has many advantageous thermochemical properties such as non toxicity, non flammability, high specific heat, good heat transfer and fluid dynamic characteristics and has a suitable liquid-vapor equilibrium temperature-pressure relationship for space heating and cooling applications.

The built-in storage water heater are compact in design, have a low cost and are easy to install. In addition, they have a good collection efficiency and satisfactory overnight thermal storage, (Sodha et al., 1979a).

A built-in storage water heater (Fig. 11.1) consists of a rectangular tank covered with 5 cm thick glass wool insulation or air gap with reflecting sheet on sides and bottom. The tank is usually made of galvanized iron (typical size of the tank being (1.12 m × 0.08 m × 0.1 m)) and encased in a mild iron or wooden box (typically 1.22 m × 0.9 m × 0.2 m). The top surface of the tank is sprayed with black board paint and is covered with one or two window glass covers (3 mm thick); with about 4 cm air gap between the absorbing surface and the glass cover. In order to avoid bulging of the tank due to enormous water pressure, braces are provided.

Cold water is fed through the inlet and hot water is withdrawn from the outlet. In case, there is no provision of piped water supply, hot water can be withdrawn by adding cold water into the inlet through a large funnel fixed at the top of the heater.

During off-sunshine hours, the major thermal losses in a built-in storage water heater are radiative and convective losses from the top surface. As such, the heater cannot provide hot water in the early morning. A simple method to improve the night time performance is to cover the glass with an adequate thickness

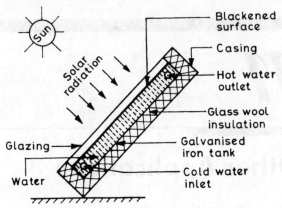

Figure 11.1 Built-in storage water heater.

of glass wool insulation or to use a reflector with an air gap. The water is allowed to heat up during the day, and the top of the tank is covered with insulation when the thermal collection reduces to zero.

EXAMPLE 11.1

Calculate convective heat transfer coefficient from blackened surface to water mass (at 20 °C) for a built-in storage water heater for an effective area of 1 m × 1 m.

Solution

The properties of water at 20 °C are (See Appendix III):
$v = 1.006 \times 10^{-6}$ m^2/s, $K = 0.5967$ W/mK and $Pr = 7.02$.
Substituting these values in Equation (2.38d), the value of Grashof number is obtained as,
$Gr = (9.8 \times 1 \times .1)/(298 \times (1.006 \times 10^{-6})^2) = 3.2495 \times 10^9$.
The value of Nusselt number can be obtained from Equation 2.41, by using the values of C and m from Table 2.2. Thus, $C = 0.15$ and $m = 1/3$.
Thus, $Nu = 425.4105$
Now, $Nu = h \, X/K$, where, X is the characteristic dimension and is given as, $(1 + 1)/2 = 1$.
Thus, $h = 253.84$ W/m^2 °C, i.e., the heat transfer coefficient from blackened surface to water mass is 253.84 W/m^2 °C.

EXAMPLE 11.2

Calculate the overall heat transfer coefficient from water to ambient through blackened surface and glass cover. Given water temperature is 70°C and heat transfer coefficient from blackened surface to water is 100 W/m^2 °C.

Solution

The heat transfer coefficient from blackened surface to glass cover (see Example 3.2) is $8.9\,\text{W/m}^2\,^\circ\text{C}$. The heat transfer coefficient from glass cover to ambient is $17.1\,\text{W/m}^2\,^\circ\text{C}$ (see Example 9.4). The overall heat transfer coefficient from water to ambient

$$U_t = \left[\frac{1}{100} + \frac{1}{8.9} + \frac{1}{17.1}\right]^{-1} = 5.53\,\text{W/m}^2\,^\circ\text{C}$$

Energy balance

In order to write down energy balance equation for built-in-storage water heater (Fig. 11.1), it is fair to assume that the temperatures of absorbing plate and water mass are same.

The energy balance will be

$$A_c[\dot{q}_{ab} - U_t(T_w - T_a)] = M_w C_w \frac{dT_w}{dt} + U_b(A_c + A_e)(T_w - T_a) \tag{11.1a}$$

The above equation is written with the help of Equations (3.23) and (3.43a) respectively. Here, A_e is side area of water heater depending on thickness of water column.

The above equation can be written as,

$$\frac{dT_w}{dt} + a\,T_w = f(t) \tag{11.1b}$$

where

$$a = \frac{A_c(U_t + U_b) + A_e U_b}{M_w C_w}$$

and

$$f(t) = \frac{A_c \dot{q}_{ab} + [A_c(U_t + U_b) + A_e U_b]T_a}{M_w C_w}$$

The solution of Equation (11.1b) can be written as

$$T_w = \frac{\overline{f(t)}}{a}[1 - \exp(-a\Delta t)] + T_{w0}\exp(-a\Delta t) \tag{11.1c}$$

or

$$\Delta t = -\frac{1}{a}\ln\left[\frac{T_w - \frac{f(t)}{a}}{T_{w0} - \frac{f(t)}{a}}\right] \tag{11.1d}$$

For detail derivation, see Section 9.9.4.

11.1.2 Shallow Solar Pond Water Heater

Shallow solar pond (SSP) is a simple and cost-effective device to harness the solar energy for industrial applications.

A shallow solar pond differs from a non-convective salt gradient solar pond (to be discussed later). The former is shallow and convective so that there is hardly any temperature stratification while the latter is of large depths and non-convective. A SSP consists of a blackened tray holding some water in it, the depth being very small. The evaporative cooling in a SSP is suppressed by covering water by means of a transparent plastic film in such a way that the film comes in contact with the top surface of water. The incident energy is absorbed at the blackened bottom and transferred to the water column due to convection. It can, thus, heat large quantities of water to appreciable temperature (Fig. 11.2(a)).

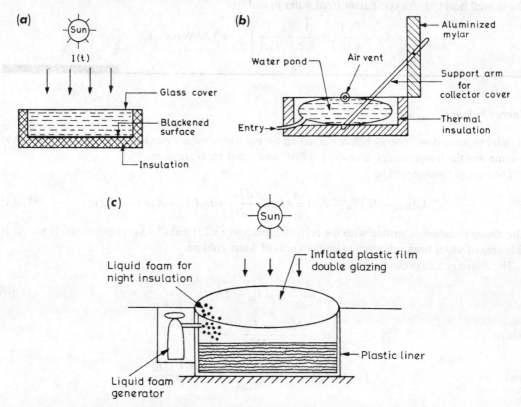

Figure 11.2   Schematic diagram of a compact SSP. (a) Without reflector (b) With reflector and (c) With liquid foam night insulation.

The solar energy collection efficiency of a SSP is directly proportional to the water depth inside, whereas the grade of thermal collection (i.e. the temperature) is inversely proportional to the water depth. In a typical design, a compact shallow solar pond consists of a pillow type of water bag, encased in a wooden box, the top being transparent, the bottom is painted black and the walls and bottom are insulated. The system has a thermally insulated cover at the top, which can be used as a booster mirror during the day, and as an insulation cover during off sunshine hours providing a means of overnight storage (Fig. 11.2(b)).

An another deep saltless SSP water heater is shown in Figure 11.2(c). A spray foam insulation has been used between glazings and between glazing and pond during low level. Due to this, there is reduction of heat losses by 50 percent. It should be noted that the spray foam used is a material normally used for fire fighting.

An efficient performance of the SSP water heater demands continuous withdrawal of hot water from the pond, to reduce the thermal losses (Sodha et al., 1981d).

Three modes of hot water withdrawal from a SSP have been suggested: batch withdrawal, closed cycle continuous flow and open cycle continuous flow. In the batch withdrawal mode, the pond is filled early in the morning and is emptied into an insulated storage reservoir in the afternoon, when the temperature is maximum. In the closed cycle mode, water is circulated continuously at a constant rate between the pond and the storage, heat may or may not be extracted from the storage by means of a heat exchanger. In this case also, the pond water is transferred to storage in the afternoon. In the open cycle mode, water at an initial temperature enters the pond and the hot water flowing through the pond is drained out continuously into a storage or for some other end use application. The pond is emptied in the evening when the heat collection reduces to zero.

Energy balance

If an inclination of glass cover of solar still (Fig. 9.3(*a*)) becomes zero (horizontal glass cover as shown in Figure 11.2(*b*)) and no space is allowed between the glass cover and water surface for evaporation, then Equations (9.41–9.43) will be used for evaluation of performance of SSP water heater. In this case h_{1w} will be only convective heat transfer coefficient between water and glass cover. Radiation and evaporative heat losses will be suppressed.

The water temperature can be determined by using Equation (9.49) as

$$T_w = \frac{\overline{f(t)}}{a}[1 - \exp(-a\Delta t)] + T_{w0}\exp(-a\Delta t) \tag{11.2}$$

The expression for parameters [a and $\overline{f(t)}$] is given in Section 9.9.3.

EXAMPLE 11.3(a)

Derive an expression for time interval Δt for Equation (11.2).

Solution

Equation (11.2) can be rewritten as

$$\exp(-a\Delta t) = \frac{T_w - \dfrac{\overline{f(t)}}{a}}{T_{w0} - \dfrac{\overline{f(t)}}{a}}$$

taking log of both sides

$$-a\Delta t = \ln\left[\frac{T_w - \dfrac{\overline{f(t)}}{a}}{T_{w0} - \dfrac{\overline{f(t)}}{a}}\right]$$

or

$$\Delta t = -\frac{1}{a}\ln\left[\frac{T_w - \dfrac{\overline{f(t)}}{a}}{T_{w0} - \dfrac{\overline{f(t)}}{a}}\right]$$

The heater can be designed to optimize the performance. The water temperature is observed to decrease with increasing water depth. The increase in depth of the tank increases the water mass and hence the heat capacity per unit area.

As seen from Figure 11.3, as the depth increases the collection efficiency increases due to the decrease of the thermal losses to the outside air. However, the rise in efficiency is fast, upto a depth of 10 cm and above that the rise in efficiency is negligible. It can, thus, be concluded that a 10 cm depth gives the optimal performance.

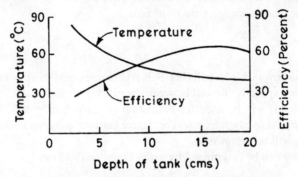

Figure 11.3 Effect of storage tank depth on storage temperature and efficiency.

In order to reduce the losses from top and bottom, we have to consider the effect of insulation thickness on the transient rise of water temperature. It has been seen that a 5.0 cm thickness of insulation is optimum for minimum heat loss from the bottom and top surfaces. Also, the effect of a reflecting sheet with an emissivity of 0.03 and an air gap of 5 cm is equivalent in performance to a 5 cm thick layer of glass wool insulation. The top reflecting surface can also be used as a booster mirror to increase the insulation on the absorber surface.

An overall efficiency of a collection-cum-storage water heater can be defined as follows:

$$\eta_{ov}(\%) = \frac{M_w C_w (T_{w(\text{max})} - T_{in})}{A_c \Sigma I_i(t) \Delta t} \times 100 \tag{11.3}$$

EXAMPLE 11.3(b)

Calculate convective heat transfer coefficient from the blackened surface to water (60°C) for SSP water heater of area 1 m × 1 m.

Solution

The properties of water at 60°C are (see Appendix III):
$\nu = 4.775 \times 10^{-7} \text{ m}^2/\text{s}$, $K = 0.6503 \text{ W/m K}$ and $Pr = 3.02$
From Equation (2.38d), Grashof number is given as,
$Gr = (9.8 \times 1 \times 0.1)/(333 \times (4.775 \times 10^{-7})^2) = 1.291 \times 10^{10}$

From Equation (2.41), Nusselt number can be calculated by using the values of C and m from Table 2.2.
Thus, $Nu = 0.15 \, (3.898 \times 10^{10})^{1/3} = 508.595$.
Now, $h = Nu \, K / X = 330.74 \, \text{W/m}^2 \, ^\circ\text{C}$.
Thus, the heat transfer coefficient from blackened surface to water is equal to $330.74 \, \text{W/m}^2 \, ^\circ\text{C}$.

EXAMPLE 11.4(a)

What will be top loss coefficient if SSP water heater is covered with 10 cm thick glass wool insulation during off-sunshine hours.

Solution

The top loss coefficient if SSP water heater is covered glass wool insulation would be given by,

$$U_t = \left[\frac{1}{h_2} + \frac{L_i}{K_i} \right]^{-1}$$

For glass wool of thickness (L_i) 10 cm, thermal conductivity (K_i) is 0.0541 W/mK.
h_2 is the heat transfer coefficient from glass wool to ambient and is given by Equation (2.49a). Its value, for $V = 3$ m/s, is 17.1 W/m^2 °C.
Thus,

$$U_t = \left[\frac{1}{17.1} + \frac{0.1}{0.0541} \right]^{-1} = 0.5247 \, \text{W/m}^2 \, ^\circ\text{C}.$$

EXAMPLE 11.4(b)

Calculate time taken for the water temperature to rise to 40°C from an initial value of 20°C for built-in storage water heater. Given $\alpha\tau = 0.8$, $\overline{I(t)} = 900 \, \text{W/m}^2$, $U_t = 5.53 \, \text{W/m}^2°\text{C}$ (Example 11.2), $U_b = 0.89 \, \text{W/m}^2°\text{C}$ (Example 3.3), $A_c = 1 \, \text{m}^2$, $A_e = 0.4 \, \text{m}^2$, $\bar{T}_a = 8.5 \, ^\circ\text{C}$, and $(MC)_w = 41900 \, \text{J/kg}°\text{C}$.

Solution

From Equation (11.1b), we have

$$\frac{\overline{f(t)}}{a} = \frac{A_c}{A_c(U_t + U_b) + A_e U_b} \propto \tau \overline{I(t)} + \bar{T}_a$$

$$= \frac{1}{(5.53 + 0.89) + .4 \times 0.89} \times 0.8 \times 900 + 8.5 = 114.757$$

and
$$a = \frac{6.775}{41900} = 1.617 \times 10^{-4}.$$

From Equation (11.1d), we get

$$\Delta t = -\frac{1}{1.617 \times 10^{-4}} \ln \left[\frac{40 - 114.77}{20 - 114.77} \right] = -\frac{1}{1.617 \times 10^{-4}} \ln \frac{74.77}{94.77}$$

$$= -\frac{1}{1.617 \times 10^{-4}} \times -0.237 = 1465.9 \sec = 24.43 \min$$

EXAMPLE 11.4(c)

Calculate Δt for SSP water heater (Fig. 11.2(b)). Given: $(\alpha\tau)_{\text{eff}} = 0.65$, $U_t = 14.6 \, \text{W/m}^2 \, {}^\circ\text{C}$, $U_b = 0.525 \, \text{W/m}^2 \, {}^\circ\text{C}$ (Example 11.4(a)), and other parameters are same as in Example 11.4(b).

Solution

From Examples 9.8 and 11.3(a), we have

$$\frac{\overline{f(t)}}{a} = \frac{0.65}{15.125} \times 900 + 8.5 = 47.18 \text{ and } a = \frac{15.125}{41900} = 3.6 \times 10^{-4}$$

Also
$$\Delta t = -\frac{1}{3.6 \times 10^{-4}} \ln \left[\frac{40 - 47.18}{20 - 47.18} \right] = -\frac{1}{3.6 \times 10^{-4}} \times -1.3312$$

$$= 3697.7 \sec = 61.6 \text{ minutes}$$

In SSP water heater Δt is more due to more heat losses from top and lower value of $(\alpha\tau)_{\text{eff}}$

11.2 NON-CONVECTIVE SOLAR POND

A solar pond refers to a stagnant (except charging and discharging operations) large body of water with black bottom and capable of collecting and storing solar energy. Practically, any water mass with black bottom can absorb solar energy but the collection efficiency is poor. The prime factor causing a lowered efficiency is the convective heat losses. The heated water at the bottom rises by convection to the top. Thus, the top most layer becomes the hottest part of the water mass and consequently the heat is lost to the environment. The pond collector will be much more effective if this convective heat dissipation is impeded. One of the method is to maintain a salt density gradient-the density increasing with depth, so that, warmer water at the bottom may not acquire low density and rise to the top by convection.

The non-convective solar pond is similar to the, earlier discussed, convective pond, but a layer of still water is used as an insulator rather than the normal glazing and air space. The three methods to accomplish the non-convecting mode of the pond and to maintain its stability are:

i. salt-stabilized,
ii. partitioned or membrane
iii. viscosity stabilized

11.2.1 Salt-stabilized Pond

A salt-stabilized solar pond (Fig. 11.4(a)) is a non-convecting body of fluid contained by an impervious bottom liner. An artificial salt solution density gradient is achieved by superimposing layers of decreasing salinity. Batches of brine can be mixed in a small evaporation pond, and then pumped into the solar pond through a horizontal diffuser floating on the surface. This is the non-convecting zone. The incident solar radiation penetrates the water surface and a fraction of it reaches the bottom after crossing the layers of varying density. This energy is trapped near the bottom due to opaqueness of water for far-infrared radiation, the relatively poor thermal conductivity of still water and the stabilizing influence of the gradient. One of the method to extract this energy is to carefully withdraw the convecting zone layer, pass it through a heat exchanger and then return the brine to the bottom of the pond. Another method is to install a heat exchanger into the non-convecting zone.

Figure 11.4(b) shows a schematic diagram of another economic prototype pond for space heating. It is contained within an approximately square excavation and earth embankment lined by 0.8 mm thick nylon reinforced black chlorinated polyethylene. The linear is a simple flat sheet of sufficient size to cover the area. Specified dimensions of the pit at the bottom and at the top of the bank are 12 m and 18 m with 45 °m bank slope. It is 3 m depth to contain 2.5 m of water.

Figure 11.4(c) shows a cross-section of the re-designed non-convective solar pond with a leak detection system and bottom insulation. The pond was re-designed at the Ohio Agricultural Research and Development Centre (OARDC Pond). The OARDC pond has post and plywood construction walls and a sand bottom with typical dimensions of $18.3 \, \text{m} \times 8.5 \, \text{m} \times 3.6 \, \text{m}$. The pond containing brine is lined with two 0.75 mm chlorinated polyethylene linears with a nylon serim. The side walls were insulated with 10 cm of polystyrene in the top 1.8 m and 5 cm in the bottom 1.8 m. The bottom surface was expected to become insulated if pond heat dried out the surrounding soil.

The bottom half of the pond has a 1.8 m convective zone of approximately 20 percent salt (sodium chloride). In the top half, a salt concentration gradient varying from the 20 percent concentration at the 1.8 m depth to zero at the surface. The top half is a non-convective zone (Short et al., 1978).

Figure 11.5 gives the variation of working temperature increment, i.e., $(T_w - T_a)$ with extraction depth Z_w for a range of output efficiencies η.

This figure can be used to predict the heat collection efficiency of an idealized pond. Instantaneous performance of a solar pond, however, cannot be defined because of the excessive thermal mass and resulting storage capacity. The figure, thus, gives average collection efficiency for an average annual insolation level of 200 W/m² for various extraction depths. The figure also suggests that for average flux intensities other than 200 W/m², the temperature rise of the pond $(T_w - T_a)$ should be directly proportional to the insolation level (I).

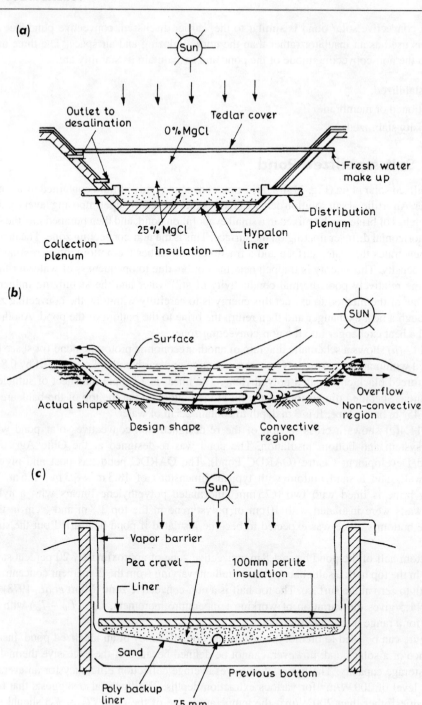

Figure 11.4 Non-convecting salt pond cross-section. (**a**) With tedlar cover, (**b**) Prototype economic, and (**c**) redesigned.

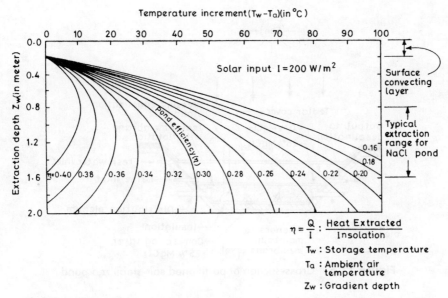

Figure 11.5 Variation of working temperature increment $(T_w - T_a)$ with extraction depth Z_w for a range of output efficiencies η.

The advantages of salt-stabilized ponds are:

i. it serves as an efficient storage,
ii. convective heat dissipation is suppressed without any additives, membrane etc.
iii. bottom insulation and top cover can be dispensed with.

Apart from these advantages, salt-stabilized solar pond have following drawbacks.

i. Salt constitutes the most expensive component of the pond.
ii. The pond suffers from energy withdrawal problems; non-convective layers are disturbed, and heat exchanger fouling may take place.
iii. Establishing the necessary salt gradient constitutes a major operational problem while maintaining it throughout the pond operation requires a desalination unit.

These factors lead to an unstable operation of the pond and the two basic concepts proposed for stable operation of the pond are:

i. Partitioning ponds by membranes.
ii. Creating a stable layer in the pond by adding thickeners to increase the viscosity instead of creating salinity gradient ponds.

11.2.2 Partitioned Salt-stabilized Pond

This pond was proposed for space heating applications. In this pond (Fig. 11.6) the convecting and non-convecting zones are separated by a transparent membrane or partition.

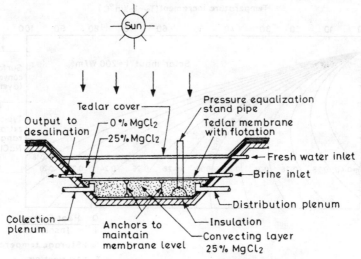

Figure 11.6   Cross-section of partitioned salt-stabilized pond.

The use of partition allows a fresh water convecting zone which can reduce the extraction problems and reduces the salt requirement.

Some of the advantages of the partitioned type ponds are:

  i. They are non-convecting.
 ii. They are collection-cum-storage ponds.
iii. Bottom insulation and top cover are not required.
 iv. Salt requirement is lower than in the case of salt stabilized pond.
  v. The non-convecting zone instabilities are suppressed.
 vi. Convecting layer being of fresh water, energy extraction is simpler
vii. In-pond heat exchanger may be practical.

The long-term operation of the pond may create the following problems:

  i. Establishing and maintaining the necessary salt gradient.
 ii. Requirement of desalination unit.
iii. Partition cost.
 iv. Partition support (pressure equalization stand pipe may be required).

11.2.3 Viscosity Stabilized Pond

This type of pond is based on a similar concept to that of the partitioned salt-stabilized pond except that in this case the non-convecting layer (Fig. 11.7) contains thickeners for stabilization rather than a salt gradient.

Polymers and detergent, oil, water gels are considered to be the required thickeners. Due to the high cost of thickeners, it has been proposed to use layers of gel or a non-convecting salt system with small amounts of gel.

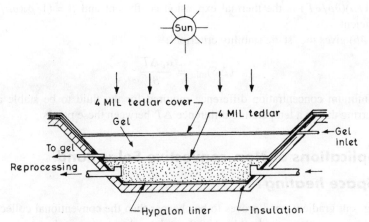

Figure 11.7 Cross-section of golled pond.

The following are the advantages of viscosity stabilized ponds:

i. It is non-convecting in nature.

ii. Bottom insulation, separate storage and salt are not required.

iii. In pond heat exchanger is practical and energy extraction is simpler, as fresh water is the convecting zone.

iv. It has reduced diffusion effect.

The pond has the following disadvantages:

i. Partition cost.

ii. High thickeners cost.

iii. Top cover indispensable, as the UV radiations may degrade the organic materials.

iv. Thickener management problems involving disposal and maintenance of a homogenous mix usually occur; the thickener may also become opaque.

11.2.4 Stability Criteria

Salt constitutes the major and most expensive component of a solar pond; thus the knowledge of the salt gradient requirement and the technology to establish and maintain it is required for efficient pond operation.

In the solar pond, the density (ρ) of the fluid is a function of salt concentration, S, and temperature, T. In the case of pond stability against vertical convection, the magnitude of the salt density gradient ($\partial \rho / \partial S$) due to salt concentration gradient ($\partial S / \partial x$) must be greater than the negative density gradient ($\partial \rho / \partial T$) produced by temperature gradient ($\partial T / \partial x$).

Considering x-axis along the vertical direction and measuring positive downwards, the equilibrium state can be expressed as

$$\frac{\partial \rho}{\partial S} \cdot \frac{\partial S}{\partial x} \geq -\frac{\partial \rho}{\partial T} \cdot \frac{\partial T}{\partial x} \tag{11.4a}$$

or

$$\frac{\partial S}{\partial x} \geq \frac{\alpha_c}{\beta} \frac{\partial T}{\partial x} \tag{11.4b}$$

where $\alpha_c = -(1/\rho)(\partial\rho/\partial T)$ is the thermal expansion coefficient and $\beta = (1/\rho)(\partial\rho/\partial S)$ is the salt expansion coefficient.

Equation (11.4b) gives the 'static stability criterion':

$$(\Delta S)_{min} = \frac{\alpha_c \Delta T}{\beta} \qquad (11.4c)$$

denoting the minimum concentration difference required for the fluid to be stable against vertical convection occurring due to a temperature difference ΔT between these points.

11.2.5 Applications of Non-convective Solar Pond

11.2.5.1 Space heating

In space heating, salt gradient pond proves to be cheaper than the conventional collector and storage systems. A pond can carry the entire heat load, without depending upon supplementary sources. A solar pond is very useful in crop drying (where a large quantity of heat is required for a short period), and heating farm buildings. Unusually low temperature heat is required for many of these applications, thus it is necessary to ensure simple and reliable operation of the pond by identifying problems and finding their practical solutions.

11.2.5.2 Green house solar pond heating system

In this case, heat is taken from the bottom of the solar pond by circulating pond brine through plastic pipe to the shell and tube heat exchanger. Brine pumping is preferred to an in-pond heat exchanger, as large in-pond heat exchange surfaces would be necessary at low pond temperatures. The heating system is so designed that when the pond is between 40 °C and 80 °C, fresh water in the tubes of the shell and tube heat exchanger is circulated to a water-to-air discharge heat exchanger in the greenhouse. When the pond is between 5 °C and 40 °C, fresh water from the tubes of the shell and tube heat exchanger is pumped through the evaporator of a heat pump to keep the temperature of the water, being delivered to greenhouse, slightly above 40 °C.

11.2.5.3 Electricity generation

Solar pond power plant uses an organic Rankine cycle convertor, utilizing a working substance like isobutane ammonia which circulates in a closed cycle.

In addition to the above applications, a non-convective solar pond can also be used for (i) salt and mineral production, (ii) solar absorption refrigeration, (iii) Rankine cycle solar engines, (iv) heating an outdoor swimming pool, and (v) agricultural drying, hot water production, distillation, industrial drying of laundry, processed food.

EXAMPLE 11.5

Calculate the total heat loss coefficient from the top of convective zone of a non-convective pond for pond temperature of 25 °C, $T_a = 15$ °C, and relative humidity of 40%.

Solution

The total heat transfer coefficient from the top of convective zone is given as, $h_1 = h_{ra} + h_{ca} + h_{ea}$. The radiative heat transfer coefficient, from Equation (3.17), can be calculated as,

$$h_{ra} = 0.9 \times 5.67 \times 10^{-8} \left[\frac{(25+273)^4 - (9+273)^4}{25 - 15} \right] = 7.97 \, \text{W/m}^2 \, ^\circ\text{C}$$

The convective heat transfer coefficient, from Equation (2.49b) (for $V = 3 \, \text{m/s}$), is

$$h_{ca} = 2.8 + 3.0 \times 3 = 11.8 \, \text{W/m}^2 \, ^\circ\text{C}.$$

The evaporative heat transfer coefficient from Equation (2.76), is

$$h_{ea} = 16.273 \times 10^{-3} \times 11.8 \times \frac{[2346.51 - 0.4 \times 1730.02]}{10} = 31.77 \, \text{W/m}^2 \, ^\circ\text{C}$$

The total heat transfer coefficient $h_1 = 7.97 + 11.8 + 31.77 = 51.54 \, \text{W/m}^2 \, ^\circ\text{C}$

11.3 SOLAR WATER HEATING SYSTEM

The conventional solar water heating system essentially consists of two units, viz. the collection unit and the storage unit. The collection unit is, usually, a blackened plate to absorb most of the solar energy incident on it. The flowing water in thermal contact with the absorber, gets heated and is transferred to the storage unit. The storage unit is a well-insulated tank, to reduce the possible heat losses. The transportation of the heated liquid from the collector to the storage unit takes place by two modes;

i. the thermosyphon mode, in which the circulation of heated water is accomplished by the natural convection, and

ii. the forced circulation mode, where a small pump is required for the flow of water.

11.3.1 Natural Circulation (Sodha and Tiwari, 1981)

The schematic of a solar water heater, with natural circulation is shown in Figure 11.8. The absorber in this case, is a set of N collectors connected in parallel (Fig. 11.8(d)). The storage tank is an insulated one with two inlets, one for the hot water from the collector and the other one to allow the cold water, from the mains, to reach the bottom of the tank without mixing with hot water. There are two outlets as well, one for the withdrawal of hot water and the other one is used to feed cold water to the collector inlet. The entire length of the connecting pipes is covered with glass wool insulation to reduce the heat loss. Solar radiation incident on the flat plate collector heat the water inside. The hot water, being less dense rises up the collector. The vacuum created by this flow is filled up by the cold water from the storage tank. Thus the upper end (2) of the collector has hot water while the lower end (1) has cold one (Fig. 11.8(a)). This hot water then enters the storage tank, from the inlet 3, from where it can be withdrawn for further use.

Figure 11.8(a) shows a view of pressure type solar water heater. In this case, an insulated storage tank is always completely filled with cold water to be heated. There should be a continuous cold water supply from the mains to get hot water at a time. This type of solar water heater can't be used for Harsh cold climatic condition, i.e., $T_a \leq 0\,°C$.

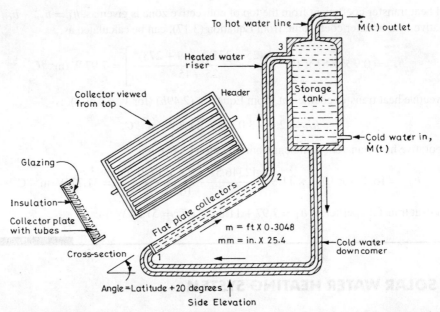

Figure 11.8(a) Schematic diagram of pressure type solar water heater with natural circulation.

For harsh cold climatic condition, there is a need of heat exchanger in a collector loop as shown in Figure 11.8(b). The position of the heat exchanger has also been shown in the same figure. In this case antifreeze liquid should be used in the collector loop. The hot water withdrawal can be made available at any time without cold water supply only by changing the position of hot water withdrawal from the top to middle of the tank (Fig. 11.8(c)).

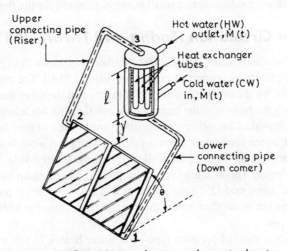

Figure 11.8(b) Heat exchanger solar water heater.

There is also a non-pressure type solar water heater as shown in Figure 11.8(c). In this case, the level of water in the tank is maintained with the help of float value. The cold water is fed at the bottom of the tank through cold water line unlike pressure type storage tank (Fig. 11.8(a)).

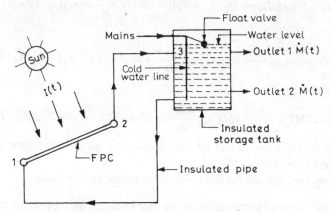

Figure 11.8(c)   Non-pressure type solar water heater.

The position of tank should be always about the upper header of the collector for natural circulation between collector panel and insulated storage tank, as shown in Figure 11.8(d).

Figure 11.8(d)   Photograph of working system.

With reference to Figure 11.8, the energy balance for the system and the tank can be written (Sodha and Tiwari, 1981).

11.3.1.1 During sunshine hour

Flat plate collector

$$A_{CN} F'[\alpha\tau\ I(t) - U_L(T_c - T_a)] - (MC)_c \frac{dT_c}{dt} = \dot{m}_2 C_w (T_2 - T_1) \qquad (11.5a)$$

Pipe

$$\dot{m}_2 C_w (T_2 - T_1) = (mC)_p \frac{dT_p}{dt} + (mC)_{wp} \frac{dT_{wp}}{dt} + \dot{m}_3 C_w (T_3 - T_2) + U_p A_p (T_p - T_a) \qquad (11.5b)$$

Storage tank

$$\dot{m}_3 C_w (T_3 - T_2) = (MC)_t \frac{dT_t}{dt} + (MC)_{wt} \frac{dT_{wt}}{dt} + U_i A_i (T_{wt} - T_a) + \dot{M}(t) C_w (T_{wt} - T_{in}) \qquad (11.5c)$$

where the subscripts c, wc, p, wp, t and wt denote collector, water in collector, pipe, water in pipe, tank and water in tank respectively.

In order to simplify the analysis, the following assumptions may be made:

i. Connecting pipes are well insulated so that the heat loss from the pipes to the ambient is negligible.

ii. Mean body temperature of the collector plate and tubes, the storage tank and the connecting pipes is equal to the mean temperature of water within them.

iii. Mean water temperature in the collector tubes, the storage tank and the connecting pipes is equal and denoted by the mean system temperature, T_m.

i.e. $T_c \approx T_{wc};\ T_p \approx T_{wp};\ T_t \approx T_{wt}$ and $T_c = T_p = T_t = T_m$

Incorporating these assumptions and simplifying Equations (11.5), gives

Whole system

$$W\frac{dT_m}{dt} + U_{LN}(T_m - T_a) + \dot{M}(t)C_w(T_m - T_{in}) = (\alpha\tau)I(t)F'A_{CN}$$

$$\underset{\substack{\text{rate of} \\ \text{heat} \\ \text{stored}}}{}\quad \underset{\text{rate of heat lost}}{}\quad \underset{\substack{\text{rate of heat lost due} \\ \text{to withdrawal of hot} \\ \text{water from tank}}}{}\quad \underset{\substack{\text{rate of solar} \\ \text{radiation absorbed} \\ \text{by the absorber}}}{} \qquad (11.6)$$

where

$$W = (mC)_c + (mC)_{wc} + (mC)_p + (mC)_{wp} + (mC)_t + (mC)_{wt}$$

$$U_{LN} = A_{CN} F' U_L + U_p A_p + U_i A_i \quad \text{and,} \quad A_{CN} = N A_C$$

Storage tank

$$W'_t\frac{dT_m}{dt} + U_T(T_m - T_a) + \dot{M}(t)C_w(T_m - T_{in}) = \dot{Q}_{useful}$$

$$\underset{\substack{\text{rate of} \\ \text{heat} \\ \text{stored}}}{}\quad \underset{\substack{\text{rate of heat} \\ \text{lost}}}{}\quad \underset{\substack{\text{rate of heat lost due} \\ \text{to withdrawal of} \\ \text{hot water}}}{}\quad \underset{\substack{\text{rate of heat collected} \\ \text{from collectors}}}{} \qquad (11.7)$$

where, $W'_t = W_t + W_{wt}$, $\dot{Q}_{useful} = \dot{m}(t)C_f(T_3 - T_5)$ and $U_T = A_T U_t$.

11.3.1.2 Off sunshine hours

Tank

$$W'_t\frac{dT'_m}{dt} + U_T(T'_m - T_a) + \dot{M}(t)C_w(T'_m - T_{in}) = 0 \qquad (11.8$$

Equation (11.6) can be rewritten in the form,

$$\frac{dT_m}{dt} + aT_m = f(t) \tag{11.9}$$

where

$$a = \frac{(U_{LN} + \dot{M}(t)C_w)}{W}$$

and

$$f(t) = \frac{[(\alpha\tau)\,I(t)F'A_{CN} + \dot{M}(t)C_wT_{in} + U_{LN}T_a]}{W}$$

The solution of Equation (11.9) with the initial condition $T_m = T_{m0}$ at $t = 0$ and $f(t) = \overline{f(t)}$ within the interval $\Delta t (0 < t < \Delta t)$, can be written as

$$T_m = \frac{\bar{f}(t)}{a}(1 - e^{-a\Delta t}) + T_{m0}e^{-a\Delta t} \tag{11.10(a)}$$

The above equation can also be solved for Δt as

$$\Delta t = -\frac{1}{a} \cdot \ln\left[\frac{T_m - \overline{\frac{f(t)}{a}}}{T_{m0} - \overline{\frac{f(t)}{a}}}\right] \tag{11.10(b)}$$

For detail, see Example 11.6(a).

The efficiency of the system can be defined in two cases:

i. Without water withdrawal

$$\eta = \frac{M_wC_w[T_m|_{t=T} - T_m|_{t=0}]}{A_{CN}\int_0^T I(t)dt} \tag{11.11a}$$

ii. With water withdrawal

$$\eta = \frac{\sum_{i=1}^n \dot{M}(t)C_w(T_{mi} - T_{in})}{A_{CN}\int_0^n (t)dt} \tag{11.11b}$$

11.3.2 Forced Circulation Water Heater, (*Sodha et al., 1982 d*)

In this case, a water pump at the inlet of collector is used to transfer the hot water available at upper header of collector to the insulated storage tank as shown in Figure 11.9(a). The collectors can also be connected in series for higher water temperature, if required. A single and double loop water heating system will be discussed in the present section. The position of storage tank becomes independent unlike in the case of natural circulation of water heater.

11.3.2.1 *Single loop (Sodha et al., 1982c)*

The single loop water heater as shown in Figure 11.9a operates continuously till the pump in switched off manually. The controller has been used in Figure 11.9(b). In this case, the pump operates only at desired temperature with help of sensor fitted at outlet of collector and storage tank. Further, a coil type heat exchanger inside storage tank with controller has been used in collector loop as shown in Figure 11.9(c). This is required for harsh cold climatic condition. Antifreeze liquid is used in a collector loop. Another type of heat exchanger in a collector loop is shown in Figure 11.9(d). This is known as *heating jacket* used outside storage tank. The maintenance of heat exchanger becomes easier.

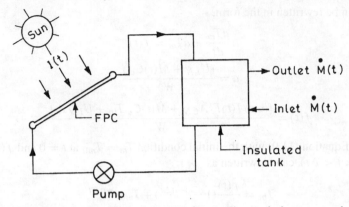

Figure 11.9(a) Schematic diagram of a simple single loop water heater.

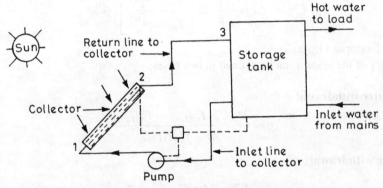

Figure 11.9(b) Schematic diagram of a single loop water heater with controller.

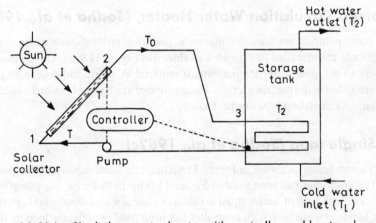

Figure 11.9(c) Single loop water heater with controller and heat exchanger.

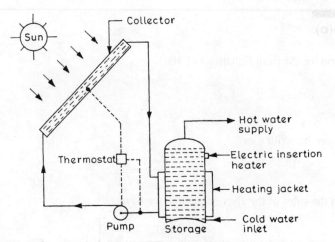

Figure 11.9(d) Single loop water heater with back-up unit and controller.

A complete set-up of domestic water heater has been shown in Figure 11.9(e). The hot water can be withdrawn for continuous supply of cold water at inlet. The non-return check valve has been used. The controller has been connected to absorber outlet and storage tank with sensor. The back-up unit has been provided outside the storage tank.

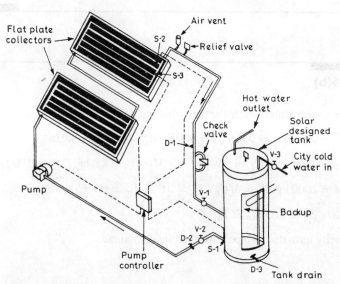

Figure 11.9(e) Schematic diagram of a single loop water heating system with back-up unit and controller. (S: Sensor, V: Gate value and D: Drainage pipe).

Referring to Figure 11.9(a), and Equation (3.88), the flow of the cold water from the storage tank to the collector is maintained by a pump (Fig. 11.9(a)). The energy balance can be written as,

$$\dot{Q}_{useful} = \dot{m}_f C_f (T_{foN} - T_{fi}) = W_t \frac{dT_m}{dt} + U_T A_t (T_m - T_a) + \dot{M}(t) C_w (T_m - T_a) \qquad (11.12)$$

EXAMPLE 11.6(a)

Derive an expression for Δt from Equation (11.10a).

Solution

Equation (11.10a) can be written as

$$e^{-a\Delta t} = \frac{T_m - \dfrac{\overline{f(t)}}{a}}{T_{m0} - \dfrac{\overline{f(t)}}{a}}$$

Taking log on both the sides of the above equation, we have

$$-a\Delta t = \ln \left[\frac{T_m - \dfrac{\overline{f(t)}}{a}}{T_{m0} - \dfrac{\overline{f(t)}}{a}} \right]$$

which gives

$$\Delta t = -\frac{1}{a} \ln \left[\frac{T_m - \dfrac{\overline{f(t)}}{a}}{T_{m0} - \dfrac{\overline{f(t)}}{a}} \right]$$

EXAMPLE 11.6(b)

Calculate 'Δt' for a solar water heater for the following parameters:

$$F' = 0.8, \ (\alpha\tau) = 0.8, \ \overline{I(t)} = 800 \ \text{W/m}^2 \, ^\circ\text{C}, \ A_{CN} = 40 \ \text{m}^2$$

$$U_{LN} = A_{CN} \, U_L, \ U_L = 6 \ \text{W/m}^2 \, ^\circ\text{C}, \ \dot{M}(t) = 72 \ \text{kg/hr}, \ C_w = 4190 \ \text{J/kg} \, ^\circ\text{C}$$

$$M = 2000 \ \text{kg}, \ W = MC_w, \ T_a = 10 \, ^\circ\text{C}, \ T_m = 60 \, ^\circ\text{C}$$

$$T_{m0} = 20 \, ^\circ\text{C} \ \text{and} \ T_{in} \cong T_a.$$

Neglect heat capacity term due to pipe, absorber, insulation etc.

Solution

From Equation (11.9), we get

$$a = \frac{6 \times 40 + 2 \times 10^{-2} \times 4190}{2000 \times 4190} = 3.8639 \times 10^{-5}$$

and

$$\overline{f(t)} = \frac{0.8 \times 0.8 \times 800 \times 40 + (6 \times 40 + 2 \times 10^{-2} \times 4190) \times 10}{2000 \times 4190} = 0.00283$$

So, from Equation (11.10*b*),

$$\Delta t = -\frac{1}{3.8639 \times 10^{-5}} \times \ln\left[\frac{60 - 73.31}{20 - 73.31}\right] = -\frac{1}{3.8639 \times 10^{-5}} \times -1.387$$

$$= 35912.12 \text{ sec} = 598.5 \text{ min} = 9.98 \text{ hrs.}$$

This is unrealistic. So the withdrawal rate should be reduced to maintain $T_m = 60\,°C$.

11.3.2.2 Double loop (Sodha et al., 1982e)

We consider a solar water heating system in which two heat exchanging loops have been used, one between collector and storage tank I and the other between storage tanks I and II with forced circulation (Fig. 11.10). The energy balance for the collector heat exchanger storage tank (I) can be written as

$$W_1 \frac{dT_1}{dt} = \dot{Q}_{\text{gain},c} - \dot{Q}_{\text{loss,I}} \tag{11.13}$$

where W_I is the thermal capacity and T_I the mean temperature of the storage tank I.

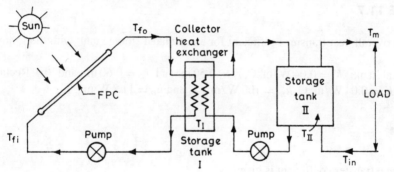

Figure 11.10 Schematic of a double loop solar water heating system.

The amount of heat energy transferred ($\dot{q}_{\text{gain},c}$) from the collector loop heat exchanger to storage tank (I), is given as

$$\dot{Q}_{\text{gain},c} = hL(\bar{T}_f - T_I) \tag{11.14}$$

or

$$\dot{Q}_{\text{gain},c} = \dot{m}C_f(1 - \exp(-\beta))(T_{f0} - T_I) \tag{11.15}$$

where

$$T_{f0} = \frac{1}{1 - \exp\{+(+\beta\beta')\}}\left[(1 - \exp(-\beta'))\left[\frac{\alpha_0\tau}{U_L}I(t) + T_a\right] + T_I(1 - \exp(-\beta'))\exp(-\beta')\right]$$

$$\beta = \frac{hL}{\dot{m}C_f} \quad \text{and} \quad \beta' = \frac{A_c U_L F'}{\dot{m}C_f}$$

The amount of heat energy transferred ($\dot{Q}_{\text{loss,I}}$) from storage tank (I) to heat exchanger of storage tank II, is given as,

$$\dot{Q}_{\text{loss,I}} = h_0 L_0(T_I - T_W) \tag{11.16}$$

or

$$\dot{Q}_{\text{loss,I}} = \dot{m}C_f(T_I - T_{II})[1 - \exp(-\beta'')] \tag{11.17}$$

where
$$\beta'' = \frac{h_0 L_0'}{\dot{m} C_f}$$

The energy balance for the storage tank (II) can be written as

$$W_{II}\frac{dT_{II}}{dt} = \dot{Q}_{loss,I} - \dot{M}(t)C_p(T_{II} - T_{in}) \tag{11.18}$$

Equations (11.13) and (11.18) can be rewritten in the form

$$\frac{dT_I}{dt} + a_1T_I + a_2T_{II} = f(t) \tag{11.19}$$

$$\frac{dT_{II}}{dt} + b_1T_I + b_2T_{II} = g(t) \tag{11.20}$$

and can be solved to get the values of T_I and T_{II}.

EXAMPLE 11.7

Calculate the overall heat transfer coefficient for an insulated storage tank with the following specifications:
(a) Cylindrical tank: $K = 0.04\,\text{W/m}^\circ\text{C}, r_i = 0.50\,\text{m}$ and $r_o = 0.60\,\text{m}$, and (b) Rectangular tank: $L = 0.10\,\text{m}, K = 0.04\,\text{W/m}^\circ\text{C}, h_w = 100\,\text{W/m}^2\,^\circ\text{C}$ and $h_o = 11.8\,\text{W/m}^2\,^\circ\text{C}$

Solution

The overall heat transfer coefficient is given as

a. for cylindrical tank (Eq. (2.34b)):

$$U = \left[\frac{1}{h_0} + \frac{r_i}{K}\ln\frac{r_o}{r_i} + \frac{1}{h_w}\right]^{-1}$$

$$= \left[\frac{1}{11.8} + \frac{0.50}{0.04}\ln\frac{0.60}{0.50} + \frac{1}{100}\right]^{-1}$$

$$= [0.0847 + 2.279 + 0.01]^{-1} = 0.42\,\text{W/m}^2\,^\circ\text{C}.$$

b. for rectangular tank (Eq. (2.33a)):

$$U = [(1/h_0) + (L_i/K_i) + (1/h_w)]^{-1} = [(1/11.8) + (0.10/0.04) + (1/100)]^{-1}$$

$$U = 0.385\,\text{W/m}^2\,^\circ\text{C}$$

11.4 HEATING OF SWIMMING POOL BY SOLAR ENERGY

Solar energy can be applied for the low temperature heating of fluids either by direct or indirect method. The direct method is generally referred to as a passive system while the indirect method is known as an active system. One of the applications of solar energy is in heating swimming pools. Outdoor swimming pools can be heated by both passive and active systems, while indoor swimming pools can be heated only by an active system.

11.4.1 Passive Heating of Swimming Pools

In this case during sunshine hours, transparent floatable plastic cover is used over the surface of the water in the swimming pool. The inner surface of the swimming pool is preferably blackened to absorb solar radiation. The solar radiation is transmitted through the plastic cover, reaches the bottom of the pool and is finally absorbed by the bottom surface. Some of the radiation is absorbed by the water of the pool, as well. Most of the absorbed solar radiation is convected to the water in the pool and the rest is lost to the ground. The loss can be minimized by using a layer of insulating material beneath the bottom surface. On receiving heat from the bottom surface, the water gets heated and moves in an upward direction due to it's low density. As the water temperature is higher than the ambient air temperature, particularly at night, convective and radiative losses occur from the water surface. In order to minimize these losses, the pool surface is covered with a waterproof material which acts as an insulating material during off-sunshine hours. As the transparent sheet must be removed when the pool is in use, there are unavoidable evaporative heat losses during this period.

11.4.2 Active Heating of Swimming Pool (*Tiwari et al., 1988*)

As explained above, the temperature of swimming pool can be increased marginally by use of a passive system. But under extreme cold climatic conditions, passive heating is not sufficient. For further increase in pool temperature, extra thermal energy is required. This can be fed in at the bottom of the pool and can maintain the temperature of the pool in a comfortable range ($\sim 20°C$). This can be done by connecting a panel of collectors to the pool either directly or through a heat exchanger. The area of collectors depending upon the capacity of the pool and the climatic conditions.

11.4.3 Description of Indoor Swimming Pool

An indoor swimming pool has certain advantages over an outdoor swimming pool, namely, (i) it is protected from dust, bird, climate etc., (ii) it becomes an integral part of the building (iii) it is easy to clean (iv) it requires less maintenance, and (v) can be used in extreme cold climatic conditions.

A schematic diagram of an indoor solar swimming pool heated by an active system is shown in Figure 11.11. There are basically three components of the system, (i) panel of collectors, (ii) circulation system consisting of pumps, valves and connecting pipes, and (iii) control system integrated with a panel of collectors and the pool water.

The cost of an active system mainly depends on the type of collectors used for heating the pool water, which, in turn, is governed by the atmospheric conditions and the availability of solar radiation. The types of collectors used for swimming pool are:

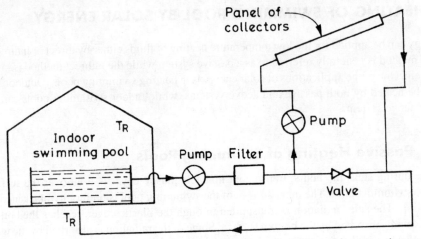

Figure 11.11   Schematic diagram of indoor swimming pool heated by an active system.

i. Unglazed collectors:
   a. Plastic panels
   b. Strip collectors
   c. Plastic pipe collectors
   d. Permanent collectors
ii. Glazed collectors:
   a. Boxed collectors
   b. Integrated collector, i.e. an integral part of the roof of a building

Unglazed collectors are preferred for heating indoor/outdoor swimming pool when the increase in pool temperature required is small, for larger requirement, however, glazed collectors are preferred. Unglazed collectors are more cost-effective than any conventional system.

In order to write down the energy balance of an indoor swimming pool integrated with a panel of collectors, the following assumptions are made:

i. Useful energy available from the panel of collectors is uniformly fed at the bottom of the pool basin.
ii. No stratification along the depth of the pool.
iii. Heat losses through connecting pipes are negligible.
iv. Panel of collectors is disconnected during off-sunshine hours.
v. Pool surface is covered with floating, waterproof, insulating cover to minimize upward convective and radiative heat losses.
vi. Evaporative heat loss is considered only when the pool is in use and it is taken out by means of an exhaust fan.
vii. Bottom heat loss is considered under a steady-state mode.
viii. Enclosure temperature is almost equal to ambient air temperature.
ix. Proper insulating material is used in the basin of the pool to avoid downward heat losses.

The energy balance without evaporation is

$$(MC)_w \frac{dT_w}{dt} + A_p U_t(T_w - T_R) + (A_p U_b + A_s U_s)(T_w - T_R) = \dot{Q}_u$$

| rate of thermal energy stored in pool water | rate of thermal energy lost from pool surface to ambient | rate of thermal energy lost through sides/bottom | rate of thermal energy available from collector panel | (11.21a) |

With evaporation

$$(MC)_w \frac{dT_w}{dt} + A_p(h_{cw} + h_{rw})(T_w - T_R) + 0.013\, h_c(P_w - \gamma P_a)A_p$$
$$+ (A_p U_p + A_s U_s)(T_w - T_R) = \dot{Q}_u$$

or

$$(MC)_w \frac{dT_w}{dt} + A_p(h_{cw} + h_{rw})(T_w - T_R) + h_e A_p(T_w - T_R)$$
$$+ (A_p U_p + A_s U_s)(T_w - T_R) = \dot{Q}_u \qquad (11.21b)$$

where, from Equation (3.70), we have

$$\dot{Q}_u = A_c F_R[(\alpha\tau)I - U_L(T_w - T_a)] \text{ without heat exchanger}$$
$$= A_c F_R[\alpha I(t) - U_L(T_w - T_a)], \text{ without glazing}$$
$$= 0 \text{ when collectors are disconnected from pool}$$

$h_e (= h_{ew})$ is given by Equation (2.76).

EXAMPLE 11.8(a)

Calculate the bottom and side heat loss coefficients for indoor swimming pool for the following given parameters:
Thickness of marble layer $(L_m) = 0.03$ m, $K_m = 2.7673$ W/mK, thickness of brick layer $(L_1) = 0.1524$ m, $K_1 = 0.6918$ W/mK, thickness of concrete layer $(L_2) = 0.1524$ m, $K_2 = 1.1069$ W/mK, wind velocity $= 0$ m/s.

Solution

If the bottom of the pond is there layered structure, i.e., zone I is the marble layer, zone II is the brick layer and zone III is the concrete layer.
The bottom loss coefficient is given as

$$U_b = \left[\frac{L_m}{K_m} + \frac{L_1}{K_1} + \frac{L_2}{K_2} + \frac{1}{h_i} \right]^{-1} = \left[\frac{0.03}{2.7673} + \frac{0.1524}{0.6918} + \frac{0.1524}{1.1069} + \frac{1}{5.7} \right]^{-1} = 1.837 \text{ W/m}^2\text{K}$$

If the side walls contain a marble layer 0.03 m thick and concrete layer 0.1 m thick, then the side loss coefficient is given as

$$U_s = [(0.03/2.7673) + (0.1/1.1069) + (1/5.7)]^{-1} = 3.615 \text{ W/m}^2\text{K}$$

EXAMPLE 11.8(b)

Calculate top loss coefficient for indoor swimming pool with and without evaporation for the following given parameters: water temperature $= 20\,°C$ and room temperature $= 15\,°C$, emissivity $= 0.9$, $\gamma = 0.8$.

Solution

The top loss coefficient without evaporation, will be equal to loss due to radiation and that due to convection. Thus, radiative heat transfer coefficient (from Eq. 3.14) is equal to,

$$h_{rw} = 0.9 \times 5.67 \times 10^{-8} \left[\frac{(20+273)^4 - (15+273)^4}{20 - 15} \right] = 5.004 \text{ W/m}^2\,°C$$

The convective heat transfer coefficient, given by Equation (2.49b), is $2.8 \text{ W/m}^2\,°C$.
Thus, the total loss coefficient is $U_t = 5.004 + 2.8 = 7.804 \text{ W/m}^2\,°C$ (without evaporation).
With evaporation, the evaporative heat loss is also considered, hence from Equation (2.76), we have,

$$h_{ew} = 0.013 \times 2.8 \times \left[\frac{2346.513 - 0.8 \times 1730.02}{20 - 15} \right] = 7.007 \text{ W/m}^2\,°C$$

The total loss coefficient is $U_t = 7.804 + 7.007 = 14.811 \text{ W/m}^2\,°C$

Equation (11.21a) without evaporation can be further rewritten as

$$\frac{dT_w}{dt} + aT_w = f(t) \tag{11.22}$$

where

$$a = \frac{A_p(U_t + U_b) + A_s U_s + A_c F_R U_L}{(MC)_w} \tag{11.22a}$$

and

$$f(t) = \frac{A_c F_R(\alpha\tau)I(t) + [A_p(U_t + U_b) + A_s U_s]T_R + A_c F_R U_L T_a}{(MC)_w} \tag{11.22b}$$

$$U_L = 7.202 \text{ W/m}^2\,°C \text{ (Example 3.2, glazed flat plate collector)}$$

$$= 5.8 + 3.8\ V \text{ (Equation (2.49a), bare absorber surface)} \tag{11.22c}$$

$$U_t = h_{rw} + h_{cw} + h_{ew} \text{ (Example 11.8(b)) with evaporation}$$

$$= h_{rw} + h_{cw}, \text{ without evaporation} \tag{11.22d}$$

The values of U_b and U_s are given in Example 11.8(a).
The solution of Equation (11.22) is

$$\Delta t = -\frac{1}{a} \ln \left[\frac{T_w - \frac{\overline{f(t)}}{a}}{T_{w0} - \frac{\overline{f(t)}}{a}} \right] \tag{11.22e}$$

EXAMPLE 11.9(a)

Calculate the value of a with and without glazing of collector for evaporation from the surface of the swimming pool.

Given parameters: $(MC)_w = 2000 \times 4190 = 8.38 \times 10^6 \, J/°C$, $A_p = 4 \, m^2$, $A_s = 4 \, m^2$, $F_R = 0.8$, $A_c = 40 \, m^2$, $\alpha = 0.9$, $\tau = 0.9$, and $V = 2 \, m/sec$.

Solution

From Equation (11.22c), $\quad U_L = 7.2 \, W/m^2 \, °C$ for glazing

$$= 13.4 \, W/m^2 \, °C \text{ for bare surface (without glazing)}$$

From Example 11.8(b) $\quad U_t = 14.8 \, W/m^2 \, °C$

From Example 11.8(a) $\quad U_b = 1.84 \, W/m^2 \, °C, U_s = 3.6 \, W/m^2 \, °C$

From Equation (11.22a)

i. with glazing

$$a = \frac{4(14.8 + 1.84) + 4 \times 3.6 + 40 \times 0.9 \times 7.2}{8.38 \times 10^6} = \frac{340.16}{8.38 \times 10^6} = 0.4059 \times 10^{-4}$$

ii. without glazing

$$a = \frac{4(14.8 + 1.84) + 4 \times 3.6 + 40 \times 0.9 \times 13.4}{8.38 \times 10^6} = \frac{563.37}{8.38 \times 10^6} = 0.6722 \times 10^{-4}$$

EXAMPLE 11.9(b)

Calculate the value of $\overline{f(t)}$ for $\overline{I}(t) = 800 \, W/m^2 \, °C$ and $\overline{T}_a = 8 \, °C$. Other parameters are same as given in Example 11.9(a).

Solution

From Equation (11.22b),

i. with glazing

$$\overline{f(t)} = \frac{40 \times 0.8 \times 0.9 \times 0.9 \times 800 + [4(14.8 + 1.84) + 4 \times 3.6] \times 15 + 40 \times .8 \times 7.2 \times 8}{8.38 \times 10^6}$$

$$= \frac{20736 + 1214.4 + 1843.2}{8.38 \times 10^6} = 2.839 \times 10^{-3}$$

ii. without glazing, in this case $\tau = 1$ and $U_L = 13.4 \, \text{W/m}^2 \, ^\circ\text{C}$

$$\overline{f(t)} = \frac{23040 + 1214.4 + 40 \times 0.8 \times 13.4 \times 8}{8.38 \times 10^6} = \frac{27684.8}{8.38 \times 10^6} = 3.304 \times 10^{-3}$$

EXAMPLE 11.9(c)

Calculate Δt for Example 11.9a and 11.9b with evaporation with and without glazing for heating the swimming pool 15 °C to 20 °C.

Solution

From Equation (11.22e)

i. with glazing

$$\Delta t = -\frac{1}{1.385 \times 10^{-4}} \ln \left[\frac{20 - \frac{2.839 \times 10^{-3}}{1.385 \times 10^{-4}}}{15 - \frac{2.839 \times 10^{-3}}{1.385 \times 10^{-4}}} \right] = -\frac{1}{1.385 \times 10^{-4}} \ln \left[\frac{20 - 20.498}{15 - 20.498} \right]$$

$$= -\frac{1}{1.385 \times 10^{-4}} \ln (0.0906) = 1.7337 \times 10^4 \text{ sec} = 288 \text{ min} = 4.8 \text{ hrs}$$

ii. without glazing

$$\Delta t = -\frac{1}{2.451 \times 10^{-4}} \ln \left[\frac{20 - \frac{3.304 \times 10^{-3}}{2.451 \times 10^{-4}}}{15 - \frac{3.304 \times 10^{-3}}{2.451 \times 10^{-4}}} \right] = -\frac{1}{2.451 \times 10^{-4}} \ln \left[\frac{20 - 13.48}{15 - 13.48} \right]$$

$$= -\frac{1}{2.451 \times 10^{-4}} \ln (4.2894) = -5941 \text{ sec}$$

The $-$ve value of Δt indicates that the swimming pool can never be heated with unglazed collector. In this case, collector is responsible for more heat loss than gain.

11.5 CONTROLLED ENVIRONMENT GREENHOUSE

Greenhouses are available in various shapes and sizes as shown in Figure 11.12(a) (Tiwari and Goyal, 1998). Essentially, a well designed greenhouse should be able to maintain a required environment for healthy growth of plants and to maximise the yield per ha. Mathematical models have been developed to describe the performance of greenhouse in terms of room and plant temperatures. Thermal model is based on basic heat and mass transfer. To facilitate the modelling procedure, a greenhouse can be considered

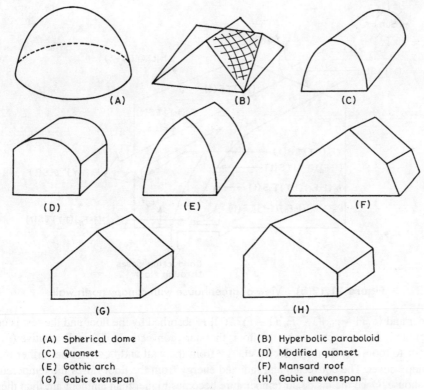

(A) Spherical dome
(C) Quonset
(E) Gothic arch
(G) Gabic evenspan

(B) Hyperbolic paraboloid
(D) Modified quonset
(F) Mansard roof
(H) Gabic unevenspan

Figure 11.12(a) View of greenhouses having different shape.

to be consist of root media, walls/roof, enclosed room air, heat capacity etc. The crop productivity depends on proper environment and more specifically on the thermal performance of the system.

As can be seen from Table 10.1, climatic condition of India has been devided into six zone based on insolation, ambient temperatures, relative humidity and rainfall. If wind is considered as another factor, then India can have another two zones mainly falling on hills. So one can have eight zone and each zone requires separate design of greenhouse.

In this section, a discussion will be covered out for a greenhouse suitable for composite climatic condition. For this zone, thermal heating/cooling are required to have suitable enclosed room air temperature for producing vegetable/crops.

11.5.1 Working Principle of a Greenhouse

Greenhouses are usually covered structures of plastic film, which is transparent to short wave length radiation and opaque to long wave length radiation. The solar radiation, $S(t)$, is incident on the canopy of greenhouse as shown in Figure 11.12(b). A fraction of solar energy $\{rS(t)\}$ is reflected back from canopy and a part of the rest radiation, $\{(1-r)S(t)\}$, is transmitted inside the greenhouse. Out of this transmitted radiation, $\{(1-r)\tau s(t)\}$, a fraction of this $\{F_n(1-r)\tau S(t)\}$, falls on north canopy cover wall which is further transmitted to atmosphere after reflection from north canopy wall (Tiwari *et al.*, 2001). The rest, $(1-F_n)(1-r)\tau S(t)$, goes to the floor of greenhouse. The magnitude of solar fraction for north wall, F_n, depends on the nth day of year and time of the day. $\{r_w(1-F_n)(1-r)\tau S(t)\}$ is reflected

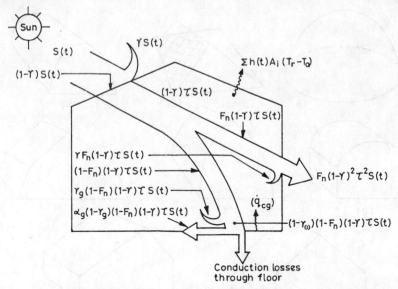

Figure 11.12(b) View of greenhouse with canopy north wall.

from the floor and $\{\alpha_g(1 - r_w)(1 - F_n)(1 - r)\tau S(t)\}$ is absorbed by the floor and the rest is conducted inside the ground. After absorption by the floor, there are convective ($\dot{q}_{cg}$) and radiative ($\dot{q}_{rg}$) losses from the floor to room air. So the reflected energy from the wall and floor can be further transmitted through canopy cover. The convected and radiated energy from the floor raises the temperature of air inside greenhouse. Once the inside air temperature becomes higher than ambient air, than there is heat losses from room air to ambient through different walls/roofs/door. This is shown in Figure 11.12(b).

The transmitted solar radiation through north canopy cover, $F_n(1 - r)^2 \tau^2 S(t)$ in Figure 11.12(b), is generally significant during the winter month. This can be retained inside the greenhouse by providing brick/opaque north wall as shown in Figure 11.12(c). In this case, incident solar radiation on north wall is absorbed by itself after reflection from the surface. After absorption, most of absorbed energy is convected and radiated to room air and rest is lost to ambient through conduction. Further, fraction (F_p) of transmitted solar energy falling on the floor, i.e., $(1 - F_n)(1 - r)\tau S(t)$ is intercepted by the plant as shown in Figure 11.12(d) and rest, i.e., $(1 - F_p)$ part is received by the floor. The magnitude of F_p depends on the status of the plant growth. There is also transpiration ($\dot{q}_{eq}$), convection ($\dot{q}_{cp}$) and radiation ($\dot{q}_{rp}$) losses from the plant to enclosed room air. The complete view of greenhouse has been shown in Figure 11.12(e). The radiative heat loss will take place from the floor and the plant to canopy cover respectively for smaller greenhouse.

11.5.2 Energy Balance

In order to write an energy balance for different component of greenhouse, the following assumptions have been made:

  i. the system is in a quasi-steady state condition,
 ii. the heat capacity of canopy cover, room air and ground has been neglected, and
iii. thermal properties of materials/air are temperature independent.

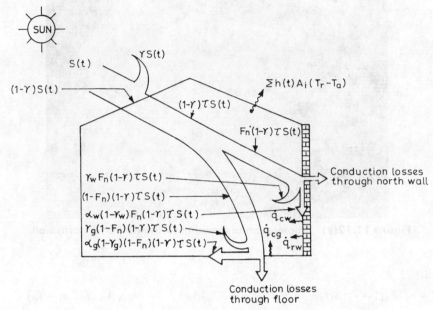

Figure 11.12(c)   View of greenhouse with brick north wall.

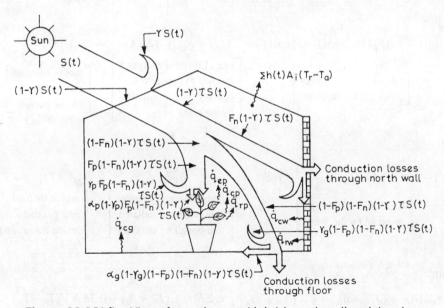

Figure 11.12(d)   View of greenhouse with brick north wall and the plant.

Figure 11.12(e)   Photograph of greenhouse showing brick north wall.

North wall:

$$\alpha_N(1 - r_w)F_n(1 - r)\tau S(t) = [\dot{q}_{cw} + \dot{q}_{rw}]A_n + h_{bN}A_N(T|_{y=0} - T_a)$$

$$\begin{bmatrix} \text{rate of thermal energy} \\ \text{absorbed by north wall} \end{bmatrix} \quad \begin{bmatrix} \text{rate of thermal energy} \\ \text{convected and radiation} \\ \text{by north wall} \end{bmatrix} \quad \begin{bmatrix} \text{rate of thermal energy} \\ \text{conducted to outside} \\ \text{air by north wall} \end{bmatrix} \quad (11.23a)$$

Plant mass:

$$\alpha_p(1 - r_p)F_p(1 - F_n)(1 - r)\tau S(t) = [\dot{q}_{ep} + \dot{q}_{cp} + \dot{q}_{rp}]A_p + M_p C_p \frac{dT_p}{dt}$$

$$\begin{bmatrix} \text{rate of thermal energy} \\ \text{absorbed by the plant} \end{bmatrix} \quad \begin{bmatrix} \text{rate of thermal energy} \\ \text{lost due to transpiration} \\ \text{convection and radiation} \\ \text{by the plant} \end{bmatrix} \quad \begin{bmatrix} \text{rate of thermal} \\ \text{energy stored} \\ \text{by the plant} \end{bmatrix}$$

$$(11.23b)$$

Floor:

$$\alpha_g(1 - r_g)(1 - F_p)(1 - F_n)(1 - r)\tau S(t) = \dot{q}_g A_g + h_g A_g(T|_{x=0} - T_\infty)$$

$$\begin{bmatrix} \text{rate of thermal energy} \\ \text{absorbed by the floor} \end{bmatrix} \quad \begin{bmatrix} \text{rate of thermal energy} \\ \text{convected and radiated} \\ \text{to room air by the floor} \end{bmatrix} \quad \begin{bmatrix} \text{rate of thermal} \\ \text{energy conducted} \\ \text{inside the ground} \end{bmatrix}$$

$$(11.23c)$$

Room air:

$$[\dot{q}_{cw} + \dot{q}_{rw}]A_N + [\dot{q}_{ep} + \dot{q}_{cp} + \dot{q}_{rp}]A_p + \dot{q}_{cg}A_g = \Sigma h(t)A_i(T_r - T_a) + 0.33\,NV(T_r - T_a) \quad (11.23d)$$

where r, r_p, r_g and r_w are reflectivity from canopy cover, plant, floor and north wall; τ is the transmittivity of canopy cover; A_i is the area of different walls/roofs; F_n is the fraction of solar radiation incident on north wall; I_i is the rate of solar energy available on different walls/roofs; α_p, α_g and α_N are absorbtivity of plant, floor and north wall respectively and h's are various heat transfer coefficients and $S(t)$ is given by $S(t) = \Sigma_{i=1}^{6} A_i I_i(t)$; $i = 1, 2, 3, 4, 5$ and 6 refer for east, west north and south walls and roofs.

$\dot{q}_{cw} + \dot{q}_{rw} = h_N(T|_{y=0} - T_r)$; $h_N = 5.7 =$ sum of convective and radiative heat transfer coefficient, W/m² °C

$\dot{q}_{ep} + \dot{q}_{cp} + \dot{q}_{rp} = h_p(T_p - T_r)$; h_p can be calculated by knowing T_p and T_r (Example 11.8b)

$\dot{q}_g = \dot{q}_{cg} + \dot{q}_{rg} = h_g(T|_{x=0} - T_r)$; $h_g = 5.7$ W/m² °C = Sum of convective and radiative heat transfer coefficient

Equations (11.23) can be written in the following form:

$$\frac{dT_p}{dt} + aT_p = f(t) \tag{11.23e}$$

The above equation can be derived after eleminating $T|_{y=0}$, $T|_{x=0}$ and T_r from Equation (11.23d) with help of Equations (11.23a), (11.23c) and (11.23d).

The room air temperature can be obtained from Equation (11.24d) after knowing T_p from Equation (11.23e).

Figures 11.12(f) and 11.12(g) shows the hourly variation of plant and room air temperature with and without north wall for a typical winter day and design parameters of greenhouse. It is observed that there is significant effect of north wall on the plant and room air temperature due to retaining of solar radiation falling on north wall by using opaque wall.

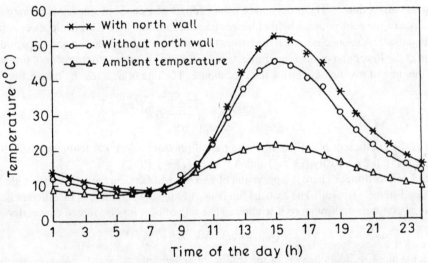

Figure 11.12(f) Hourly variation of room air temperature.

The fluctuation of room air temperature (T_r) plays an important role for survival of plants, hence there is a need to define thermal load levelling (TLL) for greenhouse as,

$$\text{TLL} = \frac{T_{r,\text{max}} - T_{r,\text{min}}}{T_{r,\text{max}} + T_{r,\text{min}}} \tag{11.24}$$

The value of TLL should have more value to smaller value of denomenator (cooling effect) and vice-versa for winter condition for given value of numerator, i.e. $T_{r,\text{max}} - T_{r,\text{min}}$ in comparison to untreated greenhouse.

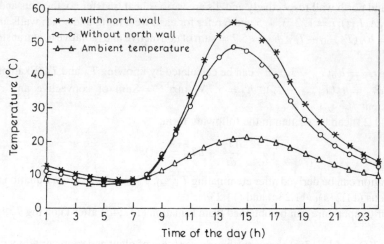

Figure 11.12(g) Hourly variation of the plant temperature.

11.5.3 Cooling

The following are the cooling arrangements available for a greenhouse:

a. Natural convection: The provision of two openings of 1 m × 1 m size is made on the south and the north sides of the roof with the help of hinges fixed at roof top. The UV plastic sheet is fixed in the frame from outside. A sponge is used at the border to avoid heat leakages during winter period. A handle is provided at the lower end of opening frame, to open the roof windows for transfer of trapped thermal energy at the time of overheating inside the greenhouse. The rate of heat loss by convection is given as

$$\dot{Q}_{\text{loss}} = C_d A_v \sqrt{\frac{2\Delta P}{\rho}}\, \Delta P \qquad (11.25a)$$

where C_d is the coefficient of diffusion, P the partial pressure (N/m^2) at temperature $T(^\circ C)$, A_v the area of vent (m^2), ρ the density (kg/m^3) and $\Delta P = P(T_r) - \gamma P(T_a)$.

b. Forced convection: There is a provision of two exhaust fans, on the left and the right side of the door, provided in the east wall. The exhaust fan is usually used when there is insufficient heat transfer due to natural convection through roof windows. The rate of heat loss by forced convection is given as

$$\dot{Q}_{\text{loss}} = 0.33\, N V (T_r - T_a) \qquad (11.25b)$$

where N is the number of air changes, V the volume of greenhouse, T_r and T_a are respectively the room and the ambient temperatures ($^\circ$C).

c. Evaporative cooling: A cooling pad is made with the help of wire mesh, wood ash and aluminum frame. This cooling pad is fixed on the west wall, opposite to the exhaust fan. Arrangement is made to sprinkle water on the cooling pad. The evaporative cooling arrangement is used during peak sunshine hour in extreme climatic condition. The air passing through the cooling pad gets cooled and then propagates inside the greenhouse by forced convection. Due to the movement of cool air, the hot air is carried away from inside the greenhouse, by forced convection. The rate of heat removal by cool water is given by

$$\dot{Q}_{\text{loss}} = 0.33\, N V (T_r - T_w) \qquad (11.26)$$

where T_w is the water temperature ($^\circ$C).

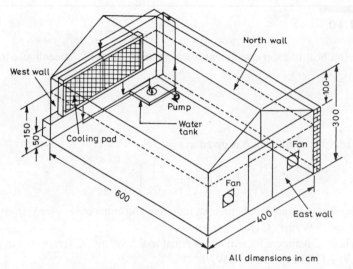

Figure 11.12(h)   Schematic diagram of greenhouse.

Figure 11.12(h) shows the schematic diagram of greenhouse incorporating the provision of forced circulation and cooling pad.

d. Misting arrangement: The water from the storage tank is pumped through filter by a water pump to a lateral pipe provided along the length of the greenhouse, at a certain height. The fogger is fitted so as to provide complete fog inside. The misting is generally used for creating high relative humidity along with cooling inside the greenhouse. Both misting and evaporative cooling arrangement are used in the extreme climatic conditions in the months of May and June.

e. Movable insulation: A movable insulation over the canopy cover is used in both winter and summer condition. In winter, it is used during night or low intensity period to avoid heat losses from inside greenhouse to outside through roofs/walls. It is used during sunshine hour during summer condition to cut-off solar radiation as shown in Figure 11.12(i).

A greenhouse provides an excellent environment to maximize the yield of vegetables, fruits and flowers.

Figure 11.12(i)   Photograph of movable insulation during sunshine hour.

EXAMPLE 11.10

Calculate the overall heat transfer coefficient from enclosed room of a greenhouse to ambient through canopy cover.

Solution

The overall heat transfer coefficient h is given as

$$h = \left[\frac{1}{h_1} + \frac{1}{h_2}\right]^{-1}$$

where h_1 is the inside heat transfer coefficient, radiative and convective, given (from Eq. 2.49a), as
$h_1 = 5.7 + 3.8 V = 5.7\,\text{W/m}^2\,^\circ\text{C}$ (for $V = 0\,\text{m/s}$).
h_2 is the outside heat transfer coefficient and is equal to $9.5\,\text{W/m}^2\,^\circ\text{C}$ (for $V = 1\,\text{m/s}$).
Thus, $h = [(1/5.7) + (1/9.5)]^{-1} = 3.56\,\text{W/m}^2\,^\circ\text{C}$.

EXAMPLE 11.11

What will be the number of air change (N) if room air is replaced three times in one hour?

Solution

Since the room air is replaced three times in one hour, according to definition of air change, number of air change is 3.

EXAMPLE 11.12

Calculate the number of air change per hour if an exhaust fan of capacity 1440 rpm has been used in greenhouse, of volume $60\,\text{m}^3$.

Solution

Let the diameter of fan $(D) = 18" = 45\,\text{cm}$ and rpm $(N') = 1440$
Thus, velocity of air $(V') = (\pi D N')/60 = 33.93\,\text{m/s}$
Cross-section area of fan, $A = (\pi D^2)/4 = 0.159\,\text{m}^2$

Thus, the volume of air sucked out $= V' \times A = 33.93 \times 0.159 = 5.3947\,\mathrm{m}^3/\mathrm{s}$
Number of air changes per second $= 5.3947/60$ and the number of air change per hour is $(5.3947/60) \times 3600 = 324$.
For two fans, number of air change is 648.

EXAMPLE 11.13

Calculate evaporative heat loss coefficient from the plant ($12\,^\circ\mathrm{C}$) to enclosed room air ($10\,^\circ\mathrm{C}$) inside greenhouse for a relative humidity of 0.4.

Solution

The evaporative heat loss coefficient, from Equation (2.76), is

$$h_{ep} = h_0 \frac{P(T_{po}) - \gamma P(T_{ro})}{T_{po} - T_{ro}}$$

where $h_o = 0.016\,h_p$, and h_p is given by Equation (2.49b), and is equal to $2.8\,\mathrm{W/m^2\,^\circ C}$ for $V = 0\,\mathrm{m/s}$. Now, $P(T_{po}) = 1433.5\,\mathrm{N/m^2}$ and $P(T_{ro}) = 1261.83\,\mathrm{N/m^2}$ and substituting in the expression for h_{ep}, gives,
$h_{ep} = 0.016 \times 2.8 \times (1433.5 - 0.4 \times 1261.83)/(12 - 10) = 20.80\,\mathrm{W/m^2\,^\circ C}$.

11.6 HEATING OF BIOGAS PLANT BY SOLAR ENERGY (*Tiwari et al., 1991*)

11.6.1 Introduction

Biogas plants have gained importance in the rural areas of India where animal dung is available in plenty and the temperature requirement for biogas production is normally met by the usual atmospheric temperature. In addition to providing gas (for cooking, lighting and small scale industries), biogas plant also produce good quality manure, a by-product from processed dung.

Biogas consists of 70 percent methane (CH_4) and 30 percent carbon dioxide (CO_2), and has a calorific value of 20 MJoules/m^3. Biogas is produced from the slurry (50% water + 50% dung) at an average temperature of about 35–40 $^\circ$C by chemical and biological process known as *anaerobic fermentation*.

The optimum temperature for maximum production of biogas from slurry is about 37 $^\circ$C. The quantity of gas production depends on the nature of dung. The gas production rate for different type of dung is shown in Table 11.1. The optimum temperature for maximum production is achieved after a number of days, referred to as *retention period*, after feeding the slurry into the digester of the system. The production of biogas starts only after the retention period. The length of the retention period can be reduced by supplying thermal energy to the system by external means, i.e. by heating slurry using either passive or active method.

TABLE 11.1 Potential gas production from different feed stocks

Type of feedstock (dung)	Gas yield per kg (m³)	Normal manure availability per animal per day	Gas yield per day (m³)
Cattle	0.036	10.00	0.36
Buffalo	0.036	15.00	0.54
Pig	0.078	2.25	0.18
Chicken	0.062	0.18	0.01
Human (adult)	0.070	0.40	0.028

On the basis of mean atmospheric temperature, during winter, India is divided into various zones as:

TABLE 11.2 Different zones in India

	Retention period (days)	Temperature	States
Zone I	30	> 25 °C	Andaman & Nicobar Islands, Andhra Pradesh, Goa, Karnataka
Zone II		20–25 °C	Kerala, Maharashtra, Pondichery, Tamil Nadu
Zone III	40	15–20 °C	Bihar, Gujarat, Haryana, Jammu, Madhya Pradesh, Orissa, Punjab, Rajasthan, Uttar Pradesh and West Bengal
Zone IV	55	10–15 °C	Himachal Pradesh, N-E States, Sikkim, Kashmir, Hill Station of UP
Zone V	not suitable for biogas plant	< 10 °C	Srinagar, Leh, Shimla etc.

Biogas is produced by the decomposition of decaying bio-mass and animal wastes, by decomposer organisms, like fungi and bacteria. The process is favored by wet, warm and dark conditions. The final stages are achieved by different species of bacteria classified as (i) aerobic or (ii) anaerobic. Aerobic bacteria flourish in the presence of oxygen while anaerobic bacteria survive in closed conditions, with no oxygen available from the environment. Being accomplished by micro-organisms, the reactions are called *fermentations*, but the term 'digestion' is often used in the anaerobic conditions, that lead to methane. The energy available from the combustion of biogas ranges between 60 and 90 percent of the dry matter heat of combustion of the input material but the gas is obtained from slurries of 50 percent water, thus efficiently providing energy. An advantage of biogas is that the digested effluent is a significantly less health hazard than the input material. Economic benefits of biogas are achieved when the digester is placed in a flow of waste material already present, for example, sewage systems, cattle shed slurries, abattoir wastes, food processing residues and municipal refuse landfill dumps. Biogas generation is used for both small and large scale operation, being particularly attractive for integrated farming.

The general equation for anaerobic digestion as:

$$C_xH_yO_z + \left(x - \frac{y}{4} - \frac{z}{2}\right)H_2O \rightarrow \left(\frac{x}{2} - \frac{y}{8} + \frac{z}{4}\right)CO_2 + \left(\frac{x}{2} + \frac{y}{8} - \frac{z}{4}\right)CH_4 \qquad (11.27)$$

For cellulose this becomes

$$(C_3H_{10}O_5)_n + n\,H_2O \rightarrow 3n\,CO_2 + 3n\,CH_4 \qquad (11.28)$$

Some organic material (e.g., lignin) and all inorganic additives do not digest adding to the bulk of the material and forming a scum which can clog the system. Generally, 95 percent of the mass of the material is water.

It is seen that digestion at higher temperature proceeds at a faster pace than at lower temperature. The optimum temperature is around 37 °C.

The biochemical processes occur in three stages, each stage being facilitated by distinct sets of anaerobic bacteria.

 i. insoluble biodegradable materials, e.g., cellulose, fats, are reduced to soluble carbohydrates and fatty acids. This takes about a day, at 25 °C.

 ii. acid forming bacteria produce mainly acetic and propionic acid. This stage also takes about a day, at 25 °C.

 iii. methane forming bacteria complete the digestion to $\sim$ 70 percent CH_4 and $\sim$ 30 percent CO_2. This takes about 14 days, at 25 °C.

The methane forming bacteria work well in mildly acidic conditions. For successful operation of digester the two basic requirements are the maintenance of constant temperature and suitable input material.

11.6.2 Classification

The biogas digester may be classified as follows:

11.6.2.1 Floating

As shown in Figure 11.13, the slurry from the inlet settlement tank reaches the digester and gets heated by absorbing the solar radiation. Biochemical process takes place and the biogas thus produced comes out through the pipe provided for gas supply. The digested fluid, meanwhile, comes out to the outlet

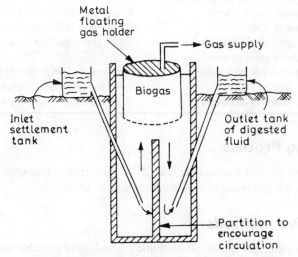

Figure 11.13 Floating gas holder digester.

tank. A partition is provided in the digester to encourage circulation. The floating gas holder provided at the top of the digester helps to keep the pressure constant, rising when the pressure is increased, to let out the biogas and lowering, when the pressure is decreased to stop the supply of biogas.

11.6.2.2 Fixed-dome

Figure 11.14 shows a fixed dome digester. Since the dome is fixed, the pressure keeps on increasing. The process is identical to the floating dome type except for the fact that in this case the pressure in the fermentation compartment is not fixed.

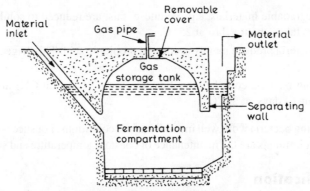

Figure 11.14 Fixed dome digester.

A comparison of the two biogas system is given in Table 11.3.

TABLE 11.3 Comparison of two types of biogas plant

Floating gas holder type	Fixed dome type
Gas is released at constant pressure	Gas is released at variable pressure
Location of defects in gas holder easy	Location of defects in dome difficult
Cost of maintenance is high	Cost of maintenance is low
Capital cost is high	Capital cost is low (for same capacity)
Movable drum does not allow the use of space for other purpose	Space above the drum can be used
Temperature is low during winter	Temperature is high during winter
Life span is short	Life span is comparatively long
Requires relatively less excavation	Requires more excavation work

11.6.3 Heating Process

In order to achieve the desired temperature, i.e., 37 °C, for greater fermentation, while keeping the retention period short, the following heating methods may be used:

11.6.3.1 Passive methods

a. Use of movable insulation during off-sunshine hours: Use of movable insulation and erection of solar canopy (greenhouse) over the dome, would be applicable to vertical floating type biogas plant.

b. Use of water heater/solar still over the dome.

c. Erection of solar canopy (green house) over the dome with a provision for movable insulation.

d. Constructing the digester with insulating material to reduce the bottom and side losses.

11.6.3.2 Active methods

a. Hot charging of the slurry: This method is most suitable for zone III and IV, where large heating is required due to the low mean ambient air temperature.

b. Integration of collectors with digester through heat exchanger (Figures 11.15 and 11.16).

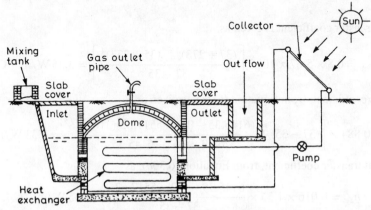

Figure 11.15 Fixed dome type biogas plant integrated with panel of collectors.

Biogas plant is integrated with a panel of collectors through a heat exchanger placed inside the digester. When the heated water from the collector passes through the heat exchanger, heat is transferred to the slurry by conduction and convection, thus raising its temperature, although, an excessive rise in the slurry temperature may lead to the death of anaerobic bacteria. Hence the slurry should be heated slowly so that anaerobic bacteria exist for microbiological process.

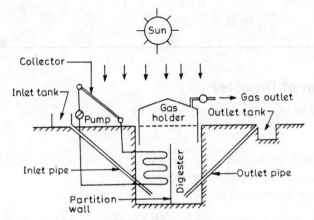

Figure 11.16 Vertical floating type biogas plant integrated with panel of collectors.

EXAMPLE 11.14

Calculate overall heat transfer coefficient from 37 °C slurry to 15 °C ambient air temperature in the case of floating type gas holder.

Solution

The air within the dome may be considered to be at 35 °C. Thus, the heat transfer coefficient from the slurry to the air within the dome would be the sum of radiative, convective and evaporative heat transfer coefficients.

Radiative heat transfer coefficient is given as,

$$h_{rs} = 0.8 \times 5.67 \times 10^{-8} \left[\frac{(37+273)^4 - (35+273)^4}{37-35} \right] = 5.35 \, \text{W/m}^2 \, °\text{C}$$

Convective loss coefficient, from Equation (6.26), is given as,

$$h_{cs} = 0.884 \times \left[(37-35) + \frac{(6145.35 - 0.9 \times 5517.62)(310)}{268.9 \times 10^3 - 6145.35} \right]^{1/3} = 1.33 \, \text{W/m}^2 \, °\text{C}$$

Evaporative heat transfer coefficient, from Equation (6.30a), as

$$h_{se} = 0.016 \times 1.33 \times \frac{6145.35 - 0.9 \times 5517.62}{37-35} = 12.55 \, \text{W/m}^2 \, °\text{C}$$

Thus, the total heat transfer coefficient from the slurry to the enclosed air is $5.35 + 1.33 + 12.55 = 19.23 \, \text{W/m}^2 \, °\text{C}$.

Now, if the thickness of the metal dome is 0.005 m and its conductivity is 62.403 W/mK, and the heat transfer coefficient from the top of the dome to ambient is given by Equation (2.49a), and is $17.1 \, \text{W/m}^2 \, °\text{C}$ (for $V = 3 \, \text{m/s}$)

Thus, the overall heat transfer coefficient is given as

$$U = \left[\frac{1}{19.23} + \frac{0.005}{62.403} + \frac{1}{17.1} \right]^{-1} = 9.04 \, \text{W/m}^2 \, °\text{C}$$

11.6.4 Design of Digester

The energy available from a biogas digester is given by

$$E = \eta H_b V_b \tag{11.29}$$

where η is the combustion efficiency of burners, boilers etc. ($\sim 60\%$), H_b the heat of combustion per unit volume of biogas and V_b the volume of biogas.

Equation (11.29) can be rewritten as,

$$E = \eta H_m f_m V_b \tag{11.30}$$

where H_m is the heat of combustion of methane and f_m the fraction of methane in the biogas.

The volume of biogas is given by

$$V_b = CM_0 \tag{11.31}$$

where C is the biogas yield per unit dry mass of whole input and M_0 is the mass of dry input.
The volume of fluid in the digester is given by

$$V_f = M_0/\rho_m \tag{11.32}$$

where ρ_m is the density of dry matter in the fluid
The volume of the digester is given by

$$V_d = \dot{V}_f t_r \tag{11.33}$$

$\dot{V}_f$ is the flow rate of the digester fluid (water) and t_r is the retention time in the digester.

EXAMPLE 11.15

Calculate (i) the volume of a biogas digester suitable for the output of six cows and (ii) the power available from the digester. The retention time is 14 days; temperature, 30 °C; dry matter consumed, 2.5 kg d^{-1}; biogas yield, 0.24 m^{3}kg^{-1}; burner efficiency, 0.6; methane proportion, 0.7; density of dry matter, 50 kg/m^3.

Solution

Given: $M_0 = (2.5 \, \text{kg} \, \text{d}^{-1}) \, (6) = 15 \, \text{kg} \, \text{d}^{-1}$
From Equation (11.32), we have
Fluid volume, $V_f = M_0/\rho_m = (15 \, \text{kg} \, \text{d}^{-1}/50 \, \text{kg} \, \text{m}^3) = 0.3 \, \text{m}^3 \text{d}^{-1}$
From Equation (11.33), we have
Digester volume, $V_d = V_f t_r = (0.3) \, (14) = 4.2 \, \text{m}^3$
From Equation (11.31), $V_b = (0.24) \, (15) = 3.6$
Equation (11.30) gives
$E = (0.6) \, (28) \, (0.7) \, (3.6) = 42.336 \, \text{MJd}^{-1} = 11.76 \, \text{kW hd}^{-1}$

EXAMPLE 11.16

Calculate top loss coefficient for fixed dome type biogas plant.

Solution

Let us assume that the thickness of the metal of dome is 0.005 m (L_c) and the conductivity (K_c) is 62.403 W/m K and that it is 1" below the ground surface. Let us also assume that the wind velocity is

3 m/s, so that the convective heat loss from the dome to the ambient is 17.1 W/m² °C (from Eq. 2.49a).
Thus, the top loss coefficient is given as,

$$U = \left[\frac{1}{h_1} + \frac{L_c}{K_c} + \frac{L_g}{K_g} + \frac{1}{h_g}\right]^{-1} = \left[\frac{1}{36.64} + \frac{0.305}{62.403} + \frac{0.025}{0.519} + \frac{1}{5.7}\right]^{-1} = 3.91 \text{ W/m}^2 \text{ °C}$$

11.7 SOLAR COOKER

Box-type solar cookers (Fig. 11.17) are suitable mainly for the boiling type of cooking. The cooking
temperature is around 100 °C. The quantities of heat required for physical and chemical changes involved
in cooking are small compared to the sensible heat of increasing food temperature and energy required
for meeting heat losses that normally occur in cooking. Thus, the speed of cooking is practically
independent of heat rate if the thermal losses are compensated, once the contents of the vessel have
been sensibly heated up to the cooking temperature (100 °C).

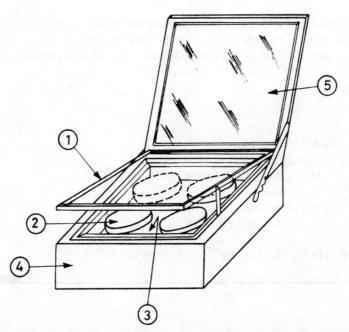

Figure 11.17 Box-type solar cooker. The components of solar cooker are:
(1) Double glass lid, (2) Aluminum cooking pots, (3) Inner tray formed out
of aluminum sheet, (4) Outer box of teakwood with glass wool insulation,
(5) Booster mirror(Mullick et al., 1987).

The complete thermal analysis of the cooker is complex due to the 3-dimensional transient heat
transfers involved. In order to find the heat loss factor, U_L, in the case of flat plate collectors, water
is circulated through the tubes at different temperatures and observations recorded in steady state. In
a solar cooker, there is no control over the temperature and the operation is transient. A quasi-steady
state is achieved when the stagnation temperature is attained.

There are two types of radiation incident on the double glass cover of the container of the solar cooker. One is the direct incident $I(t)$ and the other is the one reflected from the mirror $\rho I'(t)$. The resultant incident radiation $I_T(t)$ can be given as the sum of the two.

In order to simplify the thermal analysis of the solar cooker system, the following assumptions are made:

i. There is no stratification.

ii. The cooking pot is in contact with the inner surface of the cooker and hence their temperatures are the same.

iii. The physical properties of the material to be cooked are the same as that of water.

The rate of energy stored in the pot is $(MC)_w(dT_w/dt)$.
The rate of net thermal energy available to the cooking pot, inside the cooker

$$\dot{Q}_u = F'A_p[(\alpha\tau)I_T(t) - U_L(T_p - T_a)] \tag{11.34}$$

Now, $T_p \approx T_w$, from the above two cases, we have

$$(MC)_w \frac{dT_w}{dt} = F'A_p[(\alpha\tau)I_T(t) - U_L(T_w - T_a)] \tag{11.35}$$

or

$$\frac{dT_w}{[(\alpha\tau)I_T(t) - U_L(T_w - T_a)]} = \frac{F'A_p}{(MC)_w}dt \tag{11.36}$$

Integration of Equation (11.36) gives

$$-\frac{1}{U_L}\ln[(\alpha\tau)I_T(t) - U_L(T_w - T_a)] = \frac{F'A_p}{(MC)_w}t + C \tag{11.37}$$

At $t = 0$, $T_w = T_{w0}$ and hence from Equation (11.37)

$$C = -(1/U_L)\ln[(\alpha\tau)I_T(t) - U_L(T_{w0} - T_a)]$$

Substituting this expression of C in Equation (11.37), we get

$$\frac{1}{U_L}\ln\left[\frac{(\alpha\tau)I_T(t) - U_L(T_w - T_a)}{(\alpha\tau)I_T(t) - U_L(T_{w0} - T_a)}\right] = \frac{-F'A_p}{(MC)_w}t \tag{11.38}$$

or

$$t = -t_0 \ln\left[\frac{(\alpha\tau) - \frac{U_L(T_w - T_a)}{I_T}}{(\alpha\tau) - \frac{U_L(T_{w0} - T_a)}{I_T}}\right] \tag{11.39}$$

where t_0 is the time constant and is given by

$$t_0 = \frac{MC}{U_L F'A_p} \tag{11.40}$$

The above equation can also be written as

$$e^{-t/t_0} = \frac{(\alpha\tau) - \frac{U_L(T_w - T_a)}{I(t)}}{(\alpha\tau) - \frac{U_L(T_{w0} - T_a)}{I_T}} \tag{11.41}$$

EXAMPLE 11.17

Calculate the time taken for water at $40\,^{\circ}\mathrm{C}$ to boil in a solar cooker with the following specifications:
$(\alpha t) = 0.7$, $F' = 0.85$, $U_L = 6\,\mathrm{W/m^2\,^{\circ}C}$, $A_p = 0.36\,\mathrm{m^2}$, $MC = 4 \times 4190\,\mathrm{J/^{\circ}C}$; $T_a = 15\,^{\circ}\mathrm{C}$, $T_{w0} = 40\,^{\circ}\mathrm{C}$, $T_w = 100\,^{\circ}\mathrm{C}$

$$I_T \text{ (with + without reflector)} = (400 + 600) = 1000\,\mathrm{W/m^2}$$

Solution

From Equation (11.40) $t_0 = (MC/U_L F' A_p) = 9129\,\mathrm{s}$
From Equation (11.39)

$$t = -9129 \ln \left\{ \frac{\left[0.7 - \frac{6(100-15)}{1000} \right]}{\left[0.7 - \frac{6(40-15)}{1000} \right]} \right\} = 9703.16\,\mathrm{s} = 161.7\,\mathrm{min} = 2.69\text{ hours.}$$

EXAMPLE 11.18

If the water in a solar cooker with the specification given in Example 11.17 takes 2 hours to boil; calculate the value of initial water temperature (T_{w0}) required.

Solution

t_0 (From Example 11.17) $= 9129\,\mathrm{s}$, $t = 2\,\mathrm{hrs.}$ (given) $= 7200\,\mathrm{s}$
Using Equation (11.39), we have,

$$7200 = -9129 \ln \left\{ \frac{\left[0.7 - \frac{6(100-15)}{1000} \right]}{\left[0.7 - \frac{6(T_{w0}-15)}{1000} \right]} \right\}$$

or
$$0.45 = \frac{0.19}{0.7 - 6\frac{(T_{w0}-15)}{1000}}$$

Solving the above equation, we get

$$T_{w0} = 61.3\,^{\circ}\mathrm{C} \approx 61\,^{\circ}\mathrm{C}$$

EXAMPLE 11.19

Calculate the minimum value of the total intensity (I_T) incident on a solar cooker with the specifications given in Example 11.17, so as to heat the water at $50\,^{\circ}\mathrm{C}$ in 3 hours.

Solution

t_0 (from Example 11.17) $= 9129$ s, $t = 3$ hours $= 10800$ s (given), $T_{w0} = 50\,^{\circ}C$
From Equation 11.39, we get

$$10800 = -9129 \ln \left[\frac{0.7 - \frac{6 \times 85}{I_T}}{0.7 - \frac{6 \times 35}{I_T}} \right]$$

$$0.306 = \frac{0.7 - \frac{510}{I_T}}{0.7 - \frac{210}{I_T}}$$

Solving the above equation, the minimum intensity required.

$$I_T = 917.5 \approx 917 \, W/m^2$$

In order to evaluate the performance of solar cooker and for comparisons between various cookers we require certain parameters that are more or less independent of the climatic variables. Two important parameters have been defined in this respect. One, the first figure of merit is the ratio of optical efficiency and the heat loss factor. Second, the second figure of merit takes into account the heat exchange efficiency factor (F') (Mullick et al., 1987).

The first figure of merit, F_1, is defined as,

$$F_1 = \frac{T_{ps} - T_{as}}{I_s} \tag{11.42}$$

where T_{ps} is the plate stagnation temperature, I_s and T_{as} are, respectively, the insolation on a horizontal surface and the ambient temperature at the time when the stagnation temperature is reached.

A lower permissible limit of the value of F_1, may be specified so as to ensure a minimum level of thermal performance. For example, if it is stipulated that F_1, should equal or exceed 0.14, in a region where solar radiation and ambient temperature are $I_s = 800 \, W/m^2$ and $T_{as} = 17\,^{\circ}C$, respectively, then we have

$$\frac{T_{ps} - T_{as}}{I_s} = 0.14$$

or

$$T_{ps} \geq 129\,^{\circ}C$$

i.e. the plate temperature would equal or exceed $129\,^{\circ}C$. By specifying a suitable minimum value of F_1 (between 0.12 and 0.16, depending upon the climate) it may be ensured that the stagnation temperature is sufficiently high so that the boiling type cooking is possible.

Replacing $\alpha\tau/U_L$ by the factor F_1, Equation (11.36) can be rewritten as

$$dt = \frac{(MC)_w dT_w}{(\alpha\tau) A_p F' \left[I_T(t) - \frac{1}{F_1}(T_w - T_a) \right]} \tag{11.43}$$

Assuming that insolation $I_T(t)$ and ambient temperature T_a are constant then the above equation can be integrated over the time interval t, during which water temperature rises from T_{w1}, to T_{w2}.

$$t = \frac{-F_1(MC)_w}{A'F(\alpha\tau)} \ln \left[\frac{I_T(t) - \frac{1}{F_1}(T_{w2} - T_a)}{I_T(t) - \frac{1}{F_1}(T_{w1} - T_a)} \right] \tag{11.44}$$

However, the insolation and ambient temperature vary, so that exact integration of Equation (11.44) may not be possible. The average values of insolation and ambient temperature may be used. When insolation varies excessively and/or when higher accuracy is desired, integration may be carried out by summation of finite differences. Summing up Equation (11.44) over small, equal temperature intervals (ΔT_w), we have

$$\Sigma \Delta t = \frac{(MC)_w \Delta T_w}{A' F(\alpha \tau)} \sum_{T_{w1}}^{T_{w2}} \left[\frac{1}{I_T(t) - \frac{1}{F_1}(T_w - T_a)} \right] \qquad (11.45)$$

where T_w is the average water temperature over an interval. Insolation may be assumed constant and equal to the average value (I_T) over the time interval.

The first figure of merit is obtained by keeping the solar cooker, without the cooking pots, in the sunshine in the morning on a clear day. The cooker plate temperature, the ambient temperature and the total insolation on a horizontal surface are noted at regular intervals of time. The observations are continued upto after solar noon and stagnation temperature is determined and first figure of merit calculated. The first figure of merit ensures that the glass covers have a good optical transmission and the cooker has a low overall heat loss factor. However, for a good performance it is equally important that there is a good heat transfer to the contents of the vessel and the heat capacity of cooker interiors is small. This implies that the system to be considered should consist of the cooker and the vessels together and tested with a "full load". The tests for the second figure of merit consist of operating the solar cooker with a full load of vessels with contents. The cooker is kept in the sunshine in the forenoon and the water temperature is allowed to rise gradually until it reaches the boiling point. Such a temperature plot is shown in Figure 11.18.

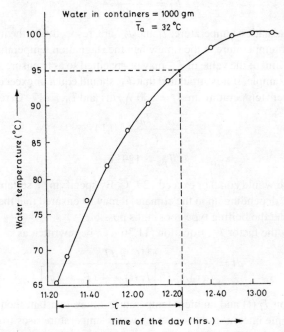

Figure 11.18 Variation of temperature of water in the vessels with time of the day (with load). (Mullick *et al.*, 1987)

The upper limit of the water temperature, T_{w2}, for the time period analyzed could have been taken as $100\,°C$, the boiling temperature, but as the water temperature approaches $100\,°C$, the rate of variation of water temperature approaches zero. Thus, there is a great uncertainty in deciding the termination point of the time interval t. The upper limit of sensible heating, T_{w2}, should be fixed in the temperature range $90–95\,°C$.

11.8 DESIGN METHOD

There are various methods for the design of active systems, which may be used according to the operating temperature range. Described here are the design methods for active systems, for which the f-chart method does not apply.

The first of the utilizability methods is monthly average hourly utilizability for flat-plate collectors (the φ method). The other methods are the generalized φ method, the φ method and the φ, f-chart method. The utilizability methods require knowledge of the collector inlet fluid temperature, which is often not known.

11.8.1 Hourly Utilizability

Utilizability is defined as the fraction of the incident solar radiation that can be converted to useful heat. It is the fraction utilized by a collector having $F_R(\alpha t) = 1$ and operating at a fixed inlet to ambient temperature difference. Although this collector has no optical losses and has a heat removal factor of unity, the utilizability is always less than 1, due to the thermal losses.

$$\dot{Q}_u = A_c F_R[(\alpha\tau)I(t) - U_L(T_i - T_a)]^+ \tag{11.46}$$

The radiation level must exceed a critical value before useful output is produced. This critical level is found by putting $\dot{Q}_u = 0$ in the above equation.

$$I_{TC} = \frac{F_R U_L(T_i - T_a)}{F_R(\alpha\tau)} \tag{11.47}$$

The useful output of the collector, in terms of the critical radiation level is

$$\dot{Q}_u = A_c F_R(\alpha\tau)(I_T - I_{TC})^+ \tag{11.48}$$

If the critical radiation level, for a particular hour for a month (N days) is constant, then the monthly average hourly collector output for this hour will be

$$\bar{Q}_u = \frac{A_c F_R(\alpha\tau)}{N} \sum_{}^{N}(I_T - I_{TC})^+ \tag{11.49}$$

The monthly average radiation in this particular hour is I_T, the average useful output can be given as

$$\bar{Q}_u = A_c F_R(\alpha\tau)\bar{I}_T\phi \tag{11.50a}$$

where the utilizability ϕ is defined as

$$\phi = \frac{1}{N} \sum_{}^{N}\frac{(I_T - T_{TC})^+}{I_T} \tag{11.50b}$$

ϕ is determined at the dimensionless critical radiation level. X_c, defined as I_{TC}/I_T, can be represented in terms of collector parameters as:

$$X_c = \frac{I_{TC}}{I_T} = \frac{F_R U_L (T_i - T_a)}{F_R (\alpha\tau)_n \frac{(\alpha\tau)}{(\alpha\tau)_n} \bar{I}_T}$$

where $(\alpha\tau)/(\alpha\tau)_n$ is determined for the mean day of the month and the appropriate hour angle.

The main advantage of hourly utilizability is in estimating the output of processes which have a critical radiation level that changes significantly through the day. The change may be due to a regular and pronounced diurnal temperature variation or due to shifts in collector inlet temperature caused by characteristics of the load.

11.8.2 Daily Utilizability

Daily utilizability is defined as the sum for a month, over all hours and all days, of the radiation on tilted surface that is above a critical level, divided by the monthly radiation.

$$\bar{\phi} = \sum_{days} \sum_{hours} \frac{(I_T - I_{TC})}{\bar{I}_T N} \qquad (11.51)$$

The expression for I_{TC} is similar to the one given earlier except that monthly average transmittance absorptance must be used in place of $(\alpha\tau)$ and T_i and T_a are the inlet and daytime temperatures for the month.

$$I_{TC} = \frac{F_R U_L (T_i - \bar{T}_a)}{F_R (\alpha\tau)} \qquad (11.52)$$

The monthly average daily useful energy gain is given by

$$\Sigma \bar{Q}_u = A_c F_R (\overline{\alpha\tau}) \bar{I}_T \bar{\phi} \qquad (11.53)$$

11.8.3 Hybrid System

Active air or liquid collectors may be used in heating of buildings in which the storage is provided in the structure of the building. This hybrid system has the advantage that the thermal losses from the receiver (the collector) can be controlled by shutdown when the radiation is below the critical level and a separate storage system is eliminated. It has the disadvantages of significant temperature swings needed in the building to provide storage and, for given building structure and allowable temperature swing, upper limits on the amount of solar energy that can be delivered to the building.

Active-collection, passive storage systems may be used in buildings that cannot accommodate large active systems. The systems are simple. Air heating collectors or liquid heating collectors with liquid-air heat exchanger may be used to supply energy that is distributed by air flow to the heated spaces. Controls are required to turn on collector fluid flow when collection is possible and to turn off the collector or otherwise dump energy when the building interior temperature attains a maximum allowable value.

The output of an active collector for a month, is

$$\Sigma Q_u = A_c \bar{I}_T F_R (\overline{\alpha\tau}) N \bar{\phi}_c = A_c F_R \bar{S} N \bar{\phi}_c \qquad (11.54)$$

where φ_c is the monthly average utilizability associated with collecting energy. The critical radiation level to be used in the determination of ϕ_c is

$$I_{TC,C} = \frac{F_R U_L (\bar{T}_i - T_a)}{F_R(\alpha\tau)} \tag{11.55}$$

$\bar{T}_i$ is the monthly average inlet temperature, that is, the building temperature during collection. For the limiting cases of zero and infinite building capacitances, T_i will be constant. For real cases, however, T_i will vary, but the variation does not significantly affect ΣQ_u.

For a building with infinite storage capacity, the monthly energy balance

$$L_{A,i} = (L - \Sigma Q_u)^+ \tag{11.56}$$

For a building with zero storage capacity, the energy will have to be dumped if the solar gain exceeds the load. The intensity of the radiation incident on collector, which just meets the building heating load, without dumping is called the *dumping critical radiation level*, $I_{TC,d}$, given by

$$I_{TC,d} = \frac{(UA)_h(\bar{T}_b - \bar{T}_a) + A_c F_R U_L(\bar{T}_i - \bar{T}_a)}{A_c F_R(\overline{\alpha\tau})} \tag{11.57}$$

where $(UA)_h$ is the overall loss coefficient-area product of the building, T_b the average building base temperature, and T_i the average building interior temperature.

Energy above $I_{TC,d}$ must be dumped and can be given as

$$Q_D = A_c F_R \bar{S} N \bar{\phi}_d \tag{11.58}$$

where f_d is the monthly average utilizability (more appropriately, unutilizability) based on $I_{TC,d}$.

The energy supplied by the collector which is useful in meeting the heating load of a zero capacitance building is the difference between the total energy collected and the energy dumped.

$$Q_{u,b} = \Sigma Q_u - Q_D = A_c F_R \bar{S} N(\bar{\phi}_c - \bar{\phi}_d) \tag{11.59}$$

The auxiliary energy required

$$L_{A,Z} = (L - \Sigma Q_u + Q_D)^+ \tag{11.60}$$

Equations (11.56) and (11.60) establish the limits of auxiliary energy requirements. The auxiliary energy requirements for a building with finite capacitance can be estimated by correlations of f, the fraction of the month's energy supplied from solar energy, with a solar-load ratio X and a storage-dump ratio Y.

$$X = A_c F_R \bar{S} N / L \tag{11.61}$$

and

$$Y = \frac{C_b \Delta T_b}{A_c F_R \bar{S} \bar{\phi}_d} = \frac{N C_b \Delta T_b}{Q_D} \tag{11.62}$$

For the month, the auxiliary energy requirement is $L_A = L(1 - f)$.

11.9 SOLAR FRACTION

A term importantly used in the concept of solar fraction is the daily auxiliary energy given as

$$\text{Auxiliary energy} = \int_0^{\text{one day}} \dot{M} C_w (T_D - T_w)^+ \tag{11.63}$$

where T_D is the temperature at which hot water is required from the system. '+' sign indicates that only positive value of $(T_D - T_w)$ is to be considered, that is, auxiliary energy is required only when the outlet temperature (T_w) from the system is less than T_D. When $T_w > T_D$, the auxiliary energy required is equal to zero. The calculations are made for an average day of each month.

The solar fraction (f) which can be used to optimize the collector size for a given requirements is defined as,

$$f = 1 - \frac{\int_0^{12 \text{ months}} \dot{M} C_w (T_D - T_w)^+ n_d dt}{\int_0^{12 \text{ months}} \dot{M} C_w (T_D - T_{in}) n_d dt} \tag{11.64}$$

In this case $T_{in} \approx T_a$. Thus

$$f = 1 - \frac{\int_0^{12 \text{ months}} \dot{M} C_w (T_D - T_w)^+ n_d dt}{\int_0^{12 \text{ months}} \dot{M} C_w (T_D - T_a) n_d dt} \tag{11.65}$$

when $T_D = T_w$, i.e. the outlet temperature from the system is the desired one, then the second term in Equation (11.65) reduces to zero and the value of f, is a maximum one. Thus, no auxiliary energy is required in this case. However, if $T_w = T_a$, i.e. the outlet temperature of water from the system is equal to the ambient temperature, the second term in the equation is one and the value of f, reduces to zero. Thus, no solar heating is taking place and auxiliary energy is required to increase the water temperature upto the desired level.

11.10 SOLAR COOLING

11.10.1 Introduction

Solar energy has been used in cooling cycles for two related purposes, (i) to provide comfort cooling, and (ii) to provide refrigeration for food preservation. Cooling is important in space conditioning of the buildings in warm climates. Cooling loads and availability of solar radiation are approximately in phase. Solar air conditioning can be done by three classes of systems: absorption cycles, desiccant cycles, and solar-mechanical processes. Within these classes the conditions may vary, use of continuous or intermittent cycles, hot or cold side energy storage, various control strategies, various temperature ranges of operation, different collectors, etc. Theoretically, the efficiencies of solar cooling processes can be immense, but practically, the temperature constraints in the operation of collectors limits them.

Cooling, like heating, is an expensive process. Reduction in cooling loads through careful building design and insulation is desirable and will be less expensive than provision of additional cooling. Good building design and construction can to an extent reduce the load on any air conditioning or heating system, but we are concerned with cooling loads that cannot be avoided by building design.

11.10.2 Solar Absorption Cooling

The basic absorption process utilizes the difference in the binding of a fluid (known as *refrigerant*) in two vessels at different temperatures; the fluid can be bound by a liquid or a solid solvent (known as *absorbent*). The underlying principle of the process can be understood as follows:

As shown in Figure 11.19, vessel I contains the refrigerant (R) bound to the absorbent (A), vessel II contains the pure refrigerant. When the two vessels are at the same temperature; with the valve between them closed, the vapor pressure in vessel II will be higher that in vessel I, i.e. $P_I < P_{II}$, when $T_I = T_{II}$. When the valve opens, the refrigerant gas flows from vessel II to I due to the existing pressure difference. The vapor pressure in I becomes higher than the equilibrium vapor pressure and the vapor condenses

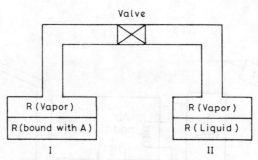

Figure 11.19 Basic absorption process.

in it. The temperature in vessel I rises, due to the heat of condensation. Reverse process taking place in vessel II reduces its temperature.

Finally, the vapor pressure tends to equalize in the two vessels, as a result of mass transfer. This causes a temperature difference between the vessels; thus $P_I = P_{II}$ and $T_I > T_{II}$. When heat is supplied to the cold vessel, the pressure equilibrium is distributed; vapor pressure in II exceeding the vapor pressure in I, which causes mass transfer and consequently, flow of heat from vessel II to I. Thus heat gets pumped from a lower temperature to a higher temperature. The absorption process may be utilized for cooling (when the cold vessel takes the heat from the surrounding or heating (when the heat developed in the warm vessel is utilized). The heat pumping ability of the device thus gets consumed and to restore it, the system must be regenerated by reversing the mass transfer and supplying the heat to I while withdrawing heat from it. In between the phases of heat pumping and regeneration, energy can be stored. Thus, an absorption process has three main characteristics, viz. cooling, heating and energy storage. The success of the absorption process depends upon the choice of suitable pair of refrigerant and absorbent.

The selection of fluids is generally based on:

i. Chemical and physical properties of fluid.
ii. Acceptability range for certain thermophysical and thermodynamic properties of the fluid.

The desirable characteristics of a refrigerant absorbent mixture for an absorption system are (i) low viscosity of the solution under operating conditions (so as to reduce the pump work), (ii) freezing point of the liquid should be lower than the lowest temperature in the cycle, (iii) good chemical thermal stability, and (iv) the components should be non-corrosive, non-toxic and non-flammable.

The thermodynamic properties of the solutions of refrigerant and absorbent are critical in the determination of its suitability for absorption systems. The combination of two should satisfy two primary thermodynamic requirements: (a) high equilibrium solubility of the refrigerant in the absorber, and (b) a larger difference in boiling points of the absorbent and refrigerant. The commonly used combinations are LiBr-H_2O and H_2O-ammonia. The absorption coolers, using solar energy, can operate in two ways: (i) by use of continuous coolers, that are similar in construction and operation to conventional gas or steam-fired units, with energy supplied to the generator from the solar collector-storage-auxiliary system, whenever cooling is required, and (ii) by use of intermittent coolers.

A possible arrangement for continuous absorption cycle is shown in Figure 11.20.

The temperature limitations of flat plate collector limit the consideration of commercial machines to lithium-bromide-water (LiBr-H_2O) systems. The principle advantage of this system are (i) water as a refrigerant has a very high latent heat of vaporization, (ii) the absorbent is non-volatile, (iii) the system operates at low pressure and hence low pumping power requirement, and (iv) the material is non-toxic and non-flammable.

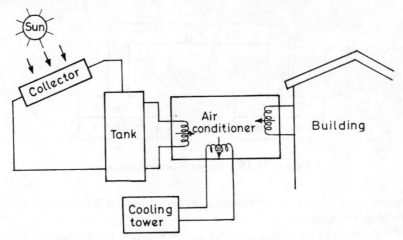

Figure 11.20   Schematic of a solar absorption air conditioning system.

LiBr-H_2O machines require cooling water to cool the absorber and condenser, and in most of the applications, a cooling tower is required. The disadvantage of using water as the refrigerant is that a water cooled condenser is required to attain temperatures corresponding to air conditioning. The temperatures corresponding to refrigeration cannot, however, be reached even with a water cooled condenser. This is due to the fact that LiBr is not sufficiently soluble in water to permit the absorber to be air cooled. The combination is corrosive to the materials of construction of the cycle (hydrogen, which is the product of the corrosion reaction, increases the system pressure, hence lowering the efficiency of the cycle). Also, during the system operation the combination has the problem of salt crystallization and cannot be used for a long period.

Ammonia-water system is mostly used for industrial absorption air conditioning and refrigeration.

11.10.3 COP of Absorption Cooling

The most common approach to solar cooling is the operation of absorption air-conditioners with energy from flat plate collector and storage systems shown in Figure 11.21.

An overall steady-state energy balance on the absorption cooler indicates that the energy supplied to the generator and evaporator must be equal to the energy removed from the machine via the coolant flowing through the absorber and condenser plus the net losses occurring to the surroundings.

$$Q_G + Q_E = Q_A + Q_C + Q_{losses} \qquad (11.66)$$

where Q_G, Q_E, Q_A, Q_C and Q_{losses} are respectively the energy supplied to generator, evaporator, the energy removed from the absorber and from condenser and the energy lost to the surroundings.

The thermal coefficient of performance *COP* is defined as the ratio of energy into the evaporator (Q_E) to the energy into the generator (Q_G).

$$COP - Q_E/Q_G \qquad (11.67)$$

It is a useful index of performance in solar cooling. The thermal *COP* (coefficient of performance) is usually in the range of 0.6 to 0.8 and the major effect of solar energy temperature variation to the generator is to vary Q_E, the cooling rate.

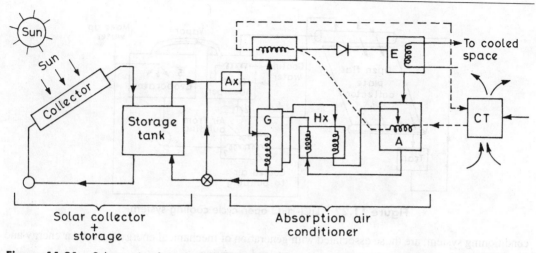

Figure 11.21 Schematic of a solar operated absorption air-conditioner with essential components of the cooler: A, absorber, B, generator, C, condenser, E, evaporator, HX, heat exchanger to recover sensible heat, CT, cooling tower, AX, the auxiliary energy source. (Duffie and Beckman, 1991)

A COP_e is the ratio of cooling to electrical energy used to provide air and liquid flows, operate controls, etc.

$$COP_e = \frac{Q_E}{\text{Electric input}} \tag{11.68}$$

The average COP over a long term can be evaluated as

$$\overline{COP} = \frac{\int Q_E dt}{Q_G dt} \tag{11.69}$$

11.10.4 Solar Desiccant Cooling and Absorption Cooling

Another type of solar airconditioners is based on open-cycle dehumidification-humidification process. They take in air from outside or from the building, dehumidify it with a solid or liquid desiccant, cool it by exchange of sensible heat, and then cool it to the desired state by evaporation. The desiccant is regenerated with solar energy. The major components of a solar desiccant cooling system are heat exchanger, heat and mass exchanger for dehumidification, and evaporative coolers.

A typical solar desiccant cooling system is shown in Figure 11.22.

The absorbing liquid is a solution of lithium chloride in water. From the absorber (1), dilute LiCl solution is transferred by a pump (2) to a heat-exchanger-distributor header (3) and then to an open flat-plate collector (4) where the water evaporates. The concentrated solution comes back to the absorber via a heat exchanger, for recovery of sensible heat. Water is cooled in the evaporator (5) with transfer of water vapor from evaporator to the absorber. The chilled water from the evaporator is sent to air-to-water heat exchanger (7) with the help of pump (6) and thus cools the building air. Provision is made to deaerate solutions, recover sensible heat, and add makeup water.

11.10.5 *Solar-mechanical Cooling*

Another cooling method involves the coupling of a solar powered Rankine-cycle engine with a conventional air conditioning system. The problems associated with solar operation of a conventional air

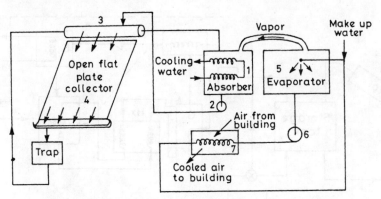

Figure 11.22 LiCl-H$_2$O open cycle cooling system.

conditioning system, are those associated with generation of mechanical energy from solar energy and adaptation of air conditioning equipment for part-load operation.

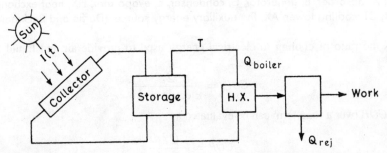

Figure 11.23 Schematic of a solar-operated Rankine-cycle cooler.

A simple Rankine-cycle cooling system is shown in Figure 11.23. Energy from the storage tank is transferred, through a heat exchanger, to a heat engine. The heat engine exchanges energy with the surroundings and produces work. As seen from Figure 11.24(a), the efficiency of the solar collector decreases as the operating temperature increases, while the efficiency of the heat engine increases with the increases of operating temperature. The overall system efficiency for conversion of solar energy to mechanical work is shown in Figure 11.24(b).

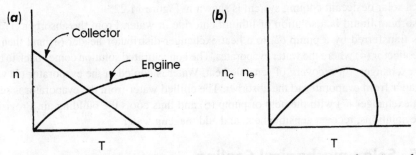

Figure 11.24 (a) Collector and power cycle efficiencies as a function of operating temperature (b) Overall system efficiency.

PROBLEMS

11.1 Write down the energy balance equation for built-in-storage and shallow solar pond water heater with hot water withdrawal.

Hint: Add hot water withdrawal term $\dot{M}C_w(T_w - T_{IN})$ on the R.H.S. of Equation (11.1a) and Equation 9.42 with $h_{1w} = h_c$.

11.2 Write down the energy balance equation for SSP water heater during off-sunshine hours.

11.3 Find out analytical expression for water temperature for SSP water heater.

Hint: Refer Section 9.9.4 and Problem 11.1.

11.4 Calculate water temperature for Problem 11.1 for given climatic and design parameters for different water depths.

a. Climatic parameters: Refer Problem 9.13a.

b. Design parameters:

$\alpha\tau = 0.65$, $h_b = 0.8\,\text{W/m}^2\,°\text{C}$, $h_3 = 100\,\text{W/m}^2\,°\text{C}$, $h_2 = 14.1\,\text{W/m}^2\,°\text{C}$, $h_1 = 50\,\text{W/m}^2\,°\text{C}$, $C_w = 4190\,\text{J/kg}\,°\text{C}$.

Hint: Select $M_w = 20, 40, 60, 80$ and $100\,\text{kg}$ and use expression for water temperature derived in Problem 11.1. Also plot η_{ov} (%) versus water depth and check the behavior of curve of Figure 11.3.

11.5 What will be the energy balance equation for built-in storage water heater during off-sunshine hours in Problem 11.1?

Hint: Use $I(t) = 0$ in Problem 11.1.

11.6 Derive the expression for water temperature in forced circulation mode with and without water withdrawal, for the following parameters:

For climatic parameters refer Problem 9.13a.

Collector area $(A_{CN}) = 10\,\text{m}^2$, $W_t = 2095 \times 10^3\,\text{J/°C}$, $U_T = 0.8\,\text{W/m}^2\,°\text{C}$, $\dot{M} = 60\,\text{kg/h}$, Hot water withdrawal duration 4 pm to 6 pm.

Hint: Use T_{foN} from Equation 3.88 and solve Equation (11.12) in a way similar to that done for a solar still in chapter 9. No water is withdrawn between 6 AM to 4 PM ($\dot{M} = 0$).

11.7 Calculate the water temperature for Problem 11.5.

Hint: See Problem 11.5.

11.8 What will be the energy balance equation for a natural circulation water heater for constant collection withdrawal ($T_m = $ constant)?

Hint: Use $T_m = T_{m0}$ constant in Equation (11.9) and $\frac{dT_{m0}}{dt} = 0$.

11.9 What is the effect of storage for constant collection of solar water heater in a natural as well as forced circulation mode of operation?

Hint: No effect.

11.10 How will you define an overall thermal efficiency for constant collection mode?

Hint: Use $T_{mi} = T_{m0}$ in Equation (11.11b).

11.11 Solve Equations (11.19) and (11.20) for T_{I} and T_{II}.

Hint: Multiply Equation (11.20) by α and add it to Equation (11.19) and assume $a_1 + \alpha b_1 = c$ and, $a_2 + \alpha b_2 = \alpha c$. Combined equation becomes first order differential equation which can be easily solved.

11.12 What will be the effect of flow rate and length of heat exchanger on $\dot{Q}_{gain,c}$ and $\dot{Q}_{loss,l}$?

Hint: Discuss Equations (11.15) and (11.17) for limiting cases, i.e., $\dot{m} \to 0$ and ∞, $L \to 0$ and ∞.

11.13 Derive an expression for time interval (Δt) required to heat indoor swimming pool from T_{w1} to T_{w2} as a function of climatic and design parameters of a collector.

Hint: Solve Equation (11.21) for $0 - \Delta t$ time interval and see Example 9.8.

11.14 Calculate the power available from a biogas digester suitable for the output of one dozen cows for 10 days retention time. The other parameters are the same as given in Example 11.15.

Hint: See Example 11.15.

11.15 Calculate the power available from a biogas digester suitable for output of one dozen animals of Table 11.1 for 14 days retention time. The other parameters are the same as given in Example 11.15.

11.16 Write down the energy balance equation for active heating biogas digester in a forced circulation mode.

Hint: Similar to Equation 11.21.

11.17 Solve the energy balance equation of Problem 11.3.

Hint: See Problem 11.3.

11.18 Calculate Δt for Problem 11.16 for different collector area for 100 kg slurry to raise the temperature from $T_{s0} = 15\,^{\circ}\mathrm{C}$ to $T_s = 38\,^{\circ}\mathrm{C}$.

Hint: $A_c = 2\,\mathrm{m}^2, 4\,\mathrm{m}^2, 6\,\mathrm{m}^2, 8\,\mathrm{m}^2$, $I(t) = 500\,\mathrm{W/m}^2$, $T_a = 15\,^{\circ}\mathrm{C}$.

11.19 Calculate the total time required to boil the water of mass 2 and 4 kg for initial temperature $40\,^{\circ}$, $60\,^{\circ}$ and $80\,^{\circ}\mathrm{C}$. The other parameters are given in Example 11.17.

Hint: Consider $T_w = 100\,^{\circ}\mathrm{C}$ and use Equation (11.39).

11.20 Calculate the total time required to boil the water from $60\,^{\circ}\mathrm{C}$ for 6 and 8 kg of water in a solar cooker. The other parameters are given in Example 11.17.

11.21 Calculate the time required to boil the water without reflector for Problem 11.20.

Hint: See Problem 11.20.

11.22 Calculate the bottom and side loss coefficient for fixed dome digester of a biogas plant.

Hint: See Example 11.8 and 11.16.

11.23 Find out the rate of air flow in terms of number of air changes.

Hint: $m = NV/3600$.

11.24 Calculate the total heat transfer coefficient from the top of convective zone for parameters of Example 11.5 for different relative humidity.

Hint: See Example 11.5 and consider $r = 10$ percent, 20 percent, 60 percent and 80 percent.

11.25 Calculate 'Δt' for different flow rate for Example 11.6b.

Hint: Repeat calculation of Example 11.6b for $\dot{M} = 18\,\mathrm{kg/hr}$ and $36\,\mathrm{kg/hr}$.

11.26 Calculate 'Δt' for different collector area for Example 11.6b.

Hint: Consider $A_{CN} = 60$ and $80\,\mathrm{m}^2$.

11.27 Calculate 'Δt' for $M_w = 500$ and $1000\,\mathrm{kg}$ for Example 11.6b.

Hint: See Problem 11.25.

11.28 Calculate the time required 'Δt' to heat the swimming pool from $15\,^{\circ}\mathrm{C}$ to $20\,^{\circ}\mathrm{C}$ for Example 11.9 for the following cases:

a. Changing heat capacity, i.e., $2.095 \times 10^6\,\mathrm{J/kg}$, 4.19×10^6 and $6.285 \times 10^6\,\mathrm{J/kg}$.

b. Collector area, i.e., $A_{CN} = 20\,\mathrm{m}^2$, $40\,\mathrm{m}^2$, $60\,\mathrm{m}^2$ and $80\,\mathrm{m}^2$ with and without evaporation for glazed collector.

Hint: See Example 11.9.

11.29 Repeat Problem 11.28 without evaporation and make your comments on outcome.

   Hint: See Problem 11.28 and the value of Δt without evaporation should be lowered.

11.30 Derive an expression for a and $f(t)$ for Equation (11.23e).

   Hint: Get $T|_{y=0}$ and $T|_{x=0}$ from Equations (11.23a) and (11.23c) respectively and substitute in Equation (11.24d). Then obtain an expression for T_r in terms of T_p only and put this in Equation (11.23a) and rearrange the equation.

11.31 Derive an expression for the plant temperature inside the greenhouse.

   Hint: Solve Equation (11.23e) (see Section 9.9.4).

11.32 Calculate evaporative (transpiration) loss for different relative humidity for Example 11.13.

   Hint: Consider $\gamma = 0.2$, 0.6, and 0.8.

CHAPTER *12*

Energy Storage

12.1 INTRODUCTION

Solar energy, being a time-dependent energy resource, must be stored and used in day-to-day life. The energy storage must be considered, keeping in view, a solar process system, the major components of which are the solar collector, storage units, conversion devices (such as air conditioners or engines), loads, auxiliary (supplemental) energy supplies, and control systems. The performance of each of these components is related to that of the other. The dependence of the collector performance on temperature makes the whole system performance sensitive to temperature.

The optimum capacity of an energy storage system depends on the expected time dependence of solar radiation availability, the nature of loads to be expected on the process, degree of reliability, the manner in which auxiliary energy is supplied, and an economic analysis weighing the relative use of solar and auxiliary energy. The solar thermal energy storage can be classified as follows:

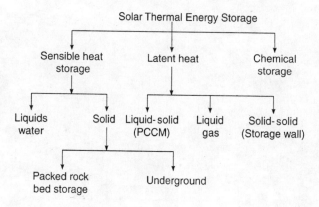

12.2 SENSIBLE HEAT STORAGE

Thermal energy may be stored as sensible heat or latent heat. Sensible heat storage systems use the heat capacity and the change in temperature of the material during the process of charging and discharging.

412

The temperature of the storage material rises when energy is absorbed and drops when energy is withdrawn. The charging and discharging operations, in a sensible heat storage system, can be expected to be completely reversible for an unlimited number of cycles, over the life-span.

The sensible heat Q gained or lost by a material in changing temperature from T_1 to T_2 is:

$$Q = M \int_{T_1}^{T_2} C_p dT = V \int_{T_1}^{T_2} \rho C_p \, dT \qquad (12.1)$$

where M is the mass (kg), C_p the specific heat (J/kg °C), ρ the density (kg/m^3) and V is volume of the material (m^3). As seen from Equation (12.1), the higher the specific heat and density of the material, more will be the energy stored in a given volume of the material. However, there are several other parameters affecting the performance of the system, viz., the operating temperature, thermal conductivity, thermal diffusivity, vapor pressure, compatibility between the storage material and the container, stability of the material at the highest temperature of the cycle and the cost of the system.

EXAMPLE 12.1

Calculate the thermal energy lost from an insulated tank filled with 1000 liter olive oil for a temperature drop of 100°C from boiling point.

Solution

From Appendix III, for olive oil (Table IIIf) $\rho = 920 \, \text{kg/m}^3$, $C = 1970 \, \text{J/kgK}$, T (boiling) $= 570 \, \text{K}$. Using Equation (12.1), we have

$$Q = 1000 \times 10^{-3} \times 920 \times 1970 \times 373 = 6.76 \times 10^8 \text{ Joules } [1\text{L} = 10^{-3} \, \text{m}^3].$$

The efficiency of thermal storage can be defined as the ratio of heat output to heat input, heat output being lower than the input by the amount of heat losses. In the case of sensible heat storage, however, the temperature of stored medium drops due to heat losses and the energy is available at a lower temperature. Thus the efficiency, in this case, can be defined as the ratio of availability of the energy discharged to the availability of the energy charged.

Sensible heat storage, on the basis of the heat storage media, may be classified as: (i) liquid media storage (ii) solid media storage and (iii) dual media storage.

12.3 LIQUID MEDIA STORAGE

Of the available liquids, water can be considered to be the most suitable liquid media for storage below 100°C. Water has the following advantages:

a. It is abundant and inexpensive.
b. Easy to handle, non-toxic and non-combustible.

c. Its flow can take place by thermosyphon action.

d. It has a high density, high specific heat, good thermal conductivity and low viscosity.

e. Can be used both as a storage medium and a working medium (thus eliminating the need of heat exchanger).

f. Charging and discharging of energy can occur simultaneously.

g. Control of a water system is variable and flexible.

Apart from above advantages, water has certain disadvantages, viz.

a. Limiting temperature range, i.e. it freezes below $0\,°C$ and boils over $100\,°C$.

b. High vapor pressure at high temperatures.

c. A corrosive medium.

d. Low surface tension, i.e. leaks easily.

e. Difficult to stratify.

Though water is the best choice in space heat systems; oils and molten salts have also been used in solar thermal power plants.

12.3.1 Well Mixed Liquid Storage

This is the mostly widely used method of sensible heat storage. The analysis presented here is for water storage but is also valid for other liquids. The transient energy balance equation for a hot liquid storage tank (Fig. 12.1) assuming the temperature in the storage tank to be uniform (well mixed storage), can be written as:

$$(MC_p)_s \frac{dT_s}{dt} = \dot{Q}_c - \dot{Q}_L - (UA)_s(T_s - T_a) \qquad (12.2)$$

where M is the mass (kg) and C_p the specific heat of liquid in the tank (J/kg $°C$), T_s and T_a are the liquid temperature ($°C$) in the tank and the ambient temperature respectively, $\dot{Q}_c$ is the rate of charging energy from the heat source (W), $\dot{Q}_L$ is the rate of heat removal by the load (W), U is the overall heat transfer coefficient between the liquid in the tank and the outside (W/m²$°C$) air and A the surface area (m²) of the storage tank.

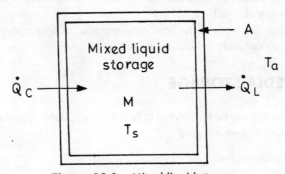

Figure 12.1   Mixed liquid storage.

If, the amount of heat addition and removal is assumed constant for a selected finite time interval, Δt, then Equation (12.2), for each time interval, can be written as

$$T_{s,\text{new}} = T_{s,\text{old}} + \frac{\Delta t}{(MC_p)_s}[\dot{Q}_c - \dot{Q}_L - (UA)_s(T_{s,\text{old}} - T_a)] \tag{12.3}$$

This equation can be used for the estimation of hourly storage water temperature if the hourly heat addition and withdrawal are known.

If solar energy is used for charging the storage, the above method gives successful results when one hour intervals are taken. Knowing the variation of the solar energy collection rate $\dot{Q}_c$ in terms of flow rate of the fluid in the collector, and the time variation of load $\dot{Q}_L$, storage temperature T_s can be determined as a function of time.

EXAMPLE 12.2

A storage tank of capacity $900\,L$ is at temperature $30\,°C$. Calculate the temperature of the tank after 2 hours, if the rate of charging from the source is 100 W. Given that the rate of heat removal by the load is zero and ambient air temperature is $12°C$. The area of storage tank is $10\,m^2$ and the overall heat transfer coefficient between liquid in the tank and the ambient is $6\,W/m^2$.

Solution

In the given problem, $\dot{Q}_L = 0$, $\dot{Q}_c = 100$ W, $T_{s,\text{old}} = 30\,°C$ $\Delta t = 2\,h$, $(MC_p) = 900 \times 4190\,J$ Substituting the above values in Equation (12.3), we get

$$T_{s,\text{new}} = 30 + \frac{2 \times 3600}{900 \times 4190}[100 - 60(30 - 12)] = 28.1\,°C.$$

12.3.1.1 *Storage tank without heat exchanger*

Without heat exchanger, the liquid flows directly into and out of storage, the solar collector inlet temperature T_{ci} is the same as temperature in the storage T_s (Fig. 12.2). Referring Equation (3.70), the energy rate is given as

$$\dot{Q}_c = A_c F_R[(\alpha\tau)I(t) - U_L(T_s - T_a)] \tag{12.4}$$

F_R is a control function given by $F_R = 1$, if the pump operates

$$= 0, \text{ otherwise}$$

It has been assumed here, that there is no drop in temperature of the fluid between the collector and the tank. The rate of heat withdrawal by the load, $\dot{Q}_L$, can be written as

$$\dot{Q}_L = (\dot{m}\,C_p)_L(T_s - T_L) \tag{12.5}$$

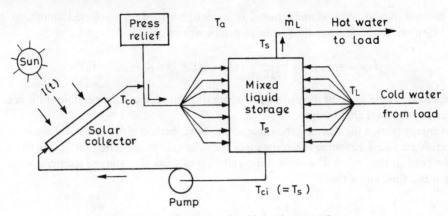

Figure 12.2   Storage tank without heat-exchanger.

where $\dot{m}_L$ is the flow rate of load stream (kg/s) and again it is assumed that there is no temperature drop between the tank and the load.

12.3.1.2 *Storage tank with heat exchanger*

A well mixed liquid storage with heat exchanger for both the collector and the load circuits is shown in Figure (12.3). In this case, the collector inlet temperature T_{ci} is not the same as the storage temperature T_s. The rate of charging is given by

$$\dot{Q}_c = (\dot{m}C_p)_c(T_{co} - T_{ci}) \tag{12.6}$$

where $(\dot{m}C_p)$ is the product of specific heat and the fluid mass flow rate through the collector.

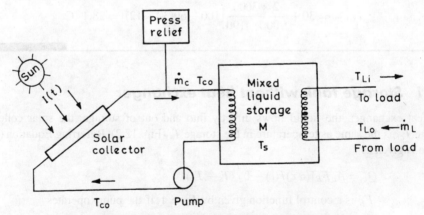

Figure 12.3   Storage tank with heat-exchanger.

The energy balance for an elemental length dx of the heat exchanger can be written as:

$$-(\dot{m}C_p)_c\frac{dT_c}{dx} = (U\,2\pi r)_c(T_c - T_s) \tag{12.7}$$

From Equation (12.7), we have

$$\frac{dT_c}{T_c - T_s} = \frac{-(U2\pi r)_c}{(\dot{m}C_p)_c} dx \qquad (12.8)$$

Integrating the above equation and using the initial condition:
At $x = 0$; $T_c = T_{c0}$, we have

$$\ln \frac{T_c - T_s}{T_{c0} - T_s} = \frac{-(U\,2\pi r)_c}{(\dot{m}C_p)_c}$$

At $x = L$; $T_c = T_{ci}$

$$\frac{T_{ci} - T_s}{T_{c0} - T_s} = \exp - \left(\frac{(UA)_c}{(\dot{m}C_p)_c} \right)$$

or

$$T_{c0} - T_{ci} = T_{c0} - T_s \left\{ 1 - \exp \left[\frac{-(UA)_c}{(\dot{m}C_p)_c} \right] \right\} \qquad (12.9)$$

Hence,

$$\dot{Q}_c = (\dot{m}C_p)_c (T_{c0} - T_{ci}) = (\dot{m}C_p)_c \left\{ 1 - \exp \left[\frac{-(UA)_c}{(\dot{m}C_p)} \right] \right\} (T_{c0} - T_s) \qquad (12.10)$$

where the factor $[1 - \exp\{-(UA)_c/(\dot{m}C_p)_c\}]$ is known as the penalty factor for the heat exchanger (Fig. 12.4a).

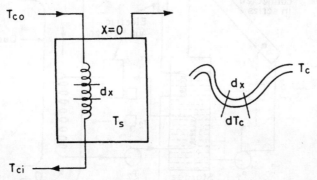

Figure 12.4(a) Cross-sectional view of collector loop heat exchanger.

Similarly, the energy withdrawn by the load is

$$\dot{Q}_L = (\dot{m}C_p)_L (T_{Li} - T_{Lo}) \qquad (12.11a)$$

and $(T_{Li} - T_{Lo})$ is obtained from

$$\frac{T_{Li} - T_{Lo}}{T_s - T_{Lo}} = \left\{ 1 - \exp \left[\frac{-(UA)_L}{(\dot{m}C_p)_L} \right] \right\} \qquad (12.11b)$$

where $(UA)_L$ is the product of surface area and the overall heat transfer coefficient for the heat exchanger between liquid in storage and load stream.

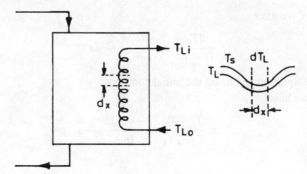

Figure 12.4(b) Cross-sectional view of load heat exchanger.

12.3.2 Space Heat and Hot Water

Figure 12.4(c) shows the configuration of the system used for providing both service of hot water and space heating. In this case, water is pumped up to the collectors, connected in series, by pump P_1 in response to a signal from the differential thermostate T_1. The differential thermostate is set at desired level. Domestic hot water is collected by a heat exchanger coil in the tank near the top.

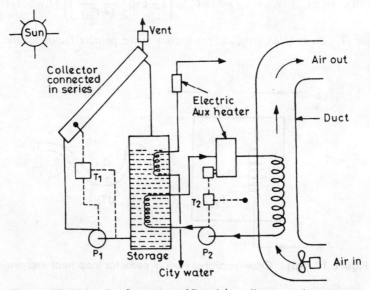

Figure 12.4(c) Configuration of flat plate collector and water storage system for space heating and domestic hot water.

An electric auxiliary hot water heater is provided to boost the temperature of the solar heated water, if required, as shown in Figure 12.4(c). The space heating provision operates in response to signals from thermostate T_2. Thermostate T_2 senses the indoor temperature and starts P_2. If the temperature of circulating hot water drop, the pump is off automatic due to thermostate. The auxiliary electric heater can be replaced by fuel-fired heaters provided an assured source of a suitable fuel is available.

If the water in the main storage tank becomes too cool to provide enough heat to the fan-coil units in the duct, the second contact on the thermostate calls for heat from the electric auxiliary heater.

EXAMPLE 12.3(a)

Derive Equation (12.11b).

Solution

Referring to Figure 12.4(b), the energy balance for an elemental length dx of the load heat exchanger is

$$(\dot{m}C_p)_L \frac{dT_L}{dx} = (U2\pi r)_L(T_s - T_L)$$

Above equation can be of the following form

$$\frac{dT_L}{T_s - T_c} = \frac{(U2\pi r)_L}{(\dot{m}C_p)_L} dx$$

Integrating the above equation with initial condition, i.e. $T_L|_{x=0} = T_{Lo}$, we have

$$\frac{T_s - T_c}{T_s - T_{Lo}} = \exp\left\{-\frac{(U2\pi r)_L x}{(\dot{m}C_p)_L}\right\}$$

At $x = L$, $T_L = T_{Li}$ and $A = 2\pi r L$

$$\frac{T_s - T_L}{T_s - T_{Lo}} = \exp\left\{-\frac{(UA)_L}{(\dot{m}C_p)_L}\right\}$$

After simplification, we get Equation (12.11b).

EXAMPLE 12.3(b)

Discuss Equation (9.11b) in the limiting cases: (i) $(UA)_L \to 0$, (ii) $(UA)_L \to \infty$ and (iii) $(\dot{m}C_p) \gg 1$.

Solution

i. Using Equation (9.11b), for $(UA)_L \to 0$, $\exp[-(UA)_L/(mC_p)_L] \to 1$, then $T_{Li} = T_{Lo}$, no heat transferred.

ii. Similarly, for $(UA)_L \to \infty$, $\exp[-(UA)_L/(mC_p)_L] \to 0$, thus, $T_{Li} = T_s$, maximum heat is transferred.

iii. For, $(\dot{m}C_p) \gg (UA)_L$, $\exp[-(UA)_L/(\dot{m}C_p)_L] \to 1$, then $T_{Li} = T_{Lo}$ no heat transferred due to large flow rate.

12.4 SOLID MEDIA STORAGE

The difficulties of high vapor pressure of water and the limitations of other liquids can be overcome by storing thermal energy as sensible heat in solids. Pebble-beds or rock-piles are generally preferred as the storage material due to their low cost. A pebble-bed consists of a bed of loosely packed rock material through which the heat transport fluid can flow. At low temperatures, air is used as the heat transport medium. The large surface area of rocks and greater contact time due to the rough path for air flow through the bed ensures good heat transfer to/form air by direct contact. The heat loss through the pebble bed (in no air flow condition) by conduction is very low, the rocks having a small surface contact with surrounding rocks and presence of air in the voids (stagnant air being a poor heat conductor). Not much insulation is thus required around the storage. Large pebble-bed storage may conveniently be placed underground. The energy stored in a packed-bed storage system depends on the thermophysical properties of the material, the rock size and shape, packing density, heat transfer fluid etc.

The following are the advantages of a packed-bed storage system.

a. Rocks are abundant and low-cost, are easy to handle, non-toxic and non-combustible.
b. High storage temperatures are possible in this case.
c. Heat exchanger can be avoided.
d. No freezing and corrosion problems

Apart from the above advantages, certain disadvantages of the system are:

a. Storage volumes are large.
b. High pressure drop.
c. Simultaneous charging and discharging is not possible.

This energy storage system is used in space heating by storage in rock-bed with air as the heat transport medium, and in the wall known as Trombe wall.

12.4.1 Packed-bed Storage

A packed-bed (pebble-bed or rock pile) storage unit utilizes the heat capacity of a bed of loosely packed particular material to store energy. A fluid, usually air, is circulated through the bed to add or remove energy.

A packed bed storage unit consists of a container, a screen to support the bed, support for the screen, and inlet and outlet ducts. Flow is maintained through the bed in one direction, during addition of heat and in the opposite direction during removal of heat. The heat, in this case, cannot be added and removed at the same time, in contrast to water storage systems.

Well-designed packed beds using rocks have several characteristics that are desirable for solar energy applications, the heat transfer coefficient between air and solid is high, which promotes thermal stratification; cost of the storage material and container are low; when there is no air flow, the conductivity of bed is low; and the pressure drop through the bed can be low.

Although there are many studies available on the heating and cooling of packed beds, the first one was done by Schumann (1929) and the equations, given below, describing a packed bed are often referred to as the Schumann model. The basic assumptions leading to this model are: one-dimensional heat flow; no axial conduction or dispersion; constant properties; no mass transfer; no temperature gradients within the solid particles and no heat loss to the environment.

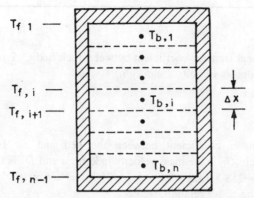

Figure 12.5 Packed bed divided into N segments.

The energy balance for fluid and bed are:

$$(\rho C_p)_f \varepsilon \frac{\partial T_f}{\partial T} = -\frac{(\dot{m} C_p)_f}{A} \frac{\partial T_f}{\partial x} + h_v (T_b - T_f) \qquad (12.12)$$

and

$$(\rho C_p)_b (1 - \varepsilon) \frac{\partial T_b}{\partial T} = h_v (T_f - T_b) \qquad (12.13)$$

where ε is the bed void fraction, h_v the volumetric heat transfer coefficient between bed and the fluid (i.e. the usual area heat transfer coefficient times the bed particulate surface area per unit bed volume), $W/m^2 \, ^\circ C$, $(\rho C_p)_f$ is heat capacity per unit volume, $J/m^3 \, ^\circ C$, of the fluid and $(\rho C_p)_b$ that of the bed.

For an air-based system the first term of Equation (12.12) can be neglected and the equation can be rewritten as,

$$\frac{\partial T_f}{\partial (X/L)} = NTU(T_b - T_f) \qquad (12.14a)$$

and

$$\frac{\partial T_b}{\partial \theta} = NTU(T_f - T_b) \qquad (12.14b)$$

where NTU, the number of transfer units, is given as

$$NTU = \frac{h_v AL}{(\dot{m} C_p)_f} \qquad (12.15)$$

and the dimensionless time

$$\theta = \frac{t(\dot{m} C_p)_f}{(\rho C_p)_f (1 - \varepsilon) AL} \qquad (12.16)$$

where A is bed cross-sectional area (m^2) and L the bed length (m). For a long term study of solar energy systems, several numerical techniques may be used. One of them is the "effectiveness — NTU" method.

As shown in Figure 12.5, the bed temperature can be considered to be uniform over Δx length of bed. The air temperature leaving bed element i is found from the relation

$$\frac{T_{f,i+1} - T_{b,i}}{T_{f,i} - T_{b,i}} = e^{-NTU(\Delta x/L)} \qquad (12.17)$$

EXAMPLE 12.4

Calculate the volumetric heat transfer coefficient between rock bed, $3.5\,m \times 3.0\,m \times 1.5\,m$, of $2\,cm$ diameter rocks and air with mass velocity $258.5\,kg/h$.

Solution

The volumetric heat transfer coefficient between the bed and air is given by the relation, $h_v = 824(G_0/D_s)^{0.92}$, where G_0 is the mass velocity in $kg/m^3 s$ and D_s is the particle diameter in m. In the given problem, $G_0 = 258.5/(3.5 \times 3.0 \times 1.5 \times 3600) = 4.56 \times 10^{-3}\,kg/m^3 s$
Substituting the values in the above relation, we get

$$h_v = 824(4.56 \times 10^{-3}/0.02)^{0.92} = 211.46\,W/m^2\,°C.$$

EXAMPLE 12.5

Calculate the number of heat transfer units, when air, with mass velocity of $1782\,kg/h$ flows through a pebble-bed of area $3.5\,m \times 3.0\,m$. Given the heat transfer coefficient between the bed and air is $225\,W/m^2\,°C$, C_p is $1006\,J/kg\,K$.

Solution

The number of heat transfer units (NTU) can be calculated from Equation (9.15) as

$$NTU = 225 \times 3.5 \times 3.0 \times 3.5/(1782 \times 1006) = 4.61 \times 10^{-3}.$$

EXAMPLE 12.6(a)

Discuss Equation (12.17) in the limiting case $\Delta x \to 0$.

Solution

From Equation (12.17), as $\Delta x \to 0$, $\exp(-NTU/N) \to 1$, thus, $T_{f,i+1} = T_{f,i}$.

EXAMPLE 12.6(b)

Derive Equation (12.17).

Solution

Refer to Figure 12.5 and elemental length Δx, the energy balance for ith element will be same as Equation (12.14a)

$$\frac{dT_f}{dx} = \frac{\text{NTU}}{L}(T_{bi} - T_f)$$

or

$$\frac{dT_f}{T_{bi} - T_f} = \frac{\text{NTU}}{L}dx$$

Integrate above equation with $T_f|_{x=0} = T_{f,i}$, then

$$\frac{T_f - T_{bi}}{T_{f,i} - T_{bi}} = \exp\left\{-\frac{\text{NTU}}{L} \cdot x\right\}$$

At $x \to \Delta x$, $T_f = T_{f,i+1}$, above equation becomes

$$\frac{T_{f,i+1} - T_{bi}}{T_{f,i} - T_{bi}} = \exp\left\{-\frac{\text{NTU}}{L} \cdot \Delta x\right\}.$$

The energy removed from the air and transferred to the bed in length ΔX will be

$$(\dot{m}C_p)_f(T_{fi} - T_{f,i+1}) = (\dot{m}C_p)_f(T_{f,i} - T_{b,i})\left(1 - e^{-\text{NTU}/N}\right). \tag{12.18}$$

where $N = L/\Delta X$.

An energy balance on the rock within region ΔX can then be expressed as

$$\frac{dT_{b,i}}{d\theta} = \eta N(T_{f,i} - T_{b,i}) \tag{12.19}$$

where η is a constant equal to $1 - e^{-\text{NTU}/N}$.

The above equation represents N differential equations for N bed temperatures. An extension of the above equation may include the energy loss to an environment at T_a. Then,

$$\frac{dT_{b,i}}{d\theta} = \eta N(T_{f,i} - T_{b,i}) + \frac{(U\Delta A)_i}{(\dot{m}C_p)_f}(T_a - T_{b,i}) \tag{12.20}$$

where $(U\Delta A)_i$ is the loss area-loss coefficient product for node i.

12.5 DUAL MEDIA STORAGE

Solid and liquid sensible heat storage materials can be combined in a number of ways. One of them is to jointly use the rock-bed and water tank: a hybrid system. Hybrid systems with water tank surrounded by rock-bed (Fig. 12.6a) have been used for solar space heating.

Another one is to incorporate rocks and oil in a single vessel. Such a system has been used in solar thermal power applications, to improve stratification and to minimize the amount of, relatively expensive, liquid.

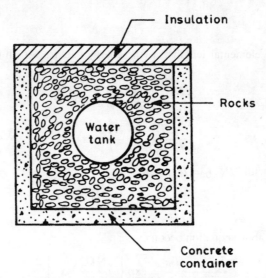

Figure 12.6(a) Hybrid sensible heat storage system.

Figure 12.6(b) shows an open flow collector mounted on a inclined south facing roof. The collector is integrated with water tank and rock-bed storage system in the basement of the building through a pump P_1. The thermostat sensor T_1 is connected to the pump P_1. The water is pumped from the bottom of the tank to a perforated pipe located at the top of the corrugated metal collector by pump P_1 as shown in Figure 12.6(b). The water is heated as it flows downward through open troughs to the collector channel at the bottom of the collector. The water drains back to the storage tank in the absence of solar radiation due to stoppage of pump P_1 after getting signal from T_1 thermostate. The rocks in contact with the tank

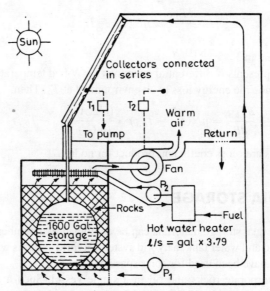

Figure 12.6(b) Configuration of water tank and rock-bed storage system in the basement of the building.

are heated by hot water in the tank. The thermal energy is spread throughout the rock bed by radiation, convection and conduction.

Cool air from the room enters the bottom of rock bed through return and the basement of the building. The fan operates after getting signal from thermostat T_2. In this case, the cool air available at bottom of rock bed passes through it. During the flow of cool air though the rock bed, it gets heated and warm air is allowed to enter into the room. There is a provision of auxilliary heating of air either by conventional hot water heater or by other fuels through a heat exchanger provided at the top of rock bed through pump P_2.

12.5.1 Ground Collector

It is well known that the ground has a large thermal capacity and hence acts as a large reservoir of solar energy. Ground, therefore, can potentially be used for solar energy collection-cum-storage purposes in low temperature ranges. A brief description and working principle of ground collector is given. The cross-sectional view of a ground collector is shown in Figure 12.7.

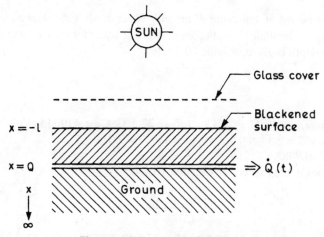

Figure 12.7 Ground collector.

The top surface of the ground is blackened and glazed. A network of pipes is laid down at some depth under the earth surface. An arrangement of water flow through a network of pipes can be made either at constant flow rate or at constant collection temperature. In this case, the solar radiation after reflection and absorption by the glass cover, is absorbed by the blackened surface. A part of the absorbed energy is convected to the ambient, through the glass cover, and the rest is conducted inside the ground. During the conduction process, the amplitude of thermal energy decreases with the depth of the ground and it becomes constant at large depths of the ground ($\sim$4 m). The temperature of ground for different surface treatment at 4 m depth has been given in Table 7.2. When the water flows through the network of pipe then the heat available at that depth is conducted to the flowing water and the flowing water gets heated. The temperature rise of the flowing water depends on the following main parameters:

  i. flow rate
 ii. length of pipes
iii. depth of network of pipe

To avoid night losses from top of the ground collector, provision of movable insulation can be made. Shelton (1975) investigated theoretically the performance of semi-spherical storage units buried under the ground (Fig. 12.8). The steady state and quasi-steady state models showed that the heat losses from the storage units to the ground are only a few percent of the heat storage capacity of the system. Further, Givoni (1977) carried out an extensive review of the various inexpensive storage mediums and analyzed different options from the view-point of their applicability under different climatic and soil conditions. During the same time, Nicholls (1977) developed some analytical models for the configurations utilizing the storage potential of the ground. Their arrangement consisted of a circular cylinder having its axis in the vertical direction and buried in the ground in such a way that the top of the cylinder was in the level of the ground. It is like a ground water well. Due to constant temperature at the depth of the well, the water inside it is warmer in winter and cooler in summer. This phenomenon is due to the annual storage effect.

EXAMPLE 12.7

A network of pipe is buried at 4 m depth of ground. Water at $40\,°C$ is allowed to pass through 10 m length of pipe, at 4 kg/s. Calculate the outlet temperature of water for the bare surface. The temperature of the ground at 4 m depth is given in Table 10.2.

Solution

In the given problem, $T_{ci} = 40°C$, $\dot{m} = 4$ kg/s, $T_s = 27.5°C$, $C_p = 4190$ J/kg $°C$
Thus, $UA/\dot{m}C_p = 6 \times 2 \times 3.14 \times 0.5 \times 10/(4 \times 4190) = 0.011$
Now, $\exp(-0.011) = 0.9888$
Substituting the values in the expression (Section 10.4.4)

$$\frac{T_{co} - T_s}{T_{ci} - T_s} = \exp\left(-\frac{UA}{\dot{m}C_p}\right)$$

$$T_{co} = 27.5 + (40 - 27.5)(0.9888) = 39.86°C$$

As seen from Figure 12.8, if the pipe is placed above the blackened surface, then it will function as a flat plate collector, discussed in detail in Chapter 3. But when the pipe is placed in the ground, the analysis will differ due to the addition of conduction term.

An earth-air tunnel discussed in Section 10.4.4 and shown in Figure 10.9 works on a similar principle. If instead of air, water is filled in the tank, the system acts as water heater/cooler storage. The efficiency of the system depends on certain parameters like:

i. diameter of the pipe or the flow rate
ii. length of the pipe
iii. depth at which tank is placed

The system can also be considered as hybrid sensible heat storage system (discussed earlier).

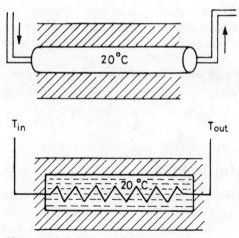

Figure 12.8 Ground water storage tank.

12.6 BASICS OF LATENT HEAT STORAGE

Phase change materials (PCM) have considerably high thermal energy storage densities as compared to the sensible heat storage materials and are able to absorb or release large quantities of energy ("latent heat") at a constant temperature by undergoing a change of phase. The following changes of phase are possible:

i. solid-gas
ii. liquid-gas
iii. solid-liquid

The above phase transformations are listed in the order of decreasing latent heat of transformation. In spite of large latent heats, the transformation solid-gas and liquid-gas are generally not employed for energy storage, since gases occupy large volumes.

In solid-solid transitions, heat is stored as the material is transformed from one crystalline form to another. Though most of the transitions have small latent heat and involve small volume changes, some transitions involve a large change in entropy as one of the states has a high disorder quite like the liquid state. Such PCM have an advantage due to minimal containment requirements.

The solid-liquid transformations include storage in salt hydrates, and employ certain inorganic salts, which are soluble in water and form crystalline salt hydrates.

Any latent heat thermal energy storage system should have at least following three components:

i. a PCM in the desired temperature range.
ii. container for holding the PCM.
iii. a suitable heat exchanger for transferring the heat from the heat source to PCM and then from PCM to the required point of use.

Latent heat storage systems are more expensive than the sensible heat storage media, viz. water and rocks. The PCMs undergo solidification and cannot generally be used as heat transfer media in a solar collector or the load. Thus a separate heat transport medium is required, with a heat exchanger in between. Many PCMs have poor thermal conductivity; others are corrosive and require special containers. Due to the high cost of latent heat storage; these system are usually used when:

a. high energy density or high volumetric energy capacity is desired,
b. the load is such that energy is required at a constant temperature; or within a small range of temperature.
c. the storage size is small.

12.6.1 Heat Transfer in PCM

A proper designing of a thermal storage system using phase change material requires details regarding heat transfer and phase change processes in PCM. The information generally required is (i) the distance of the melting front from the heated face, (ii) the temperature distribution, and (iii) the amount of heat stored as sensible heat and latent heat. The involved heat conduction problems are difficult to solve due to variable properties and moving boundaries. The analysis, however, is much simplified, by the assumption that the properties of any given phase are independent of temperature. Moreover, the convection effect (arising due to the density differences between the solid and liquid phases and the buoyancy forces produced by density differences due to temperature variations in the liquid) and the effect due to super cooling of liquid are generally not considered.

The simplest form of the heat transfer problem with phase change from liquid to solid is the freezing of upper layers of water in a pond when ambient air temperatures are below $0\,°C$. The ice formed on the surface, being poor conductor of heat reduces the rate of cooling of the water below.

12.6.2 Thermal Analysis of Freezing of Top of Ponds

A thermal analysis of freezing of top of pond (Fig. 12.9) is given below:

For an approximate solution of this problem certain assumptions are made:

i. $T_m = T_1$
ii. $R_1 = (1/h_1) \approx 0$, i.e. resistance to heat transfer on the liquid side is considered negligible
iii. heat capacity of ice is negligible in comparison to its latent heat of melting

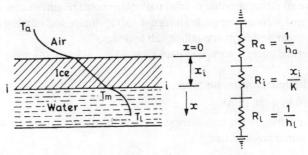

Figure 12.9(a) Diagram of a pond.

iv. thermal properties of ice are assumed constant

v. T_a and h_a are constant

The rate of heat transfer from ice-water interface to ambient can be given as:

$$\dot{q}' = \left[\frac{1}{h_0} + \frac{x_i}{K_i}\right]^{-1}(T_m - T_a) \tag{12.21}$$

The rate of ice formation, at the interface, per unit area is given by,

$$\dot{q}' = H_0\rho\frac{dx_i}{dt} \tag{12.22}$$

where H_0 and ρ are respectively the latent heat (J/kg) and density (kg/m^3) of ice.

Equations (12.21) and (12.22) give

$$H_0\rho\frac{dx_i}{dt} = \left[\frac{1}{h_0} + \frac{x_i}{K_1}\right]^{-1}(T_m - T_a)$$

or

$$\left[\frac{1}{h_0} + \frac{x_i}{K_1}\right]dx_i = \frac{T_m - T_a}{\rho H_0}dt \tag{12.23}$$

Let us assume,

$$x_i^* = \frac{h_0 x_i}{K_i} \quad \text{and} \quad t^* = \frac{T_m - T_a}{\rho H_0 K_i}t \tag{12.24}$$

Substitution of x_i^* and t^* from Equation (12.24) in (12.23) and rearrangement of the various terms, gives

$$(1 + x_i^*)dx_i^* = dt^* \tag{12.25}$$

Integration of Equation (12.25) along with the initial condition $x_i = 0$ at $t = 0$, and $x_i^* = 0$ at $t^* = 0$, gives

$$x_i^2 + 2x_i^* - 2t^* = 0$$

which yields

$$x_i^* = -1 \pm \sqrt{1 + 2t^*}$$

Since x_i^* is not negative, hence

$$x_i^* = -1 + \sqrt{1 + 2t^*} \tag{12.26}$$

The above equation can be used to determine the thickness of ice formed as a function of time.

The variation of normalized thickness of ice with normalized time is shown in Figure 12.9(b). It is clear that the thickness increases with increase of time. It saturates after some time due to poor heat transfer.

12.6.3 Analysis of Phase Change Material (PCM)

Given below is a steady-state analysis for estimating the behavior of phase change process. Let us consider a PCM slab (Fig. 12.10(a)) which is subjected to solar intensity $I(t)$. The energy balance at

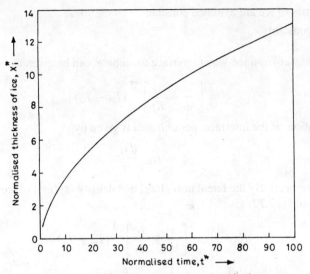

Figure 12.9(b) Variation x_i^* with t^*.

different boundaries are:

$$\alpha \tau \bar{I}(t) = -K_1 \frac{\partial T_1}{\partial x}\bigg|_{x=-L_1} + U_t(T|_{x=-L_1} - T_s) \tag{12.27}$$

$$T_1(x = 0) = T_s(x = 0) = T_0 \tag{12.28}$$

$$-K_1 \frac{\partial T_1}{\partial x}\bigg|_{x=0} = -K_s \frac{\partial T_s}{\partial x}\bigg|_{x=0} + \dot{q} \tag{12.29}$$

$$-K_s \frac{\partial T_s}{\partial x}\bigg|_{x=L_2} = h_2(T_s|_{x=L_2} - T_s) \tag{12.30}$$

Here $T_1 = A_1 x + B_1$ and $T_s = A_2 x + B_2$ $\qquad$ (12.31)

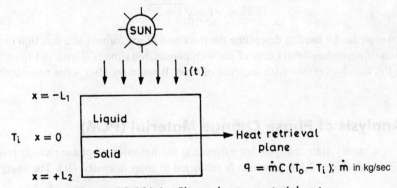

Figure 12.10(a)   Phase change material system.

Substituting the values of T_1 and T_s from Equation (12.31) in Equations (12.27) to (12.30), gives

$$A_1(K_1 + U_t L_1) - U_t B_l = -\alpha \tau \bar{I}(t) - U_t T_a \qquad (12.32)$$

$$B_1 = B_2 = T_0 \qquad (12.33)$$

$$-K_1 A_1 - \dot{m} C B_1 + K_s A_2 = -\dot{m} C T_i \qquad (12.34)$$

$$A_2(K_s + h_2 L_2) + h_2 B_2 = h_2 T_a \qquad (12.35)$$

Equations (12.32) to (12.35) can be solved for A_1, B_1, A_2 and B_2. After knowing these constants, an average temperature of PCM can be obtained as

$$\bar{T} = \frac{1}{L} \left[\int_{-L_1}^{0} T_1(x)dx + \int_{0}^{L_2} T_s(x)dx \right] \qquad (12.36)$$

Then, the effective thermal properties of PCM can be obtained from the expression:

$$C_{\text{eff}} = C_s + (H_0/(\bar{T} - T_a))(L_1/L) \qquad (12.37)$$

where C_s is the specific heat of the solid phase of PCM and H_0 is the latent heat of fusion. Similarly, the expression for the effective thermal conductivity K_{eff} can be written as

$$K_{\text{eff}} = K_s \left(1 - \frac{L_1}{L}\right) + K_1 \frac{L_1}{L} \qquad \text{where} \quad L = L_1 + L_2 \qquad (12.38)$$

The air-based solar heating system with phase change energy storage (PCES) has been shown in Figure 12.10(b). The system can be used for domestic hot water as well as space heating. There is a provision of direct space heating by byepassing PCES system. The collectors are connected in series. There is a co-axial counter current heat exchanger. The heat exchanger is connected to water tank and phase change energy system as shown in Figure 12.10(b). There are three pumps. One pump (P_l)

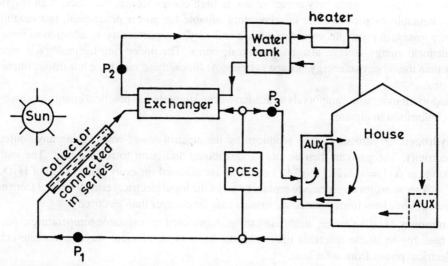

Figure 12.10(b)   Phase change energy storage (PCES) system for space heating.

operates for water heating. Pump (P_1) operates for collector loop through PCES for thermal energy storage. In the absence of solar energy, stored thermal energy in PCES can be used for space heating through pump (P_3). Pump (P_3) is used between heat exchanger and water tank.

Properties of some of the salt hydrates as PCMs is given in Table 12.1.

TABLE 12.1   Properties of salt-hydrates as PCMs

Material	Melting heat point($^\circ$C)	Specific fusion (kJ/kg)	Thermal heat conductivity		
			Solid	Liquid	W/mK
$Mg(NO_3)_2 \cdot 6H_2O$	89.9	167	1.84	2.51	0.490 at 95°C
$MgCl_2 \cdot 6H_2O$	115.0	165	1.72	2.82	0.570 at 120°C
$Zn(NO_3)_2 \cdot 6H_2O$	36.1	147	1.34	2.26	0.464 at 39.9°C
$Na_2 S_2O_3 \cdot 5H_2O$	48.5	210	1.46	2.38	0.57 at 40°C
$CaCl_2 \cdot 6H_2O$	29.7	170	1.46	2.13	0.540 at 38.7°C
$Na_2SO_4 \cdot 10H_2O$	32.4	241	1.76	3.30	0.544

12.7 CHEMICAL STORAGE

In this concept the energy is stored in the form of heat of chemical reactions which are often of larger magnitude than the latent heat storage. The idea of storing solar energy by the utilization of chemical reactions is not new. Nature has been storing energy by the use of chemical reactions in photosynthesis. Chemical energy storage is a two step process:

i. Storage or endothermic mode in which energy is absorbed to either break or rearrange the chemical bonds, producing in the process more energetic species which are stored.

ii. Discharge or exothermic mode in which the reaction is reversed producing thermal energy and regenerating the starting material.

Chemical energy storage is very attractive due to high energy storage densities, high reaction temperature attainable by present solar concentrators suitable for power production, fast reaction rates, low energy losses, in principle, an unlimited life and easy transportability. In addition to these advantages, chemical energy storage also has certain demerits. The underlying technology is much more complex than the other two energy storage systems. At times, these may have hazardous impact on the environment.

The important inorganic compounds which have been suggested as practical energy stores, by means of their combustion in air, are:

i. **Hydrogen**   Hydrogen can be produced by the electrolysis of water, using any source of DC electricity. The gas can then be stored, distributed and burnt to release heat. The enthalpy of change is $\Delta H = -242\,kJ\,mol^{-1}$, i.e. 242 kJ are released for every mole (18 g) of H_2O formed.

   A more promising idea is to replace some of the input electrical energy by heat from a cheaper source. Thus heat from solar concentrators may be cheaper than electricity.

ii. **Ammonia**   Unlike water, ammonia can be dissociated at realizable temperatures. Along with a heat engine, these reactions may form the basis of an efficient way to generate continuous electrical power from solar heat.

$$N_2 + 3H_2 \rightleftharpoons 2NH_3$$

EXAMPLE 12.8

A small well-insulated passive solar house requires an average internal heat supply of 2 kW. This will maintain an internal temperature of 25°C. A hot water store is decided to be built in a rectangular tank whose top forms the floor of the house, of area 250 m^2. The heating must be adequate for 120 days as all the heat loss from the tank passes by conduction through the floor, and as the water cools from an initial 75 °C to a final 45°C, calculate

i. volume of the tank.

ii. energy density of storage.

Solution

i. Heat required $= (2\,\text{kW})(120\,\text{days})(24\,\text{h day}^{-1})(3.6\,\text{MJ kWh}^{-1}) = 20736\,\text{MJ}$

$$\text{Volume of water} = \frac{20736(\text{MJ})}{(1000\,\text{kg m}^{-3})(4200\,\text{Jkg}^{-1}{}^{\circ}\text{C}^{-1})(25^{\circ}\text{C})} = 197\,\text{m}^3 \text{ as } Q = V \cdot \rho \cdot C_p \cdot \Delta T$$

Depth of tank $= 197/250 = 0.79\,\text{m}$

ii. Energy density of the used storage above 45°C $= (20736)/197 = 105.26 \approx 105\,\text{MJ m}^{-3}$

Energy density above ambient house temperature at 25°C $= 105 \times 2 = 210\,\text{MJ m}^{-3}$

PROBLEMS

12.1 Derive an expression for total energy lost/gain for constant thermal and physical properties of solid/liquid storage material.

Hint Integrate Equation (12.1) with constant ρ and C.

12.2 Calculate the temperature of paraffin for 75 J energy lost from boiled oil.

Hint Use Problem 12.1 and Appendix III (Table III).

12.3 Calculate (UA) for an insulated 100 litre cylindrical tank having length 1 m filled with acetic acid, if the temperature of oil drops from 300 °C to 150 °C in one hour.

Hint $(UA)(T_2 - T_1) \times 3600 = Q$ (Eq. (12.1)).

12.4 Derive an expression for an overall heat loss coefficient in terms of physical properties of an insulated storage tank.

Hint $(UA)(T_2 - T_1) \times 3600 = Q$ (Eq. (12.1)), for constant physical properties.

12.5 Calculate boiling temperature of sea water for a given Q and T_1.

Hint $Q = MC(T_2 - T_1)$.

12.6 Calculate the temperature drop when 8×10^{11} J thermal energy is lost from an insulated tank filled with 1500 liter olive oil.

Hint See Example 12.1.

12.7 Calculate the time required for the temperature of a storage tank filled with liquid of mass 1000 kg at a temperature of 35°C to change by 5°C, if there is no charging from the source and the rate of heat removal by the load is zero. Given: ambient air temperature is 15°C, area of storage is 15 m², the overall heat transfer coefficient between the liquid in the tank and ambient is 7 W/m² °C and $C_f = 4190$ J/kg °C.

Hint See Example 9.2.

12.8 Water, at 50 °C, flows at a rate of 5 kg/s, through a 8 m long pipe buried at a depth of 4 m in the ground. Calculate the outlet temperature for: (i) wetted surface, and (ii) glazed surface.

Hint See Example 12.7.

12.9 Calculate normalized thickness of ice (x_i^*) for 1 m thick ice and normalised time for 2 hr.

Hint Use Equation (12.24). Given $k_i = 2.21$ W/m °C, $\rho = 920$ kg/m³, $H_0 = 333.7$ kJ/kg, $T_m - T_a = 5$°C, $h_0 = 5.7$ W/m² °C.

12.10 Find out t^* for $x^* = 1$.

Hint Use Equation 12.26.

12.11 Plot the variation of effective specific heat C_{eff} and thermal conductivity K_{eff} with L_1/L.

Hint Use Equations (12.37) and (12.38).

CHAPTER *13*

Photovoltaic Systems

13.1 INTRODUCTION

Photovoltaic generation of power is caused by radiation that separate positive and negative charge carriers in absorbing material. In the presence of an electric field, these charges can produce a current for use in an external circuit. Such fields exist permanently at junctions or inhomogeneities in materials as 'built-in' electric fields and provide the required EMF for useful power production.

Junction devices are usually known as photovoltaic cells or solar cells, although, the term is a misnomer in the sense that it is the current that is produced by the radiation photons and not the 'voltage'. The cell itself provides the source of EMF. It is to be noted that photoelectric devices are electrical current sources driven by a flux of radiation.

A majority of photovoltaic cells are silicon semiconductor junction devices. Thus, in order to study the photovoltaic cells we should have an understanding of the basics of the semi-conductors; a brief description of which follows in the subsequent sections.

A solar cell constitutes the basic unit of a PV generator which, in turn, is the main component of a solar generator. A PV generator is the total system consisting of all PV-modules which are connected in series or parallel with each other.

Solids can be divided into three categories, on the basis of electricity conduction through them. They are: conductors, semi-conductors and insulators. The gap between the valence band and the conduction band (forbidden energy band) in the case of insulators ($h\nu < E_g$, where h is the Planck's constant and ν the frequency), is very large. Thus it is not possible for the electrons in the valence band to reach the conduction band, hence there is no conduction of current. In the case of semi-conductor ($h\nu > E_g$), the gap is moderate and the electrons in valence band may acquire energy sufficient enough for them to cross the forbidden (Fig. 13.1) region. While, in the case of conductors ($E_g \approx 0$), no forbidden gap exists and electron can easily move to the conduction band.

The semiconductor can again be divided into two categories: intrinsic and extrinsic. Intrinsic (pure) semi-conductors have Fermi-level in the middle of the conduction and valence band. In this case the densities of free electrons in conduction and free holes in valence band are equal $n = p = n_i$ and each is proportional to $\exp(-E_g/2kT)$.

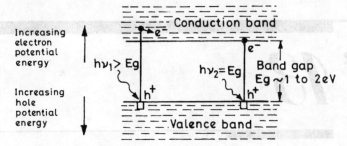

Figure 13.1 Semiconductor band structure of intrinsic material. Photon absorption $h\nu < E_g$, no photoelectric absorption. $h\nu_1 - E_g$, excess energy dissipated as heat. $h\nu_2 = E_g$, photon energy equals band gap.

EXAMPLE 13.1

Determine the band gap in a silicon crystal at 40 °C.

Solution

The variation of band gap with temperature is given by the relation:

$$E_g(T) = E_g(0) - \frac{aT^2}{T + b}$$

where a and b for different materials are as follows:

Material	$E_g(0)$	a	b
Silicon (Si)	1.16 eV	7×10^{-4} eVK^{-1}	1100 K
Gallium arsenide (GaAs)	1.52 eV	5.8×10^{-4} eVK^{-1}	300 K

Substituting the appropriate values in the above equation, we get

$$E_g(T) = 1.16 - \frac{7 \times 10^{-4} \times (313)^2}{313 + 1100} = 1.11 \text{ eV}.$$

13.2 DOPING

In order to increase the conductivity of intrinsic semiconductors, controlled quantities of specific impurity ions are added to the intrinsic semiconductor to produce doped (extrinsic) semiconductors. Impurity ions of valency less than the semiconductor enter the semiconductor lattice and become electron acceptor sites, that trap free electrons. These traps have an energy level within the band gap, but near the valence band. The absence of free electrons produce positively charged states called holes, that also move through the material as free carriers. Such a material is called p type material, having holes as majority carriers and electrons as minority carriers. If impurity ions of a valency greater than that of the semiconductor are added then n type material results which has electrons as majority carriers and holes as minority carriers. Both p and n type extrinsic semiconductors have higher electrical conductivity than the intrinsic basic material.

13.3 FERMI LEVEL

The Fermi level is the apparent energy level within the forbidden band gap from which majority carriers (electrons in n-type and holes in p-type) are excited to become charge carriers. The probability for the majority carrier excitation varies as $\exp[-e\varphi/(kT)]$, where e is the charge of the electron and hole, and φ the electric potential difference between the Fermi level and the valence or conduction band, T the temperature (K) and k the Boltzmann constant, $(= 1.38 \times 10^{-23}$ J/K$)$. For n-type material,

$$E_F = E_C + kT \ln\frac{N_D}{N_C} \tag{13.1}$$

where E_F is the Fermi-energy level, E_C the conduction band energy; k the Boltzmann constant; N_D the donor concentration and N_C the effective density of states in conduction band, and is constant at fixed temperature T. For p type material,

$$E_F = E_V - kT \ln\frac{N_A}{N_V} \tag{13.2}$$

where E_V is the valence band energy, N_A the acceptor ion concentration and N_V the effective density of states in the valence band.

EXAMPLE 13.2

Calculate the shift in Fermi energy level in a silicon crystal doped with a V group impurity of concentration 10^{15} cm^3. Given: the effective density of states in the conduction band is 2.82×10^{-19} cm^3 and the band gap is 1.1 eV; room temperature is 27 °C.

Solution

From Equation 13.1, we have

$$E_F = E_C + kT \ln(N_D/N_C)$$

If the valence band is taken as the reference level, then $E_C = 1.1$ eV. Substitution of the values gives

$$E_F = 1.1 + (1.38 \times 10^{-23}/1.6 \times 10^{-19}) \times 300 \ \ln(10^{15}/2.82 \times 10^{-19})$$

$$= 1.1 + 2.00 = 3.1 \text{ eV}$$

The shift is $3.1 - 0.55 = 2.55$ eV.

13.4 p-n JUNCTION

The basic requirement for photovoltaic energy conversion is an electronic asymmetry in the semiconductor structure known as junction. When n- and p-type semiconductors are brought in contact, then electrons from n-region near the junction would flow to the p-type semiconductor leaving behind a layer which is positively charged. Similarly holes will diffuse in the opposite direction leaving behind a negatively charged layer. A steady state is finally reached, resulting in a junction, which contains practically no mobile charges, hence the name depletion region.

The p-n junction (Figure 13.2) may be connected to a battery in two ways: (i) in forward bias (Figure 13.3(a)), the positive conventional circuit current passes from p to n material across a reduced band potential difference V_B (ii) in reverse bias (Figure 13.3(b)), the conventional positive current has an increased band potential difference V_B to overcome.

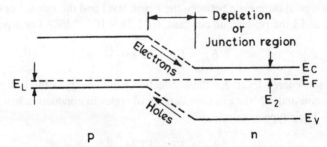

Figure 13.2 Energy levels in a p-n junction.

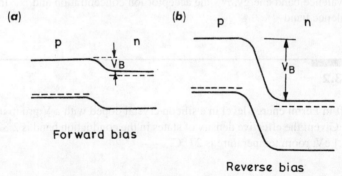

Figure 13.3 Energy levels in a p-n junction with (**a**) forward bias and (**b**) reverse bias.

Thermally or otherwise generated electrons and holes recombine after a typical relaxation time τ, having moved a typical diffusion length L through the lattice. In intrinsic material the relaxation time can be long, $\tau \sim 1$ s, but for commercial doped materials relaxation time are much shorter, $\tau \sim 10^{-2}$ to 10^{-8} s.

Electrons and holes may be generated thermally or by light, and become carriers in the material (Figure 13.4). Minority carriers in the depletion region, are pulled across electrostatically down their respective potential gradients. The minority carriers that cross the region become majority carriers in the adjacent layer. The passage of these carriers causes the generation current, I_g, which is mainly controlled by temperature in a given junction without illumination.

In an isolated junction, there can be no overall imbalance of current across the depletion region. Thus, a reverse recombination current I_r of equal magnitude occurs from the bulk material, this restores

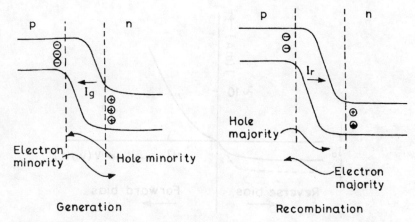

Figure 13.4 Generation and recombination currents at p-n junction.

the normal internal electric field. The band potential V_B is slightly reduced by I_r. The recombination current I_r can be varied by external bias as explained earlier (Fig. 13.5).

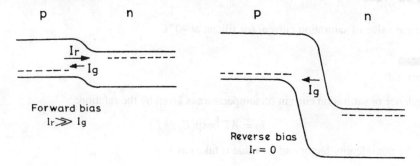

Figure 13.5 Generation and recombination currents with external bias.

13.5 *p-n* JUNCTION CHARACTERISTICS

The *p-n* junction characteristics are given in Figure 13.6. With no external bias ($V = 0$).

$$I_r = I_g \tag{13.3}$$

with a forward bias of voltage V, the recombination current becomes an increased forward current.

$$I_r = I_g \exp(eV/(kT)) - 1 \tag{13.4}$$

The total current (with no illumination) is,

$$I_D = I_r - I_g = I_g [\exp(eV/kT) - 1] \tag{13.5}$$

The above equation is the shockley equation and can be written as,

$$I_D = I_0 [\exp(eV/kT) - 1] \tag{13.6}$$

where $I_0(= I_g)$ is the saturation current under reverse bias, before avalanche breakdown occurs. It is also known as leakage or diffusion current. For good solar cells $I_0 \sim 10^{-8}$ Am^{-2}. Its value increases with temperature (Fig. 13.6, dotted curve).

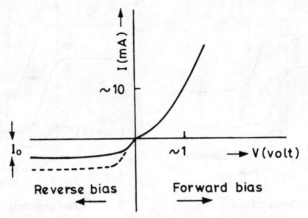

Figure 13.6   p-n junction dark characteristics.

EXAMPLE 13.3

Determine the value of saturation current for silicon at 40 °C.

Solution

The dependence of saturation current on temperature is given by the relation:

$$I_0 = AT^3 \exp(E_g/kT)$$

Here, A is the non-ideality factor and its value is taken as 1

$$E_g = 1.11 eV = 1.11 \times 1.6 \times 10^{-19} \text{ J}.$$

Substituting the known values in the above equation, we get

$$I_0 = (40 + 273)^3 \exp\left(-\frac{1.11 \times 1.6 \times 10^{-19}}{1.38 \times 10^{-23} \times 313}\right) = 4.26 \times 10^{-11} \text{Am}^{-2}.$$

EXAMPLE 13.4

Determine the value of dark current in the limiting case $V \to 0$.

Solution

From Equation (13.6), as $V \to 0$, $\exp(eV/kT) \to 1$ and hence dark current $I_D \to 0$.

13.6 PHOTOVOLTAIC EFFECT

When the solar cell (p-n junction) is illuminated, electron-holes pairs are generated, and acted upon by the internal electric fields, resulting in a photo current (I_L). The generated photocurrent flows in a direction opposite to the forward dark current. Even in the absence of external applied voltage, this photocurrent continues to flow, and is measured as the short circuit current (I_{sc}). This current depends linearly on the light intensity, because absorption of more light results in additional electrons to flow in the internal electric field force.

The overall cell current I is determined by subtracting the light induced current I_L from the diode dark current I_D.

$$I = I_D - I_L \qquad (13.7)$$

Then

$$I = I_0 \left[\exp\left(\frac{eV}{kT}\right) - 1 \right] - I_L \qquad (13.8)$$

This phenomenon is called the photovoltaic effect.

EXAMPLE 13.5(a)

Determine the value of the overall cell current in the limiting case $V \to 0$.

Solution

From Equation 13.8, as $V \to 0$, $\exp(eV/kT) \to 1$. Hence, $I \to -I_L$.

EXAMPLE 13.5(b)

Find out voltage for zero overall cell current.

Solution

Substituting $I = 0$ in Equation (13.8), we get

$$I_o \left[\exp\left(\frac{eV}{kT}\right) - 1 \right] - I_L = 0$$

or

$$\exp\left(\frac{eV}{kT}\right) = \frac{I_L}{I_o} + 1$$

or

$$V = \frac{kT}{e} \log\left[\frac{I_L}{I_o} + 1 \right]$$

13.7 PHOTOVOLTAIC MATERIAL

The solar cells are made of various material and with different structure in order to reduce the cost and achieve maximum efficiency. There are various types of solar cell material, the single crystal, poly crystalline and amorphous silicon, compound thin film material, and other semiconductor absorbing layer which give highly efficient cells for specialized applications.

Crystalline silicon cells are most popular, though they are expensive. The amorphous silicon thin film solar cell are less expensive. The amorphous silicon layer is used with both hydrogen and fluorine incorporated in the structure. These a-Si:F:H alloy have been produced by the glow discharge decomposition of SiF_4 in the presence of hydrogen. The efficiency of a-Si module is about 6–8 percent.

A variety of compound semiconductor can also be used to manufacture thin film solar cells. These compound material are $CuInSe_2$, CdS, CdTe, Cu_2S, InP. The $CuInSe_2$ solar cell stability appears to be excellent. The combination of different band gap material in the tandem configurations lead to photovoltaic generator of much higher efficiencies.

13.7.1 Single Crystal Solar Cell

Single-crystalline solar cells made from high-purity material (solar grade) show excellent efficiencies and long-term stability.

Figure 13.7 shows the diagram of a silicon cell. The electric current generated in the semiconductor is extracted by contact to the front and rear of the cell. The cell is covered with thin layer of dielectric material, the antireflecting coating or ARC (to minimize the reflection from the top surface). The total series resistance of the cell can be expressed as:

$$R_s = R_{cp} + R_{bp} + R_{cn} - R_{bn} \qquad (13.9)$$

where R_{cp} is the metal contact to p-type semiconductor resistance; R_{bp} the bulk p-type resistance (bulk of p-type region is where most of electron/hole pairs are generated by the absorption of light and where

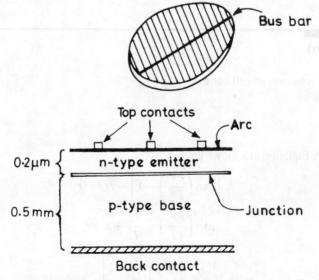

Figure 13.7 The structure of single crystal solar cell.

minority carrier (electron) are transported by diffusion and partially lost by recombination); R_{cn} the contact to n-type semiconductor resistance and R_{bn} the bulk n-type resistance. The idealized junction current is given as

$$I = I_0 \left[\exp \frac{e(V - IR_s)}{kT} - 1 \right] \tag{13.10}$$

In addition, a shunt path may exist for current flow across the junction due to surface effect or poor junction region. This alternate path for current constitutes a shunt resistance R_p across the junction. Then,

$$I = I_L - I_0 \left[\exp \left(\frac{e(V - IR_s)}{AkT} \right) - 1 \right] - \left(\frac{V - IR_s}{R_p} \right) \tag{13.11}$$

where A is an empirical non-ideality factor and is usually 1.

EXAMPLE 13.5(c)

What is the condition for zero idealised junction current ($I = 0$).

Solution

Substitute $I = 0$ in Equation (13.10), yield

$$\exp \frac{eV}{kT} = 1 \Rightarrow V = 0.$$

There are certain parameters to be mentioned in the I–V characteristics of a solar cell. They are:
Short circuit current (I_{sc}): It is the light generated current I_L.

$$V_{oc} = \frac{kT}{e} \ln \left(\frac{I_L}{I_o} + 1 \right) \tag{13.12}$$

Open circuit voltage (V_{oc}): It is obtained by setting $I = 0$ in Equation (13.8) (see Example 13.5(b)).
Both I_L and I_0 depend on the structure of solar cells. No power is generated under short or open circuit. The power output is defined as,

$$P_{out} = V_{out} \times I_{out}$$

The maximum power P_{max} provided by the device is achieved at a point on the characteristics, where the product IV is maximum.
Thus

$$P_{max} = I_{max} V_{max} \tag{13.13}$$

$$\text{Fill factor (FF)} = \frac{P_{max}}{V_{oc} I_{sc}} \tag{13.14}$$

The fill factor, also known as the curve factor, is a measure of sharpness of the knee in I–V curve. It indicates how well a junction was made in the cell and how low the series resistance has been made. It can be lowered by the presence of series resistance and tends to be higher whenever the open circuit voltage is high. It's maximum value in Si is 0.88. The maximum possible output can be given as,

$$P_{max} = V_{oc} I_{sc}\, FF \qquad (13.15)$$

The solar cell power conversion efficiency can be given as,

$$\eta = \frac{P_{max}}{P_{in}} = \frac{V_{oc} I_{sc}\, FF}{\text{Incident solar radiation} \times \text{Area of solar cell}} \qquad (13.16)$$

EXAMPLE 13.6

Calculate fill factor for a solar cell which has the following parameters:
$V_{oc} = 0.2\,\text{V}$, $I_{sc} = -5.5\,\text{mA}$, $V_{max} = 0.125\,\text{V}$, $I_{max} = -3\,\text{mA}$.

Solution

Substituting the appropriate values in Equation 13.14, we get

$$\text{Fill factor} = \frac{V_{max} I_{max}}{V_{oc} I_{sc}} = \frac{0.125 \times 3}{0.21 \times 5.5} = 0.324.$$

EXAMPLE 13.7

Calculate the maximum power at an intensity of 200 W/m². Given:
$V_{oc} = 0.24\,\text{V}$, $I_{sc} = -9\,\text{mA}$, $V_{max} = 0.14\,\text{V}$ and $I_{max} = -6\,\text{mA}$.

Solution

From Equation (13.13), we have

$$P_{max} = 0.14 \times (-6) = -0.84\,\text{mW}$$

13.7.1.1 I–V characteristics

The current equation for a solar cells is given by Equation 13.10 and shown in Figure 13.8. For a good solar cell, the series resistance, R_s, should be very small and the shunt (parallel) resistance, R_p, should be very large. For commercial solar cells, R_p is much greater than the forward resistance of a diode so that it can be neglected and only R_s is of interest. The following are few of the characteristics parameters that have been discussed.

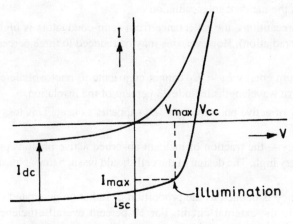

Figure 13.8 *I–V characteristics of a solar cell.*

The optimum load resistance $R_L(P_{max}) = R_{pmax}$ is connected, if the *PV* generator is able to deliver maximum power.

$$P_{max} = V_{Pmax} \, I_{Pmax} \tag{13.17}$$

and
$$R_{Pmax} = \frac{V_{Pmax}}{I_{Pmax}} \tag{13.18}$$

The efficiency is defined as,
$$\eta = P/\phi \tag{13.19}$$

$P = V \times I$, is the power delivered by *PV* generator.
$f = I \times A$, is the solar radiation falling on the *PV* generator.
I_T is the solar intensity and A is the surface area irradiated.

EXAMPLE 13.8

Calculate the cell efficiency of the cell described in Example 10.7, if its area is $4 \, cm^2$.

Solution

Cell efficiency = output/input = $(0.14 \times 6 \times 10^{-3})/(200 \times 4 \times 10^{-4}) = 0.0105 = 1.05\%$

13.7.1.2 *Limits to cell efficiency*

Photovoltaic cells are limited in efficiency by many losses; some of these are avoidable but others are intrinsic to the system and may be described as follows:

i. the electric current leaves the top surface by a web of metal contacts arranged to reduce series resistance losses in the surface. These contacts have a finite area and thus cover part of the active surface and block the incident solar radiation.

ii. without special precautions, the reflectance from semi-conductors is high (about 40 percent of the incident solar radiation). However, this may be reduced to three percent or less by the use of thin film surface.

iii. photons of quantum energy $hv < E_g$ cannot contribute to photovoltaic current generation. For silicon, the inactive wavelengths include 23 percent of the insolation.

iv. the excess energy of active photons $(hv - E_g)$ appears as heat. This loss is about 33 percent of the insolation.

v. quantum efficiency — the fraction of incident absorbed active photons producing electron-hole pairs is usually very high. The design of the cell should be such that at least 95 percent absorption takes place.

vi. collection efficiency is defined as the proportion of radiation generated electron-hole pairs that produce current in the external circuit. For 10 percent overall efficiency cells, the collection efficiency factor is usually about 0.7. Increasing this to about 0.9 would produce >20 percent overall efficiency cells.

vii. each absorbed photon produces electron-hole pairs with an electric potential difference of E_g/e (1.1 V in Si). However, only a part (V_B) of this potential is available for the EMF of an external circuit. The voltage factor F_v is equal to eV_B/E_g. The missing EMF occurs because in the open circuit the Fermi level across the junction equates at the dopant n and p levels and not at the displaced conduction to valence band levels. Increased dopant concentration increases F_v. The loss due to voltage factor is about 20 percent of the insolation.

viii. the solar cell $I-V$ characteristics is strongly influenced by the $p-n$ diode characteristics. Thus as the solar cell (Fig. 13.9(a)) output is raised towards V_{oc} the diode becomes increasingly forward biased, so increasing the internal recombination current I_r across the junction. This necessary behavior is treated as a fundamental loss in the system. The loss due to curve factor is about four percent of the insolation.

ix. in practice, the cell characteristics does not follow Equation (10.8) and is better represented as,

$$I = I_0 \exp\left[\frac{eV}{AkT} - 1\right] - I_L$$

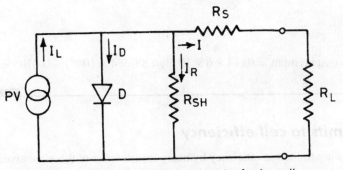

Figure 13.9(a) Equivalent circuit of solar cell.

The factor A results from increased recombination in the junction and tends to change V_{oc} and I_o. Thus, in general, optimum output would occur if $A = 1$. Within the cell, recombination is lessened if:

a. diffusion paths are long (>50 to 100 μm in Si), this requires long minority carrier lifetimes

b. junction is near the top surface (within 0.15 μm)

c. material has few defects other than the dopant

13.7.1.3 Determination of R_s

For determination of R_s, the I–V curves at the same temperature but for two different solar intensities I_{T1} and T_{T2} are plotted (Fig. 13.9(b)). A point A is selected on higher intensity curve corresponding to a voltage slightly greater than V_{Pmax}.

Thus, $$I = I_{sc1} - I(A) \qquad (13.20)$$

Next a point B is selected on the lower intensity curve.

$$I(B) = I_{sc2} - I$$

The voltage difference corresponding to the voltages at A and B is

$$\Delta V = V(B) - V(A)$$

or, $$R_{s1} = \frac{\Delta V}{I_{sc1} - I_{sc2}} \qquad (13.21)$$

This process can be repeated to obtain other values of R_s's and the mean of these values gives R_s.

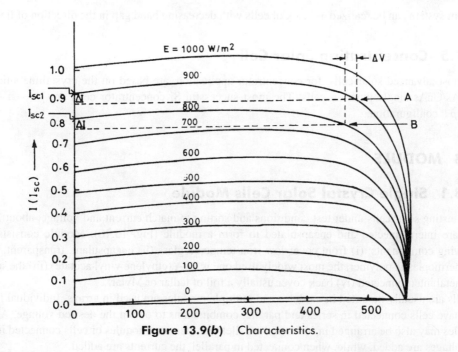

Figure 13.9(b) Characteristics.

13.7.1.4 Determination of R_p

R_p can be determined from the slope of I–V curve at the short circuit point.

$$\left.\frac{dI}{dV}\right|_{v=0} = -\frac{1}{R_p} \tag{13.22}$$

13.7.2 Thin Film Solar Cell

Thin film solar cell are efficient for large scale photovoltaic energy conversion. This not only reduces the semiconductor material required but is also beneficial for production of large area module.

The semiconductor material for thin film solar cells should have a high absorption coefficient ($\alpha > 10^4$ cm^{-1}). Two groups of material meet this requirement.

  i. Compound semiconductor with direct band gap and poly crystalline structure.
  ii. Amorphous semiconductor.

13.7.3 Amorphous Si Solar Cells (a-SiH)

Hydrogenated amorphous silicon film represent extremely suitable material for the solar cell mainly due to its optical properties. Only a thin film of about 0.7 μm thickness absorbs a large fraction of the incident solar radiation due to high absorption coefficient. The optical band gap of pure a-SiH is well matched with solar spectrum.

13.7.4 Tandem Solar Cell

Tandem system can be realized as stack of cells with decreasing band gap in the direction of light path.

13.7.5 Concentrating Solar Cell

The most advanced solar cells, for concentrator application, are based on the crystalline silicon and AlGaAs/GaAs single junction cells. The most successful Si concentrator cells are $p^+ - n - n^+$ or $n^+ - p^+$ configuration.

13.8 MODULE

13.8.1 Single Crystal Solar Cells Module

After testing solar cells under test conditions and sorting to match current and voltage, about 36 solar cells are interconnected and encapsulated to form a module (Fig. 13.10). Module consists of the following components: (i) front cover low iron tempered glass (ii) encapsulant, transparent, insulating, thermoplastic polymer, the most widely used one is EVA (ethylene vinyl acetate) (iii) the solar cell and metal interconnected (iv) back cover usually a foil of tedlar or Mylar.

Cells are usually mounted in modules and multiple modules are used in arrays. Individual modules may have cells connected in series and parallel combinations to obtain the desired voltage. Arrays of modules may also be arranged in series and parallel. For identical modules or cells connected in series, the voltages are added, while, when connected in parallel, the currents are added.

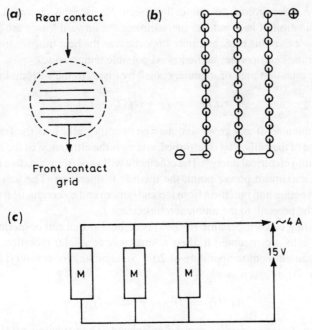

Figure 13.10 Typical arrangements of commercial Si solar cells (**a**) cell (**b**) module of 33 cells (**c**) array.

Photovoltaic generators, Figure 13.11, may be used to drive machines such as electric pumps, refrigerators, and other devices. *PV* array mounted on the roof-top offer the possibility of large scale power generations in decentralized medium size grid-connected units. The *PV* system supply the electricity need of the building, feed surplus electricity need of the building, feed surplus electricity to the grid, to earn the revenue and draws electricity from the grid at low insolation.

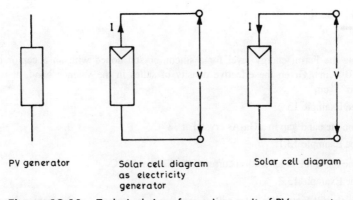

Figure 13.11 Technical signs for various unit of PV generator.

13.9 CELL TEMPERATURE

The temperature of operation of a PV module can be determined by an energy balance. The solar energy absorbed by a module is converted partly into thermal energy and partly into electrical energy.

The electrical energy is removed from the cell through the external circuit. The thermal energy is dissipated by a combination of heat transfer mechanisms; the upward losses are the same as discussed in Chapter 3. Back losses, in this case, are more important, as the heat transfer from the module should be maximized so that the cell operates at the lowest possible temperature.

An energy balance on a unit area of module, cooled by losses to the surroundings can be written as

$$\tau\alpha\, I_T = \eta_c I_T + U_L(T_c - T_a) \tag{13.23}$$

where τ is the transmittance of any cover that may be over the cells, α is the fraction of the radiation incident on the surface of the cells, that is absorbed, and η_c is the efficiency, of the module, of conversion of incident radiation into electrical energy. The efficiency will vary from zero to a maximum, depending on how close to the maximum power point, the module is operating. The loss coefficient, U_L, will include losses by convection and radiation from top and bottom and by conduction through any mounting framework that may be present, to the ambient temperature T_a.

The nominal operating cell temperature (NOCT) is defined as that cell or module temperature which is reached when the cells are mounted in their normal way at a solar radiation level of 800 W/m², a wind speed of 1 m/s, an ambient temperature of 20 °C, and no load operation (i.e. with $\eta_c = 0$). From Equation (13.23), $\tau\alpha/U_L$ is given as

$$\alpha\tau/U_L = (T_{C,NOCT} - T_a)/I_{T,NOCT} \tag{13.24}$$

Knowing T_a, $I_{T,NOCT}$, $T_{C,NOCT}$, $\tau\alpha/U_L$ can be calculated. Then treating $\tau\alpha/U_L$ as a constant, the temperature at any other condition can be found from the relation:

$$T_c = T_a + (I_T \tau\alpha/U_L)(1 - \eta_c/\tau\alpha) \tag{13.25}$$

PROBLEMS

13.1 Determine the Fermi energy level for a silicon crystal doped with an acceptor impurity of concentration 10^{17} cm³. Given the effective density of states in the valence band, at room temperature, is 1.04×10^{-19} cm³.

   Hint See Example 13.2.

13.2 Determine the band gap in a Ga-As crystal at 38 °C.

   Hint See Example 13.1.

13.3 Determine the value of saturation current for silicon at 32 °C.

   Hint See Example 13.3.

13.4 Calculate the fill factor if a solar cell of area 4 cm² is irradiated with an intensity of 100 W/m². Given $V_{oc} = 0.24$ V, $I_{sc} = -10$ mA, $V_{max} = 0.14$ V, $I_{max} = -6.5$ mA. Also calculate R_{OP}.

   Hint Use Equation (10.14) and use $R_{OP} = V_m/I_m$.

13.5 What will be solar cell current if dark and light induced current are equal.

   Hint Use Equation (13.7).

13.6 What will be the acceptor ion concentration at $-273\,°C$.

    Hint See Example 13.1.

13.7 What will be the acceptor ion concentration for extrinsic p-type material ($E_F = E_V$).

    Hint Use Equation (13.2).

13.8 Calculate dark current for a solar cell for reverse and forward bias mode and verify the results with Figure 13.6.

    Hint Use Equation (10.6) for different V in reverse and forward bias mode, for a given room temperature.

13.9 Calculate fill factor for a given solar cell for a solar intensity of $300\,W/m^2$.

    Hint Use Figure 13.9(b) and Equations (13.12) and (13.14).

13.10 Draw the curve between efficiency of a solar cell and solar intensity for Figure (13.9(b)).

    Hint Use Equation (13.19).

13.11 Calculate R_{op} for the solar cell given in Example 13.6.

    Hint $R_{op} = V_{max}/I_{max}$.

13.12 What should be the acceptor ion concentration, for the same shift in Fermi level, for a given p type material, at different temperatures.

    Hint Use Equation (10.2) and vary T between 273 and 300 K.

13.13 Calculate the band gap for silicon (Si) and gallium arsenide (Ga As) with different temperature.

    Hint See Example 13.1.

13.14 Find out temperature for zero band gap for silicon and gallium arsenide.

    Hint Put $E_g(T) = 0$ in Equation of Example 13.1.

13.15 Plot variation of Fermi-energy level for n-type and p-type materials with concentration of dipping material.

    Hint Use Equations (13.1) and (13.2), respectively.

13.16 Calculate saturation current with different temperatures.

    Hint See Example 13.3.

13.17 How does dark current vary with potential 'V'.

    Hint See Figure 13.6.

13.18 Calculate overall cell current (I) with Voltage (V) and show the variation.

    Hint Use Equation (13.8).

CHAPTER 14

Economic Analysis

14.1 INTRODUCTION

Techno-economic analysis is the area of engineering where engineering judgement and experience are utilised in the application of scientific principles and techniques to problems of project cost control, profitability analysis, planning, scheduling and optimisation of operational research etc. This study covers a wide range of topics such as time value of money, maintenance, organisational structures, integrated projects control, quality and resource management, life cycle and risk analysis etc.

Close estimation becomes crucial and critical with the advancement of technology and society to remain competitive. An estimate based on over design may be too high to sustain, whereas that based on under design may be successful for a while but again it is not possible to sustain. An effective economic analysis can be made by the knowledge of cost analysis, which can be done by the aid of cash flow diagrams and some other methods. The subsequent sections deal with their study.

14.2 COST ANALYSIS

14.2.1 Capital Recovery Factor

Let P be the present amount invested at zero ($n = 0$) time at the interest rate of i per year and S_n future value at the end of n years as shown below.

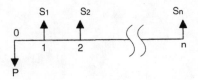

Then, at the end of one year the time value of P will be

$$S_1 = P + iP = P(1 + i)$$

At the end of second year P becomes

$$S_2 = S_1 + iS_1 = P + iP + i(P + iP)$$
$$= P(1 + i) + Pi(1 + i) = P(1 + i)^2$$

Similarly, at the end of third and nth year will be

$$S_3 = P(1+i)^3 \quad \text{and} \quad S_n = P(1+i)^n$$

Assuming S_n to be S, then S_n can be written as

$$S = P(1+i)^n$$

Here, $S > P$ for $i > 0$, compound interest law

$$S = P\, F_{PS} \tag{14.1a}$$

Future value = (present value). (Compound-interest factor)

---

where, F_{PS} is more completely designated as $F_{PS,i,n}$ where, i is the rate of interest and n is the number of years, and

$$F_{PS,i,n} = (1+i)^n \tag{14.1b}$$

is known as the *compound interest factor* or *future value factor* which converts P into S (Table 14.1).

EXAMPLE 14.1

If Rs. 20,000 compounds to Rs. 28,240 in 4 years what will be the rate of return?

Solution

By using Equation (14.1a), $S = P(1+i)^n$ and substituting S = Rs. 28,240; P = Rs. 20,000 and $n = 4$, we get,

$$28,240 = 20,000(1+i)^4 \text{ or, } (1+i)^4 = 1.412$$

Solving the above equation, we get $i = 0.09$ or nine percent per year.

EXAMPLE 14.2

How long will it take for money to double if compounded annually at 10 per cent per year?

Solution

Let us assume that the money doubles in n years, then $S = 2P$.
Using Equation (14.1a) and substituting $S = 2P$, as mentioned earlier, we get

$$2P = P(1+0.10)^n$$

$$2 = (1+0.10)^n$$

Solving the above equation, we get

$$\log 2 = n \log 1.1, \text{ i.e. } n = 7.3 \text{ years}$$

The money will be doubled in 7.3 years.

If one year is divided into p equal units of period, then n becomes np and i become i/p which is the rate of return per unit period.

Substitution of these values in Equation (14.1a), gives

$$S = P\left(1 + \frac{i}{p}\right)^{np}$$

This can be written as

$$S = P\left[\left(1 + \frac{i}{p}\right)^{p}\right]^{n} \tag{14.2}$$

where the expression $(1 + i/p)^p$ can be expressed as

$$\left(1 + \frac{i}{p}\right)^{p} = 1 + \text{effective rate of return}$$

or

$$\text{Effective rate of return} = \left(1 + \frac{i}{p}\right)^{p} - 1 = i \text{ for } p = 1$$
$$> i \text{ for } p > 1 \tag{14.3}$$

EXAMPLE 14.3

Calculate the effective rate of return for 10 percent interest for $p = 5$ and $p = 12$.

Solution

From Equation (14.3), we have

$$\text{Effective rate of return} = \left(1 + \frac{i}{p}\right)^{p} - 1$$

$$\text{For } p = 5; \text{ the effective rate of return} = \left(1 + \frac{0.10}{5}\right)^{5} - 1 = (1.02)^5 - 1 = 0.104$$

$$\text{For } p = 12; \text{ the effective of return} = \left(1 + \frac{0.10}{12}\right)^{12} - 1 = 0.1047$$

For simple interest,

$$S = P(1 + ni) = P + (iP)n \tag{14.4}$$

EXAMPLE 14.4

A student borrows Rs. 2000 and returns Rs. 2100 at the end of six months. What was the rate of interest paid by the student?

Solution

Here, $S = 2100$, $P = 2000$ and $n = 6/12$

$$2100 = 2000 \left(1 + \frac{6}{12} i \right)$$

Substituting these values in Equation (14.4), we get, $1.05 = 1 + 0.5i$, $i = 0.10$ or 10 percent.

EXAMPLE 14.5(a)

A person borrows Rs. 10,000 at 10 percent for four years and four months. Calculate the money paid considering compound interest.

Solution

Using Equation (14.4), we have

$$S = 10,000 \left[1 + \frac{4}{12}(0.10) \right] = \text{Rs. } 10,333$$

for 4 months, and Rs.10,333 become P for another 4 years. For compound interest

$$S = P F_{PS,10\%,4}$$

Using Equation (14.1a), we get

$$S = 10,333(1.4641) = \text{Rs. } 15,129$$

Equation (14.1a) can be rewritten as

$$P = S/(1+i)^n$$

or

$$P = S(1+i)^{-n}$$

This shows that the future amount (at nth year) is reduced when converted against the calendar to the present value (at zeroth time), assuming i to be positive.

$$P = S F_{SP} \qquad (14.5a)$$

Present value = (future value). (present value factor)

where F_{SP} (table 14.1) is designated as $F_{SP,i,n}$ and is given by

$$F_{SP,i,n} = \frac{1}{(1+i)^n} = (1+i)^{-n} \qquad (14.5b)$$

From Equations (14.1b) and (14.5b), F_{PS} and F_{SP} can be related as

$$F_{PS}.F_{SP} = 1$$

From Equations (14.1a) and (14.5a) the future value S of initial investment P, and vice versa, can be calculated

$$S = P(1+i)^n, \text{ moving with the calendar}$$

and $\qquad P = S(1+i)^{-n}, \text{ moving against the calendar}$

Above equations can be combined as

$$A_{t2} = A_{t1}(1+i)^n \qquad (14.6)$$

Amount at time 2 = Amount at time 1 (Compound-interest operator)

Here, n is positive with the calendar, and negative against the calendar Equation (14.6) is referred to as the *time-value conversion relationship*.

14.2.2 Unacost

In solving engineering economic problems it is convenient to diagram expenditures and receipts as vertical lines positioned along a horizontal line representing time. Expenditures and receipts can point in opposite directions. By using this concept, a uniform annual amount will be discussed.

Consider a uniform end-of-year annual amount R (unacost) for a period of n years. The diagram for this is as shown,

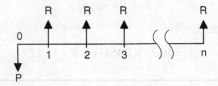

where R is referred to as unacost.

Let P be a single present value at initial time (i.e. $n = 0$). Then by using Equation (14.5a), we get

$$P = R\left[\frac{1}{1+i} + \frac{1}{(1+i)^2} + \cdots\cdots + \frac{1}{(1+i)^n}\right] \qquad (14.7)$$

Further

$$P = R\sum_{1}^{n}\frac{1}{(1+i)^n}$$

Equation (14.7) is a geometric series which has $1/(1+i)$, as the first term and $1/(1+i)$ as the ratio of n successive terms.

The summation of geometric series in Equation (14.7) can be evaluated as,

$$\sum_{1}^{n} \frac{1}{(1+i)^n} = \frac{1}{1+i} \frac{1-\left(\frac{1}{1+i}\right)^n}{1-\frac{1}{1+i}} = \frac{(1+i)^n - 1}{i(1+i)^n}$$

Then Equation (14.7) becomes,

$$P = R\frac{(1+i)^n - 1}{i(1+i)^n}$$

or,

$$P = RF_{RP,i,n} \tag{14.8a}$$

> **Present value = (unacost). (unacost present value factor)**

where,

$$F_{RP,i,n} = \frac{(1+i)^n - 1}{i(1+i)^n} \tag{14.8b}$$

which is referred to as the equal-payment series present value factor or annuity present value factor (Table 14.1).

Equation (14.8a) can also be written as,

$$R = P\frac{i(1+i)^n}{(1+i)^n - 1}$$

$$R = PF_{PR} \tag{14.9a}$$

> **Unacost = (present value). (capital recovery factor)**

where, F_{PR} (Table 14.1) is designated as

$$F_{PR,i,n} = \frac{F_{PR,i,n}(1+i)}{(1+i)^n - 1} \tag{14.9b}$$

It is also known by capital recovery factor in short, as CRF.

From Equations (14.8b) and (14.9b), it can be seen that,

$$F_{RP,i,n} = \frac{1}{F_{PR,i,n}}$$

14.2.3 Sinking Fund Factor

The future value, S, at the end of n years can also be converted into a uniform end-of-year annual amount R as shown below:

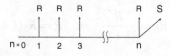

Equation (14.9a) can be expressed in terms of S, by using Equation (14.5a) as,

$$R = S(1+i)^{-n} \cdot \frac{i(1+i)^n}{(1+i)^n - 1}$$

or

$$R = S\frac{i}{(1+i)^n - 1}$$

$$R = S.F_{SR,i,n}. \tag{14.10a}$$

> ## Unacost = (future amount). (sinking fund factor)

where

$$F_{SR,i,n}. = i/[(1+i)^n - 1] \tag{14.10b}$$

It is referred to as the sinking fund factor (SFF) (Table 14.1)
Equation (14.10a) can be rewritten as,

$$S = R\frac{(1+i)^n - 1}{i}$$

$$S = RF_{RS,i,n}. \tag{14.11a}$$

> ## Future amount = (unacost). (equal payment series future value factor)

From Equations (14.10a) and (14.11a), it is concluded that F_{SR} and F_{RS} have reciprocal relationship, i.e.

$$F_{SR} = 1/F_{RS} \tag{14.11b}$$

A uniform end-of-year annual amount R is related with P and S as given in Equation (14.8a and 14.11a). Similarly, a uniform beginning of year annual amount R_b (say) can be derived in terms of P and S as follows:
The R and R_b have the following cash flow diagram:

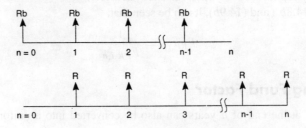

With the help of Equatioin (14.5a), R and R_b are related as

$$R_b = \frac{R}{(1+i)} \quad \text{or} \quad R = R_b(1+i) \tag{14.11c}$$

EXAMPLE 14.5(b)

Derive an expression for R_b in terms of P and S.

Solution

Substitute the expression of R from Equation (14.6b) into Equation (14.9a)

$$R_b(1+i) = PF_{PR,i,n}$$

so

$$R_b = \frac{P}{(1+i)} . F_{PR,i,n}$$

Similarly, from Equation (14.10a)

$$R_b = \frac{S}{(1+i)} F_{SR,i,n}$$

The values of various conversion factors with number of years for a given rate of interest have been given in the Table 14.1

TABLE 14.1   The values of conversion factors

			$i = 0.04$				
n	F_{PS}	F_{SP}	F_{RP}	F_{PR}	F_{RS}	F_{SR}	F_{PK}
1	1.04	0.962	0.962	1.04	1	1	26
2	1.082	0.925	1.886	0.53	2.04	0.49	13.255
3	1.125	0.889	2.775	0.36	3.122	0.32	9.009
4	1.17	0.855	3.63	0.275	4.246	0.235	6.887
5	1.217	0.822	4.452	0.225	5.416	0.185	5.616
6	1.265	0.79	5.242	0.191	6.633	0.151	4.769
7	1.316	0.76	6.002	0.167	7.898	0.127	4.165
8	1.369	0.731	6.733	0.149	9.214	0.109	3.713
9	1.423	0.703	7.435	0.134	10.583	0.094	3.362
10	1.48	0.676	8.111	0.123	12.006	0.083	3.082
11	1.539	0.65	8.76	0.114	13.486	0.074	2.854
12	1.601	0.625	9.385	0.107	15.026	0.067	2.664
13	1.665	0.601	9.986	0.1	16.627	0.06	2.504
14	1.732	0.577	10.563	0.095	18.292	0.055	2.367
15	1.801	0.555	11.118	0.09	20.024	0.05	2.249
16	1.873	0.534	11.652	0.086	21.825	0.046	2.146
17	1.948	0.513	12.166	0.082	23.697	0.042	2.055
18	2.026	0.494	12.659	0.079	25.645	0.039	1.975
19	2.107	0.475	13.134	0.076	27.671	0.036	1.903
20	2.191	0.456	13.59	0.074	29.778	0.034	1.84

			$i = 0.06$				
n	F_{PS}	F_{SP}	F_{RP}	F_{PR}	F_{RS}	F_{SR}	F_{PK}
1	1.06	0.943	0.943	1.06	1	1	17.667
2	1.124	0.89	1.833	0.545	2.06	0.485	9.091
3	1.191	0.84	2.673	0.374	3.184	0.314	6.235
4	1.262	0.792	3.465	0.289	4.375	0.229	4.81
5	1.338	0.747	4.212	0.237	5.637	0.177	3.957
6	1.419	0.705	4.917	0.203	6.975	0.143	3.389
7	1.504	0.665	5.582	0.179	8.394	0.119	2.986
8	1.594	0.627	6.21	0.161	9.897	0.101	2.684
9	1.689	0.592	6.802	0.147	11.491	0.087	2.45
10	1.791	0.558	7.36	0.136	13.181	0.076	2.264
11	1.898	0.527	7.887	0.127	14.972	0.067	2.113
12	2.012	0.497	8.384	0.119	16.87	0.059	1.988
13	2.133	0.469	8.853	0.113	18.882	0.053	1.883
14	2.261	0.442	9.295	0.108	21.015	0.048	1.793
15	2.397	0.417	9.712	0.103	23.276	0.043	1.716
16	2.54	0.394	10.106	0.099	25.672	0.039	1.649
17	2.693	0.371	10.477	0.095	28.213	0.035	1.591
18	2.854	0.35	10.828	0.092	30.906	0.032	1.539
19	3.026	0.331	11.158	0.09	33.76	0.03	1.494
20	3.207	0.312	11.47	0.087	36.786	0.027	1.453

			$i = 0.08$				
n	F_{PS}	F_{SP}	F_{RP}	F_{PR}	F_{RS}	F_{SR}	F_{PK}
1	1.08	0.926	0.926	1.08	1	1	13.5
2	1.166	0.857	1.783	0.561	2.08	0.481	7.01
3	1.26	0.794	2.577	0.388	3.246	0.308	4.85
4	1.36	0.735	3.312	0.302	4.506	0.222	3.774
5	1.469	0.681	3.993	0.25	5.867	0.17	3.131
6	1.587	0.63	4.623	0.216	7.336	0.136	2.704
7	1.714	0.583	5.206	0.192	8.923	0.112	2.401
8	1.851	0.54	5.747	0.174	10.637	0.094	2.175
9	1.999	0.5	6.247	0.16	12.488	0.08	2.001
10	2.159	0.463	6.71	0.149	14.487	0.069	1.863
11	2.332	0.429	7.139	0.14	16.646	0.06	1.751
12	2.518	0.397	7.536	0.133	18.977	0.053	1.659
13	2.72	0.368	7.904	0.127	21.495	0.047	1.582
14	2.937	0.34	8.244	0.121	24.215	0.041	1.516
15	3.172	0.315	8.559	0.117	27.152	0.037	1.46
16	3.426	0.292	8.851	0.113	30.324	0.033	1.412
17	3.7	0.27	9.122	0.11	33.75	0.03	1.37
18	3.996	0.25	9.372	0.107	37.45	0.027	1.334
19	4.316	0.232	9.604	0.104	41.446	0.024	1.302
20	4.661	0.215	9.818	0.102	45.762	0.022	1.273

			$i = 0.10$				
n	F_{PS}	F_{SP}	F_{RP}	F_{PR}	F_{RS}	F_{SR}	F_{PK}
1	1.1	0.909	0.909	1.1	1	1	11
2	1.21	0.826	1.736	0.576	2.1	0.476	5.762
3	1.331	0.751	2.487	0.402	3.31	0.302	4.021
4	1.464	0.683	3.17	0.315	4.641	0.215	3.155
5	1.611	0.621	3.791	0.264	6.105	0.164	2.638
6	1.772	0.564	4.355	0.23	7.716	0.13	2.296
7	1.949	0.513	4.868	0.205	9.487	0.105	2.054
8	2.144	0.467	5.335	0.187	11.436	0.087	1.874
9	2.358	0.424	5.759	0.174	13.579	0.074	1.736
10	2.594	0.386	6.145	0.163	15.937	0.063	1.627
11	2.853	0.35	6.495	0.154	18.531	0.054	1.54
12	3.138	0.319	6.814	0.147	21.384	0.047	1.468
13	3.452	0.29	7.103	0.141	24.523	0.041	1.408
14	3.797	0.263	7.367	0.136·	27.975	0.036	1.357
15	4.177	0.239	7.606	0.131	31.772	0.031	1.315
16	4.595	0.218	7.824	0.128	35.95	0.028	1.278
17	5.054	0.198	8.022	0.125	40.545	0.025	1.247
18	5.56	0.18	8.201	0.122	45.599	0.022	1.219
19	6.116	0.164	8.365	0.12	51.159	0.02	1.195
20	6.728	0.149	8.514	0.117	57.275	0.017	1.175

			$i = 0.12$				
n	F_{PS}	F_{SP}	F_{RP}	F_{PR}	F_{RS}	F_{SR}	F_{PK}
1	1.12	0.893	0.893	1.12	1	1	9.333
2	1.254	0.797	1.69	0.592	2.12	0.472	4.931
3	1.405	0.712	2.402	0.416	3.374	0.296	3.47
4	1.574	0.636	3.037	0.329	4.779	0.209	2.744
5	1.762	0.567	3.605	0.277	6.353	0.157	2.312
6	1.974	0.507	4.111	0.243	8.115	0.123	2.027
7	2.211	0.452	4.564	0.219	10.089	0.099	1.826
8	2.476	0.404	4.968	0.201	12.3	0.081	1.678
9	2.773	0.361	5.328	0.188	14.776	0.068	1.564
10	3.106	0.322	5.65	0.177	17.549	0.057	1.475
11	3.479	0.287	5.938	0.168	20.655	0.048	1.403
12	3.896	0.257	6.194	0.161	24.133	0.041	1.345
13	4.363	0.229	6.424	0.156	28.029	0.036	1.297
14	4.887	0.205	6.628	0.151	32.393	0.031	1.257
15	5.474	0.183	6.811	0.147	37.28	0.027	1.224
16	6.13	0.163	6.974	0.143	42.753	0.023	1.195
17	6.866	0.146	7.12	0.14	48.884	0.02	1.17
18	7.69	0.13	7.25	0.138	55.75	0.018	1.149
19	8.613	0.116	7.366	0.136	63.44	0.016	1.131
20	9.646	0.104	7.469	0.134	72.052	0.014	1.116

			$i = 0.14$				
n	F_{PS}	F_{SP}	F_{RP}	F_{PR}	F_{RS}	F_{SR}	F_{PK}
1	1.14	0.877	0.877	1.14	1	1	8.143
2	1.3	0.769	1.647	0.607	2.14	0.467	4.338
3	1.482	0.675	2.322	0.431	3.44	0.291	3.077
4	1.689	0.592	2.914	0.343	4.921	0.203	2.451
5	1.925	0.519	3.433	0.291	6.61	0.151	2.081
6	2.195	0.456	3.889	0.257	8.536	0.117	1.837
7	2.502	0.4	4.288	0.233	10.73	0.093	1.666
8	2.853	0.351	4.639	0.216	13.233	0.076	1.54
9	3.252	0.308	4.946	0.202	16.085	0.062	1.444
10	3.707	0.27	5.216	0.192	19.337	0.052	1.369
11	4.226	0.237	5.453	0.183	23.045	0.043	1.31
12	4.818	0.208	5.66	0.177	27.271	0.037	1.262
13	5.492	0.182	5.842	0.171	32.089	0.031	1.223
14	6.261	0.16	6.002	0.167	37.581	0.027	1.19
15	7.138	0.14	6.142	0.163	43.842	0.023	1.163
16	8.137	0.123	6.265	0.16	50.98	0.02	1.14
17	9.276	0.108	6.373	0.157	59.118	0.017	1.121
18	10.575	0.095	6.467	0.155	68.394	0.015	1.104
19	12.056	0.083	6.55	0.153	78.969	0.013	1.09
20	13.743	0.073	6.623	0.151	91.025	0.011	1.078

			$i = 0.16$				
n	F_{PS}	F_{SP}	F_{RP}	F_{PR}	F_{RS}	F_{SR}	F_{PK}
1	1.16	0.862	0.862	1.16	1	1	7.25
2	1.346	0.743	1.605	0.623	2.16	0.463	3.894
3	1.561	0.641	2.246	0.445	3.506	0.285	2.783
4	1.811	0.552	2.798	0.357	5.066	0.197	2.234
5	2.1	0.476	3.274	0.305	6.877	0.145	1.909
6	2.436	0.41	3.685	0.271	8.977	0.111	1.696
7	2.826	0.354	4.039	0.248	11.414	0.088	1.548
8	3.278	0.305	4.344	0.23	14.24	0.07	1.439
9	3.803	0.263	4.607	0.217	17.519	0.057	1.357
10	4.411	0.227	4.833	0.207	21.321	0.047	1.293
11	5.117	0.195	5.029	0.199	25.733	0.039	1.243
12·	5.936	0.168	5.197	0.192	30.85	0.032	1.203
13	6.886	0.145	5.342	0.187	36.786	0.027	1.17
14	7.988	0.125	5.468	0.183	43.672	0.023	1.143
15	9.266	0.108	5.575	0.179	51.659	0.019	1.121
16	10.748	0.093	5.668	0.176	60.925	0.016	1.103
17	12.468	0.08	5.749	0.174	71.673	0.014	1.087
18	14.463	0.069	5.818	0.172	84.141	0.012	1.074
19	16.777	0.06	5.877	0.17	98.603	0.01	1.063
20	19.461	0.051	5.929	0.169	115.38	0.009	1.054

			$i = 0.18$				
n	F_{PS}	F_{SP}	F_{RP}	F_{PR}	F_{RS}	F_{SR}	F_{PK}
1	1.18	0.847	0.847	1.18	1	1	6.556
2	1.392	0.718	1.566	0.639	2.18	0.459	3.548
3	1.643	0.609	2.174	0.46	3.572	0.28	2.555
4	1.939	0.516	2.69	0.372	5.215	0.192	2.065
5	2.288	0.437	3.127	0.32	7.154	0.14	1.777
6	2.7	0.37	3.498	0.286	9.442	0.106	1.588
7	3.185	0.314	3.812	0.262	12.142	0.082	1.458
8	3.759	0.266	4.078	0.245	15.327	0.065	1.362
9	4.435	0.225	4.303	0.232	19.086	0.052	1.291
10	5.234	0.191	4.494	0.223	23.521	0.043	1.236
11	6.176	0.162	4.656	0.215	28.755	0.035	1.193
12	7.288	0.137	4.793	0.209	34.931	0.029	1.159
13	8.599	0.116	4.91	0.204	42.219	0.024	1.132
14	10.147	0.099	5.008	0.2	50.818	0.02	1.109
15	11.974	0.084	5.092	0.196	60.965	0.016	1.091
16	14.129	0.071	5.162	0.194	72.939	0.014	1.076
17	16.672	0.06	5.222	0.191	87.068	0.011	1.064
18	19.673	0.051	5.273	0.19	103.74	0.01	1.054
19	23.214	0.043	5.316	0.188	123.414	0.008	1.045
20	27.393	0.037	5.353	0.187	146.628	0.007	1.038

			$i = 0.20$				
n	F_{PS}	F_{SP}	F_{RP}	F_{PR}	F_{RS}	F_{SR}	F_{PK}
1	1.2	0.833	0.833	1.2	1	1	6
2	1.44	0.694	1.528	0.655	2.2	0.455	3.273
3	1.728	0.579	2.106	0.475	3.64	0.275	2.374
4	2.074	0.482	2.589	0.386	5.368	0.186	1.931
5	2.488	0.402	2.991	0.334	7.442	0.134	1.672
6	2.986	0.335	3.326	0.301	9.93	0.101	1.504
7	3.583	0.279	3.605	0.277	12.916	0.077	1.387
8	4.3	0.233	3.837	0.261	16.499	0.061	1.303
9	5.16	0.194	4.031	0.248	20.799	0.048	1.24
10	6.192	0.162	4.192	0.239	25.959	0.039	1.193
11	7.43	0.135	4.327	0.231	32.15	0.031	1.156
12	8.916	0.112	4.439	0.225	39.581	0.025	1.126
13	10.699	0.093	4.533	0.221	48.497	0.021	1.103
14	12.839	0.078	4.611	0.217	59.196	0.017	1.084
15	15.407	0.065	4.675	0.214	72.035	0.014	1.069
16	18.488	0.054	4.73	0.211	87.442	0.011	1.057
17	22.186	0.045	4.775	0.209	105.931	0.009	1.047
18	26.623	0.038	4.812	0.208	128.117	0.008	1.039
19	31.948	0.031	4.843	0.206	154.74	0.006	1.032
20	38.338	0.026	4.87	0.205	186.688	0.005	1.027

14.3 CASH-FLOW DIAGRAMS

A cash flow diagram is simply a graphical representation of cash flows drawn on a time scale.

$$\text{Net cash flow} = \text{receipts} - \text{disbursements} \qquad (14.12)$$

As discussed earlier, Equation (14.12) can be represented as follows:

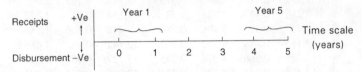

In the above cash flow diagram, a uniform end-of-year annual amount (R) will be considered at the end of each year of time scale. The above cash flow diagram will be used in the following examples.

EXAMPLE 14.6

A person decides to spend Rs. 3,000 on first, second, third and fourth year on energy efficient equipment and agrees to set aside a certain amount now and each year thereafter until the fourth year. If the contribution forms an arithmetical progression for all years increasing by 20 percent after the first year; calculate his first contribution if money is worth 10 percent.

Solution

Let us assume that the first contribution is x. The cash flow diagram can be shown as:

$$
\begin{array}{ccccc}
x & 1.2x & 1.4x & 1.6x & 1.8x \\
\uparrow & \uparrow & \uparrow & \uparrow & \uparrow \\
0 & 1 & 2 & 3 & 4
\end{array}
$$

Consider two years from now as the focal point. Now using time-value conversion relation (Equation (11.6)) in above cash-flow diagram, we get

$$x(1.10)^{-2} + 1.2x(1.10)^{-1} + 1.4x(1.10)^0 + 1.6x(1.10)^1 + 1.8x(1.10)^2 = 3000(1.10)^{-1}$$
$$+ 3000(1.10)^0 + 3000(1.10)^1 + 3000(1.10)^2$$

$$7.2553x = 3000 \times 4.2191$$

$$x = 1744.56$$

The first contribution would be Rs. 1744.56.

EXAMPLE 14.7

A person wants a down payment of Rs. 2,000 on a solar water heating system of amount Rs. 10,000. An annual end-of-year payment (R) of Rs. 1174.11 is required for 12 years. However, the person elects to pay Rs. 1000 yearly and a balance payment at the end. Find the balance payment if money is worth 10 percent interest.

Solution

Let X is the balance payment. The cash-flow diagram is

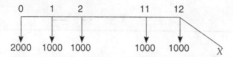

Using cash-flow diagram and Equations (14.5a) and (14.8a), we can write

$$10,000 = 2000 + 1000 F_{RP,10\%,12} + X F_{SP,10\%,12}$$
$$= 2000 + 1000(6.8137) + X(0.31863)$$
$$X = 3723.10$$

The balance payment is Rs. 3723.10

EXAMPLE 14.8

Find the equivalent present value P of the following series of receipt (Rs.) as of the end of the fourth year if money is worth 10 percent per year.

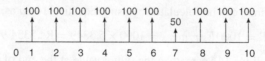

Solution

The above series can be written, after adding and subtracting 50 at 7th year, as follows:

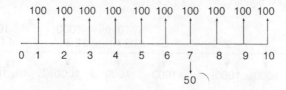

For the above cash-flow diagram, $R = 100$ and $S = 50$. By using Equations (14.5a) and (14.8a), the present value

$$P = 100F_{RP,10\%,10} - 50F_{SP,10\%,7}$$

$$= 100(6.1438) - 50(0.3855) = 595.10 \text{ (zero time)}$$

Moving 4 years with the calendar and using Equation (14.1a), the equivalent present value at 4th year will be

$$595.10F_{PS,6\%,3} = 595.10 \times 2.5937 = 1543.52 \text{ (Rs)}$$

EXAMPLE 14.9

A person plans to create a forborne annuity by depositing Rs. 1000 at the end of the year, for 8 years. He wants to withdraw the money at the end of 14 years from now to buy energy-efficient domestic appliances. Find the accumulated value at the end of the fourteenth year, if money is worth 10 percent per year.

Solution

Let X be the amount available at 14th year which can be considered as receipt. The cash flow diagram for the payment is,

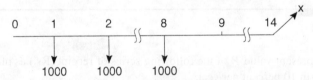

The present value (zero time) can be calculated by using Equation 14.8a as,

$$P = 1000F_{RP,10\%,8} = 1000 \times 5.3349 = \text{Rs.}5334.90$$

If this amount is deposited for 14 years, then the future value at the end of 14 years (Equation 14.1a) will be

$$S = 5334.90 \times F_{PS,10\%14} = 5334.90(3.7975) = \text{Rs.}20,259.00$$

The above cash flow diagram can also be drawn by considering Rs. 1000 paid for 14 years less Rs. 1000 paid as annuity for the last 6 years.

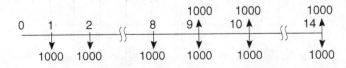

By using Equation (14.11), we get

$$S = 1000\left[\frac{(1.10)^{14} - 1}{0.10} - \frac{(1.10)^6 - 1}{0.10}\right] = 1000[27.975 - 7.7156]$$

or $\qquad S = \text{Rs. } 20,259.00$

It is clear that one can use one of the above methods.

14.4 COST COMPARISONS WITH EQUAL DURATION

In this section, a uniform expense is referred to as a uniform end-of-year cost.

EXAMPLE 14.10

Two solar water heating systems have the following cost comparison. Which system is more economical if the money is worth 10 percent per year.

Economic Components	System (A)	System (B)
First cost (Rs.)	30,000	15,000
Uniform end-of-year maintenance per year (Rs.)	2,000	5,000
Overhaul, end of the third year (Rs.)	–	3,500
Salvage value (Rs.)	4,000	1,000
Life of the system (years)	5	5
Benefit from quality control as a uniform end-of-year amount per year (Rs.)	1,000	–

Solution

Cash-flow diagram for each system have been shown as follows:

System A

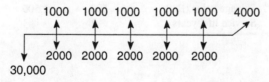

System B

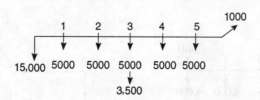

The present value of the costs for system A can be obtained by using, Equations (14.5a) and (14.8a) as

$$P_{AS} = 30,000 + (2000 - 1000)F_{RP,10\%,5} - 4000F_{SP,10\%,5} = 30,000 + 1000(3.7908)$$

$$-4000(0.62092) = \text{Rs}.31,307.12$$

The present value of the costs for system B can be obtained by using, Equations (11.5) and (11.8) as follows:

$$P_{BS} = 15,000 + 5000F_{RP,10\%,5} + 3500F_{SP,10\%3} - 1000F_{SP,10\%,5}$$

$$= 15,000 + 5,000 \times 3.7908 + 3500 \times 0.75131 - 1000 \times 0.62092$$

$$= 15,000 + 18,954 + 2629.55 - 620.92 = \text{Rs}.35,962.63$$

From above calculations, it is clear that system A is more economical than system B.

14.5 COST COMPARISONS WITH UNEQUAL DURATION

If two energy efficient systems have different duration of lives, a fair comparison can be made only on the basis of equal duration. One of the methods for comparison is to compare single present value of costs on the basis of a common denominator of their service lives.

14.5.1 Single Present Value Method (First Method)

EXAMPLE 14.11

Two solar water heating systems have the following cost comparison. Which system is more economical if the money is worth 10 percent per year.

Cost Components	System (A)	System (B)
First cost (Rs.)	20,000	30,000
Uniform end-of-year maintenance (Rs.)	4,000	3,000
Salvage value (Rs.)	500	1,500
Service life, years	2	3

Solution

The cash-flow diagram for both systems are first reduced to single present value of the cost.

System A

System B

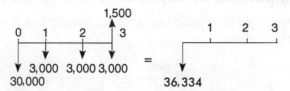

The simplified diagrams are now repeated to obtain six year duration. Note that the present value of system A is 26,529 as of its time of installation.

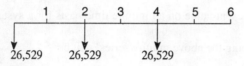

Similarly, the present value of system B is Rs. 36,334 at the time of installation. The cash-flow diagram for six year duration is,

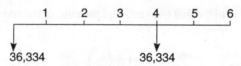

The present value of each of the preceding diagrams at 10 percent per year is

$$P_{A6} = 26,529 + 26,529 F_{SP,10\%,2} + 26,529 F_{SP,10\%,4}$$

$$= 26,529 + 26,529(1.10)^{-2} + 26,529(1.10)^{-4}$$

$$= 26,529 + 21,924.79 + 18,119.66 = \text{Rs.} 66,573.45$$

Similarly, $\qquad P_{B6} = \text{Rs.} 63,632.27$

The ratio of cost is $\qquad \dfrac{P_{A6}}{P_{B6}} = \dfrac{66,573.45}{63,632.27} = 1.0462$

Thus, system B is more economical than system A.

14.5.2 Cost Comparison by Annual Cost Method (Second Method)

In this case, uniform end-of-year annual amount will be calculated by using Equation (14.9a) for $P_{A2} =$ Rs. 26,529 and, $P_{B3} =$ Rs. 36,334 of Example 14.11

The unacost for two systems are

$$R_A = P_{A2} F_{PR,10\%,2} = 26,529(0.57619) = \text{Rs.} 15,285.74$$

and $\qquad R_B = 36,334 F_{PR,10\%,3} = 36,334 \times (0.40211) = \text{Rs.} 14,610.26$

The ratio of cost is $\qquad \dfrac{R_A}{R_B} = \dfrac{15,285.74}{14610.26} = 1.0462$

The system B is more economical than system A as concluded earlier.

14.5.3 Cost Comparison by Capitalised Cost (Third Method)

If P_n is equivalent present value of a system lasting n year, then the present value on an infinite time basis can be represented as follows:

Capitalised cost is the present value on an infinite time basis for a system costing P_n and lasting n years.

The present value, replacing the above infinity series, will be

$$K = P_n \sum_{x-0}^{\infty} \frac{1}{(1+i)^{xn}} = P_n \left[1 + \frac{1}{(1+i)^n} + \frac{1}{(1+i)^{2n}} + \cdots \right] \qquad (14.13)$$

This is a geometric series with the first term as 1 and the ratio of the consecutive terms as $1/(1+i)^n$. Its summation is given by,

$$\sum \frac{1}{(1+i)^{xn}} = 1 \frac{1 - \left(\frac{1}{(1+i)^n} \right)^{\infty}}{1 - \frac{1}{(1+i)^n}} = \frac{(1+i)^n}{(1+i)^n - 1}$$

Equation (14.13) becomes,

$$K = P_n \, F_{PK,i,n} \qquad (14.14a)$$

Capitalised cost = (Present value basis n years duration) (capitalised cost factor)

---

where K is the capitalised cost and $F_{PK,i,n}$ the factor that converts a present value to capitalised cost (Table 14.1), also known as capitalised cost factor and is given by

$$F_{PK,i,n} = \frac{(1+i)^n}{(1+i)^n - 1} \qquad (14.14b)$$

From Equations (14.9b), $F_{PR,i,n}$ is given as

$$F_{PR,i,n} = \frac{i(1+i)^n}{(1+i)^n - 1} \qquad (14.15)$$

Comparing Equations (14.15) and (14.14b), $F_{PR,i,n,}$, $F_{PK,i,n}$, gives following relationship

$$F_{PR,i,n} = i \, F_{PK,i,n} \qquad (14.16)$$

Capital recovery factor = rate of return × (capitalised cost factor)

Similarly, from Equations (14.9a) and (14.14a), R and K are related as,

$$R = iK \qquad (14.17)$$

or

$$\boxed{\textbf{Unacost = rate of return} \times \textbf{(capitalised cost factor)}}$$

In this case, we solve Example 14.11 by using the capitalised cost method.
From Example 14.11, we have, P_{A2} = Rs. 26,529 and P_{B3} = Rs. 36,334.
By using Equation (14.14a), we get

$$K_A = P_{A2}, F_{PK,10\%,2} = 26,529 \times (5.7619) = \text{Rs.}152,857.45$$

$$K_B = 36,334 F_{PK,10\%,3} = 36,334 \times (4.0211) = \text{Rs.}146,102.65$$

The ratio of cost is

$$\frac{K_A}{K_B} = \frac{152,857.45}{146,102.65} = 1.0462$$

It is clear from the above calculation that the results obtained are the same as in the earlier solution. From this calculation, we can also conclude that the system B is more economical than system A.

14.5.4 Fourth Method

As a matter of fact, it is possible to convert a present value P_{n1}, of n_1 years duration to an equivalent present value P_{n2} of n_2 years duration.

Hence, applying Equation (14.14) gives,

$$P_{n1} F_{PR,i,n1} = P_{n2} F_{PR,i,n2}$$

$$P_{n2} = P_{n1} \frac{F_{PR,i,n}}{F_{PR,i,n2}} \qquad (14.18)$$

As discussed earlier

$$P_{A2} = \text{2-year duration} = \text{Rs. 26,529}$$

$$P_{B3} = \text{3-year duration} = \text{Rs. 36,334}$$

Convert the present value of system B to an equivalent value for 2 years duration using Equation (14.18)

$$P_{B2} = P_{B3} \frac{F_{PR,10\%,3}}{F_{PR,10\%,2}} = 36,334 \frac{0.40211}{0.57619} = \text{Rs.}25,356.68$$

The ratio of cost

$$\frac{P_{A2}}{P_{B2}} = \frac{26529}{25356} = 1.0462$$

Further, the result of cost ratio is the same as that obtained by various methods discussed earlier. Hence, system B is more economical.

14.6 PAYBACK TIME/PAYMENT TIME/PAYBACK PERIOD

Profitability is a measure of the total income for a project compared to the total outlay. Money going into the project is taken as negative and money coming back from the project as positive. Payout time is one of the criteria for profitability which has been discussed in this section.

14.6.1 Payout Time without Interest

It is the time required to reduce the investment to zero. The working capital is not considered in evaluating payout time without interest.

EXAMPLE 14.12

Find the payout time for the following data given in Table 14.2.
Here, depreciation and tax are defined as follows:

Depreciation: An expenditure that decreases in value with time, must be apportioned over its life. The term used to describe this loss in value is depreciation. If there is any tax benefit due to depreciation in value, the benefit can be considered profit.

$$\text{Taxes} = (\text{Income} - \text{Deduction}) \text{ Tax rate} \tag{14.19}$$

Cummulative cash flow of Table 14.2 has been given in Table 14.3, and Figure 14.1,
The pay back time is between 2–3 years, i.e. 2.4 years.

TABLE 14.2 Data for various profit and cash flow

Time (end year)	After tax profit (Rs.)	Tax benefit due to depreciation (Rs.)	Cash flow = Profit + Tax benefit due to depreciation
0	−10,000	0	−10,000
1	2,750	2,000	4,750
2	2,000	2,000	4,000
3	1,300	2,000	3,300
4	700	2,000	2,700
5	0	2,000	2,000

TABLE 14.3
Cummulative cash flow

Time (end year)	Cummulative cash flow (B)
0	−10,000
1	−5,250
2	−1,250
3	2,050
4	4,750
5	6,750

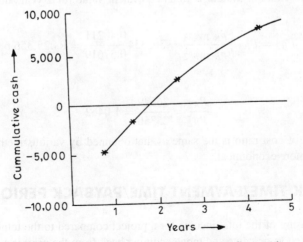

Figure 14.1 Variation of cummulative cash flow with years.

14.6.2 Payout Time with Interest

This is the method which allows for a return on investment and is subject to variations. The variation includes an interest on working capital.

EXAMPLE 14.13

Using the data given in Table 14.2, find the payout time with interest on remaining investment at 10 percent per year.

Here, an investment for the year has been computed by the following relation:

Investment for the year = Investment for the previous year

$$-(\text{Cash flow}-\text{Interest rate for previous year}) \qquad (14.20)$$

The cummulative cash flow of Table 14.2, while considering interest rate and using Equation (14.20), has been given in Table 14.4. The cummulative net cash flow can also be drawn against years similar to Figure 14.1. Based on Table 14.4, it can be inferred that the payback period is about 2.95 years which is higher due to the interest rate, as expected.

In Examples 14.12 and 14.13, simple methods have been used to calculate payout time (payback period). Now an analytical method will be discussed for evaluating payout time.

TABLE 14.4 Cummulative cash flow with interest

(1)	(2)	(3)	(4)	(5)	(6)
		Interest on	Cash flow		
End	Investment	investment	after interest	Cash flow	Cummulative
year	for year	for year	(4)–(3)	net	cash flow
0			−10,000		−10,000
1	10,000	1,000	4,750	3,750	−6,250
2	6,250	625	4,000	3,375	−2,870
3	2,875	287.5	3,300	3,012.5	137.5

14.6.3 Analytical Expression for Payout Time

The payback period (n), the number of years necessary to exactly recover the initial investment P, is computed by summing the annual cash-flow values and estimating n through the relation:

$$0 = -\text{initial investment} + \text{sum of annual cash flows}$$

or

$$0 = -P + \sum_{t=1}^{n} CF_t(F_{SP,i\%,t}) \qquad (14.21)$$

where CF_t is the net cash flow at the end of year t. If cash flow are same each year, F_{RP} factor may be used in the relation:

$$0 = -P + CF_t(F_{RP,i\%,n}) \qquad (14.22)$$

i.e. after n years, the cash flow will recover the investment and a return of i percent. If the expected retention period (life) of the asset/project is less than n years, then investment is not advisable.

Considering i to be zero, Equation (14.22) becomes,

$$0 = -P + \sum_{t=1}^{n} CF_t \qquad (14.23)$$

and if CF_t values are assumed equal, then,

$$n = P/CF \qquad (14.24)$$

There is little value in techno-economic study for n computed from Equations (14.23) and (14.24). When $i\% > 0$ is used to estimate n, the results incorporate the risk considered in the project undertaken.

EXAMPLE 14.14

A solar water heating system purchased for Rs. 18,000 is expected to generate annual revenues of Rs. 3000, and have salvage value of Rs. 3000 at any time during 10 years of anticipated ownership. If a 15 percent per year required return is imposed on the purchase, compute the payback period.

Solution

The cash flow for each year is Rs. 3000 (P) with an additional revenue of Rs. 3000 in year n. The cash flow diagram has been shown below:

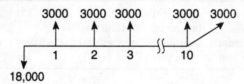

After using Equations (14.21) and (14.22) for above cash flow, one gets,

$$0 = -P + CF_t(F_{RP,15\%,n}) + SV(F_{SP,15\%,n})$$

or

$$0 = -18,000 + 3000\, F_{RP,15\%,n} + 3000\, F_{SP,15\%,n}$$

The resulting payout time can be evaluated after further using Equations (14.8b) and (14.5b) for F_{RP} and F_{SP} and we get $n = 15.3$ years which is not economical with such high interest. For $i = 0$, Equation (14.23) can be used and we get,

$$0 = -18,000 + n(3,000) + 3,000$$

The resulting payout time is 5 years, which is most economical, without interest rate.

EXAMPLE 14.15

Two solar water heating systems have the cash flow as given below:
Calculate the payout time for both the systems.

Economic components	System I	System II
Present worth (P), Rs.	16,000	8,000
Net income per year (Rs.)	4,000	1,000 (year 1–5)
		3,000 (year 6–15)
Maximum life (n), years	7	15
Interest rate (i), percentage	15	15

Solution

The cash flow diagram for both the systems are:

For system I

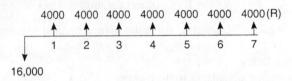

Using Equation (14.22) in above cash flow diagram, and assuming n as the payout time, we have

$$0 = -16,000 + 4000 + F_{RP,15\%,n}$$

which gives $n = 6.57$ years which is less than 7 years.

For system II

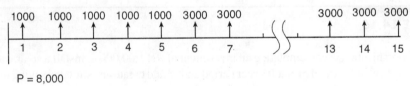

Equivalent cash flow diagram for n years as a payout time is,

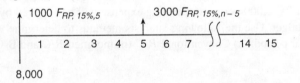

Now, $3000\, F_{RP,15\%,n-5}$ can be converted into present worth by using Equation (14.5b) and we get,

$$0 = -8000 + 1000\, F_{RP,15\%,5} + (3000\, F_{RP,15\%,n-5})F_{SP,15\%,n}$$

which yield $n = 9.52$ years, less than expected life.

14.7 BENEFIT-COST ANALYSIS

Benefit-cost ratio (B/C ratio) is a tool to select the right project based on advantage versus disadvantage analysis. A project is considered to be attractive when the benefits derived from its execution exceed its associated costs.

The conventional B/C ratio is calculated as

$$B/C = (\text{Benefits} - \text{Disbenefits})/\text{Cost} = (B - D)/C \qquad (14.25)$$

The modified B/C ratio, which is gaining support includes operation and maintenance $(0 \& M)$ costs in the numerator and treats them in a manner similar to disbenefits, and is given by,

$$B/C = \frac{\text{Benefits} - \text{Disbenefits} - O \& M \text{ Cost}}{\text{Initial investment}} \qquad (14.26)$$

The salvage value can also be considered in the denominator.
The B/C ratio influences the decision on the project approval.

If,
$$B/C > 1, \qquad \text{accept the project}$$
$$B/C < 1, . \qquad \text{reject the project.}$$

Thus, in case of mutually exclusive projects, B/C ratio gives a method to compare them against each other.
Some definitions are given below.

Benefits (B): Benefits are the advantages to the owner.
Disbenefits (D): When the project under consideration involves disadvantages to the owner.
Costs: The anticipated expenditures for construction, operation, maintenance etc.
Owner: Public: One who incurs the costs as the government.

EXAMPLE 14.16

A non-profit organization is contemplating an investment of Rs. 1,00,000 to install a solar water heating system. The grant would extend over a 10-year period and would create an estimated saving of Rs.20,000 per year. The organization uses a rate of return of 6 percent per year on all grant investments. An estimated Rs. 4,000 a year would have to be released, from other sources, for expenses. In order to make this program successful, a Rs. 2,000 per year operating expense will be incurred by the organization from it's regular O & M budget. Use the following analysis methods to determine whether the program is justified over a 10-year period: (a) Conventional B/C (b) modified B/C and B-C analysis.

Solution

The definition using an equivalent- annual-worth basis are:

  Benefits : Rs. 20,000 per year
  Investment cost : 1,00,000 $F_{PR,6\%,10}$ = Rs. 13587 per year
  O & M cost : Rs. 2,000 per year
  Disbenefits : Rs. 4,000 per year

a. Conventional Method

$$B/C = (20,000 - 4,000)/(13587 + 2,000) = 1.026$$

This project is justified since $B/C > 1$

b. Modified Method

This separates the investment and O & M cost.

Modified $B/C = (20000 - 4000 - 2000)/13587$

c. $B - C = 16,000 - 13587 = 2413$

Since $B - C > 0$, the investment is justified.

EXAMPLE 14.17

Cost analysis of built-in storage water heater:

a. Initial cost of the material used: See following table for 100 liter capacity:

Material	Cost (Rs.)
Steel structure (for box and cover material)	1,000
Glass	170
Insulation	150
Paint	170
Stand, bucket, frame etc.	550
Labor	700
Total	2,740

b. Salvage values: We may reuse various system components e.g., body structure, glass, insulation etc., even after the useful life of the system is over. Therefore, the salvage value is estimated to be 40 percent of the initial cost.

Salvage value of built-in storage water heater = Rs. 1096

c. Maintenance cost: This may be estimated as follows:

Annual maintenance cost $= 0.15 \times$ Annual first cost.

Useful life of the system $(n) = 10$ years and $r = 0.12$.

Solution

The CRF and SFF can be obtained as follows:

$$\text{CRF} = \frac{r(1+r)^n}{(1+r)^n - 1} = 0.177$$

$$\text{SFF (sinking fund factor)} = \frac{r}{(1+r)^n - 1} = 0.057$$

$$P = 2740$$

$$S \text{ (salvage value)} = 1096$$

$$\text{Annual first cost} = \text{Rs. (CRF)}.P = \text{Rs. } (0.177).2740 = 484.98 \approx \text{Rs. } 485$$

$$\text{Annual salvage value} = \text{(SFF)}S = \text{Rs. } (0.057 \times 1096) = \text{Rs. } 62.50$$

$$\text{Annual maintenance cost} = \text{Rs. } 0.15 \times 485 = \text{Rs. } 72.75$$

$$\text{Hence, annual cost/m}^2 = \text{Rs. } ((72.75 - 62.5) + 485) = \text{Rs. } (495.25)$$

Average daily insolation at $45°$
$$\text{Inclination of the absorber} = 5.8768 \text{ kWh/m}^2\text{day}$$

$$\text{Annual total insolation} = 5.8768 \times 365 = 2145 \text{ kW h/m}^2$$

Useful energy (as the efficiency of

$$\text{the system is about 70 percent)} = 2145 \times 0.7 = 1501 \text{ kWh/m}^2$$

$$\text{Annual cost/kWh} = \text{Rs.0.33}$$

EXAMPLE 14.18

Cost analysis of a mounted single basin solar still.

a. Initial cost of the material used is given in the following table:

Material	Cost (in Rs.)
Steel and aluminum structures	1150
Glass	180
Rubber material	40
Paint	75
Insulation	45
Labor	600
Total	2090

The yearly average yield of the still $= 2$ liters/m^2 day.

Solution

Assuming the reuse of various components even after the useful life of the system is over, the salvage value can be estimated to be 35 percent of the initial cost.

Salvage value of solar still $=$ Rs. 731.50
From Example (11.17),

CRF $= 0.1769$	$P =$ Rs. 2090
SFF $= 0.0569$	$S =$ Rs. 731.50

Annual first cost $= 0.1769 \times 2090 = $ Rs. 369.72

Annual salvage value $= $ Rs. $0.0569 \times 731.50 = $ Rs. 41.62

Annual maintenance cost $= $ Rs. $0.15 \times 369.72 = $ Rs. 55.46

Annual cost/m$^2 = $ Rs. $(369.72 + 55.46 - 41.62) = $ Rs. 383.56

Annual yield of the still $= 2 \times 365 = 730$ liters

Annual useful energy $= 730 \times 0.65 = 474.5$ kWh

where 0.65 kWh/kg is the latent heat of vaporisation.

Annual cost/kg $= 369.72/730 = $ Rs. 0.506

Annual cost/kWh $= 369.72/474.5 = $ Rs. 0.78

b. If the yield from the still is, say, 5 liters/m^2 day, then the annual yield $= 1825$ liters.
Consequently, annual useful energy $= 1825 \times 0.65 = 1186.25$ kW/h
Then, annual cost /kg $= $ Rs. 0.20 and annual cost/kWh $= $ Rs. 0.31.

14.8 EFFECT OF DEPRECIATION

Let us define the following terms to be used frequently:

Initial cost (C_i):   Also referred as **first cost** or **initial value** or **single amount**. It is the installed cost of the system. The cost includes the purchase price, delivery and installation fee and other depreciable direct cost (defined later) incurred to ready the asset for use.

Salvage value (C_{sal}):   It is the expected market value at the end of useful life of the asset. It is negative if dismantling cost or carrying away cost are anticipated. It can be zero also. For example, the window glass has zero salvage value.

Depreciation (C_d):   An expenditure that decreases in value with time. This must be apportioned over its lifetime. The term used to describe this loss in value is known as *depreciation*.

$$C_d = C_i - C_{\text{sal}}$$

Book value (B):   It represents the remaining undepreciated investment on corporate books. It can be obtained after the total amount of annual depreciation charges to date has been subtracted from the first cost (present value/initial cost). The book value is usually determined at end of each year.

Depreciation rate (D_t):   It is the fraction of first cost removed through depreciation from corporate book. This rate may be the same, i.e. straight-line (SL) rate or different for each year of the recovery period as shown in Figure 14.2.

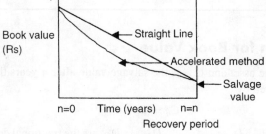

Figure 14.2 Variation of book value (Rs) with time (n) showing depreciation rate.

Mathematically, it can be written for straight-line (SL) depreciation as follows:

$$D_t = \frac{C_i - C_{sal}}{n} \tag{14.27}$$

The book value at nth year can be expressed as

$$B_n = C_i - n D_t \tag{14.28}$$

Fractional depreciation $\left(\frac{1}{n}\right)$: It is the ratio of depreciation rate to depreciation.

Recovery period (n): It is the life of the asset (in years) for depreciation and tax purpose. It is also referred as *expected life* of asset in years.

Market value: It is the actual amount that could be obtained after selling the asset in the open market. For example, (i) the market value of a commercial building tends to increase with period in the open market but the book value will decrease as depreciation charges are taken in to account and (ii) an electronic equipment(computer system) may have a market value much lower than book value due to the rapid change of technology.

EXAMPLE 14.19(a)

Calculate the depreciation rate and the book value of the asset having first cost of Rs. 60,000 and salvage value of Rs. 10,000 after 5 years.

Solution

The depreciation rate can be calculated from Equation (14.27) as

$$D_t = \frac{60,000 - 10,000}{5} = \text{Rs. } 10,000 \text{ per year for 5 years}$$

The book value at the end of each year will be

$$B_1 = 60,000 - 1 \times 10,000 = \text{Rs. } 50,000$$
$$B_2 = 60,000 - 2 \times 10,000 = \text{Rs. } 40,000$$
$$B_3 = 60,000 - 3 \times 10,000 = \text{Rs. } 30,000$$
$$B_4 = 60,000 - 4 \times 10,000 = \text{Rs. } 20,000$$

and

$$B_5 = 60,000 - 5 \times 10,000 = \text{Rs. } 10,000$$

14.8.1 Expression for Book Value

If C_i is an initial cost of the asset and C_{sal} is the salvage value after n years then total depreciation or depreciable first cost is given by

$$C_d = C_i - C_{sal} \tag{14.29a}$$

Let $D_{f1}, D_{f2}, D_{f3}, D_{f4}, D_{f5}, D_{f6}, \ldots\ldots\ldots, D_{fn-1}, D_{fn}$ are the fractional depreciation for each year, then depreciation for mth year will be,

$$D_m = D_{fm} . C_d \tag{14.29b}$$

The book value of the asset at end of the mth year can be obtained by subtracting the accumulated depreciation expense to that time from the original value of the asset, i.e.

$$\boxed{\textbf{Book value = Initial cost (first cost) -- Accumulated cost}}$$

or,
$$B_m = C_i - C_d \Sigma D_{fi} \qquad (14.30a)$$

Subtracting and adding C_d in the right side of the above equation, we get

$$B_m = C_i - C_d + C_d - C_d \Sigma D_{fi} = C_{sal} + C_d [1 - \Sigma(D_{fi})] \qquad (14.30b)$$

The book value may bear no relation to the resale value.

14.8.2 Straight Line Depreciation

If fractional depreciation is same for all years, i.e. $D_{f1} = D_{f2} = D_{f3} = D_{f4} = D_{f5} = D_{f6}, \ldots \ldots = D_{fn-1} = D_{fn}$, or $D_f = 1/n$, then
Depreciation for mth year, $D_m = C_d/n$
Accumulated depreciation upto mth year becomes, $\Sigma D_m = C_d m/n$
 The book value at end of mth year will be

$$B_m = C_i - C_d \Sigma D_{fi} = C_i - C_d(m/n)$$
$$= C_i - C_d + C_d - C_d(m/n) = C_{sal} + C_d[1 - (m/n)] \qquad (14.30c)$$

The above result can be summarised as follows:

$$\boxed{\textbf{Book value = Salvage value + Future depreciation}}$$

Depreciation remaining for future years from mth year to nth year is

$$C_d \sum_m^n D_{f,m} = C_d \sum_{m=m}^{m=n} \frac{1}{n} = C_d \frac{n-m}{n} = C_d \left[1 - \frac{m}{n}\right]$$

Present value of Re. 1 of depreciation (C_d = Re. 1) is,

$$F_{SLP,i,n} = \frac{1}{n}\left[\frac{1}{1+i} + \frac{1}{(1+i)^2} + \frac{1}{(1+i)^3} + \cdots + \frac{1}{(1+i)^n}\right]$$
$$= \frac{1}{n}\frac{(1+i)^n - 1}{i(1+i)^n} = \frac{1}{n}F_{RP,i,n} \qquad (14.31)$$

The flow chart of above series is shown below:

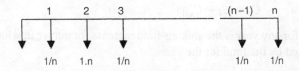

EXAMPLE 14.19(b)

Derive the relation $F_{SLP,i,n} = \frac{1}{ni\,F_{PK}}$.

Solution

We know from the above derivation that

$$F_{SLP,i,n} = \frac{1}{n}F_{RP,i,n} = \frac{1}{n}\frac{(1+i)^n - 1}{i(1+i)^n}$$

Also

$$F_{PK,i,n} = \frac{(1+i)^n}{(1+i)^n - 1}$$

Substituting $F_{PK,i,n}$ in the expression for $F_{SLP,i,n}$, we get

$$F_{SLP,i,n} = \frac{1}{ni\,F_{PK,i,n}}$$

14.8.3 Sinking Fund Depreciation

Suppose an annual deposit (Rs.) is made at end of each year to a sinking fund to restore the depreciable value at the end of n years.

The sinking fund depreciation (SF) is explained in the figure and flow chart given below:

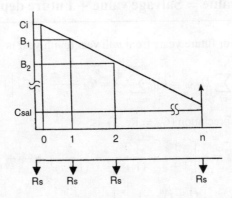

The annual deposit (Rs.) can be expressed for a known first cost C_i and salvage value C_{sal} as

$$\text{Rs.} = C_d F_{SP,i,n} F_{PR,i,n} = C_d F_{SR,i,n} = C_d \frac{i}{(1+i)^n - 1} \tag{14.32}$$

where

$$C_d = C_i - C_{sal}$$

The depreciation for any year is the sinking-fund increase for that year, which is the deposit for year plus an interest earned by the fund for the year.

14.9 COST COMPARISON AFTER TAXES

14.9.1 Without Depreciation

If i is the rate of return before taxes and t the tax rate, then the rate of return r after taxes will be

$$r = i(1-t)$$

Before taxes, the flow chart is

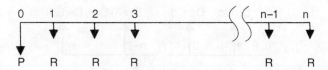

In this case, an investment compounds at a rate i and first cost P can be expressed in terms of unacost (R) as

$$P = RF_{RP,i,n} = R\frac{(1+i)^n - 1}{i(1+i)^n}$$

After taxes, the flow chart is

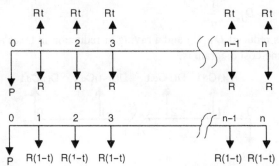

or

In this case, it will compound at a rate r and an expression for first cost P can be written as

$$P = \frac{R(1-t)}{(1+r)} + \frac{R(1-t)}{(1+r)^2} + \frac{R(1-t)}{(1+r)^3} + \cdots + \frac{R(1-t)}{(1+r)^n}$$

$$= R(1-t)\frac{(1+r)^n - 1}{r(1+r)^n} = R(1-t)F_{RP,r,n} \tag{14.33}$$

From the above equation, an expression for unacost after taxes can be expressed as

$$R = \frac{P}{(1-t)}\frac{r(1+r)^n}{(1+r)^n - 1} = \frac{P}{(1-t)}F_{PR,r,n}$$

Also

$$R = \frac{S}{(1-t)}\frac{r}{(1+r)^n - 1} = \frac{S}{(1-t)}F_{SR,r,n}$$

Here

$$S = P(1+r)^n.$$

14.9.2 With Depreciation

Consider C_i as the initial cost of an article that lasts n years with the salvage value of C_{sal}. The depreciable cost will be

$$C_d = C_i - C_{sal}$$

There is no tax consideration at the time of purchase of an article.

Let $D_{f1}, D_{f2}, D_{f3}, \ldots D_{f,n-1}, D_{fn}$ are the fractional depreciation for each year as assumed in the earlier section, then time-cost diagram without tax will be

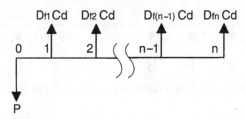

Here $\quad C_d = D_{f1}C_d + D_{f2}C_d + \cdots D_{fn}C_d$

$$= D_f C_d + D_f C_d + C_d + \cdots + D_f C_d \text{ for same fractional depreciation}$$

Also, $\quad D_{f1} = D_{f2} = D_{f_n} = \dfrac{1}{n}$

Now, the taxable base is reduced $D_{f1}C_d$ and a saving or reduction in taxes amounting to $D_{f1}C_d.t$ is realised. In this case, time-cost diagram is,

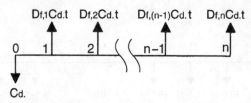

By using the above time-cost diagram and assuming $D_{f,1} = D_{f,2} = \ldots . D_{f,(n-1)} = D_{f,n} = D_f = 1/n$, the present value is

$$P = C_d - D_f C_d t \left[\frac{1}{1+r} + \frac{1}{(1+r)^2} + \ldots + \frac{1}{(1+r)^n} \right]$$

$$= C_d - C_d t \frac{1}{n} \left[\frac{1}{1+r} + \frac{1}{(1+r)^2} + \ldots + \frac{1}{(1+r)^n} \right]$$

$$= C_d - C_d t F_{SLP,r,n}$$

or, $\qquad P = C_d(1 - t F_{SLP,r,n}) \hfill (14.34a)$

After knowing an expression for present value one can also write the expressions for unacost R and capitalised cost K as follows:

$$R = P F_{PR,r,n} = C_d(1 - t F_{SLP,r,n}) F_{PR,r,n} \hfill (14.34b)$$

and, $\qquad K = P F_{PK,r,n} = C_d(1 - t F_{SLP,r,n}) F_{PK,r,n} \hfill (14.34c)$

Here, it is important to note that an expression for conversion factor from straight-line depreciation to the present value with tax is given by

$$F_{SLP,r,n} = \frac{1}{n} \frac{(1+r)^n - 1}{r(1+r)^n}$$

EXAMPLE 14.20

Calculate the conversion factor from straight-line depreciation to the present value with and without tax for 10 percent rate of return and 30 percent tax for a period of five years.

Solution

The conversion factor from straight-line depreciation to the present value is given by
a. without tax

$$F_{SLP,i,n} = \frac{1}{5} \frac{(1+0.10)^5 - 1}{0.10(1+0.10)^5} = 0.7581$$

b. with tax. Here, i should be replaced by $r = i(1-t) = 0.10(1-0.30) = 0.07$ in the above calculation

$$F_{SLP,r,n} = \frac{1}{5} \frac{(1+0.07)^5 - 1}{0.07(1+0.07)^5} = 0.8200.$$

EXAMPLE 14.21

Derive an expression for the present value P for a uniform end-of-year cost R occurring simultaneously with the tax instant t.

Solution

The uniform end-of-year cost after taxes at the end of each year $= R(1-t)$.
The $R(1-t)$ will be same up to n years at end of each year.
In order to obtain the expression for P, one has to convert unacost $R(1-P)$ in to present value as follows:

$$P = R(1-t)F_{RP,r,n}$$

or

$$P = R(1-t)\frac{(1+r)^n - 1}{r(1+r)^n}.$$

EXAMPLE 14.22

Derive an expression for the present value P for a given salvage C_{sal} value at end of nth year by treating as a non-depreciable first cost, an expense.

Solution

The C_{sal}, a non-depreciable expense is invested now and fully recovered at the end of n years with no tax consideration. The present value

$$P = C_{sal} - \frac{C_{sal}}{(1+r)^n} = C_{sal}\frac{(1+r)^n - 1}{(1+r)^n}$$

PROBLEMS

14.1 Calculate future (F_{PS}) and present (F_{SP}) value factor for a given number of years for 10 percent rate of interest and show that $F_{PS} \cdot F_{SP} = 1$ for each case.

Hint Use Equations (14.1b) and (14.5b) for $n = 0, 2, 4, 6, 8, 10$ and 12.

14.2 Calculate the effective rate of return for different value of p for 10 percent rate of interest.

Hint Use Equation (14.3) for $p = 1, 2, 3, 4, 6$.

14.3 Calculate capital recovery (F_{PR}) and sinking fund (F_{SR}) factors for different number of years ($n = 1, 5, 10, 15,$ and 20) for a given rate of interest ($r = 0.05, 0.10, 0.15,$ and 0.20 percentages).

Hint Use Equation (14.8b).

14.4 Prove that $F_{SR} \cdot F_{RS} = 1$.

Hint Use Equations (14.10b) and (14.11b) respectively.

14.5 A solar cooker purchased for Rs. 1200 is expected to generate annual revenues of Rs.150 and have salvage value of Rs. 400 at the end of 15 years. If a 18 percent per year required return is imposed on the purchase, compute the payback period.

Hint See Example 14.14 and solve the problem with cash flow diagram.

14.6 Find the equivalent present value (p) at the end of fifth year if money is worth 12 percent per year for the following series:

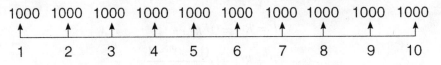

Hint See Example 14.8.

14.7 Two swimming pool have been heated by solar water heating systems which have the following cost comparison. Find out which system is more economical if the money is worth 12 percent per year.

Economic components	System I	System II
First cost (Rs.)	60,000	30,000
Uniform end-of-year maintenance per year (Rs.)	3,500	7,000
Overall, end of the fifth year (Rs.)	3,000	10,000
Salvage value (Rs.)	10,500	2,500
Life of the system(years)	10	10

Draw cash-flow diagram of both the systems.

Hint See Example 14.10.

14.8 Two solar distillation plant of capacity 100 liters per day have been constructed by using concrete/brick/cement, and fibre re-enforced plastic (FRP) materials. These distillation plants have the following cost comparison:

Cost of the components	System A	System B
First cost (Rs.)	1,00,000	2,50,000
Uniform end-of-year maintenance (Rs.)	15,000	20,000
Salvage value (Rs.)	20,000	–
Service life (years)	20 years	30 years

By using cash-flow diagram, find out which system is more economical if the money is worth 12 percent per year.

Hint See Example 14.11.

14.9 Solve the Problem 14.8 by using capitalised cost method.

Hint See Section 14.5.2.

14.10 Draw the curve between F_{PR} and n for different values of r of Problem 14.3.

APPENDIX I

Conversion of units

i. Length, m

1 yard = 3 ft = 36 inches = 0.9144 m

1 m = 39.3701 inch = 3.280839 ft = 1.093613 yd = 1650763.73 wavelength

1 ft = 0.3048 m; 1 inch = 2.54 cm = 25.4 mm

1 mil = 2.54×10^{-3} cm

ii. Area, m^2

1 ft^2 = 0.0929 m^2, 1 in^2 = 6.452 cm^2 = 0.00064516 m^2

1 cm^2 = 10^{-4} m^2 = 10.764×10^{-4} ft^2 = 0.1550 $inch^2$

1 ha = 10,000 m^2

iii. Volume, m^3

1 ft^3 = 0.02832 m^3 = 28.3168 l

1 in^3 = 16.39 cm^3 = 1.639×10^{-2} l

1 yd^3 = 0.764555 m^3 = 7.646×10^2 l

1 UK gallon = 4.54609 l

1 US gallon = 3.785 l = 0.1337 ft^3

1 m^3 = 1.000×10^6 cm^3 = 2.642×10^2 US gallons = 10^3 l

1 l = 10^{-3} m^3

1 fluid ounce = 28.41 cm^3

iv. Mass, kg

1 kg = 2.20462 lb = 0.068522 slug

1 ton (short) = 2000 lb (pounds) = 907.184 kg

1 ton (long) = 1016.05 kg

$1 \text{ lb} = 16 \text{ oz (ounces)} = 0.4536 \text{ kg}$

$1 \text{ oz} = 28.3495 \text{ g}$

$1 \text{ quintal} = 100 \text{ kg}$

v. Density and specific volumes, kg/m^3, m^3/kg

$1 \text{ lb/ft}^3 = 16.0185 \text{ kg/m}^3 = 5.787 \times 10^{-4} \text{ lb/in}^3$

$1 \text{ g/cm}^3 = 10^3 \text{ kg/m}^3 = 62.43 \text{ lb/ft}^3$

$1 \text{ lb/ft}^3 = 0.016 \text{ g/cm}^3 = 16 \text{ kg/m}^3$

$1 \text{ ft}^3 \text{ (air)} = 0.08009 \text{ lb} = 36.5 \text{ g at N.T.P.}$

$1 \text{ gallon/lb} = 0.010 \text{ cm}^3/\text{kg}$

vi. Pressure, Pa

$1 \text{ lb/ft}^2 = 4.88 \text{ kg/m}^2 = 47.88 \text{ Pa}$

$1 \text{ lb/in}^2 = 702.7 \text{ kg/m}^2 = 51.71 \text{ mm Hg} = 6.894757 \times 10^3 \text{ Pa (Pascal)} = 6.894757 \times 10^3 \text{ N/m}^2$

$1 \text{ atm} = 1.013 \times 10^5 \text{ N/m}^2 = 760 \text{ mm Hg} = 101.325 \text{ kPa}$

$1 \text{ in H}_2\text{O} = 2.491 \times 10^2 \text{ N/m}^2 = 248.8 \text{ Pa} = 0.036 \text{ lb/in}^2$

$1 \text{ bar} = 0.987 \text{ atm} = 1.000 \times 10^6 \text{ dynes/cm}^2 = 1.020 \text{ kg f/cm}^2 = 14.50 \text{ lb f/in}^2$
$\phantom{1 \text{ bar}} = 10^5 \text{ N (Newton/ m}^2) = 100 \text{ kPa}$

$1 \text{ torr (mm Hg } 0\,^\circ\text{C)} = 133 \text{ Pa}$

$1 \text{ Pascal (Pa)} = 1 \text{ N/m}^2 = 1.89476 \text{ kg}$

$1 \text{ inch of Hg} = 3.377 \text{ kPa} = 0.489 \text{ lb/in}^2$

vii. Velocity, m/s

$1 \text{ ft/s} = 0.3041 \text{ m/s}$

$1 \text{ mile/h} = 0.447 \text{ m/s} = 1.4667 \text{ ft/s} = 0.8690 \text{ knots}$

$1 \text{ km/h} = 0.2778 \text{ m/s}$

$1 \text{ ft/min} = 0.00508 \text{ m/s}$

viii. Force, N

$1 \text{ N (Newton)} = 10^5 \text{ dynes} = 0.22481 \text{ lb wt} = 0.224 \text{ lb f}$

$1 \text{ pdl (poundal)} = 0.138255 \text{ N (Newton)} = 13.83 \text{ dynes} = 14.10 \text{ g f}$

$1 \text{ lb f (i.e. wt of 1 lb mass)} = 4.448222 \text{ N} = 444.8222 \text{ dynes}$

$1 \text{ ton} = 9.964 \times 10^3 \text{ N}$

$1 \text{ bar} = 10^5 \text{ Pa (Pascal)}$

$1 \text{ ft of H}_2\text{O} = 2.950 \times 10^{-2} \text{ atm} = 9.807 \times 10^3 \text{ N/m}^2$

$1 \text{ in H}_2\text{O} = 249.089 \text{ Pa}$

$1 \text{ mm H}_2\text{O} = 9.80665 \text{ Pa}$

$1 \text{ dyne} = 1.020 \times 10^{-6} \text{ kg f} = 2.2481 \times 10^{-6} \text{ lb f} = 7.2330 \times 10^{-5} \text{ pdl} = 10^{-5} \text{ N}$

$1 \text{ mm of Hg} = 133.3 \text{ Pa}$

$1 \text{ atm} = 1 \text{ kg f/cm}^2 = 98.0665 \text{ k Pa}$

ix. Mass flow rate and discharge, $kg/s, m^3/s$

    $1 \text{ lb/s} = 0.4536 \text{ kg/s}$

    $1 \text{ ft}^3/\text{min} = 0.4720 \text{ l/s} = 4.179 \times 10^{-4} \text{ m}^3/\text{s}$

    $1 \text{ m}^3/\text{s} = 3.6 \times 10^6 \text{ l/h}$

    $1 \text{ g/cm}^3 = 10^3 \text{ kg/m}^3$

    $1 \text{ lb/h ft}^2 = 0.001356 \text{ kg/s m}^2$

    $1 \text{ lb/ft}^3 = 16.2 \text{ kg/m}^2$

    $1 \text{ litre/s (l/s)} = 10^{-3} \text{ m}^3/\text{s}$

x. Energy, J

    $1 \text{ cal} = 4.187 \text{ J (Joules)}$

    $1 \text{ kcal} = 3.97 \text{ Btu} = 12 \times 10^{-4} \text{ kWh} = 4.187 \times 10^3 \text{ J}$

    $1 \text{ Watt} = 1.0 \text{ J/s}$

    $1 \text{ Btu} = 0.252 \text{ kcal} = 2.93 \times 10^{-4} \text{ kWh} = 1.022 \times 10^3 \text{ J}$

    $1 \text{ hp} = 632.34 \text{ kcal} = 0.736 \text{ kW}$

    $1 \text{ kWh} = 3.6 \times 10^6 \text{ J}$

    $1 \text{ J} = 2.390 \times 10^{-4} \text{ kcal} = 2.778 \times 10^{-4} \text{ Wh}$

    $1 \text{ kWh} = 860 \text{ kcal} = 3413 \text{ Btu}$

    $1 \text{ erg} = 1.0 \times 10^{-7} \text{ J} = 1.0 \times 10^{-7} \text{ Nm} = 1.0 \text{ dyne cm}$

    $1 \text{ J} = 1 \text{ Ws} = 1 \text{ Nm}$

    $1 \text{ eV} = 1.602 \times 10^{-19} \text{ J}$

    $1 \text{ kW} = 1000 \text{ W} = 1 \text{ unit}$

    $1 \text{ GJ} = 10^9 \text{ J}$

xi. Power, Watt (J/s)

    $1 \text{ Btu/h} = 0.293071 \text{ W} = 0.252 \text{ kcal/h}$

    $1 \text{ Btu/h} = 1.163 \text{ W} = 3.97 \text{ Btu/h}$

    $1 \text{ W} = 1.0 \text{ J/s} = 1.341 \times 10^{-3} \text{ hp} = 0.0569 \text{ Btu/min} = 0.01433 \text{ kcal/min}$

    $1 \text{ hp (F.P.S.)} = 550 \text{ ft lb f/s} = 746 \text{ W} = 596 \text{ kcal/h} = 1.015 \text{ hp (M.K.S.)}$

    $1 \text{ hp (M.K.S.)} = 75 \text{ mm kg f/s} = 0.17569 \text{ kcal/s} = 735.3 \text{ W}$

    $1 \text{ W/ft}^2 = 10.76 \text{ W/m}^2$

    $1 \text{ ton} = 3.5 \text{ kW}$

xii. Specific heat, J/kg °C

    $1 \text{ Btu/lb °F} = 1.0 \text{ kcal/kg °C} = 4.187 \times 10^3 \text{ J/kg °C}$

    $1 \text{ Btu/lb} = 2.326 \text{ kJ/kg}$

xiii. Temperature, °C and K used in SI

    $T \text{ °C} = (5/9) (T \text{ °F} + 40) - 40$

    $T \text{ °F} = (9/5) (T \text{ °C} + 40) - 40$

$$T \, ^\circ R = 460 + T \, ^\circ F$$
$$T \, ^\circ K = (5/9) \, T \, ^\circ R$$
$$T \, ^\circ C = T \, ^\circ F / 1.8 = (5/9) \, T \, ^\circ F$$

xiv. Rate of heat flow per unit area or heat flux, W/m²

1 Btu/ft² h = 2.713 kcal/m² h = 3.1552 W/m²

1 kcal/m² h = 0.3690 Btu/ft² h = 1.163 W/m² = 27.78 × 10⁻⁶ cal/s cm²

1 cal/cm² min = 221.4 Btu/ft² h

1 W/ft² = 10.76 W/m²

1 W/m² = 0.86 kcal/h m² = 0.23901 × 10⁻⁴ cal/s cm² = 0.317 Btu/h ft²

1 Btu/h ft = 0.96128 W/m

xv. Heat transfer coefficient, W/m² °C

1 Btu/ft²h °F = 4.882 kcal/m²h °C = 1.3571 × 10⁻⁴ cal/cm²s °C

1 Btu/ft²h °F = 5.678 W/m² °C

1 kcal/m²h °C = 0.2048 Btu/ft²h °F = 1.163 W/m² °C

1 W/m²K = 2.3901 × 10⁻⁵ cal/cm²sK = 1.7611 × 10⁻¹ Btu/ft² °F = 0.86 kcal/m²h °C

xvi. Thermal conductivity, W/m °C

1 Btu/ft h °F = 1.488 kcal/m h °C = 1.73073 W/m °C

1 kcal/m h °C = 0.6720 Btu/ft h °F = 1.1631 W/m °C

1 Btu in/ft²h °F = 0.124 kcal/mh °C = 0.144228 W/m °C

1 Btu/in h °F = 17.88 kcal/mh °C

1 cal/cm s °F = 4.187 × 10² W/m °C = 242 Btu/h ft °F

1 W/cm °C = 57.79 Btu/h ft °F

xvii. Angle, rad

2π rad = 360 degrees

1 degree = 0.0174533 rad = 60′ (angle)

1 minute = 0.290888 × 10⁻³ rad = 60 seconds

1 second = 4.84814 × 10⁻⁶ rad

1 degree = 4 min (360 degree = 24 × 60 min, time)

xviii. Illumination

1 lx (lux) = 1.0 lm (lumen)/m²

1 lm/ft² = 1.0 foot candle

1 foot candle = 10.7639 lx

xix. Time, h

1 week = 7 days = 168 h = 10080 minutes = 604800 s

1 mean solar day = 1440 minute = 86400 s

1 calender year = 365 days = 8760 h = 5.256×10^5 minutes
1 tropical mean solar year = 365.2422 days
1 sidereal year = 365.2564 days (mean solar)
1 s (second) = 9.192631770×10^9 Hertz (Hz)

xx. Concentration, kg/m^3 and g/m^3
1 g/l = 1 kg/m^3
1 lb/ft^3 = 6.236 kg/m^3

xxi. Diffusivity, m^2/s
1 ft^2/h = 25.81×10^{-6} m^2/s

APPENDIX //

The geographical location of some radiation stations in India is shown in Figure I.

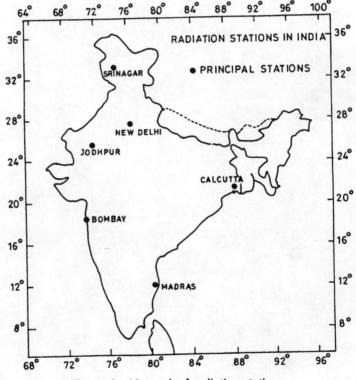

Figure I Network of radiation stations.

The hourly variation of ambient air temperature and solar intensity for different months for variou stations in India have been given below in tables (1 kW = 1000 W).

TABLE I Monthly solar radiation (kW/m²) on horizontal surface and ambient air temperature for different places in India

(a) CALCUTTA: (i) monthly solar radiation (kW/m²)

Hour/months		6	7	8	9	10	11	12	13	14	15	16	17	18
Jan./Nov.	total	0.100	0.124	0.310	0.490	0.629	0.716	0.746	0.716	0.629	0.490	0.310	0.124	0.100
	diffuse	0.100	0.117	0.141	0.156	0.165	0.168	0.170	0.168	0.165	0.156	0.148	0.111	0.100
Feb./Oct.	total	0.100	0.165	0.374	0.563	0.707	0.798	0.829	0.798	0.707	0.563	0.374	0.165	0.100
	diffuse	0.100	0.122	0.147	0.161	0.168	0.172	0.174	0.172	0.168	0.161	0.147	0.122	0.100
March/Sep.	total	0.100	0.237	0.458	0.653	0.801	0.896	0.928	0.896	0.801	0.653	0.458	0.237	0.100
	diffuse	0.100	0.133	0.154	0.166	0.173	0.177	0.178	0.177	0.173	0.166	0.154	0.133	0.100
April/Aug.	total	0.115	0.314	0.534	0.727	0.875	0.967	0.997	0.967	0.875	0.727	0.536	0.314	0.115
	diffuse	0.108	0.142	0.159	0.169	0.178	0.179	0.180	0.179	0.176	0.169	0.159	0.142	0.108
May/July	total	0.150	0.366	0.580	0.762	0.904	0.991	1.022	0.991	0.904	0.762	0.580	0.366	0.150
	diffuse	0.118	0.147	0.162	0.171	0.177	0.180	0.181	0.180	0.177	0.171	0.162	0.147	0.118
June	total	0.173	0.385	0.594	0.771	0.910	0.994	1.025	0.994	0.910	0.771	0.594	0.385	0.173
	diffuse	0.123	0.148	0.163	0.171	0.177	0.180	0.181	0.180	0.177	0.171	0.163	0.148	0.123
Dec.	total	0.100	0.117	0.289	0.465	0.603	0.688	0.717	0.688	0.603	0.465	0.289	0.117	0.100
	diffuse	0.100	0.109	0.139	0.154	0.163	0.167	0.169	0.167	0.163	0.154	0.139	0.109	0.100

(a) CALCUTTA: (ii) Mean hourly ambient temperature (°C)

Hour/months	1	2	3	4	5	6	7	8	9	10	11	12	13	14	15	16	17	18	19	20	21	22	23	24
Jan.	15.3	14.7	14.1	13.7	13.6	13.9	14.5	15.7	17.4	19.4	21.7	23.8	25.3	26.4	26.8	26.4	25.5	24.0	22.3	20.6	19.1	17.8	16.8	16.0
Feb.	18.2	17.5	17.0	16.6	16.5	16.8	17.4	18.6	20.3	22.2	24.4	26.5	28.1	29.1	29.5	29.1	28.2	26.8	25.1	23.4	22.0	20.7	19.6	18.8
March	23.2	22.5	22.0	21.6	21.5	21.8	22.4	23.5	25.2	27.1	29.3	31.4	32.9	33.9	34.3	33.9	33.0	31.6	29.9	28.3	26.9	25.6	24.6	23.8
April	26.5	25.9	25.5	25.1	25.0	25.2	25.8	26.8	28.3	30.0	31.9	33.7	35.1	36.0	36.3	36.0	35.2	33.9	32.5	31.0	29.7	28.6	27.7	27.0
May	27.7	27.2	26.9	26.6	26.5	26.7	27.2	28.0	29.2	30.6	32.2	33.7	34.8	35.5	35.8	35.5	34.9	33.8	32.6	31.4	30.4	29.5	28.7	28.2
June	27.7	27.3	27.0	26.8	26.7	26.8	27.2	27.9	28.8	30.0	31.2	32.4	33.3	33.9	34.1	33.9	33.4	32.5	31.6	30.6	29.8	29.1	28.5	28.0
July	27.0	26.8	26.5	26.4	26.3	26.4	26.7	27.2	28.0	28.8	29.8	30.7	31.4	31.8	32.0	31.8	31.4	30.8	30.1	29.3	28.7	28.1	27.7	27.3
August	27.0	26.8	26.5	26.4	26.3	26.4	26.7	27.2	28.0	28.8	29.8	30.7	31.4	31.8	32.0	31.8	31.4	30.8	30.1	29.3	28.7	28.1	27.6	27.2
Sept.	26.9	26.6	26.3	26.2	26.1	26.2	26.5	27.1	27.9	28.8	29.9	30.9	31.6	32.1	32.3	31.8	31.0	30.2	29.4	28.1	28.7	28.1	25.8	25.3
Oct.	24.9	24.5	24.2	24.0	23.9	24.1	24.5	25.2	26.2	27.4	28.7	30.0	30.9	31.6	31.8	31.6	31.0	30.1	29.1	28.1	27.2	26.4	25.8	25.3
Nov.	19.8	19.3	18.8	18.5	18.4	18.6	19.2	20.2	21.6	23.3	25.2	26.9	28.3	29.2	29.5	29.2	28.4	27.2	25.7	24.3	23.1	22.0	21.1	20.4
Dec.	15.9	15.2	14.7	14.3	14.2	14.5	15.1	16.2	17.9	19.8	22.0	24.1	25.6	26.6	27.0	26.6	25.7	24.3	22.6	21.0	19.6	18.3	17.3	16.5

(b) JODHPUR: (i) Monthly solar radiation (kW/m²)

Hour/months		6	7	8	9	10	11	12	13	14	15	16	17	18
Jan./	total	0.100	0.114	0.278	0.450	0.586	0.670	0.698	0.670	0.586	0.450	0.218	0.114	0.100
Nov.	diffuse	0.100	0.107	0.138	0.153	0.163	0.166	0.168	0.166	0.163	0.153	0.138	0.107	0.100
Feb./	total	0.100	0.148	0.347	0.530	0.670	0.757	0.788	0.757	0.670	0.530	0.347	0.148	0.100
Oct.	diffuse	0.100	0.118	0.145	0.159	0.166	0.170	0.172	0.170	0.166	0.159	0.145	0.118	0.100
Mar./	total	0.100	0.227	0.442	0.630	0.774	0.866	0.892	0.866	0.774	0.630	0.442	0.227	0.100
Sep.	diffuse	0.100	0.132	0.152	0.165	0.171	0.175	0.177	0.175	0.171	0.165	0.152	0.132	0.100
April/	total	0.119	0.316	0.533	0.717	0.861	0.951	0.980	0.951	0.861	0.717	0.533	0.316	0.119
Aug.	diffuse	0.109	0.142	0.159	0.169	0.175	0.179	0.180	0.179	0.175	0.169	0.159	0.142	0.109
May/	total	0.166	0.378	0.586	0.763	0.901	0.985	1.016	0.985	0.901	0.763	0.586	0.378	0.166
July	diffuse	0.122	0.147	0.163	0.171	0.177	0.180	0.181	0.180	0.177	0.171	0.163	0.147	0.122
	total	0.196	0.402	0.605	0.777	0.911	0.993	1.023	0.993	0.911	0.777	0.605	0.402	0.192
June	diffuse	0.128	0.149	0.164	0.172	0.178	0.180	0.181	0.180	0.178	0.172	0.164	0.149	0.128
	total	0.100	0.108	0.253	0.423	0.557	0.640	0.668	0.640	0.557	0.423	0.253	0.108	0.100
Dec.	diffuse	0.100	0.105	0.135	0.151	0.161	0.165	0.166	0.165	0.161	0.151	0.135	0.105	0.100

(b) JODHPUR: (ii) Monthly hourly ambient temperature (°C)

Hour/months	1	2	3	4	5	6	7	8	9	10	11	12	13	14	15	16	17	18	19	20	21	22	23	24
Jan.	11.5	10.7	10.1	9.7	9.5	9.8	10.6	11.9	13.9	16.1	18.7	21.1	22.9	24.1	24.6	24.1	23.1	21.4	19.5	17.5	15.8	14.3	13.1	12.2
Feb.	14.1	13.3	12.6	12.2	12.0	12.3	13.1	14.5	16.6	19.0	21.7	24.2	26.2	27.4	27.9	27.4	26.3	24.6	22.5	20.4	18.7	17.1	15.8	14.9
March	19.2	18.4	17.7	17.3	17.1	17.4	18.2	19.7	21.8	24.2	27.0	29.6	31.5	32.8	33.3	32.8	31.7	29.9	27.8	25.7	23.9	22.3	21.0	20.0
April	24.5	23.7	23.0	22.6	22.4	22.7	23.5	24.9	27.0	29.4	31.1	34.6	36.6	37.8	38.3	37.8	36.7	35.0	32.9	30.8	29.1	27.5	26.2	25.3
May	29.2	28.4	27.9	27.4	27.3	27.6	28.3	29.6	31.4	33.6	36.0	38.3	40.0	41.2	41.6	41.2	40.2	38.6	36.7	34.9	33.3	31.9	30.7	29.9
June	30.0	29.4	29.0	28.6	28.5	28.7	29.3	30.4	31.9	33.6	35.6	38.3	40.0	39.8	40.1	39.8	38.7	37.7	36.2	34.6	33.4	32.2	31.3	30.6
July	27.7	27.4	27.1	26.9	26.8	26.9	27.3	27.9	28.8	29.8	31.0	32.1	32.9	33.5	33.7	33.5	33.0	32.3	31.4	30.5	29.7	29.0	28.5	28.0
August	26.2	25.8	25.5	25.3	25.2	25.4	25.8	26.5	27.5	28.7	30.1	31.4	32.3	33.0	33.2	33.0	32.4	31.5	30.5	29.4	28.6	27.8	27.1	26.6
Sept.	25.5	24.9	24.5	24.2	24.1	24.3	24.8	25.8	27.2	28.8	30.6	32.3	33.5	34.4	34.7	34.1	33.6	32.5	31.1	29.7	28.6	27.5	26.6	26.0
Oct.	21.7	20.9	20.2	19.8	19.6	19.9	20.7	22.2	24.3	26.7	28.8	32.0	33.9	35.2	35.7	35.2	34.1	32.3	30.2	28.1	26.4	24.8	23.5	22.5
Nov.	16.2	15.3	14.6	14.1	13.9	14.2	15.1	16.7	19.0	21.6	24.6	27.4	29.5	30.9	31.4	30.9	29.6	27.7	25.4	23.2	21.2	19.5	18.1	17.0
Dec.	12.8	12.0	11.3	10.9	10.7	11.0	11.8	13.3	15.3	17.7	20.5	23.0	24.9	26.2	26.7	26.2	25.1	23.3	21.3	19.2	17.4	15.8	14.5	13.6

(c) MADRAS: (i) Monthly solar radiation (kW/m²)

Hour/ months		6	7	8	9	10	11	12	13	14	15	16	17	18
Jan./	total	0.100	0.178	0.390	0.582	0.728	0.820	0.852	0.820	0.728	0.582	0.390	0.178	0.100
Nov.	diffuse	0.100	0.124	0.148	0.162	0.169	0.173	0.175	0.173	0.169	0.162	0.148	0.124	0.100
Feb./	total	0.100	0.212	0.436	0.636	0.788	0.885	0.918	0.885	0.788	0.635	0.436	0.212	0.100
Oct.	diffuse	0.100	0.130	0.152	0.165	0.172	0.177	0.178	0.177	0.172	0.165	0.152	0.130	0.100
March/	total	0.100	0.257	0.492	0.696	0.854	0.953	0.983	0.953	0.854	0.696	0.492	0.257	0.100
Sep.	diffuse	0.100	0.136	0.156	0.168	0.175	0.179	0.180	0.179	0.175	0.168	0.156	0.136	0.100
April/	total	0.107	0.301	0.536	0.736	0.893	0.988	1.023	0.988	0.893	0.736	0.535	0.301	0.107
Aug.	diffuse	0.105	0.141	0.159	0.169	0.177	0.180	0.181	0.180	0.177	0.169	0.159	0.141	0.105
May/	total	0.199	0.326	0.552	0.745	0.893	0.986	1.020	0.986	0.895	0.745	0.552	0.326	0.119
July	diffuse	0.109	0.143	0.160	0.170	0.177	0.180	0.181	0.180	0.177	0.170	0.160	0.143	0.109
June	total	0.125	0.335	0.554	0.742	0.888	0.978	1.010	0.979	0.888	0.742	0.554	0.335	0.125
	diffuse	0.111	0.144	0.160	0.170	0.177	0.180	0.181	0.180	0.177	0.170	0.160	0.144	0.111
Dec.	total	0.100	0.166	0.374	0.563	0.706	0.797	0.828	0.797	0.706	0.563	0.374	0.166	0.100
	diffuse	0.100	0.122	0.147	0.161	0.168	0.172	0.174	0.172	0.168	0.161	0.147	0.122	0.100

(c) MADRAS: (ii) Mean hourly ambient temperature (°C)

Hour/ months	1	2	3	4	5	6	7	8	9	10	11	12	13	14	15	16	17	18	19	20	21	22	23	24
Jan.	21.4	21.0	20.6	20.4	20.3	20.5	20.9	21.7	22.8	24.0	25.5	26.8	27.9	28.5	28.8	28.5	27.9	27.0	25.9	24.8	23.9	23.0	22.3	21.8
Feb.	22.3	21.9	21.5	21.2	21.1	21.3	21.8	22.6	23.9	25.3	26.9	28.4	29.6	30.3	30.6	30.3	29.6	28.6	27.4	26.1	25.1	24.1	23.4	22.8
March	24.3	23.9	23.5	23.2	23.1	23.3	23.8	24.6	25.9	27.3	29.0	30.5	31.6	32.4	32.7	32.4	31.7	30.7	29.4	28.2	27.1	26.2	25.4	24.8
April	27.2	26.7	26.4	26.1	26.0	26.2	26.6	27.4	28.6	29.9	31.4	32.9	33.9	34.6	34.9	34.6	34.0	33.0	31.9	30.7	29.7	28.8	28.1	27.6
May	29.1	28.6	28.2	27.9	27.8	28.2	28.5	29.4	30.6	32.1	33.8	35.3	36.5	37.3	37.6	37.3	36.6	35.5	34.3	33.0	31.9	30.9	30.2	29.6
June	28.9	28.4	28.0	27.7	27.6	27.8	28.3	29.2	30.4	31.9	33.5	35.1	36.2	37.0	37.3	37.0	36.3	35.3	34.0	32.7	31.7	30.7	29.9	29.3
July	27.5	27.0	26.7	26.4	26.3	26.5	26.9	27.7	28.9	30.2	31.7	33.2	34.2	34.9	35.2	34.9	34.3	33.3	32.2	31.0	30.0	29.1	28.4	27.9
August	26.9	26.5	26.1	25.9	25.8	26.0	26.4	27.2	28.3	29.6	31.1	32.5	33.5	34.2	34.5	34.2	33.6	32.7	31.5	30.4	29.5	28.6	27.9	27.4
Sept.	26.5	26.1	25.7	25.5	25.4	25.6	26.0	26.8	27.9	29.1	30.6	31.9	33.0	33.6	33.9	33.6	33.1	32.1	31.0	29.9	29.0	28.1	27.4	26.9
Oct.	25.4	25.0	24.7	24.5	24.4	24.5	24.9	25.6	26.5	27.7	28.9	30.1	31.0	31.6	31.8	31.6	31.1	30.2	29.3	28.3	27.5	26.8	25.2	25.7
Nov.	23.4	23.0	22.8	22.6	22.5	22.6	23.0	23.6	24.4	25.4	26.6	27.7	28.5	29.0	29.2	29.0	28.5	27.8	26.9	26.1	25.3	24.6	24.1	23.7
Dec.	21.9	21.6	21.3	21.1	21.0	21.1	21.5	22.2	23.1	24.2	25.4	26.5	27.4	28.0	28.2	28.0	27.5	26.7	25.8	24.8	24.0	23.3	22.7	22.3

(d) NEW DELHI: (i) Monthly solar radiation (kW/m²)

Hour/ months		6	7	8	9	10	11	12	13	14	15	16	17	18
Jan./	total	0.100	0.108	0.256	0.424	0.558	0.641	0.669	0.641	0.558	0.424	0.256	0.108	0.100
Nov.	diffuse	0.100	0.105	0.136	0.151	0.161	0.165	0.166	0.165	0.161	0.151	0.136	0.105	0.100
Feb./	total	0.100	0.138	0.329	0.508	0.646	0.732	0.761	0.731	0.646	0.508	0.329	0.138	0.100
Oct.	diffuse	0.100	0.115	0.143	0.157	0.165	0.169	0.171	0.169	0.165	0.157	0.143	0.115	0.100
March/	total	0.100	0.222	0.431	0.616	0.756	0.846	0.877	0.846	0.756	0.616	0.431	0.222	0.100
Sep.	diffuse	0.100	0.131	0.151	0.164	0.170	0.174	0.176	0.174	0.170	0.164	0.151	0.131	0.121
April/	total	0.121	0.317	0.529	0.710	0.851	0.939	0.693	0.939	0.851	0.710	0.529	0.317	0.110
Aug.	diffuse	0.110	0.142	0.159	0.168	0.175	0.178	0.179	0.178	0.175	0.168	0.159	0.142	0.177
May/	total	0.177	0.382	0.589	0.762	0.898	0.980	1.011	0.980	0.898	0.762	0.589	0.384	0.124
July	diffuse	0.124	0.148	0.163	0.171	0.177	0.180	0.181	0.180	0.177	0.171	0.163	0.118	0.207
June	total	0.207	0.411	0.610	0.779	0.911	0.990	1.020	0.990	0.911	0.779	0.610	0.411	0.130
	diffuse	0.130	0.150	0.164	0.172	0.177	0.180	0.181	0.180	0.177	0.172	0.164	0.158	0.100
Dec.	total	0.100	0.103	0.232	0.397	0.528	0.610	0.637	0.610	0.528	0.397	0.232	0.103	0.100
	diffuse	0.100	0.103	0.133	0.149	0.159	0.164	0.165	0.164	0.159	0.149	0.133	0.103	0.100

(d) NEW DELHI: (ii) Mean hourly ambient temperature (°C)

Hour/ months	1	2	3	4	5	6	7	8	9	10	11	12	13	14	15	16	17	18	19	20	21	22	23	24
Jan.	9.1	8.4	7.9	7.4	7.3	7.6	8.3	9.5	11.4	13.5	15.8	18.1	19.8	20.9	21.3	20.9	19.9	18.4	16.5	14.7	13.2	11.8	10.7	9.8
Feb.	11.9	11.2	10.6	10.2	10.1	10.4	11.0	12.3	14.0	16.0	18.3	20.5	22.1	23.2	23.6	23.2	22.2	20.8	19.0	17.3	15.8	14.4	13.3	12.5
March	17.1	16.3	15.7	15.3	15.1	15.4	16.2	17.5	19.5	21.7	24.3	26.7	28.5	29.7	30.2	29.7	28.7	27.0	25.1	23.1	21.4	19.9	18.7	17.8
April	23.0	22.2	21.6	21.2	21.0	21.3	22.1	23.4	25.4	27.7	30.3	32.7	34.5	35.7	36.2	35.7	34.7	33.0	31.0	29.1	27.4	25.9	24.6	23.7
May	28.4	27.7	27.2	26.7	26.6	26.9	27.6	28.8	30.6	32.7	35.1	37.3	39.0	40.1	40.5	40.1	39.1	37.6	35.8	34.0	32.4	31.0	29.9	29.1
June	30.2	29.6	29.1	28.8	28.7	28.9	29.5	30.5	31.9	33.6	35.5	37.3	38.7	39.6	39.9	39.6	38.8	37.5	36.1	34.6	33.4	32.3	31.4	30.7
July	28.3	27.8	27.5	27.3	27.2	27.4	27.8	28.5	29.5	30.8	32.1	33.4	34.4	35.1	35.3	35.1	34.5	33.6	32.5	31.5	30.6	29.8	29.1	28.7
August	27.1	26.7	26.4	26.2	26.1	26.3	26.6	27.3	28.3	29.4	30.7	32.0	32.9	33.5	33.7	33.5	32.9	32.1	31.1	30.1	29.3	28.5	27.9	27.5
Sept.	25.8	25.4	25.0	24.7	24.6	24.8	25.3	26.1	27.4	28.8	30.4	31.9	33.1	33.8	34.1	33.8	33.1	32.1	30.9	29.6	28.6	27.6	26.9	26.3
Oct.	20.6	19.9	19.3	18.8	18.7	19.0	19.7	21.0	22.9	25.0	27.5	29.8	31.5	32.7	33.1	32.7	31.7	30.1	28.2	26.3	24.7	23.3	22.2	21.3
Nov.	14.0	13.2	12.5	12.0	11.8	12.1	13.0	14.5	16.7	19.2	22.1	24.8	26.8	28.2	28.7	28.2	27.0	25.2	23.0	20.8	18.9	17.2	15.9	14.8
Dec.	11.3	10.0	9.0	8.3	8.0	8.5	9.8	12.1	15.4	19.2	23.5	27.6	30.6	32.6	33.4	32.6	30.9	28.1	24.8	21.5	18.7	16.1	14.1	12.4

(e) SRINAGAR: (i) Monthly solar radiation (kW/m²)

Hour/ months		6	7	8	9	10	11	12	13	14	15	16	17	18
Jan./ Nov.	total	0.100	0.100	0.207	0.363	0.486	0.566	0.593	0.566	0.486	0.363	0.207	0.100	0.100
	diffuse	0.100	0.100	0.130	0.146	0.156	0.161	0.163	0.161	0.156	0.146	0.130	0.100	0.100
Feb./ Oct.	total	0.100	0.120	0.287	0.453	0.584	0.665	0.692	0.665	0.584	0.453	0.287	0.120	0.100
	diffuse	0.100	0.110	0.139	0.153	0.162	0.166	0.167	0.166	0.162	0.153	0.139	0.110	0.100
March/ Sep.	total	0.100	0.207	0.401	0.576	0.708	0.793	0.821	0.793	0.708	0.576	0.401	0.207	0.100
	diffuse	0.100	0.130	0.149	0.162	0.168	0.172	0.173	0.172	0.168	0.162	0.149	0.130	0.100
April/ Aug.	total	0.127	0.318	0.518	0.689	0.821	0.904	0.933	0.904	0.821	0.689	0.518	0.318	0.127
	diffuse	0.112	0.142	0.158	0.167	0.173	0.177	0.178	0.177	0.173	0.167	0.158	0.142	0.112
May/ July	total	0.202	0.398	0.592	0.755	0.883	0.963	0.987	0.963	0.883	0.755	0.592	0.398	0.202
	diffuse	0.129	0.149	0.163	0.170	0.177	0.179	0.180	0.179	0.177	0.170	0.163	0.149	0.129
June	total	0.236	0.432	0.619	0.779	0.903	0.979	1.006	0.979	0.903	0.779	0.619	0.432	0.236
	diffuse	0.133	0.152	0.164	0.172	0.177	0.188	0.181	0.180	0.177	0.172	0.164	0.152	0.133
Dec.	total	0.100	0.100	0.182	0.332	0.454	0.532	0.558	0.532	0.454	0.332	0.182	0.100	0.100
	diffuse	0.100	0.100	0.125	0.144	0.153	0.159	0.161	0.159	0.153	0.144	0.125	0.100	0.100

(e) SRINAGAR: (ii) Mean hourly ambient temperature (°C)

Hour/ months	1	2	3	4	5	6	7	8	9	10	11	12	13	14	15	16	17	18	19	20	21	22	23	24
Jan.	-1.4	-1.8	-2.0	-2.2	-2.3	-2.2	-1.8	-1.2	-0.4	0.6	1.8	2.9	3.7	4.2	4.4	4.2	3.7	3.0	2.1	1.3	0.5	-0.2	-0.7	-1.1
Feb.	0.3	-0.1	-0.5	-0.7	-0.8	-0.6	-0.2	0.6	1.7	3.0	4.5	5.9	6.9	7.6	7.9	7.6	7.0	6.1	4.9	3.8	2.9	2.0	1.3	0.8
March	4.8	4.3	3.9	3.6	3.5	3.7	4.2	5.1	6.4	7.9	9.5	11.1	12.3	13.1	13.4	13.1	12.4	11.3	10.0	8.7	7.7	6.7	5.9	5.3
April	8.9	8.4	7.9	7.5	7.4	7.6	8.2	9.3	10.9	12.6	14.7	16.6	18.0	18.9	19.3	18.9	18.1	16.8	15.3	13.7	12.4	11.2	10.3	9.5
May	12.9	12.3	11.7	11.3	11.2	11.5	12.1	13.3	15.1	17.1	19.4	21.5	23.1	24.2	24.6	24.2	23.3	21.8	20.0	18.3	16.8	15.5	14.4	13.6
June	16.3	15.6	15.0	14.5	14.4	14.7	15.4	16.7	18.6	20.8	23.3	25.6	27.4	28.6	29.0	28.6	27.5	25.9	24.0	22.1	20.5	19.1	17.9	17.0
July	20.0	19.4	18.9	18.5	18.4	18.6	19.3	20.4	22.0	23.9	26.0	27.9	29.4	30.4	30.8	30.4	29.6	28.2	26.6	25.0	23.6	22.4	21.4	20.6
August	19.5	18.9	18.4	18.0	17.9	18.1	18.7	19.8	21.4	23.2	25.2	27.1	28.6	29.5	29.9	29.5	28.7	27.4	25.8	24.3	22.9	21.7	20.8	20.1
Sept.	14.7	13.9	13.3	12.9	12.7	13.0	13.8	15.2	17.2	19.6	22.2	24.7	26.6	27.8	28.3	27.8	26.7	25.0	23.0	21.0	19.3	17.7	16.4	15.5
Oct.	7.9	7.1	6.4	5.9	5.7	6.0	6.9	8.4	10.6	13.1	16.0	18.7	20.7	22.1	22.6	22.1	20.9	19.1	16.9	14.7	12.8	11.1	9.8	8.7
Nov.	1.9	-1.0	-1.4	-1.7	-1.8	-1.6	-1.1	-0.1	1.3	2.9	4.7	6.4	7.6	8.5	8.8	8.5	7.7	6.6	5.2	3.8	2.7	1.6	0.7	0.1
Dec.	-0.4	-1.0	-1.4	-1.7	-1.8	-1.6	-1.1	-0.1	1.3	2.9	4.7	6.4	7.6	8.5	8.8	8.5	7.7	6.6	5.2	3.8	2.7	1.6	0.7	0.1

Solar radiation on surface of different orientations (kW/m²) in the months of January and June in New Delhi

Hour		6	7	8	9	10	11	12	13	14	15	16	17	18
Surface														
January														
Horizontal	total	0.100	0.108	0.256	0.424	0.558	0.641	0.669	0.641	0.558	0.424	0.256	0.108	0.100
	diffuse	0.100	0.105	0.136	0.151	0.161	0.165	0.166	0.165	0.161	0.151	0.136	0.105	0.100
South	total	0.060	0.096	0.363	0.539	0.652	0.717	0.738	0.717	0.652	0.539	0.363	0.096	0.060
	diffuse	0.050	0.053	0.068	0.075	0.080	0.083	0.083	0.083	0.080	0.075	0.068	0.053	0.005
June														
Horizontal	total	0.207	0.411	0.610	0.779	0.911	0.990	1.020	0.990	0.911	0.779	0.610	0.411	0.207
	diffuse	0.130	0.150	0.164	0.172	0.177	0.180	0.181	0.180	0.177	0.172	0.164	0.150	0.130
South	total	0.065	0.075	0.082	0.086	0.211	0.258	0.274	0.258	0.211	0.086	0.082	0.075	0.065
	diffuse	0.065	0.075	0.082	0.086	0.089	0.090	0.090	0.090	0.089	0.086	0.082	0.075	0.065

APPENDIX

Properties of Materials

TABLE III(a) Properties of metals

	Properties at 20 °C			
Metal	ρ kg/m³	C_p kJ/kg °C	K W/m °C	α m²/s × 10⁻⁵
Aluminum:				
Pure	2707	0.896	204	8.418
Al-Si (Silumin, copper bearing)				
86% Al, 1% Cu	2659	0.867	137	5.933
Lead	11400	0.1298	34.87	7.311
Iron:				
Pure	7897	0.452	73	2.034
Steel				
C ≈ 1.5% (C-Carbon steel)	7753	0.486	63	0.970
Copper:				
Pure	8954	0.3831	386	11.234
Aluminum bronze				
95% Cu, 5% Al	8666	0.410	383	2.330
Bronze				
75% Cu, 25% Sn	8666	0.343	326	0.859
Red brass				
85% Cu, 9% Sn, 6% Zn	8714	0.385	61	1.804
Brass				
70% Cu, 30% Zn	8600	0.877	85	3.412
German silver				
62% Cu, 15% Ni, 22% Zn	8618	0.394	24.9	0.733
Constantan				
60% Cu, 40% Ni	8922	0.410	22.7	0.612
Magnesium, pure	1746	1.013	171	9.708
Nickel, pure	8906	0.4459	90	2.266
Silver:				
Purest	10524	0.2340	419	17.004
Pure (99.9%)	10524	0.2340	407	16.563
Tin, pure	7304	0.2265	64	3.884
Tungsten	19350	0.1344	163	6.271
Zinc, pure	7144	0.3843	112.2	4.106

TABLE III(b)   Properties of non-metals

Non-metals	Temperature °C	K W/m K	ρ kg/m^3	C kJ/kg K	α m^2/sec. $\times 10^{-7}$
Asbestos	50	0.08	470	–	–
Building brick	20	0.69	1600	0.84	5.2
Common face		1.32	2000	–	–
Concrete, cinder	23	0.76	–	–	–
Stone 1-2-4 mix	20	1.37	1900–2300	0.88	8.2–6.8
Cork board	30	0.043	160	1.88	2–5.3
Cork, regranulated	32	0.045	45–120	1.88	2–5.3
Cotton wool	–	0.018	1500	–	–
Glass, window	20	0.78 (avg)	2700	0.84	3.4
Borosilicate	30–75	1.09	2200		
Ground	32	0.52	2050	1.88	1.35
Glass wool	23	0.38	24	0.7	22.6
Paraffin wax	–	0.25	900	2900	–
Perspex	–	0.20	1190	1500	–
Plaster, gypsum	20	0.48	1440	0.84	4.0
Rubber	–	0.15	910	–	–
Stone:					
Granite		1.73–3.98	2640	0.82	8–18
Limestone	100–300	1.26–1.33	2500	0.90	5.6–5.9
Marble		2.07–2.94	2500–2700	0.80	10–13.6
Sandstone	40	1.83	2160–2300	0.71	11.2–11.9
Saw dust	23	0.059	–	–	–
Sand dry	–	0.039	–	–	–
Soil dry	–	0.014	–	–	–
Soil wet	–	0.042	–	–	–
Tiles clay	–	0.85	1900	–	–
Tiles concrete	–	1.10	2100	–	–
Wood (across the grain):					
Fir	23	0.11	420	2.72	0.96
Maple or Oak	30	0.166	540	2.4	1.28
Yellow pine	23	0.147	640	2.8	0.82
Wood shavings	23	0.059	–	–	–
Wood chip board	–	0.15	800	–	–

TABLE III(c)   Physical properties of some other materials

S.No.	Material	Density (ρ) kg/m^3	Thermal conductivity, K W/m °C	Specific heat, C J/kg °C
1.	Air	11.177	40.026	1006
2.	Alumina	3800	29.0	800
3.	Aluminum	41–45	211	0.946
4.	Asphalt	1700	0.50	1000
5.	Brick	1700	0.84	800
6.	Carbon dioxide	1.979	0.145	871
7.	Cement	1700	0.80	670
8.	Clay	1458	1.28	879
9.	Concrete	2400	0.1	1130
10.	Copper	8795	385	–
11.	Cork	240	0.04	2050
12.	Cotton wool	1522	–	1335
13.	Fibre board	300	0.057	1000
14.	Glass-crown	2600	1.0	670
	Window	2350	0.816	712
	Wool	50	0.042	670
15.	Ice	920	2.21	1930
16.	Iron	7870	80	106
17.	Lime stone	2180	1.5	–
18.	Oxygen	1.301	0.027	920
19.	Plaster-board	950	0.16	840
20.	Polyesterene-expanded	25	0.033	1380
21.	P.V.C.- rigid foam	25–80	0.035–0.041	–
	- rigid sheet	1350	0.16	–
22.	Saw dust	188	0.57	–
23.	Thermocole	22	0.03	–
24.	Timber	600	0.14	1210
25.	Turpentine	870	0.136	1760
26.	Water (H_2O)	998	0.591	4190
	(Sea)	1025	–	3900
	(Vapour)	0.586	0.025	2060
27.	Wood wool	500	0.10	1000

TABLE III(d) Properties of air at atmospheric pressure (the values of μ, K, C_p and Pr are not strongly pressure-dependent and may be used over a fairly wide range of pressures)

T K	ρ kg/m^3	C_p kJ/ kg K	μ kg/m-sec $\times 10^{-5}$	ν m^2/s $\times 10^{-6}$	K W/mK $\times 10^{-3}$	α m^2/sec $\times 10^{-5}$	Pr
100	3.6010	1.0259	0.6924	1.923	9.239	0.2501	0.770
150	2.3675	1.0092	1.0283	4.343	13.726	0.5745	0.753
200	1.7684	1.0054	1.3289	7.490	18.074	1.017	0.739
250	1.4128	1.0046	1.488	9.49	22.26	1.3161	0.722
300	1.1774	1.0050	1.983	15.68	26.22	2.216	0.708
350	0.9980	1.0083	2.075	20.76	30.00	2.983	0.697
400	0.8826	1.0134	2.286	25.90	33.62	3.760	0.689

TABLE III(e) Properties of water (saturated liquid)

°F	°C	C_p kJ/kg °C	ρ kg/m^3	μ kg/m sec	K W/m C	Pr	$\nu = \frac{\mu}{\rho}$ m^2/s
32	0	4.225	999.8	1.79×10^{-3}	0.566	13.25	1.79×10^{-6}
40	4.44	4.208	999.8	1.55×10^{-3}	0.575	11.35	1.55×10^{-6}
50	10	4.195	999.2	1.31×10^{-3}	0.585	9.40	1.31×10^{-6}
60	15.56	4.186	998.6	1.12×10^{-3}	0.595	7.88	1.12×10^{-6}
70	21.11	4.179	997.4	9.8×10^{-4}	0.604	6.78	9.82×10^{-7}
80	26.67	4.179	995.8	8.6×10^{-4}	0.614	5.85	8.63×10^{-7}
90	32.22	4.174	994.9	7.65×10^{-4}	0.623	5.12	7.69×10^{-7}
100	37.78	4.174	993.0	6.82×10^{-4}	0.630	4.53	6.87×10^{-7}
110	43.33	4.174	990.6	6.16×10^{-4}	0.637	4.04	6.22×10^{-7}
120	48.89	4.174	988.8	5.62×10^{-4}	0.644	3.64	5.68×10^{-7}
130	54.44	4.179	985.7	5.13×10^{-4}	0.649	3.30	5.20×10^{-7}
140	60	4.179	983.3	4.71×10^{-4}	0.654	3.01	4.63×10^{-7}
150	65.55	4.183	980.3	4.3×10^{-4}	0.659	2.73	4.39×10^{-7}
160	71.11	4.186	977.3	4.01×10^{-4}	0.665	2.53	4.10×10^{-7}
170	76.67	4.191	973.7	3.72×10^{-4}	0.668	2.33	3.82×10^{-7}
180	82.22	4.195	970.2	3.47×10^{-4}	0.673	2.16	3.58×10^{-7}
190	87.78	4.199	966.7	3.27×10^{-4}	0.675	2.03	3.38×10^{-7}
200	93.33	4.204	963.2	3.06×10^{-4}	0.678	1.90	3.18×10^{-7}
210	104.4	4.216	955.1	2.67×10^{-4}	0.684	1.66	2.79×10^{-7}

TABLE III(f) Physical properties of some liquid

Liquid	Formula	Density kg/m³	Viscosity $\times 10^3$ Ns/m²	Melting point K	Boiling point K	Thermal conductivity W/mK	Specific heat J/kg K	Latent heat $\times 10^4$ J/kg	Heat of vaporisation $\times 10^4$ J/kg	Coefficient of cubical expansion $\times 10^{-4}$ s/K
Acetic acid	$C_2H_4O_2$	1049	1.219	290	391	0.180	1960	18.1	39	10.7
Acetone	C_3H_6O	780	0.329	178	330	0.161	2210	8.2	52	14.3
Ammonia	NH_3	665	–	–	–	0.558	4606	–	–	(at $-20\,°C$)
Benzene	C_6H_6	879	0.647	279	353	0.140	1700	12.7	40	12.2
Chloroform	$CHCl_3$	1489	0.569	210	334	0.121	960	7.9	2.5	12.7
Crude oil	–	800	1.379	–	–	0.155	–	–	–	9.0
Ethyl alcohol	C_2H_6O	789	1.197	156	352	0.177	2500	10.4	85	10.8
Glycerine	$C_3H_8O_3$	1262	1495	293	563	0.270	2400	19.9	83	4.7
Mercury	Hg	13546	1.552	234	630	7.6	140	1.17	29	1.82
Olive oil	–	920	85	–	570	0.17	1970	–	–	7.0
Paraffin oil	–	800	1000	–	–	0.15	2130	–	–	–
Turpentine	–	870	1.49	263	429	0.136	1760	–	29	9.7
Sea water	–	1020	1.02	264	377	–	3900	33.0	–	–
Water	H_2O	998	1.00	273	373	0.591	4190	33.4	226	2.1

TABLE III(g) Physical properties of humid air and constant

S. No.	Physical constants	Units	Expression $[T_i(°C) = (T_w + T_g)/2]$	Reference
1.	Specific heat capacity, C	J/kg K	$999.2 + 0.1434\,T_i + 1.101$ $\times 10^{-4}\,T_i^2 - 6.7581 \times 10^{-8}\,T_i^3$	Kyokai (1978)
2.	Thermal conductivity, K	W/m K	$0.0244 + 0.7673 \times 10^{-4}\,T_i$	"
3.	Viscosity, μ	N.s/m^2 (kg/m sec)	$1.718 \times 10^{-5} + 4.620 \times 10^{-8}\,T_i$	"
4.	Diffusivity, α	m^2/s	$7.7255 \times 10^{-10}(T_i + 2730)^{1.83}$	"
5.	Density, ρ	kg/m^3	$353.44/(T_i + 273.15)$	Toyama *et al.* (1987)
6.	Expansion factor, β'	K^{-1}	$1/(T_i + 273.15)$	Standard
7.	Acceleration due to gravity, g	m/s^2	9.81	Standard

TABLE III(h) Absorptivity of various surfaces for sun's ray

Surfaces	Absorptivity	Surfaces	Absorptivity
Metals		Surface (*Cont.*)	
Polished aluminium/copper	0.26	Bright aluminium	0.30
New galvanised iron	0.66	Flat white	0.25
Old galvanised iron	0.89	Yellow	0.48
Polished iron	0.45	Bronze	0.50
Oxidised rusty iron	0.38	Silver	0.52
Roofs		Dark aluminium	0.63
Asphalt	0.89	Bright red	0.65
White asbestos cement	0.59	Brown	0.70
Cooper sheeting	0.64	Light green	0.73
Uncoloured roofing tile	0.67	Medium red	0.74
Red roofing tile	0.72	Medium green	0.85
Galvanised iron, clean	0.77	Dark green	0.95
Brown roofing tile	0.87	Blue/black	0.97
Galvanised iron, dirty	0.89	Surroundings	
Black roofing tile	0.92	Sea/lake water	0.29
Walls		Snow	0.30
White/yellow brick tile	0.30	Grass	0.80
White stone	0.40	Light coloured grass	0.55
Cream brick tile	0.50	Light green shiny leaves	0.75
Burl brick tile	0.60	Sand gray	0.82
Concrete/red brick tile	0.70	Rock	0.84
Red sand line brick	0.72	Green leaf	0.85
White sand stone	0.76	Earth (black ploughed field)	0.92
Stone rubble	0.80	White leaves	0.20
Blue brick tile	0.88	Yellow leaves	0.58
Surface		Aluminium foil	0.39
White paint	0.12–0.26	Unpainted wood	0.60
Whitewash/glossy white	0.21		

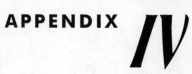
TABLE IV Extinction coefficient of methylmethacrylate plastic for five wavelength regions

j	Wavelength (μm)	Extinction coefficient μ_j (m^{-1})	Emissive power ($\Delta E_{bj}/E_b$)
1	$0 < \lambda < 0.36$	∞	0.0667
2	$0.36 \leq \lambda < 1.06$	2.378	0.6615
3	$1.06 \leq \lambda < 1.30$	12.53	0.0923
4	$1.30 \leq \lambda < 1.60$	31.0	0.0664
5	$1.60 \leq \lambda < \infty$	∞	0.1132

Physical properties of methylmethacrylate plastic (thermal trap) are as follows:

$K = 0.164\,\text{W/m}\,^\circ\text{C}$, $\rho = 1201.0\,\text{kg/m}^3$

$C = 1466.0\,\text{J/kg}\,^\circ\text{C}$, $\gamma = 0.04$

(γ is the coefficient of reflectivity)

TABLE V(a)   Expressions for saturated vapour pressure as function of temperature (°C)

S. No.	$P(N/m^2)$	Range of $T(°C)$	Reference
1.	$2.21 \times 10^{-\left[\frac{Y}{T}\left(\frac{a+bY+CY^3}{1+dY}\right)\right]}$ where, $Y = 647.27 - T_0$, $a = 3.2438$, $b = 5.8683 \times 10^{-3}$, $C = 1.17024 \times 10^{-8}$, $d = 2.1879 \times 10^{-3}$ and $T_0 = T + 273$	10–90	Keenan & Keyes (1936)
2.	$6.893 \times 10^3 \exp[54.63 - 12301.6\,T\,T_0^{-1} - 5.17 \ln T_0]$ where, $T_0 = 1.8\,T + 491.69$	10–90	Brooker *et al.* (1978)
3.	$\exp\left[25.317 - \frac{5144}{(T+273.15)}\right]$	10–90	Fernandez and Chargoy (1990)
4.	$133.29 \times 10^{\left[0.662 + \frac{7.5(T_0-273)}{T_0-35}\right]}$	10–90	Palancz (1984)
5.	$133.32 \exp\left[18.6686 - \frac{4030.1824}{(T+235.15)}\right]$	10–90	Kyokai (1978)

The variation of partial pressure calculated by using expressions above has been shown in Figure 9.2 and the numerical values are given in Table V(b)

TABLE V(b)   Steam table for saturation vapour pressure

Temp. °C	P (N/m^2)	Temp. °C	P (N/m^2)	Temp. °C	P (N/m^2)
0.0	610.8	31.0	4491	61.0	20860
1.0	656.6	32.0	4743	62.0	21840
2.0	705.5	33.0	5029	63.0	22860
3.0	757.55	34.0	5318	64.0	23710
4.0	812.0	35.0	5622	65.0	25010
5.0	871.8	36.0	5940	66.0	26150
6.0	934.5	37.0	6274	67.0	27330
7.0	1001.2	38.0	6624	68.0	28560
8.0	1072.0	39.0	6991	69.0	29840
9.0	1147.2	40.0	7375	70.0	31160
10.0	1227.0	41.0	7777	71.0	32530
11.0	1311.6	42.0	8198	72.0	33960
12.0	1401.4	43.0	8639	73.0	35430
13.0	1496.5	44.0	9100	74.0	36960
14.0	1597.3	45.0	9583	75.0˙	38550
15.0	1703.9	46.0	10086	76.0	40190
16.0	1816.8	47.0	10612	77.0	41890
17.0	1936.2	48.0	11162	78.0	43650
18.0	2062	49.0	11736	79.0	45470
19.0	2190	50.0	12335	80.0	47360
20.0	2337	51.0	12961	81.0	49310
21.0	2485	52.0	13613	82.0	49310
22.0	2642	53.0	14340	83.0	53420
23.0	2808	54.0	15002	84.0	55570
24.0	2982	55.0	15641	85.0	57800
25.0	3166	56.0	16511	86.0	60110
26.0	3360	57.0	17313	87.0	62490
27.0	3564	58.0	18147	88.0	64950
28.0	3778	59.0	19016	89.0	67490
29.0	4004	60.0	19920	90.0	70110
30.0	4241				

APPENDIX VI

TABLE VI Heating values of various combustibles and their conversion efficiencies

Fuel	Heating value (kJ/kg), H_F	Conversion efficiency of device, η_F
Coal coke	29000	70
Wood	15000	60
Straw	14000–16000	60
Gasoline	43000	80
Kerosene	42000	80
Methane (Natural gas)	50000	80
Biogas (60% methane)	20000	80
Electricity	–	95

Bibliography

Abdel-Khalik, S.I., Heat Removal Factor for a Flat-Plate Solar Collector with a Serpentine Tube, *Solar Energy*, **18**, 59 (1976).

Abu-Qudais, Md., Abu-Huleh, B., and Othman, O.N., Experimental study and numerical simulation of a solar still using an external condenser, *Energy*, **21**, 851 (1996).

Aderibigbe, D.A., Optimal Cover Inclination for Maximum Yearly Average Productivity in Roof-Type Solar Still, *Private Communication to A.I. Kudish* (1985).

Aggarwal, S., Computer Based Thermal Modelling of Advance Solar Distillation System, *Ph.D. Thesis IIT Delhi* (1998).

Akinsete, V.A. and Duru, C.U., A cheap method of improving the performance of roof type solar still, *Solar Energy*, **23**(3), 271 (1979).

Albright, L.D., Sieginer, I., Marsh, L.S., and Oko, A., In situ thermal calibration of unventilated greenhouses, *J. Agric. Engg. Res.*, **31**, 265–281 (1985).

Anwar, S.I. and Tiwari, G.N., Evaluation of convective heat transfer coefficient in crop drying under open sun drying, *Energy Conversion and Management*, **42**, 627 (2001).

ASHRAE, *Handbook of Fundamentals American Society of Heating Refrigerating and Air-Conditioning Engineers*, New York (1967, 1974, 1977).

Baibutaev, K.B., Achilov, B.M., and Ramaeva, G., Effect of Salt Concentration on the Evaporation Process in Solar Stills, *Geliotekhnika*, **6**(2), 83 (1970).

Baker, L.K., Film Heat Transfer Coefficient in solar Collector Tubes at Low Reynolds Number, *Solar Energy*, **11**, 78 (1967).

Balcomb, J.D., Barley, D., McFarland, R.D., Perry, J., Wray, W.D., and Noll, S., *Passive Solar Design Handbook*, vol. 2, US Department of Energy, Washington, DC, Report No. DOE/CS-0127/2 (1980).

Balcomb, J.D., Jones, R.W., Kosiewicz, C.E., Lazarus, G.S., McFarland, R.D., and Wray, W.O., *Passive Solar Design Handbook*, Vol. 3, American Solar Energy Society, Boulder CO (1983).

Balcomb, J.D., Jones, R.W., McFarland, R.D., and Wray, W.O., Expanding the SLR method, *Passive Solar J.*, **1**, 67 (1983).

Bansal, N.K. and Minke, G., *Climatic Zones and Rural Housing in India*, Part I of the Indo-German Project on Passive Space Conditioning (1988).

Bansal, P.K., Solar Ponds and other Collector/Storage Systems, Ph.D., Thesis, I.I.T., Delhi (1981).

Bansal, N.K., Hauser, G., and Munke, G., Passive Building Design, Elsevier Science B.V. (1994).

Bansal, P.K. and Kaushik, N.D., Salt Gradient Stabilized Solar Pond Collector, *Energy Conversion and Management*, **21**, 81–95 (1981).

Baranov, V.K., Geliotekhnika, **2**(3) 11 (1966).

Baum, V.A., Bayaramov, R.B., and Maievsky, Y.M., The Solar Still in the Desert, *Proceedings of International Solar Energy Congress*, Melbourne, 426 (1970).

Bayley, F.J., An Analysis of Turbulent Free Convection Heat Transfer, *Proc. Inst. Mech. Engg.*, **169**(20), 361 (1955).

Beekley, D.C. and Mather, G.R. (Jr.), Analysis and Experimental Test of a High Performance Evacuated Tubular Collector, DOE/NASA, CR-150 874 (1978).

Beekley, D.C. and Mather, G.R. (Jr.), Analysis and Experimental Test of a High Performance Evacuated Tubular Collector, DOE/NASA, CR-/50874 (1978).

Blank, Leland, Tarquin, T., and Antony, J., *Engineering Economy*, III Edition, McGraw-Hill Int. Editions (1989).

Bliss, R.W., Solar House Heat-A Panel, *Proc. World Symp. App. Sol. Energy*, Phoenix, 151 (1955).

Bliss, R.W., The Derivations of Several Plate Efficiency Factors Useful in the Design of Flat Plate Solar Heat Collectors, *Solar Energy*, **3**, 55 (1959).

Bowen, I.S., The Ratio of Heat Loss by Conduction and by Evaporation for any Water Surface, *The Physical Review*, **27** (2nd series), 779 (1926).

Brooker, D.B., Bakker Arkama, F.W., and Hall, C.W., Drying Cereal Grains, The AVI Publishing Co. Inc., West Port, Connecticut, USA (1978).

Brooker, D.B., Bakker-Arkema, F.W., and Hall, C.W., Drying and Storage of Grain and Oil Seeds, AVI Book Co, New York (1992).

Brown, C.K. and Gauvin, W.H., Combined Free and Forced Convection, I, II, *Can. J. Chem. Engg*, **43**(6), 306 (1965).

Buchberg, H., Laude, O.H., and Edwards, D.K., Performance Characteristics of Rectangular Honeycomb Solar Thermal Converter, *Solar Energy*, **13**, 193 (1971).

Buchberg, H., Catton I., And Edwards D.K., Natural Convection in Enclosed Space: A Review of Application in Solar Energy Collection, *Journal of Heat Transfer*, **98**, 182 (1976).

Caouris, Y. *et al.*, A Novel Solar Collector, *Solar Energy*, **21**, 157 (1978).

Chandra, R., Goel, V.K., and Raychaudhuri, B.C., Performance Comparison of Two-pass Modified Reverse Flat-plate Collector with Conventional Flat-plate Collectors, *Energy Conversion and Management*, **23**, 177 (1983).

Charters, W.W.S. and Window, B.C., Solar Collector Design and Testing, *Search* **9**(4) (1978).

Chauhan, R.S. and Kadambi, V., Performance of a Collector-cum-Storage Type of Solar Water Heater, *Solar Energy*, **18**, 327 (1976).

Chiou, J.P., El-wakil, M.H., and Duffie, J.A., A Slit and Expanded Aluminium Foil Matrix Solar Collector, *Solar Energy*, **9**, 73 (1975).

Churchill, S.W. and Ozoe, H., Correlations for Laminar Forced Convection in Flow Over and Isothermal Flat Plate and in Developing and Fully Developed Flow in an Isothermal Tube, *J. Heat Transfer*, **95**, 46 (1973).

Clark, J.A. and Schrader, J.A., An Improved Model for Steady State Heat and Mass Transfer within a Solar Still, ASME paper No. 85 W A SOL-2, Miami Beach, FL (1985a).

Clark, J.A., Schrader, J.A., Graysm, A., and Abu-Ilijleh B., Wind Speed, Insolation and Insulation Effects on Steady-State Solar Still Performance, *Solar Energy* (1985b).

Clifton, J.V. and Chapman, A.J., Natural convection of finite size horizontal plate, *Int. J. Heat Mass Transfer*, **12**, 1573 (1969).

Close, D.J., The Performance of Solar Water Heaters with Natural Circulation, *Solar Energy*, **6**, 33 (1962).

Collares-Perira, M. and Rabl, A., The Average Distribution of Solar Radiation-Correlations Between Diffuse and Hemispherical and Between Daily and Hourly Insolation Values, *Solar Energy*, **22**, 155 (1979).

Cooper, P.I., Digital Simulation of Transient Solar Still Processes, *Solar Energy*, **12**, 313 (1969a).

Cooper, P.I., The Absorption of Solar Energy Radiation in Solar Stills, *Solar Energy*, **12**, 333 (1969b).

Cooper, P.I., The Transient Analysis of Glass Covered Solar Still, Ph.D. Thesis, University of Western Australia, Australia (1970).

Cooper, P.I., Some Factors Affecting the Absorption of Solar Radiation in Solar Still, *Solar Energy*, **13**, 373 (1972).

Cooper, P.I., Digital Simulation of Experimental Solar Still Data, *Solar Energy*, **14**, 451 (1973a).

Cooper, P.I., Maximum Efficiency of a Single Effect Solar Stills, *Solar Energy*, **15**, 205 (1973b).

De Graf, J.G. and Held, E.F.M., Vander, *Applied Sci. Res. Section A*, **3**, 393 (1953).

Delyannis, A. and Delyanis, E., Solar Distillation Plant of High Capacity, 4th Int. Symp. on Fresh Water from Sea, **4**, 487 (1973).

Delyannis, E.E. and Delyannis, A., Economics of Solar Stills, *Desalination*, **52**, 167 (1985).

Dhiman, N.K., Analysis of Solar Collector-cum-Storage Water Heaters Ph.D. Thesis (IITD) (1983).

Diamante, L.M. and Munro, P.A., Mathematical Modeling of the Thin Layer Solar Drying of Sweet Potato Slices, *Solar Energy*, **51**(4), 271 (1993)

Dickinson, W.C., Clark, A.K., Day J.A., and Woultere, L.F., The Shallow Solar Pond Energy Conversion System, *Solar Energy*, **18**, 3 (1976b).

Dithus, F.W. and Boelter, L.M.K., Univ. Calif. (Berkeley) Pub. Engg., **2**, 443 (1930).

Duffie, J.A. and Beckman, W.A., *Solar Engineering of Thermal Processes*, John Wiley and Sons, New York (1991).

Dunkle, R.V., Solar Water Distillation, The Roof Type Still and a Multiple Effect Diffusion Still, International Developments in Heat Transfer, A.S.M.E., *Proc. International Heat Transfer, Part V*. University of Colorado, 895 (1961).

Dunkle, R.V., Christie, E.A., and Cooper, P.I., A Method of Measuring Sky Radiation, *Proc. Int. Solar Energy Society-Silver Jubilee*, Int. Congress, Atlanta, Georgia (1979).

Edwards, D.K., Suppression of Cellular Convection by Lateral Walls, *Trans. ASME, J. Heat Transfer*, **91**, 145 (1969).

El-Sayed, M.M., Effects of Parametric conditions on the Performance of an Ideal Diffusion Still, *Applied Energy*, **22**, 187 (1986).

Elwell, D.L., Short, T.H., and Badger, P.C., Stability Criteria for Solar (Thermal Saline) Ponds. *Proc. Annual Meeting American Section of SES*, Orlando, Florida (1977).

Faith, H.E.S., Solar Distillation, Mech. Engg., Faculty of Engg., Alexandria University, Alexandria, Egypt, (Personal Communication), (1998).

Fernandez, J. and Chargoy, N., Multistage, Indirectly Heated solar Still, *Solar Energy*, **44**(4), 215 (1990).

Frick, B., Some New Consideration about Solar Stills, *Proceedings of International Solar Energy Congress*, Melbourne, 395 (1970).

Fuji, T. and Imura, H., Natural convection heat transfer from a plate with arbitrary inclination, *Int. J. Heat Mass Transfer*, **15**, 755 (1972).

Garg, H.P., Mullick, S.C., and Bhargawa, A.K., *Solar Thermal Energy Storage*, D. Reidel Publishing Company (1985).

Garg, H.P., Treatise on Solar Energy, *Vol. 1: Fundamentals of Solar Energy*, John Wiley and Sons, New York (1982).

Garg, H.P. and Rakesh Kumar, Thermal modeling and performance prediction of drying processes under open sun drying, Http://www.kenes.com/Ises.Abstract/Htm/0458.htm(2000).

Givoni, B., Solar heating and night radiation cooling by a roof radiation trap, *Energy and Building*, **1**, 141 (1977).

Goel, V.K., Chandra, R., and Raychaudhuri, B.C., A study on the Performance of a Two-absorber Reverse Flat-plate Collector, *Energy Conversion and Management*, **27**, 335 (1987).

Gomkale, S.D., Solar Distillation, CSMCRI, Bhavnagar, India (private communication) (1969, 1980).

Gopinathan, K.K., Computating the Monthly Mean Daily Diffuse Radiation from Clearness Index and Percent Possible Sunshine, *Solar Energy*, **41**, 379–385 (1988).

Govind and Tiwari, G.N., Economic Analysis of Some Solar Energy Systems, *Energy Conversion and Management*, **24**, 131 (1984).

Goyal, R.K. and Tiwari, G.N., Performance of a reverse flat plate absorber cabinet dryer: a new concept, *Energy conversion and Management*, **40**, 385 (1999a).

Goyal, R.K. and Tiwari, G.N., Effect of thermal storage on the performance of deep bed drying, *Ambient Energy*, **20**(3), 125 (1999b).

Greenwold, M.L. and Mc Hugh, T.K., *Practical Solar Energy Technology*, Prentice-Hall, Inc., Engle wood clifts, New Jersey (1985).

Gupta, C.L. and Garg, H.P., System Design in Solar Water Heaters with Natural Circulation, *Solar Energy*, **12**, 103 (1968).

Hamid, Y.H. and Beckman, W.A., Performance of Air-Cooled Radiatively Heated Screen Matrices, *Trans. ASME., J. Engg. for Power*, **93**, 221 (1971).

Harding, J., Apparatus for Solar Distillation, *Proceedings of Institute of Civil Engineers*, **73**, 281 (1883).

Haskins, D. *et al.*, Passive Solar Buildings, Report Sandia Report 79-0824 (1979).

Hausen, H., Darstellung des Wärmeuberganges in Rohren durch verallgemeinerte Potenzbezeiehungen, VDI Z., no. 4, p 91 (1943).

Headly and Springer, A Natural Convection Crop Dryer, Presented at Int. Cong., The *Sun in the Service of Mankind*, 2–6, July, Paris, France (1973).

Held, E.F.M., Vander, *Warmetechiek*, **2**, 54 (1931).

Hirschmann, J.R. and Roefler, S.K., Thermal Inertia of Solar Stills and its Influence on Performance. Proceedings of International, *Solar Energy Congress*, Melbourne, 402 (1970).

Holland, K.G.T., Unny, T.E., and Konicek, L., Free Convection Heat Transfer Across Inclined Air Layers, *J. of Heat Transfer*, **98**, 1819 (1976).

Hollands, K.G.T., Directional Selectivity, Emittance and Absorptance Properties of Vee-corrugated Specular Surfaces, *Solar Energy*, **7**, 108 (1963).

Holman, J.P., *Heat Transfer*, 7th Edition McGraw-Hill Inc. (UK) Ltd. (1992).

Hottel, W.C. and Woertz, B.B., Performance of Flat Plate Solar Heat Exchangers, *Trans. of American Society of Mechanical Engineers*, **14**, 91 (1942).

Hottel, H.C. and Whillier, A., Evaluation of Flat-Plate Collector Performance, *Trans. of The Conference on the Use of Solar Energy*, 2, p. 74, University of Arizona Press (1958).

Huang, B.J. and Lu, J.H., Performance Test of Solar Collector with Intermittent, *Solar Energy*, **28**, 413 (1982).

Jakob, M. and Gupta, P.C., Chemical Engineering Progress, *Symposium Series*, **9**(50), 15 (1954).

Joshi, J.C. and Shrestha, J.N., Report of Second Workshop on Renewable Energy Curriculum held at Asian & Pacific Centre for Transfer of Technology, New Delhi (29th April to 3rd May, 1996).

Kays, W.M. and Grawford, M.E., *Convective Heat and Mass Transfer*, 2nd Edition, McGraw-Hill, New York (1980).

Kaushik, S.C., Tiwari, G.N., and Nayak. J.K., *Thermal Control in Passive Buildings*, IBT Publishers and Distributors, Delhi (1988).

Kaushik, N.D. and Bansal P.K., Transient Behaviour of Salt Gradient Stabilized Shallow Solar Ponds, *Applied Energy*, **10**(1), 63 (1981).

Keeman, J.H. and Keyes, F.G., *Thermodynamic Properties of Steam*, Wiley, New York, 14 (1936).

Klein, S.A., Calculation of Monthly Average Insolation on Tilted Surfaces, *Solar Energy*, **19**, 325 (1977).

Komilov, O.S., Kakharov, S.K., Rasulov, I.G., and Nazarov, M.R., Optimization of temperature condition in solar drying units by mathematical modeling, *Geliotekhnika*, **28**(5), 73 (1992).

Kooi, C.F., The Steady State Salt Gradient Solar Pond, *Solar Energy*, **23**, 37 (1979).

Kothari, D.P. and Sharma, D.K., Laboratory Manual for Energy Engineering Development Cell, Centre for Energy Studies, IIT, Delhi (1996).

Kreith, F. and Kreider, J.F., *Principles of Solar Engineering*, McGraw-Hill Book Company (1980).

Kreith, F., Lof, G.O.G., Rabl, A., and Winston, R., Solar Collectors for Low and Intermediate Temperature Applications, *Prog. Energy Combustion Science*, **6**, 1 (1980).

Kumar, S. and Tiwari, G.N., Performance evaluation of an active solar distillation system, *Energy*, **21** 805 (1996a).

Kumar, S. and Tiwari, G.N., Estimation of convecting mass transfer in solar distillation system, *Solar Energy*, **57**, 459 (1996b).

Kudish, A.I. and Wolf, D., Compact Shallow Solar Pond Hot Water Heater, *Solar Energy*, **21**, 317 (1978).

Kuzay, T.M., Malik, M.A.S., and Boer K.W., Solar Collectors of Solar One. *Proc. Workshop Solar Collectors Heating Cooling Buildings*, NSR (RANN), 99 (1974).

Kyokai, Kagaku Kogaku, ed: Kagaku Kogaku Binran, Maruzen, Tokyo (1978).

Lansing, F.L. and Clarke, V.A., High performance porous flat plate solar collectors, *J. of Energy*, **4**, 685–694 (1979).

Lansing, F.L., Clarke, V., and Reynold, R., A High Performance Porous Flat Plate Solar Collector, *Energy*, **4**, 685 (1979).

Liu, B.Y.H. and Jordan, R.C., The Interrelationship and Characteristic Distribution of Direct, Diffuse and Total Solar Radiation, *Solar Energy*, **4**, (3) (1960).

Liu, B.Y.H. and Jordan, R.C., Daily Insolation on surfaces tilted towards equator, *ASHRAE Journal*, **3**(10), 53 (1962).

Löf, G.O.G, Fundamental Problems in Solar Distillation, *Solar Energy*, **5**, 35 (1961).

Löf, G.O.G., Eilbing, J.A., and Bloemer, J.W., Energy Balances in Solar Distillation, Am. Inst. *Chem. Engrs.* **7**, 641 (1961a).

Löf, G.O.G., El-wakil, M.M., and Duffie, J.A., The Performance of a Colorado Solar House, U.N. Conference on New Sources of Energy, Rome (1961b).

Löf, G.O.G. *et al.*, Residential Heating with Solar Heated Air-the Colorado House, *ASHRAE*-77 (1963).

Luft, W. and Nemer, M.A., The Soleras Solar Energy Desalination Project, *Sun World*, **6**(1), 10 (1982).

Lugani, Neeraj, Performance Studies of Various Designs of Buildings by Using Different Heating and Cooling Concepts, Ph.D. Thesis, (IITD) (1995).

Lunde, P.J., *Solar Thermal Engineering*, Chapter 7, John Wiley & Sons, New York, pp 240–310 (1980).

Madhusudan, M., Tiwari, G.N., Hrizhikeshan, D.S., and Sehgal, H.K., Optimization of Heat Losses in Normal and Reverse Flat Plate Configuration, *Energy Conversion and Management*, **21**, 191 (1981).

Malik, M.A.S., Tiwari, G.N., Kumar, A., and Sodha, M.S., *Solar Distillation*, Pergamon Press, N.Y. (1982).

Malik, M.A.S. and Buelow, F.H., Heat Transfer Characteristics of a Solar Dryer, UNESCO Congress. *The Sun in the Service of Mankind*, Paris, paper, V. 25 (1973).

Malik, M.A.S. and Buelow, F.H., Hydrodynamic and Heat Transfer Characteristics of a Heated Air Duct, *Helio-Technique and Development*, **2**, 3 (1975).

Malik, M.A.S. and Tran, V.V., A Simplified Mathematical Model for Predicting the Nocturnal Output of a Solar Still, *Solar Energy*, **14**, 271 (1973).

McAdams, W.C., *Heat Transmission*, 3rd Edition, McGraw-Hill, New York (1954).

Meinel, A.B., Concentrating Collector, Chapter 9 in *Solar Energy Engineering* Edited by A.A.M. Sayigh. Academic Press, New York (1977).

Meinel, A.B., Unpublished Data, Helio Associated, Inc. Tucson, Arizona (1973).

Meinel, A.B. and Meinel, M.P., Applied Solar Energy, An Introduction (chapter 1), Addison-Wesley Publishing Company (1977).

Meyer, B.A., El-Wakil, M.M., and Mitchell, J.W., Natural convection heat transfer in small and moderate aspect ratio enclosures, in Thermal Storage and Heat Transfer in Solar Energy Systems (F.Kreith, R. Boehm, J. Mitchell and R.Bannerot, eds.), *American Society of Mechanical Engineers*, New York (1978).

Morgan, V.T., The Overall Convective Heat Transfer from Smooth Circular Cylinders, *Advances in Heat Transfer* (T.F Irvine and J.P Hartnett, eds.), vol.11, Academic Press, Inc., New York (1975).

Morrison, G.L. and Ranatunga, D.B.J., Thermosyphon Circulation in Solar Collectors, *Solar Energy*, **24**, 191 (1980b).

Morse, R.N. and Read, W.R.W., A Rational Basis for the Engineering Development of a Solar Still, *Solar Energy*, **12**, 5 (1968).

Morse, E.L., Warming and Ventilating Apartments by Sun's Rays, U.S. Patent, 246-626 (1881).

Mukherjee, K. and Tiwari, G.N., Economic Analysis for Various Designs of Conventional Solar Stills, *Energy Conversion and Management*, **26**, 155 (1986).

Mull, W. and Reiher, M., Gesundh-Ing.beihoffe, *Reihe*, **1**, 28 (1930).

Mullick, S.C., Kandpal, T.C., and Saxena, A.K., Thermal Test Procedure for Box-type Solar Cookers, *Solar Energy*, **39**(4), 353–360 (1987).

Nayak, J.K., Srivastava, A., Singh, U., and Sodha, M.S., The relative Performance of Different Approaches to the Passive Cooling of Roofs, *Building and Environment*, **17**(2) (1982).

Nayak, J.K., Tiwari, G.N., and Sodha, M.S., Periodic Theory of Solar Still, *Int. J. of Energy Research*, **4**, 41 (1980).

Nevers, N. De, *Air Pollution Control Engineering*, McGraw-Hill International Editions (1995).

Nicholls, R.L., Optimal proportioning of an insulated earth cylinder for storage of solar heat, *Solar Energy*, **19**, 711 (1977).

Norton Brian, Solar Energy Thermal Technology , Springer-Verlag, London (1991).

Norton, Brian, Solar Energy Thermal Technology, Springer-Verlag, New York (1992).

Nusselt, W., Der Wärmeaustausch zwischenWand und Wasser im Rohr, *Forsch. Geb. Ingenieu rwes*, 2 (1931).

Okeke, C.E., Egarievwe, S.U., and Animalu, A.O.E., Effects of Coal and Charcoal on Solar-Still Performance, *Energy*, **15**(11), 1071 (1990).

Ong. K.S., A Finite Difference Method to Evaluate the Thermal Performance of Solar Water Heater, *Solar Energy*, **18**, 137 (1974).

Oonk, R.L., Jones, D.E., and Cole-Apkel, B.E., Calculation of Performance of N-Collectors in Series from Test Data on Single Collectors, *Solar Energy*, 23–25 (1979).

Orgill, J.F. and K.G.T., Hollands, Correlation Equation for Hourly Diffuse Radiation on a Horizontal Surface, *Solar Energy*, **19**, 357 (1977).

Page, J.K., The Estimation of Monthly Mean Values of Daily Total Short Wave Radiation on Vertical and Inclined Surfaces from Sunshine Records for Latitudes 40 °N–40 °S, *Proceedings of the UN Conference on New Sources of Energy*, **4**, 378 (1964).

Palanz, B., Analysis of solar-dehumidifying drying, *Int. J. Heat and Mass Transfer*, **27**(5), 647 (1984).

Pandey, G.C., Effect of Dyes on Solar Water Distillation, *Proc. Int. Solar Energy Congress*, Brighton, Pergamon Press (1981b).

Parker, B.F., Derivation of Efficiency and Loss Factors for Solar Air Heaters, *Solar Energy*, **26**, 27 (1981).

Patil, B.G. and Ward, G.T., Simulation of solar air drying of rapeseed, *Solar Energy*, **43**(5), 305 (1989).

Phillips, W.F., The Effect of Axial Heat Conduction on Collector Heat Removal Factor, *Solar Energy*, **23**, 187 (1979).

Pitts, Donald R., and Sissiom, Leighton E., *1000 Solved Problems in Heat Transfer*, McGraw-Hill (1991).

Rabl, A., Comparison of Solar Concentrators, *Solar Energy*, **18**, 93 (1975).

Rabl, A., Optical and Thermal Properties of Compound Parabolic Concentrators, *Solar Energy*, **18**, 497 (1976).

Rabl, A. and Winston, R., Ideal Concentrators for Finite Sources and Restricted Exitangles, *Appl. Opt.* **15**, 497 (1976).

Rabl, A., *Active Solar Collectors and Their Applications*, Oxford University Press, New York and Oxford (1985).

Raj Kamal, Maheshwari, K.P., and Sawhney, R.L., *Solar Energy and Energy Conservation*, Wiley Eastern Limited (1992).

Ranjan, V., Dhiman, N.K., and Tiwari, G.N., Performance of suspended Flat Plate Air Heater, *Energy Conversion and Management*, **23**, 4 (1983).

Rao, V.S.V.B., Singh, U., and Tiwari, G.N., Transient Analysis of Double Basin Solar Still, *Energy Conversion and Management*, **23**(2), 83 (1983).

Rapp, Donald, *Solar Energy*, Prentice-Hall Inc, Engle Wood Cliffs, New Jersey (1981).

Russell, J.L., Technical Report, Gulf General Atomics corporation (1974).

Satcunanathan, S., A Crop Dryer Utilizing a Two Pass Solar Air Heater, Presented at Int. Cong., 2-6 July, Paris, France (1979).

Satcunanathan, S. and Deonarine, S., A Two Pass Solar Air Heater, *Solar Energy*, **15**, 41 (1973).

Sayigh, A.A.M., editor: *Solar Energy Engineering*, Academic Press, New York (1977).

Selcuk, M.K., Solar Air Heaters and their Applications, Chapter 8 in *Solar Energy Engineering*, Edited by A.A.M. Sayigh, Academic Press, New York (1977).

Selcuk, K., Thermal and Economic Analysis of the Overlapped Glass Plate Solar Air Heater, *Solar Energy*, **13**, 165 (1971).

Seth, S.P., Sodha, M.S., and Seth, A.K., The use of Thermal Trap for Increasing Solar Gains through a Roof or a Wall, *Applied Energy*, **10**, 141 (1982).

Sharma, V.K., Colangelo, A., and Spagna, G., Experimental performance of an indirect type solar fruit and vegitable dryer, *Energy Conversion and Management*, **34**, 293 (1993).

Shelton, J., Underground storage of heat in solar heating systems, *Solar Energy*, **17**, 137 (1975).

Shoemaker, M.J., Notes on a Solar Collector with Unique Air Permeable Media, *Solar Energy*, **5**, 138 (1962).

Shore, R., A Self Inflating Movable Insulation System Conference, **2**, 305 (1978).

Short, T.H., Badger, P.C., and Roller, W.L., A solar pond polystyrene bead system for heating and insulating greenhouse, *Acts Horticulture*, **87**, 291 (1978).

Shukla, S.N. and Tiwari, G.N., Transient Analysis of Forced Circulation Solar Water Heating System with Collectors in Series, *Energy Conversion and Management*, **23**, 77 (1983).

Singh, U., Analysis of some Solar Passive Concepts Ph.D. Thesis, Indian Institute of Technology, New Delhi (1982).

Singh, D., Bharadwaj, S.S., and Bansal, N.K., Thermal Performance of Matrix Air Heater. *Int. J. of Energy Research*, **6**, 103 (1982).

Sodha, M.S., Nayak, J.K., Kaushik, S.C., Sabherwal, S.P., and Malik, M.A.S., Performance of a Collector/Storage Solar Water Heater, *Energy Conversion*, **19**, 41 (1979a).

Sodha, M.S., Srivastava, Alok and Tiwari G.N., Thermal Performance of Double Hollow Wall/Roof, *Energy Research*, **3**, 349 (1979b).

Sodha, M.S., Kumar, A., Singh, U., and Tiwari, G.N., Transient Performance of Solar Still, *Energy Conversion*, **20**(3), 191 (1980a).

Sodha, M.S., Kumar, A., Tiwari, G.N., and Pandey, G.C., Effect of Dye on Thermal Performance of Solar Still, *Applied Energy*, **7**, 147 (1980b).

Sodha, M.S., Nayak, J.K., Tiwari, G.N., and Kumar, A., Double Basin Solar Still, *Energy Conversion*, **20**(1), 23 (1980c).

Sodha, M.S. and Tiwari G.N., Analysis of Natural Circulation of Solar Water Heating Systems, *Energy Conversion and Management*, **21**, 283 (1981).

Sodha, M.S., Bansal, P.K., and Kaushik, S.C., Simple Transient Thermal Model for Solar Collector/Storage Water heaters, *Int. J. Energy Research*, **5**(1), 95 (1981a).

Sodha, M.S., Kumar, A., Tiwari, G.N., and Tyagi, R.C., Simple Multi-wick Solar Still: Analysis and Performance, *Solar Energy*, **26**, 127 (1981b).

Sodha, M.S., Nayak, J.K., Singh, U., and Tiwari, G.N., Thermal Performance of Solar Still, *J. of Energy*, **5**, 331 (1981c).

Sodha, M.S., Tiwari, G.N., and Nayak, J.K., Shallow Solar Pond Water Heater: An Analytical Study, *Energy Conversion and Management*, **21**, 137 (1981d).

Sodha, M.S., Bansal, N.K., and Singh, D., Analysis of a Nonporous Double Flow Solar Air Heater, *Applied Energy*, **12**, 251 (1982a).

Sodha, M.S., Bapeshwar Rao, V.S.V., and Tiwari, G.N., Performance of Forced Circulation N-Water Heating Systems in Series, *Energy Research*, **7**, 2 (1982b).

Sodha, M.S., Shukla, S.N., and Tiwari, G.N., Transient Analysis of Closed Loop Solar Water heating System, *Energy Conversion and Management*, **22**(2), 155 (1982c).

Sodha, M.S., Shukla, S.N., and Tiwari, G.N., Transient Analysis of Forced Circulation Solar Water Heating System, *Energy Conversion and Management*, **22**(91), 55 (1982d).

Sodha, M.S., Tiwari, G.N., and Shukla, S.N., Thermal Modelling of Hot Water System, Chapter III in book *Review of Renewable Energy Resources* Edited by Sodha, M.S., Mathur, S.S., and Malik, M.A.S., Wiley Eastern India Ltd. (1982e).

Sodha, M.S., Shukla, S.N., and Tiwari, G.N., Transient Analysis of a Natural Circulation Solar Water Heater with a Heat Exchanger, *J. of Energy*, **7**, 107 (1983).

Sodha, M.S., Mathur, S.S., and Malik, M.A.S., *Reviews of Renewable Energy Resources*, Wiley Eastern Limited (1984).

Sodha, M.S., Dang, A., Bansal, P.K., and Sharma, S.B., An analytical and experimental study of open sun drying and a cabinet type dryer, *Energy Conversion and Management*, **25**, 263 (1985).

Sodha, M.S., Bansal, N.K., Kumar, A., Bansal, P.K., and Malik, M.A.S., *Solar Passive Building and Design*, Pergamon Press (1986).

Soliman, H.S., Effect of Wind on Solar Distillation, *Solar Energy*, **13**, 403 (1972).

Sparrow, E.M. and Ansari, M.A., A Refutation of King's Rule for Multi-Dimensional External Natural Convection, *Int. J. Heat Mass Transfer*, **26**, 1357 (1983).

Steward, W.G., A Concentrating Solar Energy System Employing a Stationary Spherical Mirror and a Movable Collector, *Proc. Solar Heating Cooling Buildings Workshop*, **24** (1973).

Suneja, S., Computer Modelling of an Inverted Solar Distillation System, Ph.D. Thesis Ch. Charan Singh University, Meerut (UP) (1997).

Suri, R.K. and Saini, J.S., Performance Prediction of Single and Double Exposure Solar Air Heater, *Solar Energy*, **12**, 525 (1969).

Swinbank, W.C., Longwave Radiation from Clear Skies, *Quart J. Roy Metorol Soc.* **89**, 539 (1963).

Tabor, H., Radiation, Convection and Conduction Coefficients in Solar Collectors, *Bulletin of the Research Council of Israel*, **60**, 155 (1958).

Tamini, A., Performance of a solar still with reflectors and black dye, *Int. Jl. of Solar and Wind Tech.*, **4**, 443 (1987).

Thekaekara, M.P., Solar Irradiance, Total and Spectral, Chapter III in *Solar Engineering*, Edited by A.A.M. Sayigh, Academic Press, Inc. New York (1977).

Tiwari, G.N., Greenhouse Technology for Controlled Environment, Narosa Publishing House, New Delhi (2003).

Tiwari, G.N., Kumar, A., and Sodha, M.S., Review on Roof Cooling by Water Evaporation, *Energy Conversion and Management*, **22**, 143 (1982).

Tiwari, G.N. and Bapeshwara Rao, V.S.V., Transient Performance of Single Basin Solar with Water Flowing Over the Glass Cover, *Desalination*, **48**(1), 101 (1983).

Tiwari, G.N., Srivastava, A., and Sharma, B.N., Transient Performance of Closed Loop Solar Water Heating System with Heat Exchanger, *Energy Research*, **7**, 289 (1983).

Tiwari, G.N. and Yadav, Y.P., Economic Analysis of Large Scale Solar Distillation Plant, *Energy Conversion and Management*, **25**, 423 (1985).

Tiwari, G.N., Singh, U., and Nayak, J.K., *Applied Thermal Energy Devices*, Kamala Kuteer Publications, Narsapur (A.P) (1985).

Tiwari G.N. and Goyal R.K., *Greenhouse Technology*, Narosa Publishing House, New Delhi (1998).

Tiwari, G.N., Gupta, S.P., Lawrence, S.A., Yadav, V.P., and Sharma, S.B., Transient performance of indoor swimming pool heating by Solar Energy, RERIC, *Int. Energy, Journal*, **10**, 7 (1988).

Tiwari, G.N., Gupta, S.P., and Lawrence, S.A., Transient analysis of solar still in the presence of dye, *Energy Conversion and Management*, **29**, 59 (1989).

Tiwari, G.N., Singh, S.K., and Srivastava, L.M., A Comparison of Active Heating of Biogas Systems for Higher Production, *Int. J. Solar Energy*, **10**, 115–125 (1991).

Tiwari, G.N., Sharma, P.K., Goyal, R.K., and Sutar, R.F., Estimation of a Efficiency Factor for a Greenhouse: A numerical and Experimental Study, *Energy and Building*, **28**, 241 (1998).

Tiwari, G.N. and Sangeeta Suneja, Solar Thermal Engineering System, Narosa Publishing House, New Delhi (1997).

Tiwari, G.N. and Suneja, S., Performance evaluation of an inverted absorber solar distillation system, *Emergy Conversion and Management.*, **39** 173 (1998).

Tiwari, G.N., Din, M., Srivastava, N.S.L., Jain, D., and Sodha, M.S., Evaluation of solar fraction for the north wall of a controlled environment greenhouse; an experimental validation, *Energy Research (In Press)* (2001).

Toyama, S., Kagaku Kikai, Gijtsu, 24, 159, Maruzen, Tokyo (1972).

Toyama, S., Aragaki, T., Salah, H.M., Murase, K., and Sando, M., Simulation of multi-effect solar still and the static characteristics, *J. of Chem. Engg.*, Japan, **20**, 473 (1987).

Trombe, F., *US Patent*, **3**, 832, 992 (1972).

Trombe, F., *Maisons Salaires Techniques de Inginiear*, **3**, 777 (1974).

Twidell, John W. and Weir, Anthony D., Renewable Energy Resources, ELBS (1987).

United Nations, Solar Distillation, Department of Economics & Social Affairs, New York (1976).

Vartanyan, A.V., Shermazanyan, Ya.T., and Arutyunyan, V.V., Analysis of the Kinematics of Heliostats. *Geliotekhnika*, **10**(3), 47 (1974).

Vasilevskis, S., Large Telescope Design Criteria, in *Telescope Design*, Edited by D.L. Crawford, Academic Press, New York (1966).

Verma, L.R., Bucklin, R.A., Eadan, J.B., and Wratten, F.T., Effect of drying air parameters on rice drying models, *Transactions of the ASAE*, **28**, 296 (1985).

Watmuff, J.H., Characters, W.W.S., and Proctor, D., Solar and Wind Induced External Coefficient for Solar Collectors, *Complex* 2 (1977).

Whillier, A., Solar Energy Collection and its Utilization for House Heating, Sc. D. Thesis, M.I.T. Michigan (1953).

Whillier, A., Black Painted Solar Air Heaters of Conventional Design, *Solar Energy*, **8**, 31 (1964).

Whillier, A. and Saluja, G., The Thermal Performance of Solar Water Heaters, *Solar Energy*, **9**(1), 21 (1965).

Wijeysundera, N.E., Ah, L.L., and Thoe, L.K., Thermal Performance Study of Two Pass Solar Air Heaters, *Solar Energy*, **28**, 306 (1982).

Williams, J. Richard, Design and Installation of Solar Heating and Hot Water Systems, *Am Arbor Science*, pp 187–229 (1983).

Winston, R., Cited by Wilson in Solar Concentrators of Novel Design, *Solar Energy*, **16**, 89 (1965).

Wong, H.Y., *Heat Transfer for Engineers*, Longman London Art, New York (1977).

Zaman, M.A. and Bala, B.K., Thin layer drying of rough rice, *Solar Energy*, **42**, 167 (1989).

Zandi, Z.M., Portable Tilted Solar Still, *Sun World*, **6**(1), 6 (1982).

Subject Index

Absorber 99
Absorber area 252
Absorptance 122
Absorption 97
Absorptivity 78, 99
Acceptance angle 252
Active solar still 302, 303, 307
Air collector 213,
    Testing 213
    Characteristic 215
Air conductance 51, 146
Air cavity 51
Air heater 101, 188, 189, 192
    Conventional 189
    Double exposure 195, 219
    Flow above the absorber 196
    Flow on both sides 201
    Honeycomb 213
    Overlapped glass plate 212
    Two pass 202
    Vee-corrugated 205
Air leakage 214, 215
Air mass 9, 11, 21, 22, 37
Air temperature 311
Albedo 2, 12
Altitude angle 17–19, 21
Angle of incidence 20, 22, 24, 122
Annual cost method 469
Aperture area 252
Atmosphere 2, 7, 9, 11–13, 29, 39
Attenuation of solar intensity 46
Auxiliary energy 347

Back (bottom) loss coefficient 113, 286, 291,
    307, 377

Beam (direct) radiation 12, 20, 22–24, 26,
    27, 38, 39
Benefit-cost analysis 475
Bio gas thermal heating 389, 410
    Passive 392
    Active 393
Bond conductance 124
Book value 479–481
Biot number 57, 58
Black body 1, 4, 39, 78, 82
Building heat loss 344
Building overall energy loss coefficient 344

Cabinet dryer 225, 240, 249
Capitalised cost 470
Capitalised cost factor 470, 471
Capitalised cost method 470
Capital recovery factor 452, 457, 470
Cash flow diagram 464, 472
Central tower receiver 259
Characteristic curve 100, 153
Characteristic dimension (length) 59, 67–69,
    116, 193
Circular fresnel lens 260
Clear day index 30, 31, 33, 35
Coefficient of performance 246, 406
Collector (see flat plate collector)
Collector efficiency factor 123–127, 146, 190, 195,
    197, 268
Collector flow factor 131, 147
Collector heat removal factor 131, 147
Compound parabolic concentrator 263
Compound interest factor 453
Compound interest operator 456

521

522 — **SOLAR ENERGY**